Addressing B5G and 6G Network Connectivity Issues in Rural Regions

Thirupathi Manickam
Christ University, India

Sriram Ananthan
Yorkville University, Canada

Subramanian Gopalakrishnan
East Point Group of Institutions, India

Chandan Chavadi
Presidency Business School, India

IGI Global
Scientific Publishing
Publishing Tomorrow's Research Today

Vice President of Editorial — Melissa Wagner
Managing Editor of Acquisitions — Mikaela Felty
Managing Editor of Book Development — Jocelynn Hessler
Production Manager — Mike Brehm
Cover Design — Phillip Shickler

Published in the United States of America by
IGI Global Scientific Publishing
701 East Chocolate Avenue
Hershey, PA, 17033, USA
Tel: 717-533-8845
Fax: 717-533-8661
Website: https://www.igi-global.com E-mail: cust@igi-global.com

Library of Congress Cataloging-in-Publication Data

Names: Manickam, Thirupathi, 1983- editor. | Ananthan, Sriram, 1983- editor. | Gopalakrishnan, Subramanian, 1964- editor. | Chavadi, Chandan, 1973- editor.
Title: Addressing B5G and 6G network connectivity issues in rural regions / edited by Thirupathi Manickam, Sriram Ananthan, Subramanian Gopalakrishnan, Chandan Chavadi.
Description: Hershey, PA : Information Science Reference [2025] | Includes bibliographical references and index. | Summary: "This book aims to provide a comprehensive examination of the current state of rural connectivity, introducing the advanced capabilities and potential applications of beyond 5G (B5G) and 6G networks. It seeks to identify the unique geographical, economic, and technical challenges impeding rural deployment and discuss innovative technologies that can facilitate network expansion in these areas. Additionally, the book explores existing policies, regulatory frameworks, and funding strategies, offering improvements and new initiatives to support broader coverage"-- Provided by publisher.
Identifiers: LCCN 2024042535 (print) | LCCN 2024042536 (ebook) | ISBN 9798369377888 (hardcover) | ISBN 9798369377895 (paperback) | ISBN 9798369377901 (ebook)
Subjects: LCSH: 6G mobile communication systems--Case studies. | 5G mobile communication systems--Case studies. | Rural telecommunication--Case studies. | Broadband communication systems--Case studies. | Digital divide--Case studies.
Classification: LCC TK5103.252 .A33 2025 (print) | LCC TK5103.252 (ebook) | DDC 384.609173/4--dc23/eng/20241129
LC record available at https://lccn.loc.gov/2024042535
LC ebook record available at https://lccn.loc.gov/2024042536

British Cataloguing in Publication Data
A Cataloguing in Publication record for this book is available from the British Library.

Table of Contents

Detailed Table of Contents

and 6G in mitigating such challenges with emphasis on their capability in providing rapid, reliable, and ubiquitous internet access in rural areas. We pay special attention to new technologies supporting such next-generation networks as more advanced antenna systems, artificial intelligence, and various spectrum-sharing strategies. Using best cases, examples, and pilot cases, we present and discuss successful pro-active strategies and use cases for B5G and 6G in rural areas. Moreover, the chapter analyses and considers the further consequences of extended connection such as the changes in the healthcare sector, education sphere and the economic advancement. We also evaluate the challenges that hinder implementation including infrastructural drawbacks, regulatory factors, and funding matters; we provide possible methods of addressing the challenge.

M. Basuvaraj, University of Allahabad, India
Keshvi Rastogi, Christ University, India
S. Chitradevi, Global Institute of Business Studies, India
D. V. VidyaShree, Christ University, India
K. M. Anu, Christ University, India

The emerging technology of the fifth-generation broadband cellular network is already ruling the market with its efficiency, lower latency, higher connectivity, and many more features. In contrast, the sixth-generation broadband cellular network is yet in its research and development stage. These technologies cannot only revolutionize the world with their features, such as high speed and enhanced cybersecurity but also empower it to reach greater heights. To understand the network requirements of the rural and under-developed areas, it is important to understand all those challenges in the way ahead. . Launching such efficient and effective technologies in rural areas would benefit the country as well as its economic growth. The large markets of these cellular networks are at constant growth and are expected to be booming in the future of the Telecom Regulatory Authority of India. (2023, September 29)..

Dankan Gowda V, Department of ECE, BMS Institute of Technology and Management, Karnataka, India
C. R. Vijay, Department of Management Studies, Dr. N.G.P. Institute of Technology, Coimbatore, India
S. V. Ramanan, Department of Electronics and Communication Engineering, PPG Institute of Technology, Coimbatore, India
K. D. V. Prasad, Symbiosis Institute of Business Management, Hyderabad, India & Symbiosis International University, Pune, India
Anil Sharma, Amity School of Engineering and Technology, Amity University, Noida, India

Realising the potential of B5G as well as 6G technologies makes potential a chance to solve the issues associated with the connectivity of the rural areas. This chapter aims at discussing elaborate plans and approaches to the deployment of B5G and 6G networks to improve on the connectivity of rural areas. We then discuss the technology enablers that rendered these networks fit for rural settings through modern communication technologies such as satellite communications, edge computing, and new spectrum management methods. Thus, by studying the presence of networks in different situations, we discuss the main practices for deploying networks and consider examples of pilot projects. Secondly, the chapter also analyses wider socio-economic effects that increased connectivity has on rural societies, including access to education, health care or employment. We also discuss the limitation that may exist in the implementation of the strategy such as infrastructural, regulatory and financial and the ways by which they could be surmounted.

In the era of transformation and technological advancement, there is one major challenge, which is connectivity. The chapter represents not only physical but digital concerns as well. Addressing the digital literacy gap is paramount today.. The chapter suggests the upbringing and support of installation of non-line-of-sight (NLOS) backhaul devices to expand internet reach in a cost efficiency and operational effectiveness. The NLOS refers to the ability to catch the internet frequency waves diffracted from the barrier in the path such, as buildings and mountains, contrasting with clear-line-of-sight (CLOS) methods. Moreover, the chapter argues that merely expanding the 5G and 6G networks to rural areas without bringing literacy, quality assurance, and involvement will not be fruitful. Overall, the spread of internet connectivity is not just an expansion but a target to be achieved with sustainable strategies.

The deployment of renewable energy technologies (RETs) in rural areas is examined in this chapter, with special attention paid to the opportunities and particular constraints that these places bring. This chapter offers a thorough review of the status of renewable energy technologies (RETs), highlighting the major governmental, technical, economic, social, cultural, and environmental obstacles that prevent RETs from being widely adopted in rural areas. Because it tackles the urgent need for sustainable energy solutions in rural areas where traditional energy access is still restricted, the research discussed in this chapter is noteworthy. Through the identification and analysis of deployment issues, this chapter adds to the larger conversation on rural electrification and sustainable development. To close the energy gap and advance global sustainability, policymakers, practitioners, and academics seeking to increase the adoption of renewable energy in underserved areas can benefit greatly from the conclusions and suggestions made.

The demand for wireless broadband communication is rapidly increasing due to the need to support more users and provide more information about more data. Ultra-wideband (UWB) technology has been developed to cover a very wide bandwidth at many frequencies using very narrow pulses in the nanosecond range and can solve this problem. UWB, a wireless technology designed to transmit information at high speeds over short distances with very low power, is an area of current interest with many applications. This chapter introduces the current UWB technology and its potential applications. It also discusses how these systems can be used without interfering with other existing communication systems. The advantages and disadvantages of these systems have also been examined in detail. Also, the main component which is the most important part of ultra-wide technology is an antenna, has been discussed here. The basic antenna design that can be used for sixth-generation system to enhance the data rates and speed has been mentioned.

The opportunity of shifting to 5G and 6G networks is momentous in dealing with inefficiency in resource management in higher learning institutions in India. This review examines the ways in which the application of these sophisticated counter-voice technologies can transform the ways that these institutions, especially universities, distribute and manage resources to support these changes for accessibility and efficiency as well as sustainability. Using URLLC and IMC connections, and the utilization of artificial intelligence, machine learning, blockchain, and similar advanced solutions, 5G and the upcoming 6G networks become capable of providing needed connection. They include dynamic resource management, fast and secure accreditation, and learning for virtual and augmented reality. Reflecting on many problems in India, such as the digital literacy gap and diversified learner needs, the manuscript underlines the factors as policy initiatives, PPP, and funding for building up the enabling environment for adopting e-learning for all.

Rural development encompasses development in social, cultural, environmental, technological and economic conditions. India introduced new land reforms by abolishing the zamindari system, distribute lands among the landless farmers and abolishing bonded labour to protect the farmers' interests. The contribution of agriculture and allied services in Gross Value Added was just 17.8% for the year 2019-20 and 18% in 2023. India's rural economic development envisages through three phases namely land reforms, green revolution, and changes due to globalization when India opened its economy in the nineties. However, a substantial gap remains in the progress trajectory, leading to a rural-urban divide. The present study is conceptualized that tries to assess the impact of digital interventions on socio-economic growth in rural communities. The authors develop an economic model which emphasizes on digitalization, technological developments and community engagement, and how all of these lead to rural development and inclusive growth in India's economy.

"Beyond Connectivity: The Impact of B5G and 6G on Rural Education Outcomes in India" explores how advanced communication technologies can enhance rural education. It addresses challenges such as teacher shortages, limited smartphone access, and digital adoption, which contribute to learning gaps. The fear of technology replacing teachers further hinders the use of digital tools in classrooms. The chapter suggests that B5G and 6G can provide high-speed internet to remote areas, improving access to cloud-based educational resources. AI-driven adaptive learning platforms can personalize education, while online training for educators fosters collaboration. Strong policies and community involvement are essential for successful implementation.

This chapter explores the transformative potential of B5G (Beyond 5G) and 6G network connectivity in promoting social inclusion and community empowerment in rural regions. It examines how advanced connectivity technologies can bridge the digital divide, enhance access to essential services such as education, healthcare, and e-governance, and foster economic opportunities through digital entrepreneurship and smart farming. The chapter incorporates global and regional case studies to illustrate the positive outcomes of B5G/6G implementations in marginalized communities, highlighting their impact on women, youth, and small businesses. Additionally, it discusses the challenges of deploying these technologies in remote areas and provides policy recommendations to ensure equitable access and sustainable development. By focusing on real-world examples, the chapter underscores the role of advanced connectivity in creating inclusive, empowered, and digitally integrated rural communities.

Mirshad Rahman, Acharya Institute of Graduate Studies, India
Bharath Booshan M. S., Acharya Institute of Graduate Studies, India
Shabista Booshan, ISBR Business School, India
S. Shabana, Bharathamatha College of Arts and Science, India
Deepika Gowda, Acharya Institute of Graduate Studies, India
S. Gowtham, Acharya Institute of Graduate Studies, India
C. G. Lakshmidevi, Mahe Co-Operative College of Higher Education and Technology, India
N. Muddasir Ahamed Khan, Acharya Institute of Graduate Studies, India

By addressing the societal demands of upcoming developments such as remote surgeries, self-driving cars, immersive environments like augmented/virtual reality (AR/VR), and other related applications, B5G and 6G networks expand on the concepts and capabilities of 5G, introducing even more sophisticated features that will provide intelligent and highly reliable communications between the devices. Small towns and rural areas, which often lack access to modern connection facilities, are also expected to see revolutionary changes as a result of all these technologies. Using a literature review methodology, this research collects and analyzes data from several sources of literature and the chapter also includes government and business publications that address the state and prospects of rural connection in India, as well as groundbreaking studies on technological innovation, digitalization, and rural development. The literature from telecommunications, public policy, and rural sociology has also been included in connection to the previously listed fields of study.

Ramya Krishnappa, B.M.S. College of Law, India
Rajiv Gurugopinath, Ramaiah College of Law, India

India, a land of villages, has over two-thirds of its population residing in rural areas, primarily dependent on agriculture. Democracy thrives at various levels, including local governance through panchayats, which play a crucial role in decentralized administration. Panchayats, constitutionally recognized, focus on social justice and economic development, covering areas like agriculture, education, and disaster management. However, they often limit "connectivity" to physical infrastructure, overlooking technological connectivity. The government has made strides in adopting modern connectivity solutions, but rural implementation faces challenges like infrastructure deficits, low awareness, and environmental concerns. While telecom decisions are centrally managed, effective implementation relies on local bodies. This paper explores the challenges panchayats face in adopting B5G and 6G technologies in rural areas and offers suggestions for overcoming these hurdles.

Foreword

The global technological landscape is evolving unprecedentedly, with Beyond 5G (B5G) and 6G networks poised to redefine how we communicate, work, and live. These advancements promise to bridge gaps in connectivity, offering transformative potential for socio-economic growth, particularly in rural and underserved regions. However, unlocking this potential requires addressing complex challenges, including infrastructural barriers, policy hurdles, and societal disparities.

In this context, *"Addressing B5G and 6G Network Connectivity Issues in Rural Regions"* is a timely and essential contribution to the field. This book meticulously examines the challenges and opportunities of deploying next-generation wireless communication technologies in rural areas. Combining theoretical insights with practical strategies offers a holistic view of how B5G and 6G networks can be leveraged to achieve inclusive growth.

The chapters in this book traverse a broad spectrum of topics. They introduce readers to the fundamentals of B5G and 6G technologies, highlighting their relevance and applications in rural contexts. The discussion then progresses to policy and regulatory frameworks, presenting actionable recommendations to overcome existing bottlenecks. Further, the book delves into technical solutions, addressing connectivity issues and exploring innovations like ultra-wideband antennas and renewable energy integration.

What makes this book particularly impactful is its focus on the human and societal aspects of technology. By examining the role of advanced connectivity in enhancing education, empowering communities, and fostering socio-economic development, the authors underline the transformative potential of B5G and 6G networks beyond mere connectivity. Including case studies and practical roadmaps further enhances its relevance for policymakers, researchers, and industry leaders.

As we stand on the threshold of a digitally connected future, the insights provided in this book serve as a guiding light for achieving equitable technological progress. It emphasises that while technological advancements are crucial, their value lies in their ability to uplift communities and reduce inequalities.

I congratulate the authors and editors on their remarkable efforts in creating a comprehensive and forward-thinking resource. I am confident that this book will inspire meaningful discourse, innovation, and action toward bridging the digital divide and ensuring that the benefits of technology reach every corner of society.

Dhanabalan Thangam
Presidency Business School, Presidency College, Bengaluru, India

Preface

The rapid evolution of wireless communication technologies has brought us to the cusp of the Beyond 5G (B5G) and 6G era, where the possibilities for transformative impacts on society are boundless. However, the reality of unequal access to these technologies, particularly in rural areas, presents a critical challenge to achieving equitable growth and development. This book, *"Addressing B5G and 6G Network Connectivity Issues in Rural Regions,"* delves into the multifaceted challenges and opportunities associated with deploying advanced communication technologies in underserved rural areas.

The chapters in this book are thoughtfully curated to guide readers through the complex interplay of technology, policy, and socio-economic development. The journey begins with an overview of B5G and 6G technologies, shedding light on their potential to revolutionize connectivity and bridge the digital divide. We then examine the unique challenges rural regions face in adopting these technologies, focusing on infrastructure, energy needs, and technical limitations.

Policy and regulatory frameworks play a pivotal role in enabling this transformation. Accordingly, dedicated chapters explore the current landscape, identify barriers, and propose strategic interventions to foster universal connectivity in rural India. These discussions are complemented by an in-depth analysis of innovative solutions, such as ultra-wideband antennas and intelligent resource management systems, that hold the promise of addressing network challenges effectively.

The book also delves into the societal impacts of these technologies, highlighting how B5G and 6G connectivity can drive social inclusion, empower communities, and revolutionize education in rural areas. Case studies and practical examples provide insights into the role of institutions like Panchayat Raj in fostering connectivity and the impact of digital marketing and economic initiatives enabled by 5G networks.

Finally, a systematic literature review offers a comprehensive synthesis of existing research, providing a solid foundation for academics, policymakers, and practitioners interested in exploring the societal impact of these groundbreaking technologies.

This book is the result of a collaborative effort by experts in wireless communication, rural development, and policy analysis. It is intended to serve as a valuable resource for researchers, engineers, policymakers, and anyone invested in leveraging technology to create an inclusive digital future. By addressing the unique challenges of rural connectivity, we aim to contribute to a vision of equitable technological progress that benefits all.

CHAPTER OVERVIEW

The book is structured into thematic sections, each addressing critical aspects of the B5G and 6G ecosystem:

Chapters 1-3 provide an introductory foundation, explaining the fundamentals of B5G and 6G technologies and their relevance to rural regions. These chapters lay the groundwork by discussing the potential benefits of these technologies and the unique opportunities they offer in transforming rural economies.

Chapters 4-5 delve into the policy and regulatory landscape, examining the frameworks necessary to support B5G and 6G deployment in underserved areas. These chapters address specific challenges, such as ensuring access to healthcare through connectivity and tailoring regulatory measures to meet rural needs.

Chapters 6-9 focus on addressing the practical connectivity challenges in rural areas. The discussion includes strategies for overcoming network infrastructure limitations, ensuring universal internet access, and integrating renewable energy solutions to sustain advanced networks in remote regions.

Chapters 10-12 highlight technological innovations essential for realizing the potential of B5G and 6G. Topics such as ultra-wideband antennas, resource management in higher education, and cutting-edge design solutions are presented to illustrate how technology can overcome rural connectivity barriers.

Chapters 13-16 explore the social and economic impacts of advanced connectivity. These chapters examine how digital interventions foster rural development, enhance education, and empower communities. They emphasize the role of B5G and 6G in driving inclusion and creating opportunities for marginalized populations.

Chapters 17-18 present case studies and strategic roadmaps for deploying B5G and 6G in rural areas. These chapters underscore the importance of local institutions, such as Panchayat Raj, in facilitating connectivity and showcase practical applications like digital marketing and resource optimization.

Chapter 19 concludes the book with a systematic literature review. This chapter synthesizes existing research, offering valuable insights into the societal impact of B5G and 6G technologies and setting the stage for future research and innovation in this domain.

Through this comprehensive exploration, the book aims to bridge the knowledge gap between technology and its application in rural contexts. By providing actionable insights, it serves as a guide for policymakers, researchers, and practitioners working towards achieving inclusive digital transformation.

Thirupathi Manickam
Christ University, India

Sriram Ananthan
Acsenda School of Management, Canada

Subramanian Gopalakrishnan
East Point Group of Institutions, India

Chandan Chavadi
Presidency Business School, India

Acknowledgement

This book, Addressing B5G and 6G Network Connectivity Issues in Rural Regions, has been a work in collaboration. I want to thank all the individuals and organisations who contributed immensely to making it possible.

First and foremost, I want to thank IGI Global Publishers, USA, for their continuous support and confidence in this project. Their devotion to advancing scholarly excellence and also supplying professional advice in the course of publication are invaluable. Working with such a great publisher has been an honour, and I appreciate that they saw this book had a life to it.

I am immensely grateful for the contribution of the chapter contributors whose expertise and dedication have made that happen. Each chapter exemplifies their vast knowledge and the provoking insights they possess on said rapid development of the B5G and 6G network connectivity area. The result is a comprehensive and authoritative resource because they have been willing to share their work and perspective.

I also thank anonymous reviewers for their very thorough evaluation and constructive feedback. Thanks to their critical insights, their contribution has resulted in an outstanding quality of this book, which is as rigorous as it is excellent. I am genuinely thankful for their time, effort, and commitment to scholarly integrity.

I want to thank CHRIST (Deemed to be University), Bangalore Yeshwantpur Campus, Bengaluru, India, for the intellectually stimulating and resourceful environment required to complete this project. The foundation of the institution's stress on academic research and innovation has always been a great inspiration. That is an order of gratitude to my colleagues and the University.

I am exceedingly grateful to present Addressing B5G and 6G Network Connectivity Issues in Rural Regions. Many people and institutions have contributed to this book, a collaborative achievement. I hope this will contribute to those working in B5G and 6G network connectivity.

Chapter 1
Introduction to B5G and 6G Technologies:
An Overview

Ravi Kumar Natashen
https://orcid.org/0000-0002-9308-7461
Presidency College, India

Ravindran Kandasamy
https://orcid.org/0000-0001-6763-0216
Presidency College, India

A. Karthikeyan
Kristu Jayanti College, India

V. Sandhya
The Oxford College of Business Management, India

ABSTRACT

With the promise of previously unheard-of improvements in connectivity, speed, and intelligent networking, the introduction of beyond 5G (B5G) and 6G technologies represents a dramatic turn in the development of telecommunications. This summary delves into the core ideas, possible uses, and significant technical breakthroughs that are propelling the development of B5G and 6G. It explores the anticipated enhancements over 5G, such as higher data rates, lower latency, better energy efficiency, and the incorporation of cutting-edge technologies like ML, AI, and the Internet of Things (IoT). The study also discusses the benefits and difficulties that come with the rollout of B5G and 6G networks, highlighting the revolutionary effects that these networks will have on a number of industries, including smart cities, transportation, and healthcare. This research attempts to clarify the important features of B5G and 6G technologies and their potential to completely transform global communication networks in the future by offering a thorough introduction.

DOI: 10.4018/979-8-3693-7788-8.ch001

INTRODUCTION

Modern civilization has been built on the swift development of telecommunications technologies, which have fueled technological innovation, social connectivity, and economic expansion. The research and development community is intensively investigating the next-generation technologies, including Beyond 5G (B5G) and 6G, as the limitations of 5G networks start to show. These developments could lead to previously unheard-of gains in network performance, which would help numerous industries undergo a digital revolution. The goal of B5G and 6G technologies is to meet the growing need for improved energy efficiency, reduced latency, and faster data throughput. Although 5G networks offer notable improvements over 4G, including quicker speeds and more dependable connections, 6G and B5G seek to push these limits even farther. 6G is anticipated to provide peak data speeds of up to 1 Tbps, which is a significant improvement over 5G networks' capabilities, according to Zhao et al. (2021).

The combination of cutting-edge technology like artificial intelligence (AI), machine learning (ML), and the Internet of Things (IoT) is a major factor driving the development of B5G and 6G. It is expected that these technologies would be essential in maximizing network efficiency and opening up hitherto unthinkable new applications. For example, network management and automation can be improved using AI and ML, resulting in more effective resource usage and higher service quality (Gupta & Jha, 2021). B5G and 6G technologies have a wide range of possible uses in industries like smart cities, transportation, and healthcare. These medical technology advancements have the potential to greatly enhance patient outcomes by enabling remote surgery and real-time health monitoring (Rappaport et al., 2019). They can facilitate the use of completely autonomous cars in the transportation sector, improving safety and easing traffic (Zhang et al., 2020). Widespread connectivity and intelligent infrastructure are advantages for smart cities, resulting in more sustainable and effective urban environments (Akyildiz et al., 2020).

While there are many opportunities for growth, implementing B5G and 6G technologies also comes with a number of difficulties. These include the requirement for large-scale infrastructure investments, the creation of new legal frameworks, and the resolution of security and privacy issues. According to Saad et al. (2020), resolving these issues will call for concerted efforts from the scientific community, industrial stakeholders, and government. We present a thorough analysis of B5G and 6G technologies in this study, looking at their underlying ideas, possible uses, and significant technological developments. We also talk about the opportunities and difficulties that come with implementing them, hoping to highlight the revolutionary potential of these next-generation networks. In addition to being a crucial enabler of the digital economy and societal developments, the expansion of wireless communication technology is also a technical advancement. Therefore, it is essential for stakeholders in a variety of fields, such as legislators, technologists, and business executives, to comprehend the direction and ramifications of B5G and 6G. This portion of the article aims to provide a comprehensive explanation of the capabilities, possible impacts, and underlying principles of B5G and 6G technologies.

Network architecture and performance have significantly evolved with the shift from 5G to 6G wireless technology. 6G seeks for extreme connection and the integration of several technologies, resulting in a more reliable and effective network, whereas 5G concentrates on increasing bandwidth and decreasing latency. The main distinctions between B5G and 6G are described in the sections that follow. The 5G network architecture is mostly based on a centralized paradigm and emphasizes large IoT connection and improved mobile broadband. Full-spectrum cell-free radio access and improved resource allocation through AI integration are made possible by the 6G architecture, which introduces a three-layer structure with super edge nodes for task-centric support (You et al., 2023) (Duan et al., 2023).

METRICS OF PERFORMANCE

5G Performance

Up to 10 Gbps data rates are possible with a latency of about 1 ms. It is anticipated that 6G performance would surpass 5G capabilities by a large margin, with Tbps-scale data rates, Kbps/Hz spectral efficiency, and microsecond latency (You et al., 2023) ("Innovative Trends in the 6G Era: A Comprehensive Survey of Architecture, Applications, Technologies, and Challenges", 2023).

Innovations in Technology:

5G technology: Make use of beam forming and huge MIMO technology.

6G Technologies: Uses cutting-edge techniques including polymorphic routing, photonics-assisted access, and terahertz communication to improve security and intelligence (Dangi et al., 2023) (Liao et al., 2023). On the other hand, although 6G promises to bring about revolutionary improvements, it also has drawbacks including deployment complexity and the requirement for new regulatory frameworks, which may hinder its uptake in comparison to the more established 5G networks.

The fundamental ideas and technical specifications of B5G and 6G have the potential to drastically alter sectors including manufacturing, healthcare, and smart cities. These developments promise improved connectivity, opening the door to creative applications that boost productivity and user experience in a variety of industries.

Impact on AI-Driven Healthcare Services: According to "6G With Smart Healthcare, Haptic Communication", 2022, 6G will enable AI-driven healthcare systems that enable remote patient monitoring and real-time diagnosis.

Advanced Applications: By getting over 5G's present constraints, technologies like telesurgery and holographic communication will become possible (Khattak et al., 2022).

Crisis Management: Better patient care during catastrophes would be possible due to faster emergency response made possible by improved connectivity ("6G With Smart Healthcare, Haptic Communication", 2022).

Impact on Manufacturing Industry 5.0 Integration: According to Das et al. (2024), 6G will facilitate the shift to Industry 5.0 by highlighting automation and customer-centric production.

Smart Cyber-Physical Systems: Automation and dependability in industrial processes will be improved by combining 6G with AI and ML (Basu et al., 2022).

Effects on Intelligent Cities Improved Connectivity: 6G will make it possible for smart city applications like driverless cars and digital sensing, raising the level of living in cities (Basu et al., 2022).

Immersion Experiences: By incorporating AR and VR into urban planning and services, these technologies will improve citizen involvement ("6G for Industry 5.0 and Smart CPS: A Journey from Challenging Hindrance to Opportunistic Future", 2022). In order to improve the capabilities of B5G and 6G networks and spur advancements in resource management, security, and network performance, artificial intelligence (AI) and machine learning (ML) are essential. By tackling the intricacies of contemporary telecommunications, its integration enables dynamic and intelligent operations. The main functions of AI and ML in these sophisticated networks are described in the sections that follow.

Network Security with AI

By detecting weaknesses and strengthening defenses against threats, AI improves security measures in 6G (Vetrithangam et al., 2024). Real-time threat identification and incident response are made possible by machine learning algorithms, which improve the network's overall security posture.

Allocating and Managing Resources: Through predictive dimensioning, AI and ML optimize resource allocation, guaranteeing equitable access and effective network resource use (Papaioannou et al., 2024). Through the analysis of traffic patterns and the dynamic adjustment of resources in response to demand, these technologies improve Quality of Service (QoS).

Astute Networking: 6G's incorporation of AI fosters intelligent networking, which enhances network performance and user experience (Deepa et al., 2024). To effectively utilize AI's capabilities in diverse situations, issues like scalability and real-time decision-making need to be resolved.

There are a number of noteworthy obstacles to the deployment of B5G and 6G networks, which fall into two categories: technical and non-technical. For these cutting-edge communication technologies to be implemented successfully, several issues must be resolved.

Challenges in Technology:

High Data speeds and Latency: 6G requires cutting-edge technology like THz band communication and AI integration in order to reach data speeds of up to terabits per second and ultra-low latency (Niranjana et al., 2024) (Ahmed et al., 2023). Significant free space path loss and atmospheric absorption are introduced when operating in the THz range, necessitating the use of cutting-edge antenna technologies and channel management techniques (Ahmed et al., 2023).

Network Complexity: As technologies like network slicing, SDN, and NFV are integrated, the network becomes more complex, necessitating advanced orchestration and management solutions (Serôdio et al., 2023) ("6G Survey on Challenges, Requirements, Applications, Key Enabling Technologies, Use Cases, AI integration issues and Security aspects", 2022).

Non-Technical Difficulties:

Legal and Standardization Issues: The deployment of new networks is made more difficult by the speed at which technology is developing, which surpasses legal frameworks ("Technological and Non-Technical Challenges Associated With 6G Networks", 2022).

Public Acceptance and Security Concerns: For broad acceptance, it is essential to guarantee user trust and solve privacy concerns ("6G Survey on Challenges, Requirements, Applications, Key Enabling Technologies, Use Cases, AI integration issues and Security aspects", 2022). Even if these obstacles are strong, they can be lessened by working together on research, modifying regulations, and creating strong security frameworks. But the speed at which technology is developing could also bring up unanticipated difficulties that are needed for constant innovation and adaptability.

By improving connectivity, data processing power, and overall system intelligence, B5G and 6G technologies have the potential to drastically change applications like remote collaboration, augmented reality, and driverless cars. These developments will promote smooth communication and instantaneous data exchange, spurring creativity in a number of industries.

Improved Connectivity for Autonomous Vehicles: 6G networks will provide extremely low latency and high data speeds, which are necessary for real-time communication between infrastructure and automobiles (Singh et. al., 2022).

AI Integration: By using adaptive learning techniques, 6G and AI together allow autonomous cars to make intelligent decisions, increasing efficiency and safety (Mizmizi et al., 2021).

Resource Utilization: Performance in dynamic situations can be optimized through real-time modifications made possible by the orchestration of cloud and network services (Shahzadi et al., 2023).

Reality Augmentation

High Bandwidth: Richer, more immersive material without latency will be possible because 6G's potential to provide data speeds of up to 1 Tbps, which will improve AR experiences (Singh et. al., 2022).

Context-Aware Services: By evaluating user context and preferences in real-time, the combination of AI with 6G can offer customized augmented reality experiences (Primmia et al., 2024).

Working Together Remotely- Smooth Communication: High-quality video and data sharing will be made possible by 6G, which is necessary for productive distant collaboration in work environments (Singh et. al., 2022).

Smart Infrastructure: More interactive and effective remote work environments will be made possible by the installation of smart sensors and AI-driven analytics (Mizmizi et al., 2021). Although B5G and 6G technologies have enormous potential, issues like security threats and moral dilemmas need to be resolved to guarantee their safe and fair use in all of these game-changing applications. (Primmia et al., 2024).

PRINCIPLES OF B5G AND 6G DEVELOPMENT

A number of significant reasons are contributing to the rapid development of technologies like B5G and 6G. First and foremost, the exponential rise in data consumption and the proliferation of linked devices necessitate the development of more advanced and efficient communication networks. There is a seven-fold expected increase in global mobile data traffic between 2017 and 2022., according to Cisco (2020). Networks with the capacity and speed to manage this increase in data demand must be in place. Furthermore, a key factor driving B5G and 6G is the incorporation of AI and ML into network operations. According to Chen et al. (2020), the utilization of these technologies results in improved efficiency and performance of communication systems through the implementation of predictive maintenance, intelligent network management, and efficient resource allocation. Additionally, it is predicted that the convergence of communication networks with other cutting-edge technologies like blockchain, edge computing, and quantum computing would lead to further breakthroughs in B5G and 6G. (Khan et al., 2020).

Capabilities and Key Technological Advancements

The capabilities that B5G and 6G networks will provide will significantly outweigh those of its predecessors. The ability to transmit data at extremely high speeds is one of the most significant developments; 6G is aiming for peak data rates of up to 1 Tbps. Higher frequency bands, like terahertz (THz) frequencies, will help with this. Although they offer more bandwidth, they also pose serious obstacles to signal penetration and propagation (Rappaport et al., 2019). The notable decrease in latency is another important development, as it is essential for real-time communication applications like remote surgery and driverless cars. As opposed to the 10 milliseconds that are usually associated with 5G networks, 6G networks are predicted to attain latencies as low as 1 millisecond (Zhang et al., 2020). Another important area for improvement is energy efficiency, with 6G technologies attempting to lower network energy usage through the use of AI-driven energy management techniques and the integration of energy-efficient hardware. This is especially crucial when it comes to sustainable growth and lowering the carbon footprint of the infrastructure supporting telecommunications (Akyildiz et al., 2020).

Potential Applications and Impact

The potential for B5G and 6G technology to revolutionize numerous industries is immense. These technologies have the potential to greatly improve the quality and accessibility of healthcare by enabling improved telemedicine services, remote monitoring, and AI-driven diagnostic tools (Rappaport et al., 2019). Ultra-reliable low-latency communication (URLLC) can help the transportation industry adopt autonomous cars and intelligent transportation systems more widely, increasing efficiency and safety (Zhang et al., 2020). Another significant area of effect is smart cities, where the integration of several IoT systems and devices—from smart grids to intelligent waste management and environmental monitoring—is made possible by B5G and 6G. This may result in better urban planning, more effective use of resources, and an improvement in the standard of living for locals (Akyildiz et al., 2020).

Modern civilization has been built on the swift development of telecommunications technologies, which have fueled technological innovation, social connectivity, and economic expansion. The research and development community is intensively investigating the next-generation technologies, including Beyond 5G (B5G) and 6G, as the limitations of 5G networks start to show. These developments could lead to previously unheard-of gains in network performance, which would help numerous industries undergo a digital revolution. In order to meet the growing need for faster data speeds, reduced latency, and increased energy efficiency, B5G and 6G technologies are planned. Although 5G networks offer notable improvements over 4G, including quicker speeds and more dependable connections, 6G and B5G seek to push these limits even farther. 6G is anticipated to provide peak data speeds of up to 1 Tbps, which is a significant improvement over 5G networks' capabilities, according to Zhao et al. (2021).

The development of B5G and 6G is mostly driven by the integration of cutting-edge technologies like artificial intelligence (AI), machine learning (ML), and the Internet of Things (IoT). These technologies are anticipated to be crucial for optimizing network efficiency and enabling previously unimaginable new applications. AI and ML, for instance, can be used to enhance network automation and management, leading to better resource consumption and increased service quality (Gupta & Jha, 2021). There are several potential applications for B5G and 6G technology in sectors such smart cities, transportation, and healthcare. These developments in medical technology, which allow for remote surgery and real-time health monitoring, have the potential to significantly improve patient outcomes (Rappaport et al., 2019). They can make it easier for fully autonomous vehicles to be used in the transportation sector, increasing traffic flow and safety (Zhang et al., 2020). Smart cities benefit from widespread connectivity and intelligent infrastructure, which creates more effective and sustainable urban environments (Akyildiz et al., 2020). Although there are plenty of development prospects, there are a lot of challenges associated with putting B5G and 6G technologies into practice. These include the need for significant expenditures in infrastructure, the development of new regulatory frameworks, and the resolution of privacy and security-related concerns. Saad et al. (2020) state that coordinated efforts from the government, industry stakeholders, and research community will be necessary to resolve these problems.

BRIEF OVERVIEW OF THE IMPORTANCE OF ADVANCED NETWORK TECHNOLOGIES

Advanced network technologies, such as B5G and 6G, are critical to supporting the burgeoning digital economy and enabling the proliferation of innovative applications. These technologies facilitate seamless connectivity, enabling real-time data exchange and interaction across various devices and platforms. This capability is essential for the development of smart cities, autonomous vehicles, telemedicine, and other applications that require reliable and high-speed communication networks. Furthermore, advanced network technologies are pivotal in driving economic growth by enhancing productivity and efficiency across multiple sectors. For instance, the implementation of IoT in manufacturing can lead to optimized production processes and reduced operational costs. In healthcare, advanced networks enable remote diagnostics and patient monitoring, improving healthcare accessibility and quality.

The fact that these technologies improve social connectedness serves as more evidence of their significance. As the use of digital communication platforms increases, modern networks make sure that people stay linked, even when they are separated by distance. In times of crisis, such natural disasters or pandemics, when strong communication networks can enable prompt coordination and response activities, this connectedness is especially important. Furthermore, cutting-edge network technologies support the creation of infrastructures that are more resilient and sustainable. Smart grid technologies and effective resource management systems can help these networks drastically cut down on energy use and their negative environmental effects. With the growing emphasis on sustainability in society, innovative network technologies play an ever-more-important role in accomplishing these goals. Modern civilization has been built on the swift development of telecommunications technologies, which have fueled technological innovation, social connectivity, and economic expansion. The research and development community is intensively investigating the next-generation technologies, including Beyond 5G (B5G) and 6G, as the limitations of 5G networks start to show. These developments could lead to previously unheard-of gains in network performance, which would help numerous industries undergo a digital revolution.

B5G and 6G technologies aim to address the increasing demand for higher data speed, lower latency, and better energy efficiency. While 5G networks provide significant advantages over 4G, such as faster speeds and more stable connections, 6G and B5G aim to push these boundaries even farther. According to Zhao et al. (2021), 6G is predicted to offer peak data speeds of up to 1 Tbps, which is a notable advance over the capabilities of 5G networks. The development of B5G and 6G is mostly driven by the integration of cutting-edge technologies like artificial intelligence (AI), machine learning (ML), and the Internet of Things (IoT). These technologies are anticipated to be crucial for optimizing network efficiency and enabling previously unimaginable new applications. AI and ML, for instance, can be used to enhance network automation and management, leading to better resource consumption and increased service quality (Gupta & Jha, 2021).

There are several potential applications for B5G and 6G technology in sectors such smart cities, transportation, and healthcare. These developments in medical technology, which allow for remote surgery and real-time health monitoring, have the potential to significantly improve patient outcomes (Rappaport et al., 2019). They can make it easier for fully autonomous vehicles to be used in the transportation sector, increasing traffic flow and safety (Zhang et al., 2020). Smart cities benefit from widespread connectivity and intelligent infrastructure, which creates more effective and sustainable urban environments (Akyildiz et al., 2020). Although there are plenty of development prospects, there are a lot of challenges associated with putting B5G and 6G technologies into practice. These include the need for significant expenditures in infrastructure, the development of new regulatory frameworks, and the resolution of privacy and security-related concerns. Saad et al. (2020) state that coordinated efforts from the government, business stakeholders, and scientific community will be necessary to resolve these problems.

PURPOSE AND SCOPE

The primary purpose of this research paper is to provide an exhaustive overview of Beyond 5G (B5G) and 6G technologies. This study aims to elucidate the key drivers, capabilities, applications, and challenges associated with these next-generation communication networks. By doing so, the paper seeks to inform and support a wide range of stakeholders, including policymakers, technologists, and business leaders, in understanding the transformative potential and implications of B5G and 6G technologies.

The following important aspects are included in the scope of this study:

Basic Concepts: This section delves into the fundamental ideas and technical developments that set B5G and 6G apart from their predecessors. It will go over the fundamentals that make these technologies special, providing the framework for comprehending their special traits and abilities.

Key Drivers: This section looks at the things that are driving the development of B5G and 6G technologies. It focuses especially on how advanced technologies like AI, ML, and the Internet of Things (IoT) are integrated. Understanding the motivation for the need for more sophisticated communication networks requires an understanding of these drivers.

Capabilities: The article examines the predicted gains in network intelligence, latency, data throughput, and energy efficiency that B5G and 6G are likely to offer. This research will show how these improvements clear the path for new opportunities while addressing the shortcomings of the current 5G networks.

Potential Applications: The revolutionary uses of B5G and 6G across a range of industries, such as smart cities, transportation, and healthcare, are covered in this section. It will demonstrate how these technologies have the power to completely transform these industries by presenting previously unheard-of chances for creativity and productivity.

Obstacles and Prospects: The study delineates the obstacles linked to the implementation of B5G and 6G technologies and investigates the chances for inventive thinking and cooperation that these obstacles afford. This section attempts to present a fair assessment of the possible obstacles and the teamwork needed to get beyond them.

Future Directions: The study offers an understanding of the research and development that will be required in the future to tackle the technological, legal, and security issues associated with the implementation of B5G and 6G. The future direction of these technologies will be discussed in this part, with a focus on the value of interdisciplinary cooperation.

HISTORICAL CONTEXT AND EVOLUTION OF MOBILE NETWORKS

Overview of 1G to 5G Technologies

From 1G to 5G, mobile networks have seen a tremendous technological journey. Each generation has brought about considerable advances in speed, capacity, and usefulness, transforming international communication.

First Generation, or 1G: Initiated in the 1980s, 1G stands for the first generation of analog cellular networks. These networks were mainly meant for voice communication and offered very few mobile telecommunication services. The Advanced Mobile Phone System (AMPS) was the most notable system because it laid the groundwork for further advancements in digital technology.

Second Generation or 2G: The definitive transition from analog to digital communication was signaled in the 1990s with the launch of 2G networks. Technologies like Code Division Multiple Access (CDMA) and the Global System for Mobile Communications (GSM) improved speech quality and enabled for more efficient use of the spectrum, paving the way for short messaging services (SMS). Moreover, compared to 1G networks, 2G networks had stronger security mechanisms.

Third Generation (3G): 3G networks made mobile internet access more feasible by introducing faster data transfer rates in the early 2000s. Technologies such as Universal Mobile Telecommunications System (UMTS) and CDMA2000 enabled multimedia messaging, video calling, and mobile internet browsing, considerably enhancing the mobile user experience, with average data speeds ranging from 200 kbps to several Mbps.

Fourth Generation, or 4G: Launched in the late 2000s and early 2010s, 4G networks yielded significant increases in data capacity and speed. With data rates ranging from 100 Mbps to 1 Gbps, Long-Term Evolution (LTE) and LTE-Advanced technologies enabled high-definition video streaming, online gaming, and other data-intensive activities. Reduced latency and improved connectivity are further benefits of 4G networks.

The most recent evolutionary stage, 5G (Fifth Generation) networks offer incredibly fast data rates, low latency, and wide-ranging connection for a wide range of devices. They were introduced in the late 2010s. 5G technology, such as millimeter-wave (mmWave) frequencies, massive MIMO (Multiple Input Multiple Output), and beamforming, offer data rates of up to 10 Gbps. In order to support cutting-edge applications like the Internet of Things (IoT), autonomous cars, and augmented/virtual reality (AR/VR), 5G represents a significant leap in mobile communications.

Transition from 5G to B5G and the Emergence of 6G

As 5G networks reach their full potential, attention gradually turns to 6G and Beyond 5G (B5G) technologies. The ongoing need for faster data speeds, lower latency, and improved network efficiency is what is causing this shift.

Beyond 5G (B5G): B5G aims to improve and expand the capabilities of current 5G technologies. It is the transitional phase between 5G and 6G. By reducing energy usage and addressing issues with coverage gaps, B5G seeks to enhance network dependability and performance. The use of new frequency bands, integration of non-terrestrial networks (such as satellites and drones), and sophisticated AI-driven network management are important areas of research. Sixth Generation, or 6G, is expected to appear around 2030 and promises revolutionary improvements over 5G and B5G. Peak data rates of up to 1

Tbps, extremely low latency in the microsecond range, and improved energy efficiency are all anticipated with 6G. In 6G networks, the incorporation of cutting-edge technologies like artificial intelligence (AI), machine learning, quantum computing, and sophisticated IoT will be essential. New applications like immersive extended reality (XR), holographic communication, and improved automation across multiple industries will be supported by these networks. The necessity for communication networks to be resilient and sustainable is another factor driving the development of 6G. In order to address the increasing demands of the digital age, researchers are looking into novel techniques to energy harvesting, effective spectrum usage, and secure communication protocols (Saad et al., 2020).

Defining B5G and 6G

Beyond 5G (B5G)

The term "Beyond 5G" (B5G) describes the advancement of wireless communication technology to a level that builds upon and expands upon the capabilities of fifth generation (5G) networks. B5G is a set of improvements and developments intended to alleviate some of the shortcomings of 5G, boost performance, and lay the groundwork for the future rollout of 6G networks rather than a single, clearly defined technology. Important traits and objectives of B5G:

Improved Performance: B5G promises to outperform 5G in terms of latency, spectrum efficiency, and data throughput. This entails combining cutting-edge modulation schemes, more effective coding techniques, and upgraded antenna technology with the optimization of current technologies.

Advanced Network Management: Using machine learning (ML) and artificial intelligence (AI) to run and manage networks more effectively. To meet the growing complexity and demands of future wireless communications, this includes adaptive network optimization, dynamic resource allocation, and predictive maintenance.

Integration of Non-Terrestrial Networks: In order to improve connectivity and offer ubiquitous coverage in rural or underserved areas, B5G aims to integrate non-terrestrial networks (NTNs), such as satellite communication systems, high-altitude platform stations (HAPS), and unmanned aerial vehicles (UAVs).

Energy Efficiency and Sustainability: Developing more energy-efficient network topologies and components, as well as investigating renewable energy sources for network infrastructure powering, are ways to address concerns about energy consumption and sustainability.

Security and Privacy: In the increasingly complicated and networked world of wireless communication, strengthening security and privacy measures is important to guard against new threats and weaknesses.

Sixth Generation (6G)

Sixth generation (6G) wireless communication technology is the next big step forward from B5G and is expected to be deployed by 2030. Revolutionary gains in data throughput, latency, connectivity, and energy economy are anticipated with 6G. The necessity for more intricate and immersive applications as well as the integration of new technology will drive these advancements. Among the primary characteristics and goals of 6G are:

Abruptly Elevated Data Rates With a maximum data throughput of one terabit per second (Tbps), 6G is expected to function significantly better than 5G and B5G. Ultra-high quality (UHD) and holographic video streaming will be seamless with this, supporting data-intensive applications like virtual reality (VR) and augmented reality (AR).

Ultra-Low Latency: 6G networks are expected to provide ultra-low latency, in the microsecond range, enabling real-time interaction and communication for applications such as driverless vehicles, industrial automation, and remote surgery.

Massive connection: 6G will make it feasible for billions of things to have a massive connection, advancing the Internet of Things (IoT) and paving the way for the development of interconnected ecosystems like smart cities and homes. This involves complex machine-to-machine (M2M) communication and the smooth integration of multiple IoT devices. connectivity.

Intelligent and Autonomous Networks: 6G will make use of quantum computing, AI, and ML to build intelligent and autonomous networks that can manage resources adaptively and optimize themselves. These networks will have the capacity to anticipate and react instantly to shifting circumstances and user needs.

Advanced Services and Applications: 6G will make it possible to use cutting-edge services and applications like digital twins, holographic communication, immersive extended reality (XR), and enhanced telepresence. Innovation in a number of industries, including healthcare, education, entertainment, and manufacturing, will be fueled by these applications.

Energy Efficiency and Sustainability: To further reduce the environmental impact of wireless networks, 6G will investigate cutting-edge methods for energy harvesting, effective spectrum usage, and green communication technologies.

Security and Privacy: To fend against ever-more-advanced cyberattacks, 6G will include cutting-edge security and privacy protections. This covers the creation of improved privacy-preserving methods, quantum-resistant encryption, and secure communication protocols.

TECHNICAL DEFINITIONS

To give a complete understanding of these state-of-the-art wireless communication systems, it is essential to define the key technological terms and concepts that underpin Beyond 5G (B5G) and 6G technologies. Peak data rate is the maximum data transfer rate that a network can achieve under ideal conditions. Peak 6G data rates are anticipated to be much higher than 5G ones, possibly exceeding 1 terabit per second (Tbps). The smooth streaming of ultra-high-definition video and data-intensive applications like augmented reality (AR) and virtual reality (VR) will be made possible by this enormous increase in data rates. The time difference between sending and receiving data is referred to as “latency”. 6G promises to enable real-time applications such as remote surgery and autonomous driving with ultra-low latency in the microsecond range. Applications that need to interact and receive data instantly would profit immensely from this decrease in latency. The spectral efficiency of a network indicates how well it uses a given spectral bandwidth. Greater spectral efficiency, which permits more data to be carried over a given bandwidth, is one of the main goals of B5G and 6G. Improved spectral efficiency will ensure that the steadily growing demand for wireless data can be satisfied without requiring an equivalent amount of additional spectrum.

Massive MIMO (Multiple Input Multiple Output) technology uses a large number of antennas at the base station to increase spectral efficiency and accelerate data rates. This technology is essential for supporting simultaneous increases in the number of users and data capacity in both B5G and 6G networks. The term "millimeter-wave" (mmWave) frequencies refers to electromagnetic waves with millimeter-wavelengths between 30 and 300 GHz. These frequencies are utilized in 5G and B5G and provide high data speeds; 6G is anticipated to investigate even higher frequency bands. 6G will also make use of ***Terahertz (THz) Frequencies***, which run from 0.1 THz to 10 THz, in order to accommodate new applications and achieve extremely high data rates.

The integration of Non-Terrestrial Networks (NTNs) into B5G and 6G networks will provide ubiquitous coverage and improved connection in remote locations. These NTNs include satellites, high-altitude platform stations (HAPS), and unmanned aerial vehicles (UAVs). With this connection, cutting-edge wireless services will be available to even the most remote areas. Antenna arrays employ beamforming, a signal processing technique, to steer signal transmission or reception in particular directions. Through the enhancement of signal quality, reduction of interference, and concentration of the signal where it is most needed, this technique increases the effectiveness and performance of wireless communication systems. By bringing processing and data storage closer to the point of need, edge computing lowers latency and boosts application performance in B5G and 6G networks. This technology improves user experience overall and facilitates real-time decision-making by processing data at the edge of the network. Technologies related to Artificial Intelligence (AI) and Machine Learning (ML) allow networks to self-manage and optimize. Predictive maintenance, dynamic resource allocation, and adaptive network optimization are made possible by these technologies, which are essential to the functioning of intelligent and adaptable B5G and 6G networks. Compared to traditional computers, quantum computing uses the ideas of quantum physics to execute complicated calculations more quickly.

It is anticipated that quantum computing would contribute to the advancement of 6G technology, specifically by improving security and opening up new application possibilities.

The term "Internet of Things" (IoT) describes a system of networked, data-exchanging, and communicating objects. Massive IoT deployments are supported by B5G and 6G networks, enabling smart homes, businesses, and cities. IoT's connectivity and intelligence will propel major developments across a number of industries. Augmented reality (AR), virtual reality (VR), and mixed reality (MR) are all included under the general term "extended reality" (XR). High data rates and low latency XR apps will be supported by 6G networks, offering engaging and immersive experiences. Digital twins are controlled, analyzed, and simulated versions of real-world objects. 6G networks will make it easier to create and maintain digital twins, which will spur efficiency and innovation in sectors like logistics, healthcare, and manufacturing. Energy harvesting is the process of obtaining and storing energy from outside sources, like ambient RF signals or the sun. In order to decrease dependency on conventional power sources and increase the sustainability of communication networks, energy harvesting techniques are being investigated for use in B5G and 6G.

The term "quantum-resistant encryption" describes cryptographic techniques created to withstand assaults by quantum computers. Quantum-resistant encryption will be incorporated into 6G networks to improve security and guarantee that data is safeguarded even as computational power grows. Real-time transmission of three-dimensional holographic pictures is known as holographic communication. With the help of 6G, holographic communication should be possible, providing immersive and dynamic experiences for telepresence, entertainment, and educational uses. A method called "network slicing" makes it possible to build several virtual networks on top of a single real network infrastructure. In B5G and 6G, network slicing allows for customized network services for various applications, offering specialized solutions for a range of customer needs.

KEY DIFFERENTIATORS FROM PREVIOUS GENERATIONS

Wireless communication technology has advanced significantly between 1G and 5G, with each generation building on the advantages of the one before it. Beyond 5G (B5G) and 6G bring significant innovations that set them apart from earlier generations while resolving current issues and expanding the limits of efficiency, performance, and capability. One of the biggest improvements in B5G and 6G is Data Rates and Bandwidth. B5G and 6G aspire to attain peak data rates of up to 1 Tbps, whereas 5G supports up to 20 Gbps and uses a wider spectrum, including mmWave frequencies (30 GHz to 300 GHz). In order to enable previously unheard-of bandwidth and data speeds, this sharp rise will make use of higher frequency bands, such as Terahertz (THz) frequencies (0.1 THz to 10 THz). Applications requiring massive amounts of data and ultra-high quality, such as immersive extended reality (XR) and holographic communication, depend on these capabilities.

Another important differentiator is latency. 5G enables real-time applications like industrial automation and autonomous driving by achieving low latency of about 1 millisecond. B5G and 6G, on the other hand, seek to achieve microsecond-level latency reduction, enabling almost instantaneous communication. For applications like remote surgery and advanced telepresence that demand instantaneous input and interaction, where even little delays can be harmful, this ultra-low latency is essential.

Improvements in device density and connectivity are essential to B5G and 6G. 5G is able to handle the expansion of the Internet of Things (IoT) by supporting up to one million devices per square kilometer. With the goal of supporting even higher device densities, B5G and 6G will make it easier for IoT devices and smart environments to proliferate. Smart homes, towns, and businesses will innovate as a result of this improved connectivity, becoming more effective and user responsive.

Significant progress has been made in the areas of network capacity and spectrum efficiency. Massive MIMO (Multiple Input Multiple Output) and beamforming are two new 5G technologies that increase network capacity and spectral efficiency. Through more sophisticated modulation schemes, effective coding techniques, and enhanced antenna technology, B5G and 6G will significantly improve these parameters. Thanks to these developments, the increasing demand for wireless data will be able to be satisfied without requiring an equal number of airwaves. With B5G and 6G, intelligence and network management are about to undergo a revolutionary change. While AI and ML were incorporated into network management and optimization in 5G, B5G and 6G will fully utilize these technologies in conjunction with quantum computing to create intelligent and autonomous networks that are capable of adaptive resource management, self-healing, and self-optimization. These networks will improve overall performance and efficiency by anticipating and reacting in real-time to changing conditions and user needs.

Energy Efficiency and Sustainability are increasingly prioritized in the development of B5G and 6G. Despite being more energy-efficient than prior generations, 5G still has issues because of the higher power requirements of sophisticated technologies. In order to reduce the negative environmental effects of wireless networks and increase their sustainability, B5G and 6G will investigate novel methods for energy harvesting, effective spectrum usage, and green communication technologies.

Improvements in Security and Privacy are essential to B5G and 6G. Although security and privacy safeguards were enhanced by 5G in comparison to 4G, new threats and vulnerabilities are still being discovered. Advanced security and privacy features will be incorporated into B5G and 6G to guard against increasingly complex cyberthreats. To provide strong security for users and data, this includes the development of secure communication protocols, quantum-resistant encryption, and improved privacy-preserving strategies.

The integration of Non-Terrestrial Networks (NTNs) is a significant accomplishment in B5G and 6G. 5G began integrating NTNs, such satellites, to extend coverage to remote areas, but B5G and 6G will fully integrate NTNs, including high-altitude platform stations (HAPS) and unmanned aerial vehicles (UAVs), to give ubiquitous coverage and increase connectivity. Even in the most isolated locations, cutting-edge wireless services will be accessible with this connection. Advanced apps and services are another domain in which B5G and 6G will excel. B5G and 6G will open up new and more sophisticated applications like digital twins, immersive XR, holographic communication, and advanced telepresence, while 5G makes it easier to use applications like enhanced mobile broadband, large machine-type communication, and ultra-reliable low-latency communication. These applications will drive innovation across several areas such as healthcare, education, entertainment, and manufacturing.

CORE PRINCIPLES AND TECHNICAL STANDARDS

A number of fundamental ideas and technological guidelines that guarantee these cutting-edge wireless communication systems live up to the high requirements of performance, efficiency, and dependability are guiding the development of B5G and 6G technologies. In order to achieve the high standards set for B5G and 6G, promote innovation, and guarantee global interoperability, these guidelines and standards are needed.

Fundamental Ideas-Ubiquitous connection: The goal of ubiquitous connection, which is the cornerstone of B5G and 6G technologies, is to enable smooth and continuous communication in a variety of settings. In order to guarantee that all areas have access to fast and dependable wireless networks, this covers urban, rural, and distant areas. This idea backs the objective of creating fair access to digital services globally and closing the digital gap.

Ultra-Low Latency: Applications requiring real-time feedback and interactivity must be supported with ultra-low latency. With latency in the microsecond range, B5G and 6G networks are able to support cutting-edge use cases like self-driving cars, remote surgery, and fully immersive virtual reality experiences. Time-sensitive applications can function successfully and efficiently thanks to this principle.

High Data Rates: In order to meet the growing demand for services and applications that rely heavily on data, high data rates are necessary. Massive IoT deployments, holographic communication, and streaming of ultra-high-definition video are all made possible by the peak data rates that B5G and 6G are expected to deliver—up to 1 Tbps. This idea guarantees consumers receive dependable and quick data transfer, improving the overall caliber of the service.

Energy Efficiency and Sustainability: For B5G and 6G networks, energy efficiency and sustainability are essential factors. By employing creative strategies like energy harvesting, effective spectrum usage, and green communication technologies, these technologies place a high priority on lowering power consumption and their negative effects on the environment. This idea guarantees the environmental sustainability and sustainability of cutting-edge wireless networks.

Security and Privacy: In B5G and 6G networks, security and privacy are critical for safeguarding users and data. Advanced security features such as secure communication protocols, quantum-resistant encryption, and privacy-preserving methods are included in these technologies. This approach protects sensitive data by providing strong defenses against cyberattacks and unlawful access.

Technical Standards and Protocols

3GPP Standards: A collaboration between telecom standards groups, the 3rd Generation Partnership Project (3GPP) develops international standards for mobile networks. B5G and 6G technologies will build upon the standards outlined in Release 17 and subsequent 3GPP releases. These standards encompass network architecture, radio access technologies, and security protocols.

IEEE Standards: Among other fields of wireless communication, the Institute of Electrical and Electronics Engineers (IEEE) develops standards for emerging technologies, Wi-Fi, and the Internet of Things. IEEE standards including IEEE 802.11 for wireless local area networks and IEEE 802.15 for wireless personal area networks will play a major role in the development and compatibility of B5G and 6G technologies. The International Telecommunication Union's Radio Communication Sector (ITU-R) provides global norms for radio communication systems, including spectrum allocation and utilization. To ensure efficient spectrum usage and worldwide harmonization, ITU-R criteria, such as those outlined in the IMT-2020 and IMT-2030 frameworks, will direct the development of B5G and 6G technologies. Several virtual networks can be created on a single physical network infrastructure thanks to a technique called network slicing. By providing customized network services for different applications, this approach guarantees efficient resource allocation and peak performance. Network slicing, a crucial element of B5G and 6G networks, will meet a range of use cases and service needs.

AI and ML Integration: The intelligent and autonomous fusion of artificial intelligence (AI) and machine learning (ML) is essential to the intelligent and autonomous management of B5G and 6G networks. Because of these advancements, wireless networks will operate more efficiently and effectively overall. Adaptive optimization, dynamic resource allocation, and predictive maintenance are made possible. Consistency and interoperability will be ensured via standards and conventions pertaining to AI and ML.

Quantum-Resistant Encryption: As quantum computing advances, traditional encryption methods could be vulnerable to quantum attacksB5G and 6G networks will use quantum-resistant encryption techniques to ensure data security in the face of exponentially increasing processing power. These encryption best practices will provide robust protection against impending cyber-attacks.

Key Innovations in B5G and 6G

The shift from 5G to 6G and B5G (Beyond 5G) technologies signals the beginning of a new chapter in wireless communication history, one filled with ground-breaking inventions. The goal of these developments is to overcome the shortcomings of the networks that are in use today and to push the limits of connectivity, speed, efficiency, and applicability.

1. Transportation in Terahertz (THz): Among the biggest advancements in 6G is the application of Terahertz (THz) frequencies, which span from 0.1 THz to 10 THz. With the massive bandwidth these frequencies provide, data rates as high as 1 Tbps are possible. Holographic communication,

ultra-high resolution video streaming, and other data-intensive applications will be made possible by THz communication, offering a user experience never before seen.

2. Advanced Antenna Technologies: Intelligent reflecting surfaces (IRS) and ultra-massive Multiple Input Multiple Output (MIMO) are two examples of the advanced antenna technologies that 6G will use. By using hundreds or even thousands of antennas at the base station, ultra-massive MIMO dramatically increases network capacity and spectral efficiency. Particularly in difficult settings, intelligent reflecting surfaces have the ability to control electromagnetic waves to increase signal strength and coverage.
3. Integration of Artificial Intelligence (AI) and Machine Learning (ML): One of the main innovations in B5G and 6G is the integration of AI and ML into network management. These technologies will allow for intelligent, self-healing, autonomous networks with predictive maintenance and self-optimization capabilities. AI-driven analytics will optimize network speed, optimize resource allocation, and improve energy efficiency while instantly adjusting to changing user demands and environmental circumstances.
4. Quantum computing and communication: B5G and 6G networks stand to be completely transformed by quantum technology. Using the concepts of quantum mechanics, quantum communication provides extremely secure data transport, offering strong defense against cyberattacks and eavesdropping. Complex problem-solving capabilities will be made possible by quantum computing, improving network security, optimization, and overall performance.
5. Holographic and Immersive Communication: 6G will enable cutting-edge uses including immersive extended reality (XR), which encompasses virtual, augmented, and mixed reality, and holographic communication. Ultra-high data speeds, minimal latency, and flawless communication are necessary for these applications. These futuristic communication technologies will become a reality with advancements in real-time data processing, sensor integration, and display technology.
6. Network Slicing and Customization: A method called network slicing makes it possible to build several virtual networks on a single physical infrastructure. This invention ensures optimal performance and resource utilization by enabling tailored network services for various applications and user groups. Network slicing capabilities will be improved by B5G and 6G, enabling a variety of use cases like industrial automation, smart cities, and autonomous driving.
7. Energy Harvesting and Green Communication: B5G and 6G technologies must prioritize sustainability. Energy harvesting innovations, such harnessing ambient energy sources (solar, wind, and radiofrequency energy), will power network devices and lessen dependency on conventional power sources. Wireless networks' negative environmental effects will be reduced by green communication technology, which will increase energy efficiency.
8. Sophisticated Security procedures: B5G and 6G will have sophisticated security procedures to safeguard user privacy and secure data in light of the growing sophistication of cyber threats. Some of the breakthroughs that will improve network security include blockchain technology, AI-driven threat detection, and encryption that is immune to quantum attacks. Strong defense against cyberattacks and unauthorized access will be offered by these procedures.
9. Integration of Non-Terrestrial Networks (NTNs): One of the main innovations of B5G and 6G will be the integration of non-terrestrial networks, such as satellites, unmanned aerial vehicles (UAVs), and high-altitude platform stations (HAPS). By providing coverage to underserved and rural places, NTNs will guarantee ubiquitous connection. Applications for remote sensing, disaster recovery, and international communication will be supported by this integration.

10. Advanced Coding and Modulation Schemes: B5G and 6G networks will have higher data throughput and spectrum efficiency because to advances in coding and modulation techniques. The performance and dependability of wireless communication will be enhanced by strategies like polar coding, orthogonal time frequency space (OTFS) modulation, and sophisticated error correction techniques, particularly in high-mobility and high-interference conditions.

ADVANCED COMMUNICATION TECHNOLOGIES IN B5G AND 6G

Advanced Coding and Modulation Schemes: Thanks to developments in coding and modulation techniques, B5G and 6G networks will be able to handle more data at a faster rate and with greater spectrum efficiency. Strategies such as polar coding, orthogonal time frequency space (OTFS) modulation, and advanced error correction techniques will improve wireless communication performance and reliability, especially in high-mobility and high-interference environments.

Terahertz (THz) Communication: In the field of wireless networks, Terahertz communication represents a major advancement. THz communication, which uses frequencies between 0.1 and 10 THz, provides previously unheard-of bandwidth, allowing for data rates of up to 1 Tbps. For data-intensive applications like holographic communication, virtual and augmented reality (VR) experiences, and streaming of extremely high-resolution video, this wide spectrum availability is perfect. THz waves are ideal for indoor applications where high-speed communication is essential and dense urban environments because of their capacity to transport massive volumes of data over short distances.

Artificial Intelligence (AI) and Machine Learning (ML) Integration

B5G and 6G technology development depends on artificial intelligence (AI) and machine learning (ML). These advanced networks will leverage AI and ML to enable intelligent, autonomous operation and enhance network performance, efficiency, and user experience. Dynamic and adaptive network resource optimization is made possible by AI-driven network management. By continuously observing and analyzing network conditions, artificial intelligence (AI) may anticipate possible problems and take action before they undermine performance. This guarantees that the network performs at its best all the time and includes load balancing, interference management, and energy efficiency improvement.

Real-time detection and mitigation of cyber threats can be achieved via machine learning algorithms, hence improving security. ML can detect anomalies that can point to malicious activities by examining user behavior and network traffic patterns, allowing for proactive defense against cyberattacks. Another use of AI in B5G and 6G networks is predictive maintenance. Artificial Intelligence (AI) can reduce downtime and operating expenses by predicting failures and scheduling repair before interruptions occur by analyzing data from network equipment. Beamforming with Massive MIMO Large arrays of antennas are used in massive multiple input multiple output (MIMO) technology to send and receive more data at once. This breakthrough is essential to B5G and 6G networks since it greatly boosts spectrum efficiency and network capacity.

Using a technique called beamforming, communications are targeted at particular people or devices as opposed to being broadcast everywhere. Particularly in densely populated locations, this concentrated broadcast lessens interference, boosts signal quality, and improves user experience. With the use of cutting-edge technology, **Intelligent Reflecting Surfaces (IRS)** can regulate electromagnetic wave propagation to increase signal strength and coverage. Through surface features like reflectivity, IRS can improve signal quality in places like interior spaces and urban canyons that are usually difficult for wireless communication to work in. A revolutionary development in safe data transfer is quantum communication. Superposition and entanglement, two concepts from quantum mechanics, can be used in quantum communication to create extremely secure communication channels that are impervious to hacking and eavesdropping. This technology is especially important for sensitive applications that need the utmost security, like government communications, financial transactions, and military activities. With the introduction of B5G and 6G, holographic communication and immersive extended reality (XR), which encompasses virtual reality (VR) and augmented reality (AR), are expected to become commonplace. For these applications to provide lifelike experiences, very high data rates, minimal latency, and seamless communication are necessary.

These futuristic communication techniques will be made possible by developments in real-time data processing, sensor integration, and display technology. These advancements will revolutionize industries like education, healthcare, entertainment, and distant work.

Network Segmentation and Personalization: With network slicing, it is possible to create several virtual networks, each suited to a particular set of application requirements, on a single physical infrastructure. For a wide range of use cases, including high-throughput mobile broadband services and low-latency industrial automation, this breakthrough allows for optimal resource allocation and performance. Network slicing capabilities will be improved by B5G and 6G, guaranteeing that networks can be dynamically and individually tailored to meet the demands of many industries and applications.

Energy Harvesting and Green Communication: By utilizing ambient energy from sources such as solar, wind, and radio frequency signals, energy harvesting technology can power network devices while lowering dependence on conventional power grids. Green communication solutions can minimize the environmental impact of wireless networks by improving their energy efficiency. The longevity of upcoming communication networks depends on these technologies.

Advanced Coding and Modulation Schemes: Data throughput and spectrum efficiency will be improved by advances in coding and modulation techniques. The performance and dependability of wireless communication will be enhanced by methods like polar coding and orthogonal time frequency space (OTFS) modulation, particularly in difficult situations with high mobility or interference.

Enhanced Network Capabilities in B5G and 6G

With the introduction of B5G (Beyond 5G) and 6G technologies, wireless communication is about to enter a revolutionary new age marked by much improved network capabilities. With new capabilities that serve a wide range of applications, from ultra-reliable low-latency communication to vast connectivity for IoT devices, these developments aim to overcome the shortcomings of existing networks.

Extremely Low Latency: Ultra-low latency is one of the key features of B5G and 6G networks. Applications that demand real-time feedback and interaction must consider latency, or the amount of time it takes for data to go from its source to its destination. In comparison to the millisecond latencies typical of 4G and 5G networks, B5G and 6G aspire to attain latencies in the region of microseconds.

For autonomous vehicles to operate safely and effectively, real-time communication between the vehicles and infrastructure must be enabled, which requires ultra-low latency. The ability for surgeons to operate remotely and receive prompt feedback is known as remote surgery, and it is essential for accuracy and patient safety. The goal of virtual reality (VR) and augmented reality (AR) is to provide smooth, immersive user experiences free from motion sickness and lag. Abundant Interconnectivity Massive connectivity, which tackles the exponential expansion in the number of connected devices, especially with the rise of the Internet of Things (IoT), is another essential feature of B5G and 6G. Compared to earlier generations, B5G and 6G networks are far more capable, supporting up to one million connected devices per square kilometer.

Strong connectivity is essential for:

Smart Cities: Enabling the communication and integration of several IoT devices for effective urban management, such as energy management, public safety, and traffic control.

Industrial Automation: Increasing productivity and efficiency in factories by enabling large sensor networks and automated systems.

Environmental Monitoring: Encouraging the extensive use of sensors to monitor environmental variables, which will improve management and aid in conservation efforts. Enhanced Mobile Broadband (eMBB) intends to meet the needs of data-intensive apps and services by offering extremely fast internet access. Peak data speeds of up to 1 Tbps will be available with B5G and 6G, guaranteeing customers a dependable and quick data transfer—crucial for applications such as enabling smooth streaming of video in ultra-high quality (4K/8K).

Holographic Communication: Providing three-dimensional, real-time interaction for personal and business purposes encouraging distant cooperation and high-quality video conferencing through telepresence and telecommuting.

Dependability and Resilience: B5G and 6G networks will also have improved resilience and dependability, guaranteeing that communication services continue to function and be reliable even in the face of unfavorable circumstances or heavy traffic volumes. This includes:

Ultra-Reliable Low-Latency Communications (URLLC): Essential for mission-critical applications, this technology ensures data transfer with extremely low error rates.

Mechanisms for redundancy and failover: Providing network continuity and disaster recovery in the event of natural disasters or component failures.

Energy Efficiency and Sustainability: When designing contemporary communication networks, energy efficiency and sustainability are becoming more and more crucial. The goal of B5G and 6G technologies is to lower power usage by using creative methods like utilizing ambient energy sources, such as sun, wind, and radiofrequency energy, to power network devices is known as energy harvesting.

Green Communication Techniques: Using energy-saving algorithms and protocols to reduce their negative effects on the environment. A sophisticated feature called "network slicing" enables the development of several virtual networks on a single physical infrastructure. Every slice provides excellent performance for various use scenarios and can be customized to meet the needs of individual applications.

Sliced networking allows for:

Personalized Service Levels: Offering distinct services for different uses, including high-bandwidth slices for streaming videos and low-latency slices for playing games.

Efficient Resource Utilization: Increasing the overall efficiency of the network by making sure that resources are distributed dynamically in response to demand. Intelligent and autonomous management of network resources will be made possible by the integration of artificial intelligence (AI) and machine learning (ML) into B5G and 6G networks.

Predictive Upkeep: Consider and resolve possible problems before they have an influence on network performance.

Dynamic Optimization: Adjust network settings on a constant basis to accommodate shifting circumstances and user needs. Security Enhancements: Use powerful pattern recognition and anomaly detection to quickly identify and neutralize cyber-attacks.

Applications and Use Cases for B5G and 6G Technologies

Beyond 5G (B5G) and 6G technologies are expected to completely transform a variety of sectors and societal roles. Innovative use cases and applications that take advantage of these cutting-edge networks' improved features—like ultra-low latency, wide connectivity, and high data rates—will be made possible by them. Some of the most important use cases and applications for B5G and 6G technologies are listed below:

Autonomous vehicles: The ultra-low latency and high reliability required for autonomous vehicles to connect in real time with infrastructure will be made available by B5G and 6G. As a result, self-driving cars will be able to operate in more safe and effective conditions, avoiding collisions and improving traffic flow.

Intelligent Transportation Systems (ITS): Smart traffic management systems that can monitor and control traffic conditions, lessen congestion, and increase the effectiveness of public transportation by gathering and analyzing data in real-time will be supported by advanced networks.

Integration of IoT with Smart Cities: B5G and 6G's huge connection will make it possible to build extensive IoT networks inside of smart cities. Applications like energy management, water distribution, trash management, and smart lighting will be supported by this, creating more sustainable and effective urban environments.

Public Safety and Surveillance: By making it easier to use high-definition cameras and sensors for emergency response and real-time surveillance, enhanced network capabilities will improve public safety and security.

Healthcare & Telemedicine-Remote Surgery: With robotic systems, surgeons can execute surgeries remotely while maintaining accurate and instantaneous control over surgical tools thanks to their ultra-low latency and great reliability.

Telemedicine: B5G and 6G will facilitate the transfer of massive medical data files, high-quality video consultations, and remote patient monitoring. This will improve access to healthcare services, especially in underserved and remote locations.

Industry 4.0 and Industrial Automation-Smart Factories: Smart sensors, robotics, and automation systems can be deployed in industrial settings thanks to the integration of advanced networks. Real-time communication between these systems results in increased productivity, decreased downtime, and better-quality output.

Predictive Maintenance: By combining AI and ML with B5G and 6G, machinery can be maintained predictably, prolonging its lifespan and preventing unplanned malfunctions.

Virtual Reality (VR) and Augmented Reality (AR): Smooth and immersive VR and AR experiences will be made possible by the high data rates and low latency of B5G and 6G. Applications for training, education, gaming, and remote collaboration will all benefit from this.

Holographic Communication: Real-time holographic communication will be made possible by advanced networks, enabling lifelike interactions in social media, teleconferences, and virtual meetings.

Ultra-High-Definition Content over Enhanced Mobile Broadband (eMBB): This technology enables users to stream content in 4K, 8K, and even higher resolutions, giving them better video quality and more engaging media experiences.

Cloud gaming: B5G and 6G will make it possible for cloud gaming services to provide high-quality graphics and low latency, so users can play demanding games without needing expensive hardware.

Sensor Networks for Disaster Management and Environmental Monitoring: The use of advanced networks will make it easier to set up sensor networks for weather, water, and air quality monitoring. Analyses and responses can be made using this data in real time.

Disaster Management: Better communication and coordination during natural disasters will allow for quicker and more efficient emergency response and recovery efforts. This is made possible by enhanced connectivity.

B5G and 6G networks will enable the use of IoT devices, drones, and autonomous machinery in agriculture, enabling precision agriculture and smart farming. By monitoring crop health, weather patterns, and soil conditions, these technologies can help improve farming techniques and boost output.

Resource Management: Precise control over pest, fertilization, and irrigation will be made possible by advanced networks, lowering resource consumption and environmental effect.

Improved Consumer Applications-Smart Homes: By integrating B5G and 6G into smart home ecosystems, better connectivity will be available for home automation, security systems, and smart appliances. This will increase convenience, security, and energy efficiency.

Wearable Devices: Newer wearables with improved features for fitness tracking, health monitoring, and augmented reality apps will be supported by advanced networks.

Quantum communication technologies will be used by B5G and 6G to create ultra-secure communication channels that guarantee data privacy and protection from cyberattacks. This will result in advanced security and privacy.

Blockchain Integration: More advanced networks will facilitate the incorporation of blockchain technology for data integrity, identity verification, and safe and transparent transactions across a range of applications, including supply chain management and finance.

SMART CITIES AND IOT IN THE CONTEXT OF B5G AND 6G TECHNOLOGIES

Beyond 5G (B5G) and 6G technologies are expected to transform urban environments by incorporating cutting-edge communication networks into smart cities and Internet of Things (IoT) architectures. Cities will be able to function more effectively, sustainably, and securely thanks to these technologies, which will provide hitherto unheard-of levels of automation, data interchange, and communication. Smart Cities use cutting edge technology to promote sustainable growth, increase operational effectiveness, and improve the quality of life for their citizens. These cities will be supported by B5G and 6G, which will allow smooth data integration and communication between diverse urban systems.

1. Intelligent Transportation Systems (ITS):
    Traffic Management: To maximize traffic flow, lessen congestion, and enhance road safety, real-time data from sensors, traffic signals, and linked cars will be examined.
    Public Transportation: With real-time updates and predictive maintenance, enhanced connection will allow for more dependable and efficient public transportation systems.
2. Energy Management:
    Smart Grids: B5G and 6G will facilitate the installation of smart grids, which may enhance energy efficiency, integrate renewable energy sources, and dynamically balance the supply and demand for energy.
    Smart Lighting: By adjusting brightness in response to pedestrian activity and environmental conditions, connected streetlights can save energy and enhance public safety.
3. Public Safety and Security:
    Surveillance Systems: B5G and 6G networks will link high-definition cameras and sensors, enabling real-time monitoring and quick incident response.
    Disaster Management: Coordinating and communicating during calamities will be made easier by advanced communication networks, which will also make disaster response and recovery activities more effective.
4. Trash Management: IoT-enabled Smart Bins trash bins can minimize environmental impact and save operational expenses by monitoring fill levels and optimizing collection routes.
    Recycling Programs: Data from linked systems will help increase recycling rates and enhance waste management.
5. Water Management:
    Smart Meters: State-of-the-art water meters will offer real-time usage data, assisting in leak detection, consumption management, and water conservation promotion.
    Flood Monitoring: IoT sensors will keep an eye on water levels and send out early alerts in the case of possible flooding, enabling prompt action.

Internet of Things (IoT)

The Internet of Things, or IoT, is a network of physical items that are outfitted with sensors, software, and other technologies to exchange data and communicate with other systems and devices over the internet. B5G and 6G will significantly expand the possibilities of the Internet of Things by enabling a higher density of linked devices and more complex and data-intensive applications.

Massive Device Connectivity: Large-scale IoT ecosystems can be deployed in urban areas thanks to B5G and 6G networks, which can handle up to a million connected devices per square kilometer.
Enhanced Data Transmission: Real-time data gathering and analysis will be possible with ultra-high-speed data transmission and low-latency communication, which is crucial for applications like smart grids and autonomous cars.

Edge Computing: By combining edge computing with B5G and 6G networks, data processing may be done closer to the source, which will save bandwidth and latency. This is critical for applications like industrial automation and healthcare that need quick reactions.
Integration of AI and Machine Learning: AI and ML will be essential in evaluating the massive volumes of data produced by Internet of Things devices, offering insights for anomaly detection, predictive maintenance, and optimal resource management.
Security and Privacy: To safeguard IoT devices and data from cyber-attacks, B5G and 6G will have cutting-edge security methods. Secure communication and data integrity will be guaranteed by improved encryption, authentication methods, and blockchain integration.

Applications and Use Cases

1.Smart Homes:
Home Automation: B5G and 6G-connected IoT devices will allow for smooth control of appliances, heating, lighting, and security systems, improving comfort and energy economy.
Health Monitoring: By continuously monitoring vital signs and giving healthcare providers access to real-time data, wearable health devices will enable proactive health management.

2. Industrial IoT (IIoT) Predictive Maintenance: By foreseeing issues and tracking performance, linked sensors in industrial equipment can increase operational efficiency and save maintenance costs.

Supply Chain Optimization: Inventory management and logistics are optimized through the use of real-time tracking and data analysis, which lowers costs and boosts productivity.

3. Precision farming in agriculture: By employing Internet of Things (IoT) sensors to track crop health, weather patterns, and soil conditions, farmers will be able to make data-driven decisions and boost yields.

Automated Machinery: Planting, fertilizing, and harvesting will be done precisely and efficiently by drones and linked tractors.

4. Healthcare:

Remote Patient Monitoring: IoT-enabled devices will enable ongoing patient health monitoring, supplying information for problem early identification and minimizing hospital visits.
Smart Medical Devices: The usage of linked medical devices for patient management, diagnosis, and treatment will be supported by advanced networks.

5. Retail:

Smart Stores: Internet of Things (IoT) technologies will make it possible for automated checkout processes, customized shopping trips, and effective inventory control.
Supply Chain Transparency: Real-time visibility into the supply chain will be made possible by improved connectivity, assuring product quality and cutting waste.

Enhanced Mobile Broadband (eMBB) in B5G and 6G Technologies

One of the key elements of B5G (Beyond 5G) and 6G technologies is enhanced mobile broadband, or eMBB. Its purpose is to offer incredibly fast internet access and seamless connection for applications that require a lot of data. By providing noticeably faster data rates, more capacity, and wider coverage than earlier mobile broadband technologies, eMBB expands their potential. The fundamentals and uses of eMBB are examined in this section in relation to B5G and 6G.

Core Aspects of eMBB

Ultra-High Data Rates: In order to facilitate quick download and upload times, B5G and 6G plan to offer peak data rates of up to 1 Tbps. Real-time cloud services, immersive digital experiences, and high-definition media streaming will all be made possible by this breakthrough in data transport capacity.

Enhanced Network Capacity: eMBB in B5G and 6G networks will significantly increase network capacity to manage the surge in data consumption and the growing number of connected devices. Sophisticated technologies, like large MIMO (Multiple Input Multiple Output) and beamforming, will be employed to improve resource allocation and maximize spectral efficiency.

Broad Coverage: B5G and 6G networks will expand coverage to ensure consistent and dependable connectivity in both urban and rural locations. This will close the digital gap and increase the number of individuals with access to high-speed internet services.

Seamless Connectivity: eMBB will offer seamless connectivity in a variety of settings, including isolated locales, densely populated areas, and situations involving high mobility, like fast trains and aircraft. This guarantees that customers receive consistent, excellent service regardless of where they are or how they move.

Uses for eMBB High-Definition Media Streaming: eMBB enables buffer-free streaming of 4K, 8K, and even higher quality media. This will improve the quality of video streaming services, making it possible for applications in the fields of entertainment, education, and business to have clearer and more detailed graphics.

Virtual Reality (VR) and Augmented Reality (AR): With its high bandwidth and low latency, eMBB will enable realistic VR and AR experiences that require real-time interaction. Interactive simulations, virtual tours, gaming, and remote training will all be transformed by this.

Cloud computing and edge computing: Users will be able to access cloud services with ease thanks to eMBB's fast data rates and low latency, which will enable them to work remotely on data-intensive jobs. Furthermore, by integrating edge computing, processing power will be closer to the user, decreasing latency and enhancing the functionality of apps like AI-driven analytics and Internet of Things systems.

Telepresence and distant Collaboration: By offering top-notch video and audio for online meetings, conferences, and cooperative projects, eMBB will improve telepresence and distant collaboration capabilities. This will lessen the need for actual travel and enable worldwide teamwork.

IoT and smart cities: eMBB's expanded network coverage and capacity will facilitate the installation of IoT devices in smart cities, opening the door for uses including traffic control, environmental monitoring, intelligent lighting, and public safety. Real-time data interchange and analysis will be made possible by the high-speed connectivity, which will enhance resource management and urban living conditions.

Enhanced Mobile Gaming: eMBB will improve mobile gaming with quicker download times, real-time multiplayer connections, and better graphics. Additionally, consumers will have easier access to cloud gaming services, which will enable them to play resource-intensive games on mobile devices without requiring expensive hardware.

Remote Work and Education: eMBB will facilitate the transition to remote work and online learning by offering dependable, fast internet connection for online courses, video conferences, and collaborative applications. This will facilitate more flexible work and learning environments and increase the efficacy of remote connections.

Healthcare & Telemedicine: eMBB will make it possible to provide top-notch telemedicine services, including large-scale medical file transmission, real-time video consultations, and remote diagnostics. This will make healthcare services more accessible, especially in rural and underdeveloped areas.

Technological Innovations Supporting eMBB

Massive MIMO and Beamforming: By using numerous antennas to send and receive more data concurrently, massive MIMO technology boosts the efficiency and capacity of networks. By directing the wireless signal toward certain users, beamforming increases signal strength and lowers interference.

Millimeter Wave (mmWave) Frequencies: To increase data rates, eMBB will make use of mmWave frequencies. Although mmWave provides more bandwidth, its penetration and range limitations force the use of small cells and sophisticated signal processing methods.

Network Slicing: This technique makes it possible to design virtual networks that are customized to meet the needs of users and certain applications. This guarantees optimal performance across a range of use cases, encompassing low-latency industrial automation and high-speed mobile broadband.

Advanced Modulation and Coding Schemes: By improving spectrum efficiency and data throughput, innovations in modulation and coding schemes will allow eMBB to provide ultra-high-speed connectivity even in difficult environments.

Ultra-Reliable Low-Latency Communication (URLLC) in B5G and 6G Technologies

A key element of B5G (Beyond 5G) and 6G technologies is Ultra-Reliable Low-Latency Communication (URLLC), which is intended to serve applications that demand very high reliability and very low latency. In the context of B5G and 6G networks, this section examines the importance, technological elements, and uses of URLLC.

Significance of URLLC: In applications where even a small delay or data loss might have serious repercussions, URLLC is indispensable. It seeks to deliver reliability levels of 99.999% or above and end-to-end latency of less than 1 millisecond. Mission-critical applications in a variety of industries, including public safety, manufacturing, transportation, and healthcare, need these strict specifications.

Technological Aspects of URLLC-Low Latency: From the physical layer to the application layer, every part of the communication process must be optimized to achieve ultra-low latency. To reduce latency, strategies including network slicing, edge computing, and sophisticated modulation algorithms are used.

High Reliability: Robust network protocols, sophisticated error correction methods, and redundancy are used to achieve high reliability. Extensive transmission channels and redundant copies of vital information provide precise delivery of information under demanding circumstances.

Network Slicing: This technique enables the development of specialized virtual networks that are adapted to the unique needs of various applications. This entails resource isolation and making sure the network can meet the strict requirements for latency and reliability for URLLC.
Edge Computing: By bringing processing and data storage closer to the consumer, edge computing minimizes the amount of distance data must travel, which lowers latency and boosts reliability.
Advanced Modulation and Coding Schemes: These schemes contribute to URLLC's low-latency and high-reliability goals by increasing data transmission efficiency and robustness.

Applications of URLLC

1. Autonomous Vehicles: For safe and effective navigation, autonomous cars need to be in constant communication with other cars, infrastructure, and control centers. For these interactions, URLLC offers the low latency and high dependability required, facilitating quick decision-making and coordination.

    Vehicle-to-Everything (V2X) Communication: URLLC facilitates V2X communication, which enables cars to share data about their location, speed, and driving circumstances. This data is essential for managing traffic and preventing collisions.
2. Industrial Automation and Manufacturing: In intelligent factories, machinery and robots must work together harmoniously, necessitating accurate and prompt communication. URLLC guarantees excellent reliability and low latency in the transmission of control commands and feedback.

    Robotics and Automation: To guarantee efficiency and safety in manufacturing processes, real-time control of robots and automated systems requires extremely low latency and great dependability.
3. Healthcare and Remote Surgery: With real-time medical applications like remote surgery made possible by URLLC, a single communication hiccup might potentially be fatal. Medical practitioners may confidently perform remote procedures thanks to URLLC's excellent reliability and low latency.

    Remote Surgery: By ensuring that their movements are precisely and instantaneously recreated, URLLC helps surgeons to use robotic tools from a distance while reducing the possibility of mistakes.
4. Augmented Reality (AR) and Virtual Reality (VR): Low latency is essential to a smooth user experience in immersive AR and VR applications. By shortening the time lag between user input and system responses, URLLC helps these apps.

    Immersive Experiences: URLLC makes sure that the virtual world responds quickly to user inputs in gaming, training, and remote collaboration, boosting the feeling of presence and involvement.
5. Public Safety and Emergency Services: For emergency agencies and first responders to efficiently coordinate their activities, they need robust communication systems. URLLC offers the dependability and low latency required to guarantee the timely delivery of vital information.

    Emergency Response: URLLC makes it possible for emergency teams to coordinate, share data, and transmit live video in real time, which enhances the efficacy and efficiency of rescue efforts.
6. Financial Services: Milliseconds can have a big impact on high-frequency trading and other financial services. By enabling incredibly quick transactions and data transfers, URLLC guarantees the smooth operation of financial systems.

High-frequency trading: By facilitating quicker transaction processing and lowering the possibility of latency-induced inconsistencies, URLLC helps the financial markets.

7. Gaming and E-Sports: Players and game servers must communicate in real time for competitive gaming and e-sports. By ensuring low latency, URLLC creates an even playing field and improves the game experience.

   Games for Real-Time Multiplayer: In order to minimize lag and guarantee that every player experiences the game in real time, URLLC offers synchronized gaming.

Challenges and Future Directions

Spectrum Availability: In order to satisfy its strict specifications, URLLC needs to have access to a wide range of frequencies. Effectively managing and allocating these frequencies is a major task.

Energy Efficiency: Higher energy consumption is frequently incurred in the pursuit of low latency and high reliability. The goal of future developments must be to maximize energy efficiency without sacrificing functionality.

Management of Interference: URLLC applications have to deal with possible interference from other networks and devices, especially in crowded metropolitan settings. Sophisticated methods of managing interference will be essential.

Scalability: Ensuring that URLLC can scale successfully while upholding its performance standards is crucial as the number of connected devices keeps increasing.

Massive Machine-Type Communication (mMTC) in B5G and 6G Technologies

Massive Machine-Type Communication (mMTC), a crucial element of B5G (Beyond 5G) and 6G technologies, aims to support the Internet of Things' (IoT) extensive connectivity requirements. This section discusses the significance, technological aspects, and applications of mMTC within the framework of B5G and 6G networks.

Importance of mMTC: By connecting billions of devices, mMTC hopes to offer scalable and effective communication solutions for a wide range of uses. mMTC is designed to enable communication between devices and facilitate automation, monitoring, and control processes across several sectors, in contrast to traditional mobile communication, which concentrates on human-to-human connection. The key to the spread of smart environments is mMTC's capacity to accommodate a large number of inexpensive, low-power devices with little to no human involvement.

Technological Aspects of mMTC-Scalability: A large number of linked devices must be managed by mMTC in a certain area. To accommodate the increased device density, sophisticated resource allocation strategies and well-designed network architecture are necessary.

Energy Efficiency: A lot of Internet of Things devices depend on batteries for long periods of operation. To increase battery life and lower energy consumption, mMTC uses low-power wide-area networks (LPWAN) and energy-efficient communication protocols.

Cost-Effectiveness: mMTC solutions need to be affordable in order to be widely used. This entails using network management that is effective and simplifying hardware to lower the cost of devices, deployment, and maintenance.

Robustness and Reliability: Although mMTC networks don't require high data rates, they nevertheless need to provide robust data transfer, especially for vital applications like industrial automation and health monitoring. To improve dependability, strategies like error correction and redundancy are used.

Lateness Tolerance: The majority of mMTC applications can work even in the presence of longer communication delays since they are latency-tolerant. This feature makes asynchronous communication techniques and energy-saving protocols possible.

Applications of mMTC

The implementation of a broad spectrum of IoT devices for smart city applications, such as waste management, traffic control, smart lighting, and environmental monitoring, is made possible by Smart Cities mMTC.

Smart Lighting: By allowing for remote control and monitoring, connected streetlights optimize energy use and maintenance schedules.

Traffic Control: Real-time traffic data is provided by IoT sensors and linked cars, allowing for dynamic traffic control and the reduction of congestion.

Agriculture: By integrating a wide range of sensors and devices that track crop health, weather, and soil conditions, mMTC facilitates precision agriculture by maximizing resource efficiency and raising production.

Soil Monitoring: Sensors gather information for targeted fertilization and irrigation by measuring the pH, moisture content, and nutrient levels of the soil.

Livestock Tracking: By monitoring the whereabouts and well-being of animals, Internet of Things devices enhance herd management and minimize losses.

Healthcare: By integrating wearable technology and medical sensors, mMTC enables remote health monitoring and management, allowing clinicians to track patient health and provide prompt interventions.

Remote Monitoring: Wearable technology keeps an eye on vital signs and notifies medical professionals of any irregularities.

Drug Adherence: Interconnected pill dispensers serve as a reminder for patients to take their prescriptions and let caregivers know when they've done so.

Industrial Automation: Real-time monitoring, predictive maintenance, and automated operations are made possible by mMTC's ability to connect sensors, machines, and control systems. This improves industrial processes. IoT sensors track the condition of equipment, anticipating malfunctions and planning repairs to reduce downtime. This is known as predictive maintenance.

Automated Operations: As a result of connected machines and robots coordinating and exchanging information, efficiency is increased and human intervention is decreased.

Energy Management: To optimize energy production and consumption, mMTC connects meters, appliances, and renewable energy sources with smart grids and energy management systems.

Smart Meters: Real-time data on energy consumption is provided by connected meters, allowing for demand response and dynamic pricing.

Renewable Integration: To maintain stability and efficiency in the grid, IoT devices keep an eye on and manage the integration of renewable energy sources.

Environmental Monitoring: By linking a network of sensors that monitor the quality of the air and water, the weather, and natural disasters, mMTC makes it possible to conduct thorough environmental monitoring.

Air Quality Monitoring: Sensors collect data for public health campaigns and regulatory compliance as well as the measurement of contaminants.

Disaster detection: Early warning systems and quick action are made possible by IoT devices that identify natural disasters like earthquakes and floods.

Technological Innovations Supporting mMTC Low-Power Wide-Area Networks (LPWAN): These networks, which are perfect for mMTC applications, offer long-range communication at low power consumption thanks to technologies like LoRa (long range) and NB-IoT (narrowband IoT).

Advanced Network Management: mMTC devices can communicate effectively and scalable thanks to methods like network slicing and edge computing, which maximize processing power and resource allocation.

Enhanced Protocols: MQTT (Message Queuing Telemetry Transport) and CoAP (Constrained Application Protocol), two protocols created especially for mMTC, allow for simple and effective device-to-device communication.

Self-Organizing Networks (SON): SON technologies lower the complexity and expense of managing large-scale IoT deployments by enabling networks to self-configure, self-optimize, and self-heal.

Focus on Rural Connectivity and Its Transformative Potential in B5G and 6G Technologies

Beyond 5G (B5G) and 6G technologies have great potential to close the digital divide, especially in underserved and rural areas. The significance, difficulties, and revolutionary possibilities of cutting-edge network technologies in augmenting rural connection are explored in this section.

The Value of Rural Internet Access In order to promote economic growth, increase access to healthcare and education, and improve the general standard of living in remote places, rural connectivity is essential. Although communication technology has advanced significantly, a large section of the world's population still lacks dependable internet connectivity, especially in rural and remote areas. By giving these neglected populations access to fast, dependable, and reasonably priced internet, B5G and 6G technologies seek to alleviate these discrepancies.

Obstacles to Rural Internet Access Geographical Barriers: The tough topography of rural and isolated places frequently makes the installation of infrastructure expensive and time-consuming. Traditional network infrastructures like fiber optics are difficult to deploy because of mountains, woods, and large distances between settlements.

Economic Restrictions: The reduced population density in rural regions raises the cost of network implementation and maintenance per user. Network providers are frequently discouraged from investing due to this economic constraint.

Limited Infrastructure: In order to support cutting-edge communication technology, rural areas frequently lack the infrastructure that is required, such as transportation networks and power grids.

Technical Restrictions: It's possible that current technologies aren't tailored to the particular needs of rural connection, such as the demand for robust performance in a variety of environmental circumstances and extended coverage.

Transformative Potential of B5G and 6G in Rural Connectivity

Extended Coverage: By utilizing high-altitude platforms, sophisticated satellite systems, and low-power wide-area networks (LPWAN), B5G and 6G technologies aim to offer larger coverage. Even the most remote and unreachable places can have connectivity thanks to these technologies.

High-Speed Broadband: Compared to earlier generations, next-generation networks offer far faster data speeds and lower latency. Because of this improved performance, high-speed internet is now available to remote populations, facilitating e-commerce, telemedicine, and online learning.

Affordability: New developments in network architecture and implementation, like edge computing and network slicing, lower operating costs and boost productivity. For customers in remote areas, this may mean more reasonably priced services.

Resilience and Reliability: The resilience and reliability of connectivity in rural areas are improved by sophisticated network architectures and technologies. Advanced error correction algorithms and self-organizing networks (SON) guarantee reliable performance even in difficult situations.

Boosting Local Economies: By allowing rural firms to access new technology, expand their market reach, and take part in the digital economy, improved connection can spur economic growth. Better data and communication services can help small-scale businesses, tourism, and agriculture.

Access to Essential Services: Rural communities can make use of telemedicine, online learning, and e-government services when they have dependable internet access. This has the potential to greatly enhance civic involvement, educational attainment, and health outcomes.

Community Development: By facilitating improved citizen cooperation, communication, and information exchange, increased connectedness promotes community development. Better digital infrastructure can foster the growth of social networks, neighborhood projects, and community groups.

Technological Innovations Supporting Rural Connectivity

Satellite Communication: High-speed internet connectivity with worldwide coverage is made possible by next-generation satellites, such as constellations in low Earth orbit (LEO). These systems eliminate the need for substantial ground infrastructure by providing connectivity to isolated and underserved locations.

High-Altitude Platforms (HAPs): HAPs, which include drones and balloons, work in the stratosphere to supplement terrestrial networks and offer wide-area coverage. They are especially helpful for getting to places that are hard to reach by conventional infrastructure.

Low-Power Wide-Area Networks (LPWAN): LPWAN technologies, such as Long Range (LoRa) and Narrowband IoT (NB-IoT), provide low-power long-range communication, which makes them perfect for sensor networks and other rural IoT applications.

Edge Computing: By moving data processing closer to the customer, edge computing lowers latency and boosts service dependability. In rural places, where major data centers could be far away, this is very helpful.

Network Slicing: This technique optimizes resource allocation and guarantees that the demands for rural connectivity are satisfied in an efficient and economical manner by enabling the development of virtual networks suited to particular requirements.

Challenges and Opportunities in B5G and 6G Technologies

There are several potential and challenges associated with the development and implementation of B5G (Beyond 5G) and 6G technologies. This section examines the numerous chances for innovation and development that these cutting-edge network technologies present, as well as the regulatory and policy issues, technological and infrastructure difficulties, and policy considerations.

Difficulties with Technology and Infrastructure Scarcity of Spectrum: A major technological obstacle is the limited availability of spectrum. New frequency bands, such as the terahertz and millimeter-wave spectrum, which are now underutilized but also subject to competition from other services and governmental restrictions, are necessary for B5G and 6G technologies to function.

Interference and Signal Propagation: Signal propagation and interference pose major issues for high-frequency signals, as those employed in terahertz and millimeter-wave communication.

Due to their short range and susceptibility to physical obstructions, these frequencies require sophisticated propagation methods and effective interference management techniques.

Energy Efficiency: As the number of connected devices rises and B5G and 6G networks demand higher data rates, energy consumption becomes a crucial concern. Sustainable development requires that all devices and network components have high energy efficiency.

Complexity of Networks: B5G and 6G networks have extremely complicated architectures that combine high-altitude platform communications, satellite communications, and terrestrial communications. Coordinating with several stakeholders and implementing advanced network management solutions are necessary to handle this complexity.

Infrastructure Deployment: There are considerable financial and logistical obstacles in deploying the required infrastructure for B5G and 6G, particularly in rural and isolated places. In these areas, creating and maintaining cutting-edge communication networks calls for significant financial outlays and creative deployment techniques.

Regulatory and Policy Considerations

Spectrum Allocation: The successful implementation of B5G and 6G technologies depends on efficient spectrum management and allocation. It is imperative for policymakers to guarantee the availability of ample spectrum and devise allocation methods that promote equitable access while reducing inter-service interference.

Data Security and Privacy: Data security and privacy are issues brought on by the enhanced connectivity and data transmission that come with B5G and 6G. It is imperative to build regulatory frameworks in order to safeguard user data and provide secure communication connections.

Standardization: To guarantee compatibility and interoperability between various networks and devices, it is crucial to develop international standards for B5G and 6G technologies. To set these standards, collaboration and cooperation on a global scale are needed. Environmental Impact: When implementing new network infrastructures, policymakers need to take the environment into account. This entails tackling the carbon footprint of network operations and encouraging the use of eco-friendly practices and technologies.

Digital Inclusion: By encouraging fair access to cutting-edge network technology, regulatory measures should endeavor to close the digital gap. This entails providing incentives for investments in underprivileged regions and endorsing programs that improve digital inclusion and literacy.

Opportunities for Innovation and Development

Advanced Communication Technologies: B5G and 6G enable the development of state-of-the-art communication technologies such as visible light, quantum, and terahertz. These technologies, with their incredible speed and security, have the power to revolutionize data transfer.

Integration of Artificial Intelligence: B5G and 6G offer a significant opportunity for the integration of artificial intelligence (AI) into network optimization and administration. AI can help with resource allocation that is intelligent, anticipate and mitigate network failures, and increase network performance.

IoT and Smart Cities: With the deployment of B5G and 6G networks, the Internet of Things (IoT) and smart cities will expand at a faster rate. Many applications, including intelligent transportation systems and smart grid management, will be made possible by improved connectivity.

Healthcare and Telemedicine: Real-time health monitoring, remote surgery, and telemedicine will all be made possible by cutting-edge network technologies. This can facilitate better access to medical care, particularly in isolated and rural places.

Industrial Automation: Smart factories, automated supply chains, and sophisticated robotics will all benefit from the innovation that B5G and 6G will bring about in this field. These developments will improve industrial operations' productivity, safety, and efficiency.

Education and Training: Better connection will transform education and training by making immersive learning experiences possible with virtual reality (VR) and augmented reality (AR), as well as by giving people access to top-notch educational resources all around the world.

Environmental Monitoring: The sophisticated environmental monitoring made possible by B5G and 6G technology will support efforts to mitigate climate change and manage natural resources. Real-time data analytics and IoT sensors can offer insightful information about trends and environmental conditions.

Implications of Future Research

There are several ramifications for future studies in B5G and 6G technologies, providing a plethora of opportunities for academic inquiry. These cutting-edge network technologies offer chances to investigate not only their technological innovations but also their wider socio-economic and ethical ramifications as they develop. Subsequent investigations ought to focus on comprehending the ways in which B5G and 6G could stimulate innovation in diverse sectors, in addition to tackling the possible exacerbation of the digital gap that these technologies may introduce. It will be imperative to look into the ethical implications of mass adoption to make sure that these developments promote fair progress rather than acting as obstacles to it. Further research might explore the lessons learnt from the shift from 1G to 5G, looking at the historical background and evolution of mobile networks. The deployment methods of B5G and 6G will benefit greatly from comparative studies of the rates of technical adoption in various countries and the associated economic effects. These kinds of research could help policymakers and business executives make better decisions, facilitating a seamless and advantageous global rollout of new network technology generations.

Since the definitions of B5G and 6G are still being developed, further study is urgently needed to clarify these ideas and come to an agreement on the salient characteristics that set these technologies apart from their forerunners. Clarifying these concepts should be the main goal of future study, as this will have an impact on the creation of industry standards and protocols for international interoperability. These initiatives will guarantee that these technologies are future-proof and can be easily incorporated into current frameworks. Another important topic for future research is the creation of technical standards and guiding principles for B5G and 6G. Since these standards are still in the early stages of development, researchers can help shape them by investigating ways to improve security and privacy measures, optimize protocols, and create standards flexible enough to keep up with quickly advancing technology. In order to guarantee that the implementation of these technologies is both technically sound and compliant with international requirements, this work is crucial.

Numerous research opportunities are presented by innovations in B5G and 6G, including as terahertz communication and AI integration. Subsequent research ought to go into the pragmatic implementations of these advancements and their capacity to revolutionize the current network infrastructure. Research on the environmental effects of these technologies should also look at sustainable deployment options that strike a balance between ecological responsibility and technical advancement. Scholars can aid in the creation of more effective and long-lasting network solutions by tackling these topics. Future research has a lot of chances because of the practical applications and use cases of B5G and 6G technologies, especially in areas like smart cities, improved mobile broadband, and rural connectivity. Academics ought to concentrate on assessing the efficiency of these technologies in a variety of fields, such as education, healthcare, and transportation, paying special emphasis to underprivileged areas. Understanding the social and economic advantages of B5G and 6G as well as spotting any unexpected repercussions from their broad use will depend heavily on the results of this research.

CONCLUSION

The developing B5G and 6G technologies have been thoroughly covered in this chapter, highlighting their importance in the development of mobile networks. In order to demonstrate the technological developments that have prepared the way for the shift to B5G and 6G, we started by looking at the historical development from 1G to 5G. By describing these next-generation networks, we emphasized the salient technical features—like the incorporation of artificial intelligence and the utilization of terahertz communication—that distinguish them from their forerunners. The basis for comprehending the basic ideas underlying these technologies was created by the conversation on technical standards and essential principles. The chapter also examined important developments that have the potential to completely transform communication networks, such as huge connectivity and ultra-low latency. The showcased use cases and applications, which included everything from improved mobile broadband and rural connectivity to smart city development, showed how B5G and 6G have the potential to revolutionize a wide range of industries. But the chapter also discussed the obstacles that need to be overcome, such as technological, infrastructure, policy, and regulatory problems, while emphasizing the chances for advancement and creativity.

REFERENCES

Ahmed, M. A., Majeed, A. S., Dawood, Y. S. (2023). A Survey of 6G Mobile Systems, Enabling Technologies, and Challenges. *International journal of electrical and electronic engineering and telecommunications 12*(1). doi:DOI: 10.18178/ijeetc.12.1.1-21

Akbar, M. S., Hussain, Z., Sheng, Q. Z., & Mukhopadhyay, S. (2022). *6G Survey on Challenges, Requirements, Applications, Key Enabling Technologies, Use Cases, AI integration issues and Security aspects*. 10.48550/arXiv.2206.00868.

Akyildiz, I. F., Kak, A., & Nie, S. (2020). 6G and beyond: The future of wireless communications systems. *IEEE Access : Practical Innovations, Open Solutions, 8*, 133995–134030. DOI: 10.1109/ACCESS.2020.3010896

Altaf Khattak, S. B., Nasralla, M. M., & Rehman, I. U. (2022). The Role of 6G Networks in Enabling Future Smart Health Services and Applications. *2022 IEEE International Smart Cities Conference (ISC2)*, Pafos, Cyprus, 2022. doi:DOI: 10.1109/ISC255366.2022.9922093

Altaf Khattak, S. B., Nasralla, M. M., & Rehman, I. U. (2022). The Role of 6G Networks in Enabling Future Smart Health Services and Applications. *2022 IEEE International Smart Cities Conference (ISC2)*, Pafos, Cyprus, 2022. doi:DOI: 10.1109/ISC255366.2022.9922093

Álvaro, V. R., Petteri, K., Silvio, M., & Harish, V. (2023). The Role of AI on 6G MAC. doi:DOI: 10.36227/techrxiv.23708310.v1

Andrews, J. G., Buzzi, S., Choi, W., Hanly, S. V., Lozano, A., Soong, A. C., & Zhang, J. C. (2014). What will 5G be? *IEEE Journal on Selected Areas in Communications, 32*(6), 1065–1082. DOI: 10.1109/JSAC.2014.2328098

ArunKumar. S., Sivakami, K., VijayaKarthik, S. V., & Deepa, S. M. (2022). 6G With Smart Healthcare, Haptic Communication. In D. Satishkumar, G. Prabhakar, & R. Anand (Eds.), *Handbook of Research on Design, Deployment, Automation, and Testing Strategies for 6G Mobile Core Network* (pp. 303-329). IGI Global Scientific Publishing. doi:DOI: 10.4018/978-1-7998-9636-4.ch015

Basu, D., Ghosh, G., & Datta, R. (2022). 6G for Industry 5.0 and Smart CPS: A Journey from Challenging Hindrance to Opportunistic Future. *2022 IEEE Silchar Subsection Conference (SILCON). Silchar, India, 2022*, 1–6. DOI: 10.1109/SILCON55242.2022.10028927

Basu, D., Ghosh, U., & Datta, R. (2022). 6G for Industry 5.0 and Smart CPS: A Journey from Challenging Hindrance to Opportunistic Future. *2022 IEEE Silchar Subsection Conference (SILCON). Silchar, India, 2022*, 1–6. DOI: 10.1109/SILCON55242.2022.10028927

Carlos, S. (2023, October 25). José, Cunha., Guillermo, Candela., Santiago, Rodriguez., Xosé, Ramón, Sousa., Frederico, Branco. (2023). 4. The 6G Ecosystem as Support for IoE and Private Networks: Vision, Requirements, and Challenges. *Future Internet, 15*(11), 348. Advance online publication. DOI: 10.3390/fi15110348

Chen, S., Wang, H., Xu, X., & Zhou, X. (2020). Artificial intelligence for 6G: Challenges and opportunities. *IEEE Communications Surveys and Tutorials, 23*(1), 40–57. DOI: 10.1109/COMST.2020.3031481

Cisco. (2020). Cisco Annual Internet Report (2018–2023) *White Paper*. https://www.cisco.com/c/en/us/solutions/collateral/executive-perspectives/annual-internet-report/white-paper-c11-741490.html

Dang, S., Amin, O., Shihada, B., & Alouini, M. S. (2020). What should 6G be? *Nature Electronics*, *3*(1), 20–29. DOI: 10.1038/s41928-019-0355-6

Das, P., Ray, S., Sadhukhan, D., & Govil, C. G. (2024). 6G Communication Technology for Industry 5.0. In Ghosh, U., Mhlanga, F., & Rawat, D. B. (Eds.), *Secure and Smart Cyber-Physical Systems*., DOI: 10.1201/9781003376712-5

Deepa, R., Sivasamy, A., Selvam, S. (2024). Intelligent networking unleashed: exploring AI and ML for enhanced 6G wireless communication systems. doi:DOI: 10.58532/V3BINC3P8CH3

Gupta, A., & Jha, R. K. (2021). A survey of 5G network: Architecture and emerging technologies. *IEEE Access : Practical Innovations, Open Solutions*, *3*, 1206–1232. DOI: 10.1109/ACCESS.2021.3059431

Hatem, I.-K., Laroui, M., Alfaqawi, M., Magnouche, A., Moungla, H., & Afifi, H. (2023). 6G-edge support of Internet of Autonomous Vehicles: A survey. *European Transactions on Telecommunications*. Advance online publication. DOI: 10.1002/ett.4918

Hosseini, S. A., Nguyen, D. T., Garcia, N., & Tran, D. C. (2021). A comprehensive survey of 6G networks: Applications, challenges, trends, and opportunities. *IEEE Access : Practical Innovations, Open Solutions*, *9*, 43253–43273. DOI: 10.1109/ACCESS.2021.3072101

Khan, L. U., Yaqoob, I., Imran, M., Han, Z., & Hong, C. S. (2020). 6G wireless systems: A comprehensive survey. *IEEE Internet of Things Journal*, *8*(1), 400–418. DOI: 10.1109/JIOT.2020.3010833

Kumar, P., Chauhan, D. S., Srivastava, P. K., & Ahmad, F. (2021). Sustainable development in 6G: Challenges and future directions. *Wireless Personal Communications*, *120*(4), 2573–2591. DOI: 10.1007/s11277-021-08582-3

Li, X., Zhang, Q., & Wang, T. (2021). Ethical implications of 6G technology: A systematic review. *Journal of Information.Communication and Ethics in Society*, *19*(4), 487–502. DOI: 10.1108/JICES-04-2021-0034

Liao, X., You, W., Lv, J., Wang, M., Zhao, Y., & He, G. (2023). 6G Polymorphic Smart Network Architecture and Key Technologies. *2023 5th International Conference on Frontiers Technology of Information and Computer (ICFTIC)*, Qiangdao, China. doi:DOI: 10.1109/ICFTIC59930.2023.10455881

Maligi, A. S., Bhavana, H. T., Sanjana, T., & Archana, H. R. (2022). Technological and Non-Technical Challenges Associated With 6G Networks. In Bharathy, A., & Alhadidi, B. (Eds.), *Challenges and Risks Involved in Deploying 6G and NextGen Networks* (pp. 167–176). IGI Global Scientific Publishing., DOI: 10.4018/978-1-6684-3804-6.ch011

Marouan, M., Brambilla, M., Tagliaferri, D., Mazzucco, C., Debbah, M., Mach, T., Simeone, R., Mandelli, S., Frascolla, V., Lombardi, R., Magarini, M., Nicoli, M., & Spagnolini, U. (2021). *6G V2X Technologies and Orchestrated Sensing for Autonomous Driving*. 10.48550/arXiv.2106.16146.

Nadeem, Q., Seddik, K. G., & Ghauri, S. A. (2022). Regulatory challenges in the adoption of 6G networks: A critical review. *IEEE Communications Standards Magazine*, *6*(1), 25–33. DOI: 10.1109/MCOMSTD.001.2000015

Niranjana, C., & Jijesh, J. J. Lakshmi, Manasa, B., & Vineeth, Kumar, P, K. (2024). Future outlook and challenges in deploying 6G. In *Futuristic Trends in Network & Communication Technologies Volume 3 Book 3*.doi:DOI: 10.58532/V3BINC3P7CH1

Papaioannou, P., Pastellas, I., Tranoris, C., Karagiorgou, S., & Denazis, S. (2024). AI-fuelled Dimensioning and Optimal Resource Allocation of 5G/6G Wireless Communication Networks. *2024 IEEE International Mediterranean Conference on Communications and Networking (MeditCom)*, Madrid, Spain. doi:DOI: 10.1109/MeditCom61057.2024.10621426

Primmia, D. R., Mahabooba, M., Karpagam, J., Sharma, K., Singh, A., & Manoj, S. (2024). The Development of 6-G Technology in Integration with AI type of Synergy. *2024 4th International Conference on Advance Computing and Innovative Technologies in Engineering (ICACITE)*, Greater Noida, India. doi:DOI: 10.1109/ICACITE60783.2024.10616538

Quy, V. K., Chehri, A., Quy, N. M., Han, N. D., & Ban, N. T. (2023). Innovative Trends in the 6G Era: A Comprehensive Survey of Architecture, Applications, Technologies, and Challenges. *IEEE Access : Practical Innovations, Open Solutions*, *11*. Advance online publication. DOI: 10.1109/ACCESS.2023.3269297

Rainer, L., Shafi, M., Shivanandan, G., Chandramouli, D., & Thiebaut, L. (2022). Perspectives on a 6G Architecture. 10.48550/arXiv.2210.03286

Ramraj, D., Choudhary, G., Dragoni, N., Lalwani, P., Khare, U., & Kundu, S. (2023). 6G Mobile Networks: Key Technologies, Directions, and Advances. *Telecom*, *4*(4). Advance online publication. DOI: 10.3390/telecom4040037

Rappaport, T. S., Xing, Y., Kanhere, O., Ju, S., Madanayake, A., Mandal, S., & Singh, A. (2019). Wireless communications and applications above 100 GHz: Opportunities and challenges for 6G and beyond. *IEEE Access : Practical Innovations, Open Solutions*, *7*, 78729–78757. DOI: 10.1109/ACCESS.2019.2921522

Saad, W., Bennis, M., & Chen, M. (2020). A vision of 6G wireless systems: Applications, trends, technologies, and open research problems. *IEEE Network*, *34*(3), 134–142. DOI: 10.1109/MNET.001.1900287

Saimler, M., İckin, S., Bernini, G., Toumi, N., Diamanti, M., Papavassiliou, S., Zivkovic, M., Akgül, Ö. U., & Khorsandi, B. M. (2024). The Role of AI Enablers in Overcoming Impairments in 6G Networks. *2024 Joint European Conference on Networks and Communications & 6G Summit (EuCNC/6G Summit)*, Antwerp, Belgium. doi:DOI: 10.1109/EuCNC/6GSummit60053.2024.10597102

Shahzadi, S., Chaudhry, N. R., & Iqbal, M. (2023). A Novel 6G Conversational Orchestration Framework for Enhancing Performance and Resource Utilization in Autonomous Vehicle Networks. *Sensors (Basel)*, *23*(17), 7366. Advance online publication. DOI: 10.3390/s23177366

Singh, M. K., Singh, R., Singh, N., & Yadav, C. S. (2022). Technologies Assisting the Paradigm Shift from 5G to 6G. In Dutta Borah, M., Singh, P., & Deka, G. C. (Eds.), *AI and Blockchain Technology in 6G Wireless Network. Blockchain Technologies*. Springer., DOI: 10.1007/978-981-19-2868-0_1

Vetrithangam, D., Komala, C. R., Sugumar, D., Rao, Y. M., Karpagavalli, C., & Kiruthiga, B. (2024). AI's Role in 6G Security Machine Learning Solutions Unveiled. In Pandey, D., Pandey, B., & Ahmad, T. (Eds.), *Security Issues and Solutions in 6G Communications and Beyond* (pp. 26–44). IGI Global Scientific Publishing., DOI: 10.4018/979-8-3693-2931-3.ch003

Viswanathan, H., & Mogensen, P. (2020). Communications in the 6G era. *IEEE Access : Practical Innovations, Open Solutions*, *8*, 57063–57074. DOI: 10.1109/ACCESS.2020.2974747

Wang, X., Ma, X., & Niu, Y. (2020). Security and privacy issues in 6G: A comprehensive survey. *IEEE Network*, *34*(4), 16–23. DOI: 10.1109/MNET.001.1900467

Xiao, D. D., Wang, X. Y., Lu, L., Shi, N. X., Liu, C., Zhang, T., & Sun, T. (2023). 6G Architecture Design: From Overall, Logical and Networking Perspective. *IEEE Communications Magazine*, *61*(7), 158–164. Advance online publication. DOI: 10.1109/MCOM.001.2200326

Yoshihisa, H. (2022). Defining 6G: Beyond the conventional boundaries of wireless communication. *International Journal of Wireless Networks*, *28*(3), 321–338. DOI: 10.1007/s11276-021-02656-0

You, X., Huang, Y., Liu, S., Wang, D., Ma, J., Zhang, C., Zhan, H., Zhang, C., Zhang, J., Liu, Z., Li, J., Zhu, M., You, J., Liu, D., Cao, Y., He, S., He, G., Yang, F., Yang, L., & Gao, W. (2023). Toward 6G TKμ Extreme Connectivity: Architecture, Key Technologies and Experiments. *IEEE Wireless Communications*, *30*(3), 86–95. Advance online publication. DOI: 10.1109/MWC.004.2200482

Zhang, Z., Xiao, Y., Ma, Z., Xiao, M., Ding, Z., Lei, X., & Li, H. (2020). 6G wireless networks: Vision, requirements, architecture, and key technologies. *IEEE Vehicular Technology Magazine*, *14*(3), 28–41. DOI: 10.1109/MVT.2020.3003744

Zhao, Z., Li, Z., & Zhang, H. (2021). The road to 6G: Ten critical issues. *IEEE Communications Magazine*, *59*(3), 68–73. DOI: 10.1109/MCOM.001.2000397

Zong, B., Fan, P., Zhang, A., Li, C., & Shen, X. (2019). Evolution of mobile communications: From 1G to 5G. *IEEE Communications Magazine*, *57*(4), 58–64. DOI: 10.1109/MCOM.2019.1800411

Chapter 2
Opportunities and Challenges of 6G Communication Technology for a Rural Economy

L. Rajeshkumar
https://orcid.org/0009-0007-8513-206X
St. Joseph's College of Engineering, OMR, Chennai, India

S. Usha
Sri Sairam Engineering College, Chennai, India

Joel Jebadurai Devapitchai
https://orcid.org/0000-0002-6947-8497
St. Joseph's College of Engineering, OMR, Chennai, India

ABSTRACT

Sixth-generation (6G) mobile network technology is poised to set new standards aimed at addressing performance demands that fifth-generation (5G) networks have yet to meet. This chapter explores how 6G communication technology, when integrated with advancements such as artificial intelligence, cloud computing, blockchain technology, and quantum computing, presents significant opportunities across various sectors, including banking, agriculture, and education, particularly within rural economies. Additionally, the chapter addresses the challenges associated with implementing 6G communication applications in these regions. By providing valuable insights, this chapter will serve as a resource for researchers and policymakers, aiding in enhancing research outcomes and informing policy recommendations that support the effective deployment of 6G technology in rural settings.

INTRODUCTION

The emergence of 6G communication technology is poised to revolutionize connectivity, especially for rural economies that have long faced challenges related to unreliable internet access. As the next evolution in wireless communication, 6G is expected to provide exceptional speed, ultra-low latency, and the ability to connect a vast array of devices simultaneously. This technological advancement is

DOI: 10.4018/979-8-3693-7788-8.ch002

particularly crucial for rural areas, where traditional communication networks often fall short, hindering access to essential services and economic opportunities. The potential benefits of 6G for rural communities are substantial. Enhanced connectivity can greatly improve access to healthcare services through telemedicine, allowing residents to consult healthcare professionals remotely and receive timely medical advice—an essential resource in regions where medical facilities are limited. In the realm of education, 6G can facilitate access to online learning resources, virtual classrooms, and training programs, empowering students and workers to enhance their skills and better prepare for a competitive job market (Linu et al., 2024). Additionally, in agriculture, the integration of Internet of Things (IoT) devices and smart sensors can enable precision farming techniques, helping farmers monitor crop health, optimize resource usage, and ultimately increase yields while minimizing environmental impact. Collectively, these advancements not only promote economic growth but also significantly enhance the quality of life for residents in rural areas.

The transition to 6G technology is accompanied by several challenges that must be addressed. A significant issue is the need for substantial infrastructure investment; building the necessary networks to support 6G—such as fiber-optic connections, cell towers, and other telecommunications equipment—requires considerable financial resources and strategic planning. Additionally, there is a pressing need to tackle the digital skills gap prevalent in rural populations, as many residents may lack the foundational knowledge to effectively utilize advanced technologies (Chauhan et al., 2024). This underscores the importance of implementing training programs that enhance digital literacy and equip individuals with the skills required to navigate the evolving digital landscape. Furthermore, as connectivity expands, concerns regarding data security and privacy become more pronounced. With the vast amounts of data generated by IoT devices and increased online interactions, rural communities must adopt robust cybersecurity measures to protect sensitive information and guard against potential threats. Ensuring data integrity and security will be crucial for building trust within these communities as they embrace new technologies.

This chapter is presented in three different sections. The first section explains the modern technologies of 6G in the rural economy. The second section deals with opportunities of 6G in the rural economy. The last and final section presents the challenges of 6G in the rural economy

APPLICATION OF MODERN TECHNOLOGIES OF 6G IN THE RURAL ECONOMY

Robotics

Robotics stands to gain immensely from the advancements brought by 6G communication technology, which promises ultra-reliable low-latency communication and massive connectivity for a wide range of devices. This enhanced connectivity enables real-time data exchange between robots and their environments, allowing for more sophisticated automation and collaboration in various applications, such as manufacturing, healthcare, and logistics. With AI and machine learning integration, robots can process and analyze data instantaneously, adapting to dynamic conditions and improving their decision-making capabilities. Moreover, 6G's ability to support simultaneous connections for numerous devices facilitates the development of swarms of robots that can work together seamlessly, enhancing efficiency and productivity in complex tasks (Sharma et.al. 2024). Overall, 6G technology will be a game-changer, transforming how robots operate and interact within increasingly interconnected ecosystems.

Artificial Intelligence

Artificial Intelligence (AI) in 6G communication technology presents a transformative opportunity for rural areas, particularly in overcoming persistent connectivity challenges and enhancing service delivery across various sectors. The ultra-reliable low-latency communication enabled by 6G allows for seamless data transmission, ensuring that even remote locations benefit from stable and high-speed connections. This reliability is crucial for optimizing network management through AI, which can dynamically allocate resources, monitor network health, and adapt to fluctuating demands, thus providing a consistent user experience.

In the realm of agriculture, this technology enables smart farming practices by facilitating real-time data analysis from IoT sensors deployed throughout fields. Farmers can monitor vital factors such as soil moisture, nutrient levels, and weather conditions, allowing for more precise and informed decision-making. This capability not only boosts crop yields but also promotes sustainable practices by optimizing resource use, thereby enhancing food security in rural regions.

Telemedicine is another area where AI and 6G can have a profound impact. By enabling reliable, high-quality video consultations, healthcare providers can reach underserved populations, delivering essential medical services without the need for travel. This can significantly improve health outcomes in rural communities, where access to specialized care is often limited (Shin et al. 2024). AI can also assist in diagnosing conditions through advanced image processing and predictive analytics, further enhancing the quality of care.

Moreover, educational resources in rural areas can be revolutionized through AI-driven adaptive learning platforms. These systems can tailor educational content to individual learning styles and paces, ensuring that students receive personalized instruction that meets their specific needs. By improving access to quality education, rural communities can empower the next generation, fostering skills that are essential for participating in a digital economy.

Machine Learning

Machine learning (ML) in 6G communication technology has the potential to significantly enhance connectivity and services in rural areas, addressing unique challenges and fostering development. With the ultra-reliable low-latency communication capabilities of 6G, ML algorithms can optimize network performance by analyzing data traffic patterns and dynamically adjusting resource allocation. This ensures stable and efficient connections, even in areas with sparse infrastructure.

In agriculture, ML can be employed to process data from IoT sensors, enabling farmers to make data-driven decisions. By analyzing variables such as soil conditions, crop health, and weather forecasts, ML models can provide actionable insights that help improve yield and reduce resource waste (Blue et al., 2024). Predictive analytics can also anticipate pest outbreaks or crop diseases, allowing for timely interventions.

Additionally, ML can enhance telemedicine services in rural communities. By analyzing patient data and treatment outcomes, machine learning algorithms can assist healthcare providers in delivering personalized care and identifying the most effective treatment plans. This capability can be especially valuable in areas with limited access to specialists, improving overall health outcomes.

In education, ML can support adaptive learning platforms that tailor content to individual students' needs and learning styles. By analyzing student performance and engagement data, these systems can recommend personalized learning paths, ensuring that students receive the support they need to succeed.

Drones

Drones are set to revolutionize various sectors with the advent of 6G communication technology, which offers ultra-reliable low-latency connectivity and massive device support. This advanced communication framework enhances real-time data transmission, enabling drones to operate with greater autonomy and efficiency in applications such as delivery services, surveillance, and environmental monitoring. The integration of AI and machine learning allows drones to optimize flight paths and make informed decisions on the fly, while seamless connectivity addresses challenges like fragmented signal reception and excessive handovers. Furthermore, the collaborative capabilities enabled by 6G facilitate swarm operations, where multiple drones can coordinate their activities in real time, enhancing their utility in smart cities and large-scale projects.

Drones are increasingly integral to various applications, including transportation, package delivery, surveillance, and media production. Given their diverse use cases, ensuring secure, reliable, and stable wireless communication is essential for the safe operation and control of these aerial vehicles (Chataut et al., 2024). The advent of 6G mobile technology offers promising AI-based solutions to tackle various challenges, particularly those related to connectivity in the skies.

One of the primary obstacles in drone operations is maintaining seamless connectivity during flight. Issues such as Fragmented Reference Signal Received Power (RSRP) can lead to radio link failures, frequent handovers, and the disruptive phenomenon known as ping-pong handovers, where a drone repeatedly switches between base stations. To ensure uninterrupted connectivity for Unmanned Aerial Vehicles (UAVs), Reinforcement Learning (RL) presents a compelling approach to developing optimal handover strategies that minimize communication disruptions.

Research indicates that traditional methods of drone communication, which typically involve connecting to the Base Station (BS) with the strongest RSRP, often result in excessive handovers. This can compromise the stability of the connection. In contrast, an RL-based solution can significantly decrease the frequency of these handovers by effectively balancing the trade-off between maximizing RSRP and minimizing handover events. By continuously learning and adapting to changing conditions, RL can enhance overall communication stability, ensuring that drones maintain a reliable connection throughout their flight.

Internet of Things

The integration of the Internet of Things (IoT) with 6G communication technology has the potential to profoundly impact rural areas, enhancing connectivity and enabling a range of innovative applications. With 6G's ultra-reliable low-latency capabilities, IoT devices can communicate seamlessly, providing real-time data exchange that is crucial for various sectors.

In agriculture, IoT sensors can monitor soil moisture, crop health, and weather conditions, allowing farmers to make informed decisions and optimize resource usage. This data-driven approach can lead to increased yields and sustainable practices addressing food security in rural communities (Jebadurai et al., 2024). For instance, precision agriculture can be implemented, where irrigation and fertilization are adjusted based on real-time sensor feedback, reducing waste and maximizing efficiency.

Moreover, IoT can enhance healthcare services in rural areas. Remote patient monitoring devices can track vital signs and health metrics, transmitting data to healthcare providers via 6G networks. This capability allows for timely interventions and personalized care, significantly improving access to healthcare in underserved regions. Telemedicine applications can flourish, enabling consultations with specialists without the need for travel, thereby bridging the healthcare gap.

In smart homes and communities, IoT devices can improve energy management and resource conservation. Smart meters and connected appliances can optimize energy consumption based on usage patterns, helping residents save on utility costs while contributing to environmental sustainability.

Internet of Everything

The integration of the Internet of Everything (IoE) with 6G communication technology has the potential to radically transform rural areas by delivering ultra-fast, low-latency connectivity that seamlessly connects people, devices, and infrastructure. This powerful combination allows rural regions to bridge the digital divide, offering access to resources and services previously limited to urban centers. In agriculture, IoE enables precision farming, where interconnected sensors monitor soil moisture, nutrient levels, and crop health in real-time, allowing farmers to make data-driven decisions and automate equipment, which in turn optimizes the use of water, pesticides, and fertilizers, improving yields and sustainability. Healthcare in rural areas stands to benefit enormously from 6G-powered IoE as well, with remote health monitoring, telemedicine, and even remote surgical procedures becoming feasible. This advanced connectivity brings critical medical services closer to communities that lack access to hospitals and specialist care, reducing healthcare inequalities. In education, IoE supported by 6G can revolutionize learning and skill development by facilitating immersive virtual and augmented reality experiences that allow rural students to participate in virtual classrooms and simulations, connecting them to global knowledge resources. Furthermore, 6G-enabled IoE systems can help manage decentralized energy sources such as solar panels and microgrids, making rural energy systems more efficient and resilient (Devapitchai et al. 2024). By incorporating environmental monitoring and early warning systems, IoE can also improve disaster preparedness and response, providing real-time data to predict natural disasters and mobilize resources rapidly in emergencies. This holistic approach to connectivity and integration positions IoE in 6G as a foundational technology for sustainable development in rural areas, fostering resilient, economically vibrant, and more inclusive communities.

Deep Learning

Deep learning in 6G communication technology has the potential to bring substantial improvements to rural areas, enhancing connectivity, automation, and local development. Deep learning, a subset of Artificial Intelligence (AI), processes and learns from vast amounts of data, making it well-suited for optimizing the high-speed, low-latency networks of 6G. In rural contexts, deep learning can power intelligent network management, where algorithms predict network demand and adjust resources dynamically, ensuring stable connectivity even in hard-to-reach areas. This adaptability allows for efficient handling of limited infrastructure, essential for providing consistent service in remote regions.

In precision agriculture, deep learning integrated with 6G can analyze data from IoT sensors, drones, and cameras to assess crop health, predict yields, and detect diseases or pests early. Farmers benefit from insights that help them use resources more efficiently and increase productivity, all while maintaining sustainable practices. Deep learning also plays a critical role in healthcare in rural areas by enabling remote diagnostics and predictive analytics. Through 6G, deep learning models can process data from wearable devices in real-time, allowing healthcare providers to monitor patients remotely, detect health issues early, and recommend interventions, making advanced care more accessible in underserved areas.

Furthermore, deep learning enables intelligent disaster management systems. In rural settings where natural disasters like floods or wildfires can be devastating, 6G-powered deep learning algorithms can analyze environmental sensor data to predict such events, issue early warnings, and coordinate emergency responses (Devapitchai et.al., 2024). This technology can also enhance education by providing adaptive learning experiences through personalized AI-powered virtual tutors and immersive simulations, overcoming challenges related to access and quality in rural schools. By embedding deep learning within 6G communication infrastructure, rural areas can become smarter, safer, and more self-sufficient, contributing to sustainable development and bridging the rural-urban digital divide.

Predictive Analytics

Predictive analytics within the 6G network ecosystem has vast potential to improve rural areas by enabling data-driven decision-making across agriculture, healthcare, infrastructure, and disaster management. Leveraging 6G's ultra-fast speeds and low latency, predictive analytics can analyze vast amounts of data from connected devices and sensors in real time, delivering insights that help rural communities address challenges and plan for the future.

In agriculture, predictive analytics powered by 6G can analyze data from soil sensors, weather stations, and satellite imagery to forecast crop yields, identify potential pest infestations, and optimize planting schedules. This enables farmers to maximize their resources and reduce crop losses, enhancing food security and economic resilience. In healthcare, predictive analytics in 6G networks can process data from wearable devices and remote health monitors to anticipate health issues and prevent disease outbreaks. For patients with chronic conditions, this real-time analysis enables proactive care management, reducing the need for emergency interventions in areas where medical facilities may be sparse.

Infrastructure management also benefits from predictive analytics in 6G, where it can monitor the condition of roads, bridges, and utilities, identifying early signs of wear or damage. This allows for timely maintenance and reduces the likelihood of costly repairs or service interruptions in remote regions. In disaster management, predictive analytics can harness environmental data to forecast events such as floods, droughts, or wildfires, triggering early warnings and allowing communities to prepare effectively (Jebadurai 2013). By integrating predictive analytics into 6G networks, rural areas can become more resilient, sustainable, and responsive, enhancing overall quality of life and driving economic growth.

Virtual Reality (VR) and Augmented Reality (AR)

Virtual Reality (VR) and Augmented Reality (AR) powered by 6G technology have transformative potential in rural areas, enabling immersive experiences that can revolutionize education, healthcare, training, and tourism. With ultra-fast speeds, high bandwidth, and low latency of 6G, VR and AR applications can operate seamlessly, delivering real-time, high-definition content even in remote areas.

In education, VR and AR can bring immersive learning experiences to rural students, allowing them to attend virtual labs, historical simulations, and global classrooms. This can bridge the gap between rural and urban educational opportunities, providing students with hands-on learning that fosters critical thinking and engagement. For vocational training, AR can be particularly impactful; for example, farmers or technicians can receive real-time, step-by-step visual guides through smart glasses or mobile devices, enabling them to learn new techniques or repair machinery without needing in-person training.

In healthcare, AR and VR supported by 6G can facilitate remote diagnostics and telemedicine. For example, doctors in urban hospitals can guide local healthcare providers through AR overlays during procedures, or patients can attend virtual consultations with specialists without leaving their communities (Raja et al., 2024). VR could also be used for physical rehabilitation or mental health therapy, providing rural patients with treatment options that would otherwise be unavailable locally.

Finally, rural tourism stands to benefit from 6G-enabled VR and AR by creating virtual tours of scenic locations, cultural sites, and festivals. This technology can attract virtual visitors globally, generating interest and potential tourism revenue for rural communities. AR-enhanced tours in person can also offer visitors interactive, informative experiences that boost engagement and appreciation for local culture. With 6G powering VR and AR, rural areas can become more connected, empowered, and economically vibrant, unlocking opportunities that enhance quality of life and foster growth.

Visible Light Communication

Visible Light Communication (VLC) in 6G technology offers a novel solution for rural areas, where conventional communication infrastructure may be limited. VLC utilizes LED light sources to transmit data, providing a secure, high-speed alternative to radio-frequency-based networks. With VLC, LED lights—installed in streetlamps, homes, or public spaces—can serve as both illumination and data transmission sources, offering efficient and sustainable connectivity options, especially where installing cell towers or fiber-optic cables is challenging.

In education, VLC can provide internet access in schools and community centers using existing LED lighting. This would allow rural students and teachers to connect to high-speed internet for digital learning, enabling access to online resources and educational content even in areas without reliable internet infrastructure. Similarly, in healthcare, VLC-enabled LEDs in clinics or mobile healthcare units could support high-speed data transfers, enabling telemedicine services, remote diagnostics, and real-time data exchange between rural and urban healthcare facilities.

Agriculture in rural areas can also benefit from VLC in 6G. For instance, greenhouses with LED lighting could simultaneously support data transmission from IoT sensors that monitor soil quality, temperature, and moisture levels, enabling farmers to make informed, data-driven decisions (Pandey et.al 2024). Additionally, VLC can enhance security and surveillance by enabling data transmission through LED streetlights, allowing real-time video streaming and monitoring in rural areas, which can improve community safety and support emergency responses.

Moreover, VLC is resistant to electromagnetic interference, making it ideal for areas where radio waves may be unreliable. It is also more secure, as light cannot pass through walls, reducing the risk of unauthorized access. With 6G-driven VLC, rural areas could experience greater digital inclusion, benefiting from connectivity that is efficient, sustainable, and adaptable to diverse environmental and infrastructural constraints.

Machine-Type Communication

Machine-type communication (MTC) in 6G technology offers transformative potential for rural areas by enabling reliable, low-latency, and high-capacity connectivity for interconnected devices. MTC, which includes both Massive MTC (mMTC) and Ultra-Reliable Low Latency Communication (URLLC), allows a vast number of devices, sensors, and machines to communicate autonomously, with minimal human intervention. In rural contexts, this capability supports applications across agriculture, healthcare, infrastructure, and environmental monitoring.

In precision agriculture, MTC can connect sensors across large fields, gathering real-time data on soil moisture, crop health, and weather conditions. 6G's ability to handle a massive number of devices per square kilometer allows farmers to monitor and control irrigation, fertilizer, and pesticide distribution precisely, reducing waste and boosting crop yields. Livestock management can also benefit from MTC, with sensors on animals monitoring health indicators and movement, providing insights for early disease detection or location tracking.

In rural healthcare, MTC can enable the widespread use of remote monitoring devices for patients, especially those with chronic conditions. Through URLLC's ultra-reliable, low-latency connections, devices like heart monitors or glucose sensors can transmit critical health data to healthcare providers in real time, enabling faster responses in emergencies and reducing the need for frequent hospital visits.

Infrastructure management is another area where MTC in 6G proves valuable. In rural regions, where maintenance can be logistically challenging, MTC-enabled sensors can monitor the condition of roads, bridges, and utilities (Sinha et al., 2024). These sensors alert maintenance teams when repairs are needed, reducing costly infrastructure failures and improving safety. Environmental monitoring is also enhanced, as MTC can support networks of sensors that detect environmental factors like soil erosion, water levels, or air quality, providing real-time data to manage resources sustainably and predict natural disasters.

MTC in 6G technology provides rural areas with a reliable, self-sustaining network that operates autonomously, supporting large-scale IoT applications and helping to bridge the urban-rural technology gap. With 6G-enabled MTC, rural regions can enhance productivity, improve public services, and foster sustainable development, making communities more connected, resilient, and economically vibrant.

Cognitive Radio

Cognitive Radio (CR) in 6G technology has the potential to significantly enhance connectivity in rural areas by intelligently managing and optimizing available spectrum resources. Cognitive radio allows devices to sense unused frequency bands and dynamically allocate these for communication, which is especially useful in rural areas where network infrastructure may be sparse, and conventional networks may struggle with bandwidth constraints. By efficiently using available frequencies, CR ensures that rural regions can achieve better connectivity without the need for extensive physical infrastructure.

In agriculture, cognitive radio can connect IoT devices and sensors on large, spread-out farms. Since cognitive radio detects and uses available frequencies without interference, it ensures reliable communication for precision farming applications, such as soil monitoring, weather data collection, and automated irrigation systems. This enables farmers to make data-driven decisions and boosts productivity even in remote areas with minimal connectivity infrastructure.

For rural healthcare, cognitive radio can support telemedicine and real-time health monitoring. By allocating necessary spectrum only when required, CR provides stable and reliable connections for remote health services, which is essential for timely consultations, diagnostics, and monitoring of patients who may be far from urban healthcare centers.

In education, CR can expand access to digital learning platforms. Schools in rural areas can benefit from cognitive radio's ability to maximize spectrum use, enabling high-quality video streaming and interactive online learning sessions for students who may otherwise have limited access to educational resources.

Furthermore, cognitive radio is highly advantageous in disaster response. In emergencies, where communication lines are often overloaded or disrupted, CR can dynamically find available frequencies, supporting real-time coordination and information sharing for emergency teams and local residents.

By integrating cognitive radio into 6G, rural areas can experience a significant improvement in connectivity and resource management (Thota et al., 2024). This capability not only expands access to essential services but also fosters innovation and economic development by making spectrum usage more efficient, affordable, and adaptable to rural needs.

Intelligent Reflective Surfaces

Intelligent Reflective Surfaces (IRS) in 6G technology represent a revolutionary advancement in enhancing connectivity, particularly in rural areas where traditional telecommunications infrastructure is often insufficient or economically unfeasible. By utilizing IRS—surfaces embedded with a multitude of small antennas capable of dynamically reflecting and manipulating wireless signals—rural communities can experience substantial improvements in signal coverage and quality. This capability is especially beneficial in landscapes characterized by natural obstructions like trees, hills, or buildings, which can interfere with direct line-of-sight communication from conventional cell towers.

The deployment of IRS is particularly advantageous because these surfaces can be integrated into existing structures, such as rooftops, lamp posts, or billboards, allowing for an efficient and cost-effective enhancement of wireless networks. Instead of investing heavily in new infrastructure, rural areas can utilize IRS to extend the range of wireless communication systems. This not only helps in delivering stronger signals to remote locations but also optimizes the overall network performance by directing and focusing signals where they are needed most, thus improving data rates and reducing latency.

Moreover, IRS facilitates the efficient allocation of bandwidth, which is crucial for supporting the increasing number of Internet of Things (IoT) devices in rural settings. In sectors like precision agriculture, where farmers rely on real-time data from sensors monitoring soil conditions, crop health, and weather patterns, IRS can ensure that these devices remain connected even in areas with previously limited network access. Similarly, in environmental monitoring, IRS can support networks of sensors that track air and water quality, helping communities address ecological concerns and comply with regulatory requirements.

In remote healthcare, IRS can enhance telemedicine capabilities by ensuring that healthcare providers can maintain stable connections for remote consultations and diagnostics, which is vital for patients living far from medical facilities. The ability to transmit high-quality video and data seamlessly can significantly improve healthcare outcomes in rural areas, where access to specialists is often lacking.

By reducing the reliance on costly infrastructure investments, IRS not only offers a scalable solution for enhancing connectivity but also contributes to greater economic viability. Enhanced connectivity leads to better access to digital services, educational resources, and e-governance initiatives, ultimately fostering economic growth and improving quality of life. In this way, IRS can play a pivotal role in bridging the digital divide, ensuring that rural areas can leverage high-speed, reliable communication essential for their development and integration into the broader digital economy (Ahad et.al. 2024). As communities adopt these advanced technologies, they will likely witness enhanced social and economic opportunities, paving the way for a more connected and sustainable future.

Cloud Computing

Cloud computing within 6G technology has the potential to dramatically improve connectivity, productivity, and access to essential services in rural areas. By leveraging 6G's ultra-low latency, high speeds, and massive connectivity, cloud computing can overcome many of the geographical and infrastructural challenges that rural regions face. Cloud-based services, accessible via 6G networks, allow rural communities to store, process, and access data remotely, reducing the need for powerful on-premises computing resources. This setup enables a wide range of applications that support agriculture, healthcare, education, and small businesses in rural areas.

In precision agriculture, for instance, cloud computing can process large volumes of data collected from IoT devices monitoring soil health, crop conditions, and weather. With 6G connectivity, farmers can access this data in real-time, receiving actionable insights for optimizing irrigation, fertilization, and harvesting, which improves yield and resource efficiency. In healthcare, cloud computing can enhance telemedicine by allowing real-time processing of data from remote monitoring devices, electronic health records, and diagnostic tools. This facilitates quicker diagnoses and continuous care for patients in rural areas, where access to healthcare facilities is often limited.

Education in rural areas also stands to benefit from cloud computing in 6G, as students and teachers can access digital classrooms, collaborative tools, and high-quality multimedia content without the need for expensive local infrastructure (Das et al., 2024). This opens up opportunities for rural students to engage with educational content from around the world, bridging the rural-urban educational gap.

In addition, cloud computing can support small businesses in rural areas by providing access to enterprise-grade software, analytics tools, and e-commerce platforms without the need for significant hardware investment. With 6G's fast and reliable connectivity, rural businesses can streamline operations, reach broader markets, and compete in the digital economy more effectively.

Edge Intelligence

Edge intelligence in 6G technology empowers rural areas by enabling local data processing and reducing the need for continuous cloud connectivity, making communities more self-sustaining and efficient. Edge-powered sensors monitor infrastructure like roads, bridges, and utilities, detecting maintenance needs or safety hazards in real-time and immediately communicating critical information to local authorities. This allows for prompt repairs, enhancing community safety and reducing long-term maintenance costs. Additionally, edge intelligence supports sectors vital to rural development, such as agriculture, healthcare, and education, by providing real-time insights directly at the network edge. By integrating AI locally, 6G's edge intelligence reduces bandwidth costs and enhances resilience in connectivity, even in remote regions. This accessibility to advanced technology fosters economic growth, improves quality of life, and strengthens community resilience, effectively bridging the rural-urban digital divide.

Edge intelligence in 6G technology is set to play a crucial role in bringing advanced data processing and AI-driven applications directly to rural areas, significantly improving connectivity, efficiency, and access to essential services. Edge intelligence combines edge computing—where data processing happens closer to the source—with artificial intelligence (AI), allowing data to be processed locally at the network edge rather than relying on distant cloud servers. When paired with 6G's ultra-low latency and high bandwidth, this setup enables real-time processing and analysis, which can address many challenges that rural communities face, such as limited infrastructure and connectivity.

In agriculture, edge intelligence enables IoT devices on farms to analyze data locally, providing insights in real-time. For instance, edge-enabled sensors can monitor soil conditions, predict weather impacts, and detect crop diseases without the need for constant connectivity to central servers. This real-time data analysis helps farmers optimize their resources, make timely decisions, and improve yields, all while using a fraction of the bandwidth required for cloud processing.

For healthcare, edge intelligence can support remote patient monitoring by processing health data locally on devices, ensuring reliable diagnostics even with intermittent internet connectivity. Vital signs can be analyzed and flagged for irregularities on-site, reducing the dependency on centralized healthcare facilities, which are often far from rural patients. This enables more immediate medical responses, allowing for faster diagnostics and the potential for preventive care in communities that would otherwise have limited access to healthcare.

In education, edge intelligence can power local educational platforms and learning tools that adapt to individual student needs. AI-driven educational content can be stored and processed locally, enabling interactive learning experiences and simulations even where connectivity is inconsistent. This means rural students can benefit from high-quality digital education without relying on continuous cloud access, bridging educational inequalities.

Edge intelligence in 6G technology empowers rural areas by enabling local data processing and reducing the need for continuous cloud connectivity, making communities more self-sustaining and efficient. Edge-powered sensors monitor infrastructure like roads, bridges, and utilities, detecting maintenance needs or safety hazards in real-time and immediately communicating critical information to local authorities (Yunus et.al. 2024). This allows for prompt repairs, enhancing community safety and reducing long-term maintenance costs. Additionally, edge intelligence supports sectors vital to rural development, such as agriculture, healthcare, and education, by providing real-time insights directly at the network edge. By integrating AI locally, 6G's edge intelligence reduces bandwidth costs and enhances resilience in connectivity, even in remote regions. This accessibility to advanced technology fosters economic growth, improves quality of life, and strengthens community resilience, effectively bridging the rural-urban digital divide.

Blockchain Technology

Blockchain technology integrated with 6G networks offers a transformative solution for rural areas, enhancing transparency, security, and efficiency across various applications such as agriculture, supply chains, healthcare, and local governance. With 6G's ultra-reliable, high-speed connectivity, blockchain's decentralized, tamper-proof ledger system can be effectively applied to track and verify transactions, records, and data, which is crucial in areas where traditional infrastructure and trusted intermediaries are often limited.

In agriculture, blockchain can trace the entire supply chain, from farm to market, ensuring transparency in production processes and fair pricing for farmers. Through 6G-enabled IoT devices, data such as crop quality, harvest dates, and storage conditions can be recorded on the blockchain in real time, allowing buyers and consumers to verify the origin and quality of products. Theis fosters trust in local produce and gives farmers more control over pricing, which can help boost their income and market competitiveness.

In healthcare, blockchain on a 6G network can secure patient records, medical histories, and telemedicine transactions, providing rural residents with a reliable way to access and share their health information. By enabling patients to securely manage and share their data, blockchain facilitates accurate, coordinated care across different healthcare providers, even in remote regions. Additionally, it offers a trustworthy system for distributing medical supplies and vaccines, reducing the risk of counterfeit medications.

Blockchain can also be used in rural finance, where it supports decentralized financial services that are accessible to individuals who may lack access to traditional banking. With 6G-powered blockchain platforms, rural residents can use secure, peer-to-peer transactions, micro-lending, and digital identity verification, which are especially valuable in areas where financial infrastructure is minimal (Kim et.al. 2024). Blockchain enables secure, cost-effective financial interactions that support economic inclusion and community resilience.

Furthermore, local governance can benefit from blockchain by improving transparency and accountability in land records, voting, and public fund distribution. Land ownership records can be securely stored on a blockchain, helping to address issues like land disputes and fraud, which are common in rural areas. Transparent voting systems on blockchain can enhance trust in elections, and 6G-powered blockchain can streamline the allocation of public resources, ensuring that funds are directed toward community priorities.

Quantum Computing

Quantum computing in 6G technology holds transformative potential for rural areas, offering groundbreaking capabilities to solve complex problems that require immense computational power. Although still an emerging field, quantum computing can enhance 6G networks by optimizing data processing and decision-making in real time. When applied in rural contexts, this advanced technology could address key challenges in agriculture, healthcare, environmental management, and education, contributing to rural economic growth and quality of life.

In agriculture, quantum-enhanced 6G can optimize crop management by rapidly analyzing vast amounts of data from IoT sensors monitoring soil, water, and climate conditions. Quantum algorithms can identify patterns in this data more efficiently than classical computers, helping farmers make precise, data-driven decisions about irrigation, fertilization, and crop rotation. This application improves productivity, resource management, and sustainability in farming, benefiting both local farmers and the food supply chain.

In healthcare, quantum computing with 6G can enable advanced medical diagnostics and personalized treatment plans. For rural areas, where healthcare facilities are limited, quantum-enabled AI models could analyze complex genetic or molecular data from remote monitoring devices, identifying health risks or personalizing treatment options more accurately. Quantum-powered 6G can also streamline telemedicine, providing real-time insights and facilitating diagnostic tools that rural practitioners and patients can access directly, reducing the need for physical visits.

Environmental management is another area that could benefit from quantum-enhanced 6G, as climate and weather modeling are data-intensive processes. Quantum computing can process environmental data from IoT sensors at a faster rate and with greater accuracy, enabling better predictions for weather events and resource use (Sharma et.al. 2024). Rural communities can benefit from this information in disaster preparedness, water resource management, and conservation efforts, making their environments safer and more sustainable.

In education, quantum-powered 6G networks can provide rural students access to cutting-edge research, simulations, and virtual laboratories, enriching STEM education and offering unique learning experiences. Quantum computing could allow for faster and more realistic simulations, enabling rural students and educators to access interactive, high-quality educational content that is otherwise unavailable due to limited local resources.

Finally, quantum-enhanced 6G can help rural economies by supporting logistics and supply chains through complex optimization and predictive modeling. Rural businesses, including small-scale farmers and artisans, can use these insights to streamline production, manage inventories, and ensure timely product delivery to broader markets, which is often a challenge in remote areas.

Opportunities of 6G in the Rural Economy

Ahad et.al (2024) present the following opportunities of 6g in the rural economy.

Industrial Automation

Industrial automation has garnered significant attention in recent years, particularly with the rise of Industry 4.0. Various sectors are working towards fully automating their manufacturing processes. While the services offered by these sectors differ, their primary goal in pursuing automation is consistent: to reduce or eliminate human involvement in production. This drive to minimize human factors stems from the potential risks and errors that can arise from human activity.

To achieve complete automation, the machines integrated into these processes must be highly reliable, capable of effectively replacing human workers. Low latency is a critical requirement for this full automation, as any delays could disrupt systems and impact thousands of connected devices. Ensuring rapid communication and response times among all modules in the system is essential for seamless operation. This level of dependable and swift data transmission can only be achieved through the capabilities of 6G mobile technology.

Enhanced Connectivity

6G technology promises to deliver ultra-high-speed connectivity and low latency, significantly enhancing communication and data transfer capabilities. This leap in connectivity is particularly crucial for rural areas, where traditional internet services often fall short, creating a persistent digital divide. With 6G, rural residents will have the opportunity to access a wealth of online services that were previously either unavailable or difficult to obtain. Educational resources can become readily accessible through high-quality online courses, virtual classrooms, and interactive learning tools, empowering students and lifelong learners to engage in educational opportunities that were once out of reach. In healthcare, 6G can revolutionize telehealth solutions by enabling real-time virtual consultations, remote patient monitoring, and timely access to medical advice, thereby addressing barriers to healthcare access caused by distance and limited local facilities. This enhanced connectivity will not only facilitate better quality of life for individuals in rural communities but also foster economic development, social inclusion, and overall community resilience, ensuring that all residents can participate fully in the digital age.

Agricultural Innovation

Agricultural innovation stands to benefit immensely from the advanced capabilities of 6G technology, particularly through its integration with the Internet of Everything (IoE). By enabling real-time data collection and analysis via smart sensors and drones, 6G empowers farmers to monitor critical factors such as soil health, weather patterns, and crop conditions with unprecedented accuracy. These smart sensors can provide continuous feedback on soil moisture levels, nutrient content, and temperature, allowing farmers to make informed decisions about irrigation and fertilization. Drones equipped with high-resolution cameras and sensors can conduct aerial surveys of fields, identifying areas that require attention and enabling targeted interventions. This level of precision agriculture not only leads to optimized resource management—ensuring that water, fertilizers, and pesticides are used efficiently—but also contributes to increased crop yields and improved overall productivity. Additionally, by reducing the reliance on chemical inputs and minimizing waste, 6G-supported agricultural practices can significantly lower the environmental impact of farming, promoting sustainability and conservation of natural resources. As a result, the adoption of 6G technology in agriculture holds the potential to transform rural economies, enhance food security, and foster a more resilient agricultural sector in the face of climate change and growing global demand for food.

Telehealth and Remote Healthcare

6G technology has the potential to enhance remote health monitoring and telemedicine services, providing advanced healthcare solutions to rural communities that often lack adequate medical facilities. With this connectivity, patients can engage in virtual consultations with healthcare professionals, enabling them to receive timely medical advice without the need for extensive travel. Additionally, remote monitoring devices can be utilized to keep track of chronic conditions, allowing healthcare providers to continuously assess patient health from a distance. This capability not only improves accessibility to healthcare services but also contributes to better overall health outcomes, ensuring that individuals in rural areas can manage their health more effectively and receive the care they need when they need it.

Education and Skill Development

6G technology facilitates access to high-quality online educational resources, virtual classrooms, and various training programs, opening up new avenues for learning. This connectivity allows rural students and workers to engage in distance learning and skills development initiatives, significantly expanding their educational opportunities. By participating in these programs, they can better prepare themselves for the demands of a rapidly evolving job market, equipping themselves with the knowledge and skills necessary to succeed in diverse professional fields. Overall, 6G empowers individuals in rural areas to enhance their education and career prospects through greater access to innovative learning platforms.

Smart Community Development

The integration of 6G technology can facilitate the creation of smart communities, where data-driven solutions significantly enhance local infrastructure and services. With the implementation of smart energy management systems, efficient waste disposal methods, and advanced transportation networks, these communities can experience improved efficiency and sustainability. Such initiatives not only streamline operations but also promote community engagement by involving residents in decision-making processes and resource management. Ultimately, the adoption of 6G can transform rural areas into interconnected, intelligent environments that support a higher quality of life for their inhabitants.

Economic Growth and Job Creation

The deployment of 6G infrastructure has the potential to drive economic development in rural areas by attracting new businesses and fostering entrepreneurship. Enhanced connectivity allows local enterprises to broaden their market reach, tap into new opportunities, and innovate their services. This increased capability can lead to job creation and promote economic diversification, helping rural communities build a more resilient and dynamic economy. By improving access to advanced technologies, 6G can play a pivotal role in transforming the economic landscape of these regions.

Environmental Monitoring and Sustainability

6G technology has the capacity to enhance sophisticated environmental monitoring systems that assess air and water quality, wildlife populations, and natural resource management. By implementing these systems, rural communities can take proactive measures to address environmental challenges and encourage sustainable practices. This proactive approach not only helps maintain a healthier ecosystem but also empowers residents to make informed decisions regarding environmental conservation and resource management, ultimately contributing to the overall well-being of their communities.

Crisis Management and Disaster Response

Improved connectivity offered by 6G can significantly strengthen rural emergency response systems by facilitating real-time communication among local authorities, first responders, and community members. This enhanced data-sharing capability enables faster response times during natural disasters, ultimately bolstering community resilience and safety. By ensuring that all parties can exchange critical information promptly, 6G technology helps rural areas better prepare for and manage emergencies, leading to more effective disaster response and recovery efforts.

CHALLENGES OF 6G IN THE RURAL ECONOMY

Gryech et.al (2024) present the challenges of 6g in the rural economy.

Infrastructure Costs

Infrastructure Costs for establishing 6G networks in rural areas are substantial and present unique challenges compared to urban deployments. Rural regions, often characterized by low population density and challenging terrain, require extensive infrastructure investment to ensure wide coverage, which includes constructing new towers, laying fiber-optic cables, and deploying advanced antennas. Unlike urban areas where high population density makes infrastructure investments more profitable, the dispersed nature of rural communities reduces the potential Return On Investment (ROI) for telecom providers. Consequently, private companies are often hesitant to undertake the costs alone, as the low user base does not immediately offset the high upfront expenditures needed to build and maintain 6G infrastructure.

Digital Literacy and Skills

Digital literacy and skills development are critical for enabling rural communities to fully benefit from the Internet of Everything (IoE) and other advanced technologies. IoE relies on a seamless integration of smart devices, sensors, and data-sharing systems that enhance connectivity and automation across various sectors, including agriculture, healthcare, education, and commerce. However, for rural residents to harness these benefits effectively, they need foundational digital literacy skills and training to operate, manage, and troubleshoot IoE-enabled devices and applications. One of the primary challenges in rural areas is the digital skills gap; many rural residents may have limited exposure to digital tools or internet-based applications due to infrastructure constraints or lack of access to affordable internet. This can create a barrier to IoE adoption, as residents may struggle with using IoE-enabled devices, managing data privacy, or troubleshooting basic technology issues. To address this, training programs focused on digital literacy are essential, equipping individuals with the foundational skills to operate IoE devices and navigate the digital landscape with confidence.

Data Security

Data security is a critical consideration in IoE networks, which generate vast amounts of data across various applications, raising significant privacy and security concerns. In remote regions with limited access to cybersecurity resources, protecting the data exchanged between IoE devices becomes particularly challenging. Sensitive information related to healthcare, agriculture, local governance, and financial transactions is transmitted across these networks, making rural communities vulnerable to data breaches, unauthorized access, and cyber-attacks. Additionally, as IoE devices are often connected to a broader network, a breach in one device can compromise the entire system, leading to data misuse or loss of control over essential functions. To address these risks, it is essential to implement security measures such as encryption, multi-factor authentication, and regular software updates on IoE devices. Furthermore, establishing cybersecurity training and resources in rural communities can empower residents to protect their data, creating a safer and more resilient IoE ecosystem that supports sustainable digital transformation in these areas.

Energy Consumption

Energy consumption is a challenge in rural areas, where consistent power supplies may be less accessible. Advanced 6G infrastructure and devices typically require significant power to operate, meaning rural areas may need alternative energy solutions to maintain continuous service.

Geographic and Environmental Factors

Geographic and environmental factors also pose significant obstacles, as rural areas may contain difficult terrain, forests, or mountainous regions that hinder signal propagation and require more specialized equipment to ensure coverage. High-frequency 6G signals are particularly susceptible to interference from physical obstructions, creating the need for additional network equipment like Intelligent Reflective Surfaces (IRS) or satellite support, which increases deployment complexity and cost.

CONCLUSION AND FUTURE DIRECTIONS OF RESEARCH

6G communication technology represents a significant advancement in connectivity, delivering high-speed internet access and robust network performance across various sectors globally, with a particular emphasis on rural areas. This next-generation technology integrates a range of sophisticated innovations, such as artificial intelligence, edge computing, and advanced IoT capabilities, enabling transformative changes in multiple sectors within the rural economy. For instance, in agriculture, 6G facilitates precision farming through real-time data analytics and smart sensors, helping farmers optimize resource use and increase crop yields. In healthcare, it supports telemedicine and remote monitoring, allowing residents in remote areas to access essential medical services (Ansi et.al 2024). Additionally, in education, 6G enables online learning platforms and virtual classrooms, empowering rural students with access to quality educational resources. By bridging the connectivity gap, 6G not only enhances the quality of life in rural communities but also drives economic growth and innovation, making it a vital component of sustainable development in these regions.

REFERENCES

Ahad, A., Jiangbina, Z., Tahir, M., Shayea, I., Sheikh, M. A., & Rasheed, F. (2024). 6G and Intelligent Healthcare: Taxonomy, technologies, open issues and future research directions. *Internet of Things : Engineering Cyber Physical Human Systems*, *25*, 101068. DOI: 10.1016/j.iot.2024.101068

Al-Ansi, A., Al-Ansi, A. M., Muthanna, A., & Koucheryavy, A. (2024). Blockchain technology integration in service migration to 6g communication networks: A comprehensive review. *Indonesian Journal of Electrical Engineering and Computer Science*, *34*(3), 1654–1664. DOI: 10.11591/ijeecs.v34.i3.pp1654-1664

Chataut, R., Nankya, M., & Akl, R. (2024). 6G networks and the AI revolution—Exploring technologies, applications, and emerging challenges. *Sensors (Basel)*, *24*(6), 1888. [REMOVED HYPERLINK FIELD]DOI: 10.3390/s24061888 PMID: 38544151

Chauhan, D., Mewada, H., Gondalia, V., Almalki, F. A., Patel, S., Modi, H., Kavaiya, S., Trivedi, Y., & Mujlid, H. M. (2024). Balancing Technological Innovation and Environmental Sustainability: A Lifecycle Analysis of 6G Wireless Communication Technology. *Sustainability (Basel)*, *16*(15), 6533. DOI: 10.3390/su16156533

Das, S. L., Prakash, O., Naidu, V. J., Das, S., Chowdhury, S., & Somassoundaram, T. *6G applications and use cases. In 6G Communication Network*. CRC Press.

Devapitchai, J. J., Dheenadhayalan, K., Rajeshkumar, L., Meena, M. R., Soundarya, M., Sowmiya, M. S., & Manickam, T. (2024). Applications of Artificial Intelligence Techniques in Modern Banking Sectors. In *Blockchain Applications for Smart Contract Technologies* (pp. 279–298). IGI Global. DOI: 10.4018/979-8-3693-1511-8.ch014

Devapitchai, J. J., Krishnapriya, S. V., Karuppiah, S. P., & Saranya, S. (2024). Using AI-Driven Decision-Making Tools in Corporate Investment Planning. In *Generative AI for Transformational Management* (pp. 137–160). IGI Global. DOI: 10.4018/979-8-3693-5578-7.ch006

Gryech, I., Vinogradov, E., Saboor, A., Bithas, P. S., Mathiopoulos, P. T., & Pollin, S. (2024). A systematic literature review on the role of UAV-enabled communications in advancing the UN's sustainable development goals. *Frontiers in Communications and Networks*, *5*, 1286073. DOI: 10.3389/frcmn.2024.1286073

Jebadurai, D. J. (2013). An overview of problems of rural entrepreneurs in India. *International Journal of Advanced Research in Management and Social Sciences*, *2*(7), 202–208.

Jebadurai, D. J., Sheela, M. V., Rajeshkumar, L., Soundarya, M., Meena, R., Manickam, T., & Manikandan, M. (2024). AI-Driven Decision-Making and Optimization in Modern Agriculture Sectors. In *Using Traditional Design Methods to Enhance AI-Driven Decision Making* (pp. 269–288). IGI Global. DOI: 10.4018/979-8-3693-0639-0.ch012

Jha, A. V., Appasani, B., Khan, M. S., Zeadally, S., & Katib, I. (2024). 6G for intelligent transportation systems: Standards, technologies, and challenges. *Telecommunication Systems*, *86*(2), 1–28. DOI: 10.1007/s11235-024-01126-5

Kim, N., Kim, G., Shim, S., Jang, S., Song, J., & Lee, B. (2024). Key Technologies for 6G-Enabled Smart Sustainable City. *Electronics (Basel), 13*(2), 268. DOI: 10.3390/electronics13020268

Liu, Y., Razman, M. R., Zakaria, S. Z. S., Ern, L. K., Hussain, A., & Chamola, V. (2024). Utilizing ubiquitous learning to foster sustainable development in rural areas: Insights from 6G technology. *Computers in Human Behavior, 161*, 108418. DOI: 10.1016/j.chb.2024.108418

Matinmikko-Blue, M., Yrjölä, S., & Ahokangas, P. (2024). Multi-perspective approach for developing sustainable 6G mobile communications. *Telecommunications Policy, 48*(2), 102640. DOI: 10.1016/j.telpol.2023.102640

Pandey, B. K., Paramashivan, M. A., Pandey, D., George, A. S., Agarwal, A., Dadheech, P., & Mahajan, D. A. (2024). A Recent Survey on 6G Vehicular Technology: Applications and Challenges. In Pandey, B. K., Pandey, D., & Ahmad, T. (Eds.), *6G Security Education and Multidisciplinary Implementation*, 337-350. IGI Global. 10.4018/979-8-3693-7421-4.ch019

Raja, S., Jebadurai, D. J., Ivan, L., Mykola, R. V., Ruslan, K., & Nadiia, P. R. (2024). Impact of Artificial Intelligence in Students' Learning Life. In *AI in Business: Opportunities and Limitations* (Vol. 2, pp. 3–17). Springer Nature Switzerland. DOI: 10.1007/978-3-031-49544-1_1

Sharma, S., Popli, R., Singh, S., Chhabra, G., Saini, G. S., Singh, M., Sandhu, A., Sharma, A., & Kumar, R. (2024). The Role of 6G Technologies in Advancing Smart City Applications: Opportunities and Challenges. *Sustainability (Basel), 16*(16), 7039. DOI: 10.3390/su16167039

Sharma, S., Popli, R., Singh, S., Chhabra, G., Saini, G. S., Singh, M., Sandhu, A., Sharma, A., & Kumar, R. (2024). The Role of 6G Technologies in Advancing Smart City Applications: Opportunities and Challenges. *Sustainability (Basel), 16*(16), 7039. DOI: 10.3390/su16167039

Shin, H., Park, S., Kim, L., Kim, J., Kim, T., Song, Y., & Lee, S. (2024). The future service scenarios of 6G telecommunications technology. *Telecommunications Policy, 48*(2), 102678. DOI: 10.1016/j.telpol.2023.102678

Sinha, P., Kandasamy, M., Shanmugam, R., & Nageswari, P. (2024). *6G communication challenges. 6G Communication Network: Architecture*. Security and Applications.

Thota, S., & Govind, T. R. (2024). 6G: Opportunities and challenges. Towards Wireless Heterogeneity in 6G Networks, 1-17.

Yunus, N. A. M., Hanapi, Z. M., & Kamarudin, S. (2024, July). 6G on the Horizon: Technologies, Requirements, Trends, and Potential Techniques. In *2024 International Conference on Electrical, Computer and Energy Technologies (ICECET* (pp. 1-6). IEEE.

KEY TERMS AND DEFINITIONS

6G: Sixth Generation.
5G: Fifth Generation.
IoT: Internet of Things.

AI: Artificial Intelligence.
ML: Machine Learning.
RSRP: Fragmented Reference Signal Received Power.
UAVs: Unmanned Aerial Vehicles.
RL: Reinforcement Learning.
BS: Base Station.
IoE: Internet of Everything.
VR: Virtual Reality.
AR: Augmented Reality.
VLC: Visible Light Communication.
MTC: Machine-Type Communication.
URLLC: Ultra-Reliable Low Latency Communication.
CR: Cognitive Radio.
IRS: Intelligent Reflective Surfaces.
RoI: Return on Investment.

Chapter 3
Policy and Regulatory Challenges in Implementing Beyond 5G and 6G Networks in Rural India:
Ensuring Right to Health

Rajiv Gurugopinath
https://orcid.org/0000-0003-3845-6778
Ramaiah College of Law, India

Bangaru Venugopal
Vel Tech Rangarajan Dr. Sagunthala R&D Institute of Science and Technology, India

ABSTRACT

The development of beyond 5G and 6G communication systems is revolutionizing connectivity, with a focus on connecting the unconnected and enhancing rural access to services, particularly in healthcare. India's government is heavily investing in 6G R&D, with organizations like TSDSI working to develop India-specific standards. This chapter analyzes India's current policy and regulatory environment, identifies barriers to implementing 6G in rural areas, and examines the impact of these networks on health and public health. A critical objective is to study the case law of Juhi Chawla v. Science and Engineering Research Board to assess the health implications of these networks in India and globally.

INTRODUCTION

The advent of first-generation cellular networks (1G) in the 1980s marked the beginning of the development of technology that enables wireless communication for the first time. In the course of the progression of cellular networks from 2G to 4G, there have been significant advancements in the fields of communications and networking. Regardless of the progress made in networking and communication technologies, it is crucial to prioritise security in order to guarantee the durability and dependability of networks. Hence, it would be beneficial for the research community to ascertain the research directions

DOI: 10.4018/979-8-3693-7788-8.ch003

pertaining to security. The proposed 6G networks. Given the conventional functions the requirements and standards for 6G have not yet been determined, but they are currently under development. There is a scarcity of books that offers security and Insights on the privacy implications of networks beyond 5G. Moreover, it is essential to systematically expand upon 5G research. Integrate current and developing research on 6G security achievement. Deploying Beyond 5G (B5G) and 6G networks in rural India has substantial policy and regulatory obstacles, especially in guaranteeing the right to health. The capacity of B5G and 6G to revolutionise healthcare through improved connectivity, velocity, and dependability is extensively documented.

These networks have the capability to facilitate innovative healthcare applications, such as remote diagnostics and telemedicine, which are of great importance for rural areas that have restricted access to healthcare (Haneya et al., 2022). It brought forth an unexpected slew of studies which were unique and required an in-depth study of the communication networks the effectiveness of Autonomous, Resilient, and Self-organizing Intelligent Software-defined Telecommunications (A-REaLiST) in low bandwidth conditions. Under normal conditions the A-REaLiST solution showed a satisfactory Quality of Experience (QoE) for remote visual examination over IP connectivity, even while using a 2G network. This was unexpected considering the often inadequate ICT infrastructure in rural India, which traditionally obstructs such applications. The unexpected discovery was the recognition of 5G networks, namely utilising the sub-6 band, as a feasible remedy for delivering rural internet coverage. The reason for this is that rural areas are sometimes neglected since mobile operators do not see a high enough Return on Investment (RoI). There are various theories surrounding these factors that are to be observed which include:

PUBLIC INTEREST THEORY OF REGULATION

The early 20th century saw the introduction of a concept derived from classical economic theory, pioneered by intellectuals such as Arthur Pigou and John Stuart Mill. This concept aimed to address market failings and enhance community well-being by government intervention. The theory posits that regulation is vital in safeguarding the public against the extremes of capitalism, particularly in instances where market forces fail to ensure optimal outcomes, such as in public utilities and essential services. The population Interest Theory regards regulation as a mechanism to safeguard and promote the welfare of the general population. The government, in its role as an impartial mediator, intervenes in the market to rectify disparities, guarantee equity, and safeguard consumers against monopolistic activities and market deficiencies. Regulations are viewed as tools for improving social well-being by allocating resources to ensure that everyone, particularly the disadvantaged and underserved, has access to vital services like healthcare and communication.

Figure 1. This figure illustrates the key developments and milestones in the evolution of Public Interest Theory, starting from its introduction in the early 20th century. The timeline captures the growing importance of government regulation in response to market failures and the rise of the welfare state (1920s-1940s). It also highlights significant periods where regulation played a key role, such as in the regulation of public utilities and essential services (1950s-1960s), efforts to address monopolies and economic disparities (1970s-1980s), and the ongoing challenges of globalization, technological progress, and evolving markets (1990s-2020s). This chronological representation emphasizes how the theory has adapted to modern economic and social realities

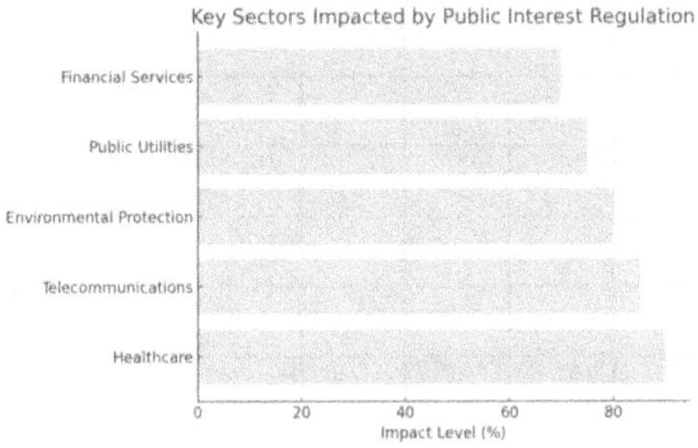

Source: Author

The basic concept of Public Interest Theory is its emphasis on the well-being of society, rather than the interests of individuals or corporations. It presupposes that the government would enact policies that favour the general population by rectifying market inefficiencies and supplying public benefits that would otherwise be insufficiently provided by the market. The philosophy prioritises the principles of openness, accountability, and the advancement of equity and social justice through regulation. The Public Interest Theory emerged as a response to the shortcomings of laissez-faire capitalism, which frequently resulted in monopolies, economic disparities, and the disregard for public well-being due to unregulated market forces. The theory has progressed in parallel with the establishment of the welfare state and the growing intricacy of contemporary economies, where the necessity for regulation in sectors like healthcare, telecommunications, and environmental protection has become more evident. In order to remain relevant, Public Interest Theory must adjust to the difficulties presented by globalisation, technological progress, and evolving markets. This necessitates striking a balance between promoting the well-being of the general people and cultivating innovation and economic expansion.

Figure 2. This chart shows the major sectors that have been significantly impacted by the application of Public Interest Theory through government regulation. The sectors include healthcare, telecommunications, environmental protection, public utilities, and financial services. The impact levels reflect how these sectors have benefited from government intervention aimed at correcting market failures, ensuring fair competition, and promoting public welfare. The chart demonstrates that sectors such as healthcare and telecommunications have seen a high level of regulatory influence, aligning with the goals of Public Interest Theory to promote social equity and protect consumers

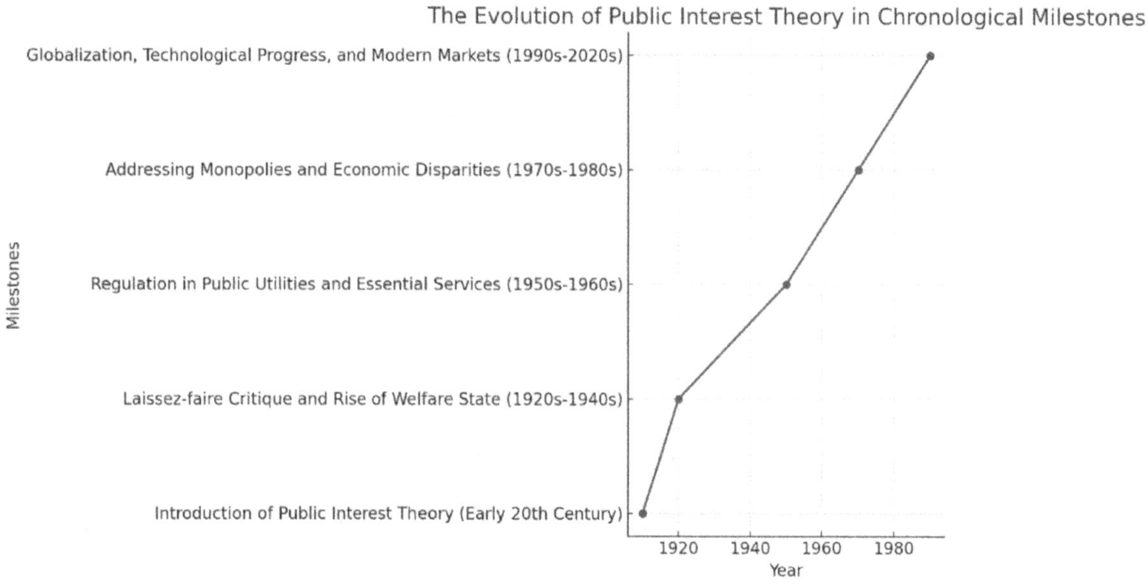

Source: Author

The Regulatory Capture Theory

The Regulatory Capture Theory, formulated by George Stigler, posits that regulatory agencies can become "captured" by the sectors they are tasked with overseeing, resulting in choices that prioritise industry interests at the expense of the public welfare. This phenomenon arises when industries wield their influence over regulators through lobbying, the practice of revolving door employment, or other similar methods. Within the framework of B5G and 6G implementation, this theory emphasises the possible dangers of telecom firms exerting influence over regulatory authorities in order to prioritise their own financial interests, perhaps to the detriment of wider public access and fairness.

Figure 3. This bar chart illustrates the core concept of the Regulatory Capture Theory, which highlights the potential influence of telecom firms on regulatory agencies. Telecom firms, represented with the highest influence level (90%), often leverage their power to shape regulations that benefit their interests. Regulatory agencies, although tasked with ensuring fair practices, are sometimes swayed by this influence (70%), potentially compromising their duty to protect public welfare. The lower influence on public welfare (40%) reflects the concern that when industry interests dominate, regulations may not prioritize the general population, leading to less equitable outcomes

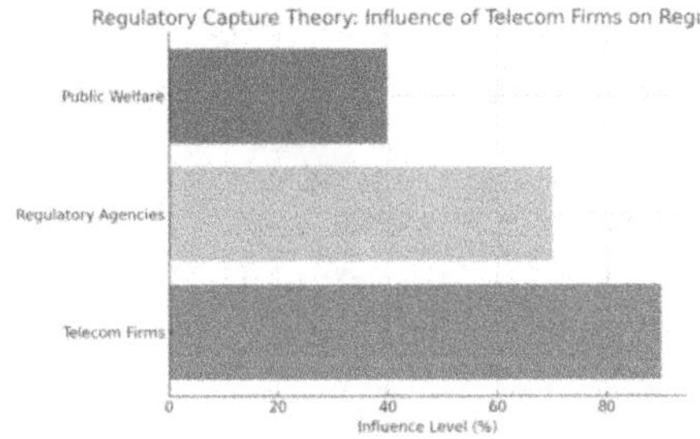

Source: Author

The Technology Acceptance Model (TAM)

The Technology Acceptance Model (TAM), developed by Fred Davis, elucidates the process by which consumers adopt and utilise technology, with a particular focus on two pivotal factors: perceived usefulness and perceived ease of use. The Technology Acceptance Model (TAM) is highly valuable for forecasting the adoption of emerging technologies such as B5G and 6G, particularly in rural regions where factors such as digital literacy and the perceived advantages of technology have a significant impact. Gaining insight into these elements can aid in formulating tactics to promote the adoption of these sophisticated networks, guaranteeing their widespread acceptance and utilisation.

Figure 4. This figure focuses on the two primary factors from the Technology Acceptance Model (TAM) that influence the adoption of new technologies: perceived usefulness and perceived ease of use. The chart shows that perceived usefulness (85%) plays a slightly more significant role than ease of use (75%) in driving adoption. In the context of B5G and 6G deployment, especially in rural areas, ensuring that people see the tangible benefits of these technologies is crucial for widespread acceptance, alongside making the technology easy to understand and operate

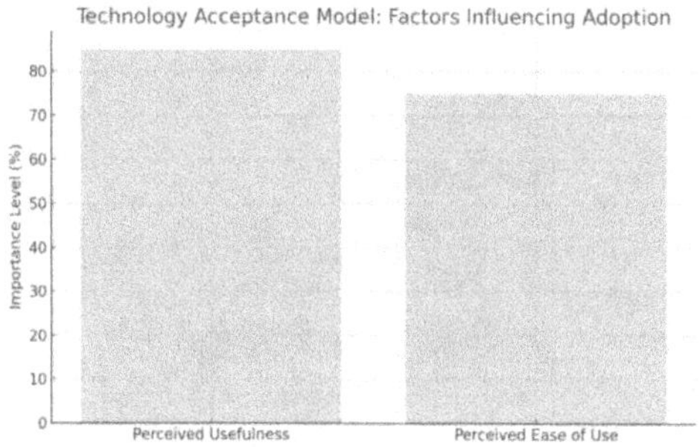

Source: Author

The Theory of Diffusion of Innovations

The Diffusion of Innovations Theory, developed by Everett Rogers, explains the process by which new ideas and technology are adopted and spread across cultures as time progresses. The theory delineates many adopter categories, including innovators, early adopters, early majority, late majority, and laggards. It elucidates how factors such as relative advantage, compatibility, and complexity impact the process of adoption. This theory is applicable to the deployment of B5G and 6G in rural areas, as it offers valuable insights on the effective introduction and adoption of these technologies, ensuring widespread usage among users.

Figure 5. This bar chart visualizes the different adopter categories from Everett Rogers' Diffusion of Innovations Theory. The percentages represent how populations adopt new technologies over time. Innovators (2.5%) are the first to adopt, followed by early adopters (13.5%). The early majority and late majority each represent 34% of adopters, while laggards (16%) are the last to adopt new technologies. Understanding these categories helps in strategizing the deployment of B5G and 6G, ensuring that the needs of each group are addressed for smooth technology diffusion, particularly in rural areas

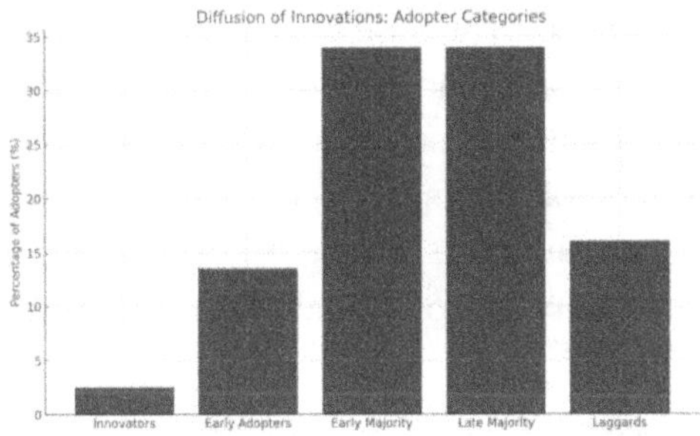

Source: Author

The Capability Approach

The Capability Approach, created by Amartya Sen and Martha Nussbaum, centres on the tangible freedoms and possibilities individuals possess to attain the lives they deem important, rather than solely their financial assets. It highlights the significance of facilitating individuals to pursue their well-being by providing them with access to vital services such as education, healthcare, and technology. Within the framework of B5G and 6G networks, the Capability Approach emphasises the importance of ensuring that these technologies augment individuals' capabilities, especially in underserved rural regions, by granting them access to services that promote their overall well-being. We have to understand the various factors surrounding the implementation of the B5G and 6G networks in rural areas:

The obstacles in enhancing rural connectivity, particularly in the context of implementing Beyond 5G (B5G) and 6G technologies, are complex and sometimes interconnected. Rural areas sometimes suffer from inadequate infrastructure, which hampers the establishment of modern connectivity. This includes the absence of essential components like fiber-optic cables, cell towers, and dependable power sources. The presence of this gap poses challenges and incurs high costs when attempting to expand contemporary telecommunications services to these regions. The exorbitant expenses associated with implementing rural connections pose a major obstacle to its economic feasibility. The expenses linked to implementing and up keeping network infrastructure in thinly populated or geographically difficult regions are frequently not balanced by the potential income, discouraging private sector investment.

Technological Challenges: The implementation of advanced technologies in remote locations frequently involves overcoming technical obstacles relating to the landscape, distance, and the current technological infrastructure, which may be obsolete or incompatible with new technology. Rural locations see a lower return on investment (ROI) for service providers due to the smaller population density and lower average income levels. Consequently, this leads to lower service adoption rates and less usage. Regulatory and policy barriers, such as insufficient or ambiguous regulatory frameworks, can impede the implementation of telecommunications infrastructure. Policies that fail to properly consider the distinct requirements of rural connectivity or that introduce administrative obstacles might exacerbate the challenges of implementing connectivity initiatives. The degree of digital literacy in rural people is a substantial obstacle to the uptake and efficient utilisation of connection services. In the absence of sufficient educational activities, the utilisation of technology remains limited, which weakens the rationale for investing in it.

Sustainability Concerns: Ensuring the long-term viability of connectivity efforts in rural areas necessitates continuous support and maintenance, which can be difficult owing to the scarcity of local skills and resources. Societal and cultural factors can pose a hurdle to the adoption of technology, as resistance may arise from adherence to cultural norms and established habits. Occasionally, there could be doubts regarding the advantages of being connected or a tendency to favour conventional lifestyles. To tackle these difficulties, a collaborative approach is needed, incorporating government policy, private sector investment, community engagement, and focused innovation. This approach aims to create sustainable and affordable solutions that address the lack of digital access in rural areas.

Figure 6. Directed acyclic graph of the various factors surrounding the policy and regulatory challenges implementing the B5G and 6G networks

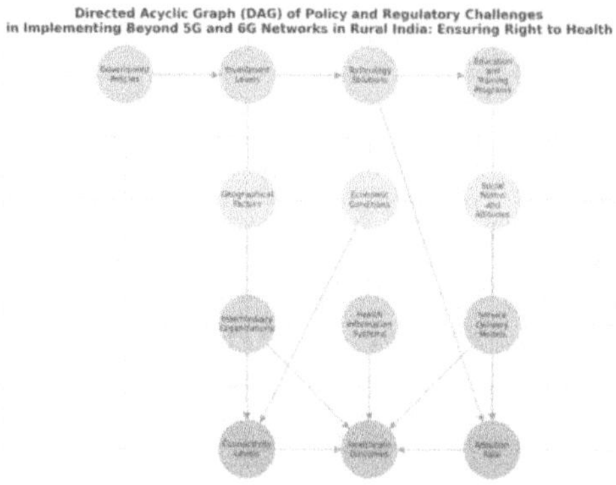

Source: Author

The unexpected importance of non-terrestrial networks (NTN) in future networks beyond 5G and 6G is noteworthy. There was a widespread understanding that terrestrial networks alone were insufficient to provide the criteria for flexibility, scalability, and coverage, thus emphasising the significance of incorporating NTNs. The surprising discovery was that implementing centralised baseband processing in distributed large Multiple Input, Multiple Output (MIMO) systems could result in excessively high backhaul data rates and pose significant processing difficulties.

As a result, Decentralised Baseband Processing (DBP) was developed to effectively handle the data. Service Level Agreements (SLAs) pertaining to the implementation of 5G technology in the healthcare sector: The idiosyncrasies and subtleties of 5G installations in healthcare exposed deficiencies in current service level agreements (SLAs). This finding was unexpected because it emphasised the importance of strong service level agreements (SLAs) to guarantee the secure and efficient incorporation of 5G technology into healthcare systems. Nevertheless, the implementation of these sophisticated networks in rural India encounters numerous obstacles. A significant obstacle is the absence of pre-existing infrastructure. Rural areas frequently experience inadequate ICT infrastructure, marked by limited bandwidth and sporadic connectivity, which impedes the successful deployment of e-health services (Bhattacharyya et al., 2019).

The challenge is exacerbated by the limited profitability for mobile operators in sparsely populated and low-income rural areas, which makes it commercially impractical to expand network coverage without significant regulatory intervention and incentives (Najjuuko et al., 2021). Furthermore, the incorporation of non-terrestrial networks (NTN) is crucial for ensuring the required extent and connection in remote regions. NTNs possess the capacity to scale, be flexible, and adapt in order to fulfil the varied requirements of rural healthcare systems. Nevertheless, the process of integrating these elements presents notable obstacles in terms of research and innovation, which include the requirement for novel technological facilitators and standards (Vanelli-Coralli et al., 2020). In order to tackle these difficulties, it is necessary to establish strong service level agreements (SLAs) that will guarantee a shared framework for the secure integration of B5G technology in healthcare. These agreements will involve all parties involved, such as device manufacturers, network providers, and regulatory bodies.

In addition, utilising advanced technologies such as distributed massive MIMO and energy-efficient solutions can assist in overcoming certain technical obstacles related to rural deployment (Gustavsson et al., 2021). In order to address the legislative obstacles for the introduction of 6G in rural India, a comprehensive strategy is required. This strategy should utilise open-source technology, blockchain, and prioritise sustainability and social effects. First and foremost, the implementation of open-source hardware and software is essential. This strategy has the potential to substantially decrease expenses and enhance availability, hence enabling the implementation of 6G networks in rural regions where economic motivations are generally less prominent. Rural communities can enhance digital inclusion by utilising free software and open hardware to construct sustainable and dependable communication networks (Towards 6G Networks for Rural Environments, 2022). Furthermore, blockchain technology has the potential to significantly contribute to resolving regulatory obstacles. Blockchain provides a distributed and open system that can establish confidence among various parties, such as legislators, service providers, and end-users (Ivan et al., 2022).

This technology has the ability to optimise the process of implementing policies, lower the expenses associated with storing data, and improve the security of data. These factors are crucial in obtaining regulatory approval and building public confidence. The inherent openness and verifiability of blockchain technology can effectively guarantee adherence to regulatory standards and facilitate a cooperative atmosphere for the implementation of 6G. Furthermore, prioritising sustainability and social effect is crucial for addressing the digital divide in rural areas. The research and standardisation efforts for 6G are giving more importance to these factors, which can be in line with regulatory objectives focused on fair service supply. It is crucial to comprehend the precise requirements of remote wireless connectivity and tackle the limitations of network design and deployment in order to successfully complete this process (6G for Bridging the Digital Divide, 2022). By emphasising sustainability as a primary motivator, 6G initiatives can secure legislative backing by showcasing their capacity to provide enduring social advantages (Chaoub et al., 2020).

The development of the sixth generation 6G mobile communication technology is inclusive and sustainable while bridging the digital gap in remote and underserved regions. Unlike its ancestors, 6G places high importance on the social impact that fuels advanced connectivity in rural areas with poor infrastructure.

Innovative network solutions like the deployment of mega cells and the integration of heterogeneous RATs will support this vision. Additionally, advancements like utilizing visible light communication will extend coverage efficiently and cost-effectively to geographically challenging areas. To reduce even further the economic inhibitors of rural connectivity, 6G is further emulating the elastic optical networks EONs. Besides repurposing some already deployed tech from urban areas, so no new massive spending is required, this route allows regions to have robust communications infrastructure that enables more individuals to access a wider variety of services digitally.

Figure 7. Mind map of the various factors surrounding the challenges in implementing the 6G networks

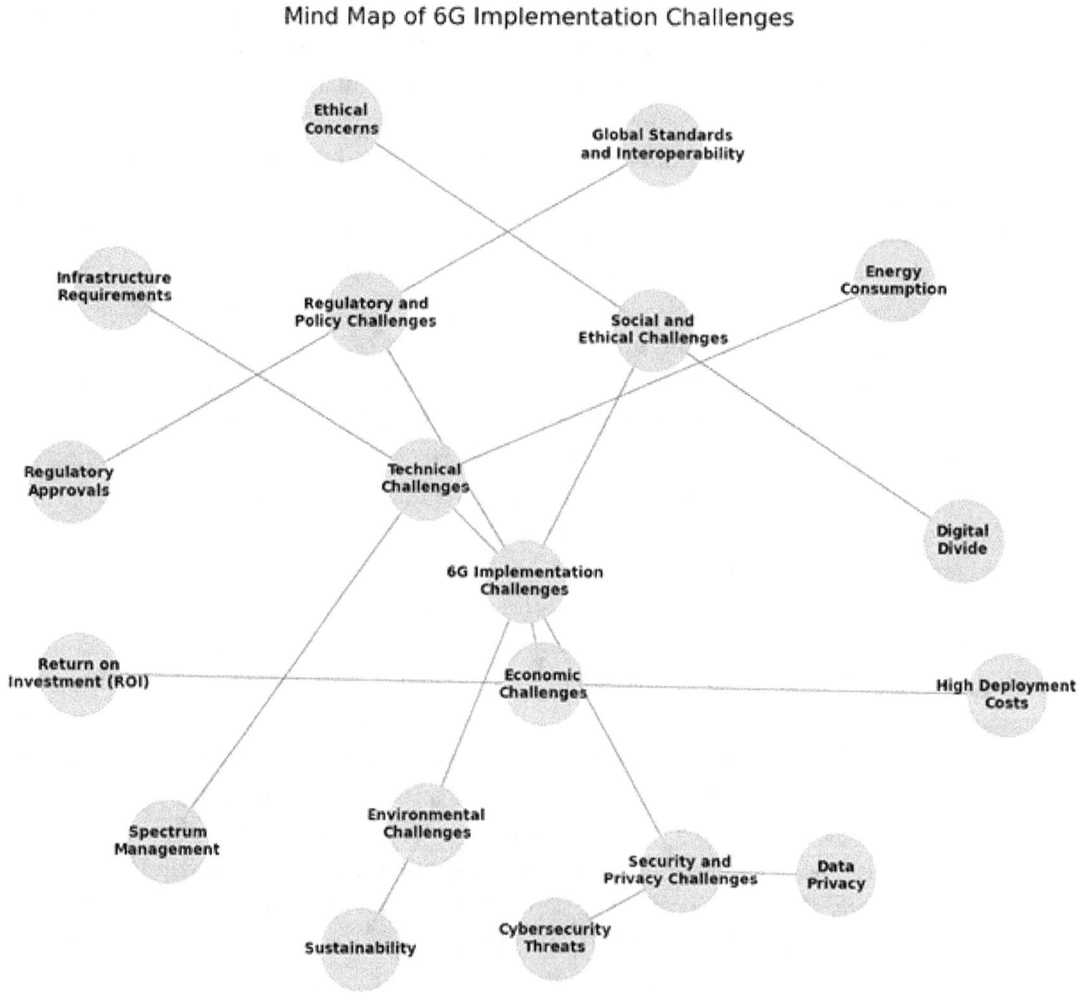

Source: Author

Strong attention to the need for intelligent maintenance, 6G marks a recognition of the unique challenges in ensuring service quality and reliability in remote, inaccessible locations, where it is most needed. Artificial intelligence has played an enabling role in self-organization and the automatic management of services. 6G further underlines the collaborative development of networks, allowing telecom operators and stakeholders to share these networks, a factor that enhances better usage of resources and increased provision of safe competitive services in rural areas, which is the overarching goal of bridging the digital divide and promoting inclusive growth.

Implementing 6G networks presents a multitude of challenges that span technical security privacy and infrastructural domains. These challenges are critical to address as 6G aims to deliver unprecedented speeds lower latencies and advanced capabilities beyond what 5G offers. One of the foremost challenges in 6G implementation is ensuring robust security and privacy. The integration of new technologies such as reconfigurable intelligent surfaces and cell free architecture introduces vulnerabilities like impersonation, attacks, side channel leaks and insider threats from rogue base stations.

Addressing these requires advanced security mechanisms like end-to-end encryption block chain and machine learning (Pandey et al., 2024). Additionally, the joint communication and sensing JCAS capabilities of 6G raise significant privacy concerns particularly regarding the handling of sensitive personally identifiable information (PII). This necessitates enhanced architectures that incorporate privacy specific features to comply with regulations like the GDPR (Dass et al., 2024). The transition from 5G to 6G involves overcoming several technological hurdles. These include achieving spectrum efficiency optimizing network parameters and deploying the necessary infrastructure to support the high demands of 6G such as terahertz communication and massive Multiple Input Multiple Output (MIMO) systems (Anjanabhargavi et al., 2024). The deployment of open radio access networks RAN also poses challenges as it requires a framework that supports automation and machine learning to manage the complexity and variability of local network needs (Gatherer et al., 2023).

Implementing network intelligence through machine learning and federated learning (FL) is crucial for 6G yet it presents its own set of challenges. These include managing communication and computing resource scarcity computing overhead and privacy issues associated with FL (Paul, 2024). The integration of AI and softwarization technologies like network function virtualization NFV and software defined networking SDN is essential for efficient resource utilization and responsive service provisioning but it requires significant advancements in network architecture and management (Serôdio et al., 2023).

The architectural evolution necessary for 6G involves innovative solutions to overcome the limitations of current 5G systems. Proposed components like the Global Service Based Architecture (GSBA) Compute Continuum Layer (CCL) and Zero Trust Layer (ZTL) aim to enhance network efficiency security and scalability (Livia et al., 2024). These innovations are critical for supporting the dynamic and demanding needs of future mobile networks. The deployment of 6G networks also faces challenges related to standardization and the need for a flexible adaptive network architecture. This includes addressing issues of latency synchronization and power goals with fixed installed equipment which requires automation and machine learning for real time adaptability (Niranjana et al., 2024).

The implementation of 6G networks is fraught with challenges that require comprehensive solutions across multiple domains. Addressing these challenges involves advancing security and privacy measures overcoming technological and infrastructural barriers integrating network intelligence and innovating network architecture and deployment strategies. These efforts are essential to realize the full potential of 6G and its transformative impact on global communication networks.

The Role of Research and Development and Organizations like Telecommunications Standards Development Society, India (TSDSI) in 6G Implementation

The development and deployment of 6G technology represent a significant leap forward in telecommunications, building upon the advances of 5G and pushing the boundaries of what is possible in wireless communication. The success of 6G will depend heavily on robust research and development (R&D) efforts and the role of standard-setting organizations like the Telecommunications Standards Development Society, India (TSDSI). These organizations are crucial in ensuring that the technology is not only cutting-edge but also accessible, secure, and beneficial to all sectors of society, particularly in a diverse and challenging landscape like India. R&D is the backbone of innovation in telecommunications, driving the discovery and development of new technologies that underpin advancements like 6G.

The transition from 5G to 6G is not merely an incremental upgrade but a transformative shift that promises to integrate advanced technologies such as artificial intelligence (AI), machine learning (ML), and the Internet of Things (IoT) into a more unified and efficient system. 6G is expected to operate at extremely high frequencies, including terahertz (THz) bands, which offer the potential for ultra-high-speed data transmission (Kulkarni et al., 2024). R&D is critical in exploring these frequencies, developing the necessary hardware, and ensuring efficient spectrum management to avoid interference and maximize the use of available bandwidth. The development of massive MIMO (multiple-input, multiple-output) technology, which involves using a large number of antennas at the base station, is essential for 6G. R&D efforts focus on improving these technologies to enhance signal strength, reduce latency, and increase the capacity of networks (Mona et al., 2023). With the increased data rates and the proliferation of connected devices, 6G networks are expected to consume more energy.

R&D in this area is crucial for developing energy-efficient technologies that minimize the environmental impact and reduce operational costs, making 6G more sustainable in the long term. As 6G networks will likely support critical infrastructure and applications, they will be attractive targets for cyberattacks. R&D efforts are needed to develop robust security protocols, including quantum-resistant encryption methods, to protect these networks from emerging threats. 6G will integrate various emerging technologies, including AI, ML, and IoT, into a coherent system. R&D is vital for developing the algorithms, data processing methods, and communication protocols necessary to ensure these technologies work together seamlessly.

In the context of India, the Telecommunications Standards Development Society, India (TSDSI) plays a pivotal role in shaping the future of telecommunications. TSDSI is an autonomous, membership-based standards development organization that develops and promotes indigenous standards for telecommunications in India, ensuring that the country's specific needs and conditions are addressed in global standards. One of TSDSI's primary functions is to develop standards that align with global benchmarks while also catering to India's unique requirements (Vankadara et al., 2024). For 6G, this means ensuring that the standards take into account India's diverse geography, population density, and varying levels of infrastructure development. TSDSI works closely with international standardization bodies like the International Telecommunication Union (ITU) and the 3rd Generation Partnership Project (3GPP) to ensure that India's voice is represented in the global standard-setting process.

This global alignment is crucial for ensuring interoperability and the global scalability of Indian technologies and solutions. TSDSI encourages the development of indigenous technologies that can be integrated into global 6G standards. This is particularly important for a country like India, which has a vast and diverse market with unique challenges. By promoting local innovations, TSDSI helps ensure that the solutions developed are not only relevant to India but can also contribute to global advancements in 6G technology. This focus on indigenous innovation also supports the broader national goals of self-reliance and economic development. TSDSI serves as a platform for collaboration among various stakeholders, including government bodies, academic institutions, and private sector companies. This collaborative approach is essential for driving R&D in 6G, as it allows for the pooling of resources, expertise, and knowledge. TSDSI facilitates joint research projects, the development of prototypes, and the testing of new technologies, all of which are crucial for the successful implementation of 6G (Okasana et al., 2023). Another key role of TSDSI is in capacity building and the dissemination of knowledge related to 6G technologies.

TSDSI organizes workshops, seminars, and training programs to educate stakeholders about the latest developments in telecommunications technology. By building the technical capacity of industry professionals, government officials, and researchers, TSDSI helps ensure that India has the human resources needed to lead in the 6G era. TSDSI plays an important role in advocating for policies and regulatory frameworks that support the deployment of 6G technologies. By providing insights and recommendations to the government, TSDSI helps shape policies that encourage innovation, protect consumer rights, and ensure the equitable distribution of technological benefits (Al-Shareeda et al., 2024). This policy advocacy is particularly important in ensuring that the regulatory environment is conducive to the rapid and widespread deployment of 6G networks.

The successful implementation of 6G in India will require a concerted effort across multiple fronts, with R&D and organizations like TSDSI playing critical roles. Through robust R&D efforts, the development of relevant standards, and the promotion of indigenous innovations, TSDSI is well-positioned to ensure that India is not just a participant but a leader in the global 6G landscape. By fostering collaboration, building capacity, and advocating for supportive policies, TSDSI can help India harness the full potential of 6G technology, driving economic growth, enhancing connectivity, and improving the quality of life for millions of people across the country.

India and Its Role in 6G Journey

India has been actively participating in the global development of 6G technology and has made several key contributions that are shaping the future of telecommunications. These contributions stem from the country's efforts in research and development, standardization, and policy initiatives, aiming to position India as a significant player in the 6G landscape (Ouyang et al., 2023). One of the significant contributions from India, particularly in the context of 5G and looking forward to 6G, is the development of the Low Mobility Large Cell (LMLC) technology. This innovation addresses the unique challenges of providing broadband connectivity in rural and remote areas. LMLC technology focuses on optimizing the performance of networks in low-mobility environments, which are common in rural regions. By enhancing connectivity over larger cells, this technology could play a critical role in ensuring that 6G networks are inclusive and capable of serving diverse geographic areas.

Indian research institutions and universities are actively involved in developing prototypes and conducting research that contributes to the 6G ecosystem. Areas of focus include terahertz communication, advanced AI-driven network management, and energy-efficient network design (Abdullahi & Al-Turjman, 2020). These research efforts are laying the groundwork for future 6G technologies that will be essential for high-speed, low-latency communication networks. TSDSI, as India's representative in global standardization bodies like the International Telecommunication Union (ITU) and the 3rd Generation Partnership Project (3GPP), has been instrumental in ensuring that India's requirements and innovations are included in the global 6G standards. Through TSDSI, India has made significant technical contributions to discussions on 6G, particularly in areas such as spectrum management, network architecture, and security protocols. TSDSI's efforts ensure that the global standards for 6G are inclusive of the needs of developing nations and emerging markets (De Alwis et al., 2022).

India has contributed to the global 6G roadmap through various research papers, technical proposals, and collaborative projects. These contributions are crucial for ensuring that the global standards consider diverse use cases and are adaptable to different socio-economic environments. India's contributions help in shaping a 6G standard that is not only technologically advanced but also practical for widespread adoption in varied environments. Although primarily focused on the immediate rollout of 5G, the NDCP 2018 sets the stage for India's involvement in 6G by emphasizing the need for research and development in next-generation communication technologies. The policy outlines India's ambition to position itself as a global leader in the development of 5G and 6G technologies, with a focus on creating an innovation-driven ecosystem. The Indian government, through bodies like the Department of Telecommunications (DoT), has been actively involved in setting up a roadmap for 6G development. The government's focus on creating a conducive environment for 6G research includes initiatives such as establishing 6G research centers, funding startups and academic institutions, and fostering public-private partnerships. This forward-looking approach ensures that India is well-prepared to contribute to and benefit from the global 6G revolution.

Indian researchers and companies are at the forefront of integrating artificial intelligence (AI) into telecommunications networks, which is expected to be a cornerstone of 6G technology. AI will play a crucial role in managing the complexity of 6G networks, optimizing network performance, and enhancing user experience through predictive analytics and autonomous network management. India has been investing in quantum computing and communication research, which is expected to influence the security frameworks of 6G networks. Quantum communication technologies promise unprecedented levels of security, which will be essential for protecting the vast amounts of data that 6G networks will handle. India's research in this area positions the country to contribute significantly to the security standards of 6G.

With sustainability becoming a critical concern, India's R&D in energy-efficient technologies is a key contribution to the 6G narrative. Indian researchers are exploring new materials, network designs, and power management techniques to reduce the energy consumption of 6G networks. This focus on sustainability ensures that 6G technology can be deployed in an environmentally responsible manner. India's contributions are also focused on ensuring that 6G technologies contribute to bridging the digital divide. The country is advocating for standards and technologies that prioritize affordability and accessibility, ensuring that 6G benefits are available to all, including those in rural and underserved areas. By focusing on inclusive growth, India is helping to shape a 6G future that is equitable and beneficial to diverse populations.

Indian academic institutions, research centers, and tech companies are engaged in global collaborations that drive 6G research. These partnerships facilitate knowledge exchange and joint development of technologies, allowing India to contribute to and benefit from the collective global expertise in 6G. The Indian government is fostering public-private partnerships to accelerate 6G development. These partnerships bring together the strengths of the private sector, such as innovation and agility, with the government's regulatory support and resources, creating a robust ecosystem for 6G research and deployment. India's contributions to 6G are multifaceted, ranging from the development of indigenous technologies to active participation in global standardization and policymaking (Vankadara et al., 2024). Through its efforts in R&D, standardization, and strategic initiatives, India is positioning itself as a key player in the global 6G landscape. The country's focus on inclusive, sustainable, and secure telecommunications technologies ensures that 6G will not only advance India's digital capabilities but also contribute significantly to the global development of next-generation communication networks (Liu et al., 2023).

India's role in the 6G journey is multifaceted, involving both challenges and opportunities as the country navigates its technological landscape. Despite a delayed 5G rollout, India is ambitiously planning to develop a nationally driven 6G network. This initiative is part of a broader strategy to leapfrog into the next generation of wireless communication, potentially positioning India as a leader in 6G technology development and deployment (Manoj, 2022). The transition from 5G to 6G is not merely an upgrade but a revolutionary shift in communication systems. 6G is expected to integrate advanced technologies such as Artificial Intelligence (AI), Machine Learning (ML), and Deep Learning (DL) to enhance network performance and enable new use cases.

These technologies will be crucial in achieving ultra-low latency and high reliability, which are essential for applications like autonomous vehicles and industrial automation. The integration of AI and ML is anticipated to transform 6G networks into intelligent, self-optimizing systems, offering unprecedented capabilities in areas such as smart cities and autonomous systems. Globally, the International Telecommunication Union (ITU) is spearheading the development of 6G standards, known as IMT-2030, which are expected to be finalized around 2030. This timeline aligns with India's aspirations to establish a robust 6G infrastructure, potentially allowing the country to contribute to and benefit from global standards and innovations.

India's journey towards 6G is also intertwined with the broader context of Industry 5.0, which emphasizes the integration of human intelligence with machine capabilities. This synergy is expected to drive innovations in smart cyber-physical systems, remote connectivity, and digital sensing, among others (Deborsi, 2022). The convergence of 6G and Industry 5.0 could enable India to address various societal challenges, such as enhancing rural connectivity and supporting smart healthcare solutions.

However, India's path to 6G is not without challenges. The country must overcome the technological and infrastructural hurdles that have historically impeded its advancements in telecommunications. Additionally, the deployment of 6G will require addressing regulatory concerns and ensuring the security and privacy of data, potentially through technologies like block chain (6G for Bridging the Digital Divide, 2022). In conclusion, India's role in the 6G journey is characterized by a strategic push towards innovation and leadership in next-generation communication technologies. By leveraging AI, ML, and other advanced technologies, India aims to not only catch up with global trends but also to set new benchmarks in wireless communication. The successful implementation of 6G in India could significantly impact various sectors, from healthcare to industrial automation, ultimately contributing to the country's digital transformation and economic growth. As India navigates this complex landscape, collaboration with global entities and adherence to international standards will be crucial in realizing its 6G ambitions.

The deployment of Beyond 5G (B5G) and 6G networks in rural India presents a transformative opportunity to enhance healthcare accessibility and uphold the right to health. The advanced capabilities of 6G, such as ultra-low latency, high bandwidth, and improved quality of service, are pivotal in supporting intelligent healthcare applications, including remote patient monitoring, telesurgery, and augmented reality-based diagnostics (Kulkarni et. Al., 2024). These technologies can bridge the healthcare gap in rural areas by enabling real-time communication and intelligent service integration, which are essential for patient-centered care and effective treatment delivery (M et al., 2021).

The integration of the Internet of Healthcare Things (IoHT) with 6G networks further enhances connectivity, allowing for seamless interaction between medical devices, healthcare professionals, and patients, thereby improving the quality of healthcare experiences (6G with Smart Healthcare Systems, 2022). Moreover, the use of open-source hardware and software in 6G networks can facilitate the creation of sustainable, reliable, and low-cost communication infrastructures in rural areas, addressing the digital divide and ensuring equitable access to healthcare services (Towards 6G Networks for Rural Environments, 2022). The deployment of 6G in rural India can thus support the realization of the right to health by providing the necessary technological infrastructure to deliver high-quality healthcare services ubiquitously, overcoming the limitations of current 5G networks and addressing challenges related to network performance and security (6G for healthcare, 2022). As 6G networks continue to evolve, they are expected to play a crucial role in enabling future smart health services, thereby transforming healthcare delivery and improving health outcomes in rural communities.

Figure 8. Policy advancements related to B5G/6G in India: This chart tracks the progress of policy advancements from 2020 to 2028 in the context of B5G and 6G deployment in rural India

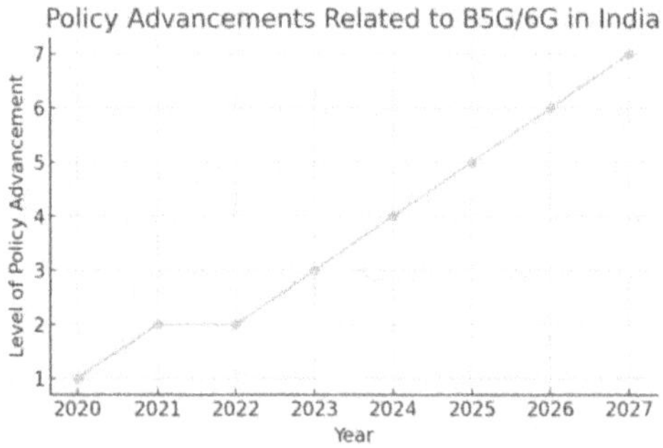

Source: Author made

Bharat 6G Vision and Future

The Bharat 6G and Bharath 6G alliance aim to propel India's 6G technology development by focusing on several key objectives and initiatives. These initiatives are designed to address the evolving demands of communication technologies and to position India as a leader in the global 6G landscape. One of the primary objectives is to establish a robust research and innovation ecosystem that fosters the development of enabling technologies for 6G, such as advanced autonomous systems, smart manufacturing, and wireless e-health, which are crucial for societal and industry-driven needs (Alonistioti et al., 2024). The alliance emphasizes the integration of cutting-edge technologies like AI, IoT, and blockchain to enhance the quality of service and enable new applications such as mixed reality and holographic communication (Saxena et al., 2024).

Furthermore, the synergy between 6G and AI is highlighted as a transformative force across various sectors, including healthcare, transportation, and education, while also addressing security and ethical challenges (Primmia et al., 2024). The operational goals of 6G include creating a holistic ICT platform that unifies communication capabilities to meet diverse requirements, with a focus on scalability, efficiency, and agility (Wang et al., 2023). The alliance also aims to develop intelligent, distributed, and programmable networks that can seamlessly integrate communication and computation to meet increasing service demands (Talwar et al., 2021).

Additionally, the alliance is committed to advancing stakeholder collaboration and harmonizing standardization efforts to empower users and developers within the 6G ecosystem (Yrjölä et al., 2023). By leveraging these strategic initiatives, the Bharat 6G and Bharath 6G alliance seek to drive innovation, address technical challenges, and ensure that India plays a pivotal role in the global 6G narrative.

India has demonstrated a strong commitment to advancing telecommunications, exemplified by the "Bharat 6G Vision" document released in March 2023, which aspires to position India as a leader in 6G technology by 2030. Furthermore, the establishment of the Bharat 6G Alliance (B6GA) aims to foster innovation and collaboration in 6G technology development. Despite these initiatives, the absence of specific policies addressing the unique challenges of rural connectivity, such as infrastructure deficits and affordability concerns, poses a significant barrier.

The Bharat 6G Vision can significantly enhance healthcare accessibility in rural areas by leveraging advanced technologies such as artificial intelligence (AI), the Internet of Things (IoT), and cloud computing, which are integral to the 6G framework. The ultra-low latency and high bandwidth capabilities of 6G networks facilitate real-time medical data analysis, remote diagnostics, and even telesurgery, thereby overcoming the geographical and infrastructural barriers that have traditionally hindered healthcare delivery in rural regions (Nayak & Patgiri, 2021). The integration of AI into healthcare systems can provide cost-effective solutions tailored to the specific needs of rural populations, improving data quality and availability, which are crucial for effective healthcare delivery (Olugboja et. al., 2024).

Figure 9. This chart compares the potential benefits of 5G technology in rural India (such as advancements in telemedicine and smart education) against the challenges that have impeded its full realization (such as high costs and lack of infrastructure). It highlights the gap between the anticipated advantages of 5G and the barriers to achieving these benefits, underscoring the difficulties in implementing advanced networks in rural regions

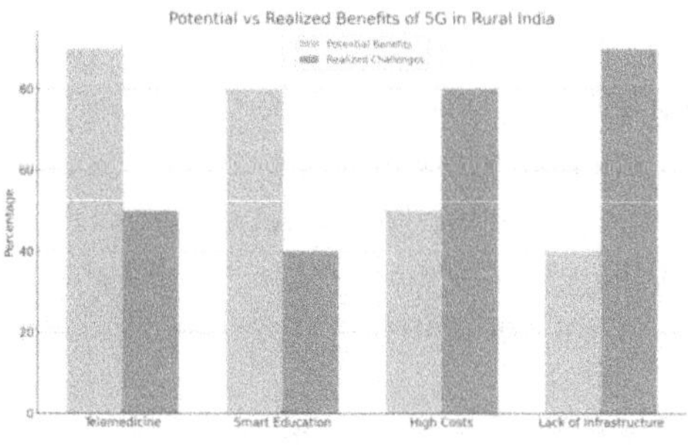

Source: Author made

Moreover, 6G's support for intelligent wearable devices and the Internet of Medical Things (IoMT) can enable continuous health monitoring and personalized care, enhancing the quality of life for rural residents. The deployment of 6G technology also promises to support immersive and haptic technologies, which can offer interactive healthcare experiences and stress management solutions, further broadening the scope of healthcare services available remotely (Mihovska et al., 2024). Additionally, the use of telemedicine and remote healthcare services, facilitated by 6G's robust connectivity, can ensure 24/7 access to healthcare professionals and resources, thus addressing the critical issue of healthcare inaccessibility in remote areas (Mamoona et al., 2023). The Bharat 6G Vision, by incorporating these advanced technologies, can thus play a pivotal role in bridging the healthcare disparity gap, ensuring equitable access to quality healthcare services for rural populations across India.

Figure 10. Infrastructure growth in rural India: The infrastructure growth chart illustrates the steady development of essential telecom infrastructure in rural areas, such as fiber optic cables and power supply systems, from 2020 to 2028

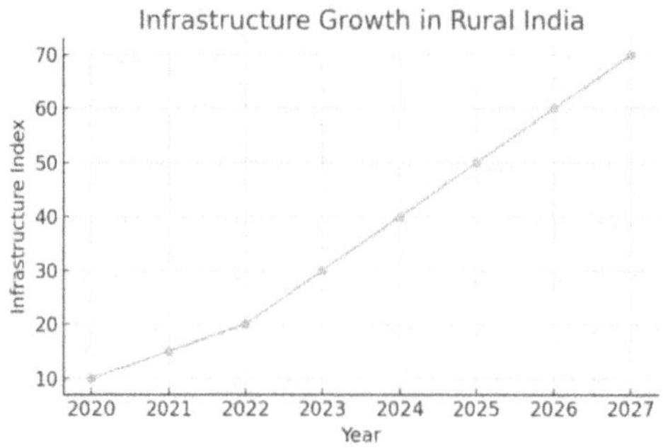

Source: Author made

Case Analysis: Juhi Chawla 5G Suit and Its Implications for B5G and 6G

The Juhi Chawla 5G suit, which raised concerns about the potential health impacts of 5G technology, highlights the broader implications for the implementation of Beyond 5G (B5G) and 6G technologies, particularly in rural India.

The suit underscores the need for careful consideration of both technological advancements and public perception in the deployment of next-generation networks. 5G technology, with its components of Enhanced Mobile Broadband, Massive Machine Type Communication, and Ultra Reliable and Low Latency Communication, has the potential to revolutionize rural education in India by enabling innovative concepts like "School on Wheels" and smart educational systems (Agrawal et al., 2022).

However, the deployment of 5G in rural areas faces significant challenges, including the lack of viable business models and the risk of not realizing its full potential due to insufficient private sector investment (Skouby, 2021).

This is compounded by the fact that rural areas are often excluded from the benefits of modern technology, as seen in the current 5G rollouts that favour urban areas (Ivan et al., 2022). The transition to B5G and 6G technologies offers opportunities to address these challenges. B5G/6G networks are designed to support high data rate applications and the integration of advanced technologies like massive multiple input multiple output (mMIMO) systems, which can efficiently manage the increasing demand for wireless connectivity.

These technologies are crucial for the development of smart cities and can potentially transform rural environments by providing sustainable, reliable, and low-cost communication networks through open-source hardware and software (Towards 6G netwroks for rural environments, 2022).

Implementation in Rural India

However, the implementation of B5G/6G in rural India must consider several factors, including demographics, economic status, and the current state of development. The high computational complexity and latency issues associated with mMIMO systems present additional challenges that need to be addressed through AI and machine learning techniques to optimize network performance.

Moreover, the experience of other countries, such as Bangladesh, highlights the importance of addressing barriers like limited technology literacy and high costs, which are also relevant to rural India. Collaborative efforts, infrastructure development, and regulatory support are essential to overcoming these challenges and ensuring the successful integration of advanced communication technologies in rural areas.

Figure 11. This bar chart outlines the main legal and procedural issues that arose during the Juhi Chawla 5G suit. It emphasizes the challenges faced by the plaintiffs, including procedural delays, incorrect court fees, lack of mandatory notices, and the disruptions caused by the public sharing of the hearing's video conferencing link. These issues played a significant role in the dismissal of the case

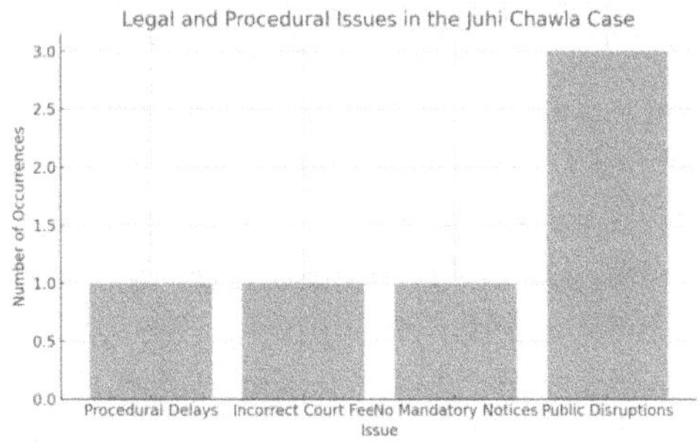

Source: Author made

Juhi Chawla & Ors. v. Science and Engineering Research Board & Ors. (CS(OS) 262/2021) in the Delhi High Court, which culminated in a judgment on June 4, 2021, serves as a significant point of discussion in the broader context of the implementation of advanced telecommunication technologies such as Beyond 5G (B5G) and 6G in India. This case brings to light several important issues, including the public's apprehensions about health impacts, the legal frameworks governing the deployment of telecommunication infrastructure, and the challenges faced by the judiciary in addressing complex scientific disputes.

The plaintiffs, led by actress and environmental activist Juhi Chawla, filed a suit against various government entities and private telecom operators, challenging the roll-out of 5G technology in India. The plaintiffs claimed that the government had not conducted sufficient research on the potential health risks associated with 5G, particularly the effects of electromagnetic field (EMF) radiation on human health. They sought a permanent injunction against the defendants, restraining them from implementing

5G technology until such research was completed. The plaintiffs invoked the precautionary principle, embedded within Article 21 of the Indian Constitution, which guarantees the right to life and personal liberty. They argued that, in the absence of conclusive studies on the safety of 5G, its roll-out could violate this fundamental right by exposing the population to potentially harmful radiation.

Figure 12. This chart provides a chronological overview of the significant events during the Juhi Chawla 5G suit in June 2021. It highlights the timeline from the filing of the case on May 31, 2021, to its dismissal on June 10, 2021. The disruptions during the virtual court hearing, the imposition of fines, and the court's final decision are also captured, reflecting the legal process's progression

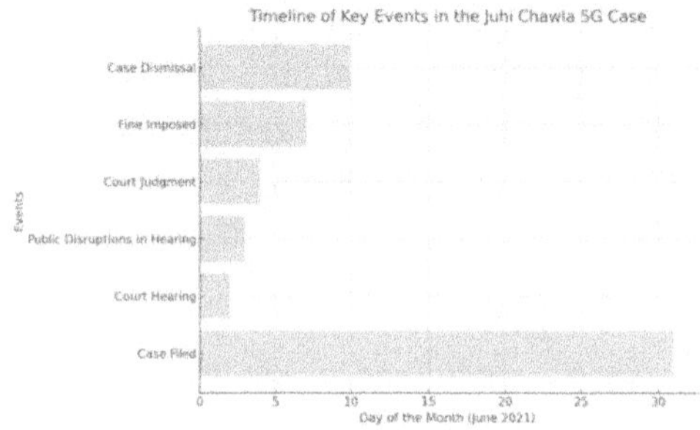

Source: Author made

The Delhi High Court, presided over by Justice J.R. Midha, dismissed the suit on multiple grounds. The court found that the suit was frivolous, poorly drafted, and an abuse of the legal process. It noted that the plaintiffs had failed to comply with several procedural requirements, such as the payment of the correct court fees and the issuance of mandatory notices under Section 80 of the Civil Procedure Code (CPC). Furthermore, the court observed that the plaintiffs had not presented any concrete evidence to support their claims about the health risks of 5G. Instead, their suit was based largely on conjecture and unfounded fears. The court also criticized the plaintiffs for circulating the video conferencing link of the court proceedings on social media, which led to repeated disruptions during the hearing. The court imposed a cost of Rs. 20 lakhs on the plaintiffs for wasting judicial time and directed them to deposit the amount with the Delhi State Legal Services Authority (DSLSA). The court also issued show cause notices to individuals who disrupted the proceedings, asking why contempt of court proceedings should not be initiated against them.

The case highlights the growing public concern about the health impacts of emerging technologies, particularly those related to telecommunications. While public concerns are legitimate and must be addressed, this case underscores the importance of grounding legal actions in solid scientific evidence. The plaintiffs' failure to present credible scientific studies weakened their case and ultimately led to its dismissal. The plaintiffs attempted to use the precautionary principle as a legal tool to halt the 5G roll-out. The precautionary principle, which advocates preventive action in the face of uncertainty, is indeed a critical component of environmental and public health law. However, its application requires a careful balance between caution and progress. The court's decision suggests that while the precautionary principle is important, it cannot be invoked without substantial evidence to back the claimed risks.

This case illustrates the challenges courts face when dealing with scientific disputes. The judiciary often lacks the expertise to independently assess complex scientific issues, relying instead on expert testimony and established scientific consensus. The dismissal of the suit reflects the court's reluctance to intervene in matters where the scientific evidence is not clear-cut or where the regulatory authorities, presumed to be the experts, have given their approval. The judgment also raises questions about the role of the judiciary in the regulation of technological advancements. Courts must strike a balance between ensuring public safety and not stifling technological progress. This case suggests that the courts are likely to defer to the expertise of regulatory bodies like the Department of Telecommunications (DoT) and the Indian Council of Medical Research (ICMR) unless there is compelling evidence of harm (Key Considerations to Achieve 5G and B5G Connectivity in Rural Areas, 2022).

As India moves towards the implementation of B5G and 6G technologies, this case sets a precedent for how similar legal challenges might be handled. The rapid deployment of these advanced networks is essential for economic development, particularly in rural areas. However, this must be done in a way that addresses public concerns and adheres to legal and procedural norms. The court's strong language regarding the abuse of the legal process is particularly noteworthy. It reflects the judiciary's frustration with frivolous litigations that drain judicial resources and delay justice. For future litigants, especially in cases involving public interest, this serves as a warning that the courts will not tolerate the misuse of legal provisions under the guise of public interest. The imposition of a Rs. 20 lakhs fine on the plaintiffs serves as a deterrent against frivolous litigation. It emphasizes the need for due diligence before approaching the courts, particularly in cases involving complex and high-stakes issues like the roll-out of national telecommunications infrastructure.

Figure 13. This flowchart illustrates how the Juhi Chawla 5G case set a judicial precedent for future litigation involving telecommunications technology in India. It highlights the court's dismissal of the case, citing insufficient evidence for the claimed health risks, and its decision not to apply the precautionary principle. This ruling suggests that future litigants must present stronger scientific evidence, and it may discourage frivolous legal actions, especially when regulatory authorities have approved technological advancements

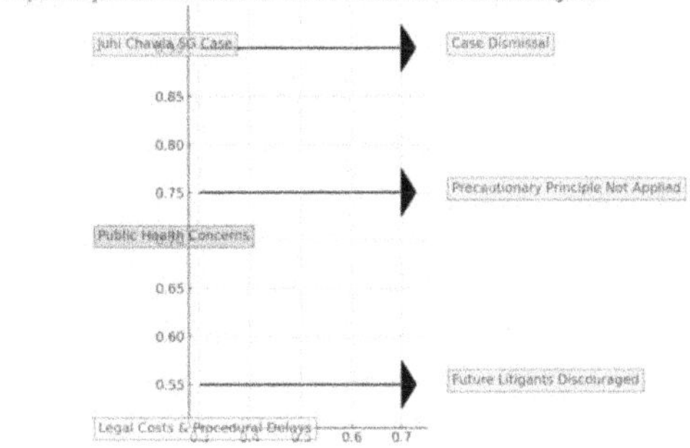

Source: Author made

The case also brings to the forefront the need for continuous and transparent regulatory oversight as new technologies are rolled out. While the court found no immediate grounds to halt 5G, the concerns raised by the plaintiffs about health effects cannot be entirely dismissed. It is crucial for regulatory bodies to conduct and communicate the results of comprehensive studies on the potential impacts of B5G and 6G technologies to reassure the public. This case serves as a critical reference point in the ongoing dialogue about the deployment of advanced telecommunications technologies in India. It underscores the necessity of grounding public policy and legal challenges in rigorous scientific research while balancing the imperatives of technological progress and public safety.

As India prepares for the widespread implementation of B5G and 6G, this case offers valuable lessons on the importance of adhering to legal norms, the potential consequences of frivolous litigation, and the need for clear communication between the government, judiciary, and the public on the benefits and risks associated with new technologies. The observations from this case will likely influence how future challenges to telecommunications advancements are handled, particularly in ensuring that the legal process is used constructively to address legitimate concerns without hindering technological and economic progress.

CONCLUSION AND WAY FORWARD

The exploration of India's focus on 6G technology, including its benefits, challenges, and the role of organizations like the Telecommunications Standards Development Society, India (TSDSI), underscores the nation's strategic commitment to positioning itself as a global leader in next-generation telecommunications. India's proactive engagement in the development and deployment of 6G is not merely a technological pursuit but a multidimensional strategy aimed at fostering economic growth, social equity, and global influence. The substantial investments in research and development, combined with the promotion of indigenous technologies such as Low Mobility Large Cell (LMLC), highlight India's focus on addressing unique national challenges, particularly the digital divide that exists between urban and rural areas.

The integration of advanced technologies like artificial intelligence, quantum computing, and IoT within the 6G framework is poised to revolutionize sectors such as healthcare, education, and smart city development, offering transformative solutions that can significantly enhance the quality of life for millions. Moreover, India's emphasis on energy-efficient network designs aligns with global sustainability goals, ensuring that the environmental impact of these technological advancements is minimized, thus contributing to a more sustainable future.

The active participation of TSDSI in global standardization processes ensures that India's needs and innovations are effectively represented in the global arena, enabling the country to influence the development of international standards that are inclusive and equitable. This involvement not only strengthens India's position in global telecommunications but also ensures that the benefits of 6G are accessible to diverse populations worldwide.

The potential economic benefits of 6G, including job creation in high-tech sectors and the growth of the digital economy, are immense, positioning India as a hub for technological innovation and entrepreneurship. The anticipated advancements in cybersecurity, driven by quantum communication research, underscore India's commitment to national security in an increasingly digital world.

The government's forward-looking policies, such as the National Digital Communications Policy (NDCP) 2018, provide a robust framework for the development of 6G, encouraging public-private partnerships and fostering an innovation-driven ecosystem. Furthermore, the focus on inclusive digital growth, which aims to bring high-speed connectivity to underserved regions, is critical for ensuring that the benefits of 6G extend to all segments of society, thus contributing to broader socio-economic development.

The international collaboration facilitated by India's participation in global research networks enhances knowledge exchange and accelerates the pace of technological advancements, positioning India at the forefront of global R&D efforts in telecommunications. As 6G technology continues to evolve, India's strategic focus on this area is likely to yield significant long-term benefits, not only for the country itself but also for the global community. The emphasis on creating a secure, sustainable, and inclusive 6G ecosystem reflects India's broader vision of leveraging technology to drive social progress and economic prosperity.

Figure 14. This figure provides a timeline of significant milestones in India's 6G development, starting from the release of the Bharat 6G Vision in 2023 and ending with the projected full-scale 6G deployment in 2028. The milestones include key events such as TSDSI's engagement in global 6G standardization, policy formulation, funding of research projects, and trial runs in rural areas. This timeline reflects India's proactive approach in shaping the global 6G landscape while addressing domestic challenges

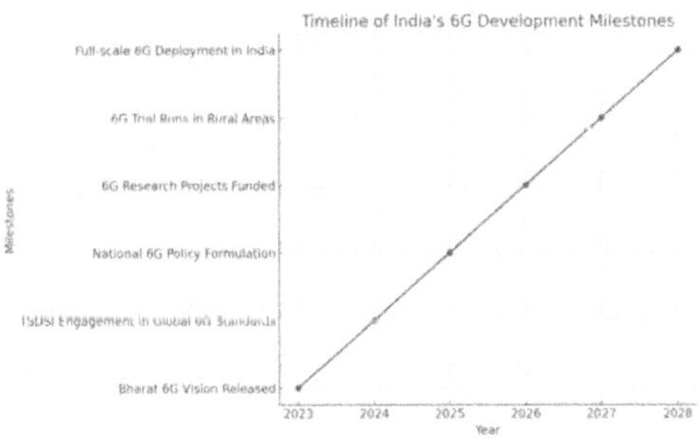

Source: Author made

The advent of 6G technology is poised to have a transformative impact on rural areas, potentially addressing many long-standing challenges related to connectivity, access to services, economic development, and social inclusion. As 6G promises ultra-fast internet speeds with extremely low latency, rural areas, which have traditionally suffered from limited or unreliable connectivity, stand to benefit significantly. This enhanced connectivity will play a critical role in bridging the digital divide, allowing rural communities to access the same digital services and opportunities as their urban counterparts. One of the most immediate impacts of 6G in rural areas will be the provision of high-speed internet, which can enable a wide range of digital services that were previously inaccessible or difficult to use due to bandwidth constraints. Affordable and efficient connectivity solutions, such as those tailored for sparsely populated areas, will make it economically viable for service providers to offer high-quality internet in rural regions, thereby reducing costs for end-users and making digital inclusion a reality.

The economic implications of 6G in rural areas are profound. With reliable and fast internet, local businesses will be able to tap into broader markets, engage in e-commerce, and offer digital services, driving the growth of local economies and creating jobs. This could also reduce the migration of youth from rural to urban areas, as they can find new opportunities within their own communities. In agriculture, 6G will enable the use of advanced technologies such as IoT, AI, and precision farming, providing farmers with real-time data on weather, soil conditions, and crop health, leading to better decision-making, increased productivity, and sustainable farming practices. The educational landscape in rural areas will also be transformed by 6G technology. With the ability to deliver high-quality educational content online, rural students will gain access to the same learning resources as those in urban areas, thereby levelling the educational playing field. E-learning platforms, virtual classrooms, and interactive learning tools will become more accessible, improving educational outcomes and providing rural students with the skills needed to thrive in a digital economy.

Moreover, 6G will facilitate virtual teacher training programs, enabling educators in rural areas to enhance their teaching methodologies and receive real-time guidance from experts, further boosting the quality of education. Healthcare in rural areas will see significant advancements through 6G technology. Telemedicine services, supported by 6G's low latency and high reliability, will enable patients in remote locations to consult with doctors, receive diagnoses, and even undergo certain treatments without the need to travel long distances. This is particularly crucial in areas where access to healthcare facilities is limited, and it can lead to improved health outcomes and reduced mortality rates.

Real-time health monitoring devices, powered by 6G, will allow for continuous tracking of patients' health, enabling timely medical interventions and better management of chronic conditions. Beyond economic and health benefits, 6G will enhance social and community development in rural areas. Technology will support richer communication experiences, such as high-definition video calls and virtual reality interactions, helping to reduce social isolation and strengthen community bonds.

Access to e-governance platforms will be greatly improved, allowing rural citizens to access essential government services, such as identity verification and financial aid, more easily and efficiently. Furthermore, 6G will play a pivotal role in the development of smart villages, where infrastructure such as electricity, water, and transportation can be managed more efficiently through connected devices and AI. This will not only improve the quality of life in rural areas but also make them more resilient to challenges such as natural disasters, with 6G enabling real-time data collection and communication for disaster management and response. Cultural preservation and digital literacy are other important aspects where 6G will make a difference. This technology will support platforms that promote and preserve local cultures, allowing rural communities to document and share their heritage online.

Digital literacy initiatives will be essential to ensure that rural populations can fully leverage the benefits of 6G, making technology an enabler rather than a barrier. However, realizing these benefits will require addressing several challenges, such as the significant investment needed to deploy 6G infrastructure in rural areas, ensuring the sustainability of this expansion, and developing policies that ensure equitable access.

The focus must be on creating inclusive digital growth that leaves no one behind, thereby preventing further entrenchment of existing inequalities. In conclusion, 6G technology has the potential to be a transformative force in rural areas, providing enhanced connectivity that drives economic growth, improves access to essential services like education and healthcare, and supports social and cultural development. With the right strategies and investments, 6G can play a crucial role in bridging the rural-urban divide, fostering inclusive growth, and improving the quality of life for millions of people living in rural regions.

The implementation of Beyond 5G (B5G) and 6G networks in rural India presents a confluence of opportunities and challenges that reflect the broader ambitions of India's technological advancements and the inherent complexities of its diverse socio-economic landscape. These next-generation networks are poised to revolutionize connectivity and access to essential services, including healthcare, education, and public welfare, particularly in underserved rural areas. However, this transformative potential is contingent upon overcoming significant policy, regulatory, and infrastructural hurdles.

The public concerns surrounding health impacts, as evidenced by cases like Juhi Chawla v. Science and Engineering Research Board & Ors., underscore the necessity of grounding technological progress in robust scientific evidence and transparent regulatory oversight. The case also highlights the judiciary's role in balancing the imperatives of public safety with the need for technological advancement. As India embarks on the 6G journey, the role of organizations like the Telecommunications Standards Development Society, India (TSDSI) becomes crucial in shaping standards that are not only globally relevant but also tailored to India's unique needs.

The case further emphasizes the importance of integrating advanced technologies such as artificial intelligence (AI), machine learning (ML), and block chain into the telecommunications framework to address challenges related to security, privacy, and network optimization. The successful deployment of B5G and 6G in rural India will require a concerted effort from multiple stakeholders, including the government, private sector, and civil society, to create an enabling environment that fosters innovation while ensuring equitable access to the benefits of these technologies. This involves not only addressing the infrastructural deficits and economic barriers that have historically impeded rural connectivity but also fostering digital literacy and community engagement to ensure widespread adoption and utilization of these advanced networks.

Moreover, the experience of other countries and the lessons from India's delayed 5G rollout underscore the need for a proactive and forward-looking approach to policy and regulation, one that anticipates and addresses potential challenges before they become significant obstacles. The case also underscores the importance of Service Level Agreements (SLAs) in the deployment of these networks, particularly in sectors like healthcare, where reliability and security are paramount. In this context, the integration of non-terrestrial networks (NTNs) and decentralized baseband processing (DBP) represents significant innovations that can help overcome the technical challenges of rural deployment.

Additionally, the emphasis on sustainability and social impact in the development of 6G technologies reflects a broader global shift towards more inclusive and responsible technological progress. This approach aligns with India's goals of bridging the digital divide and promoting inclusive growth, particularly in rural areas that have traditionally been left behind in the digital revolution. The incorporation of open-source technologies and block chain further enhances the potential for creating cost-effective and secure networks that can be widely adopted across diverse geographies. Ultimately, the success of B5G and 6G in India will hinge on the ability to navigate the complex interplay of technological innovation, regulatory frameworks, and public perception, ensuring that the deployment of these networks contributes meaningfully to the broader goals of sustainable development and social equity.

Figure 15. This bar chart shows the projected economic and social benefits of 6G deployment in rural India, including job creation, improved healthcare access, better education opportunities, local business growth, and advancements in smart agriculture. Each category is represented by its projected impact percentage, highlighting how 6G could transform rural communities by providing access to high-speed connectivity and digital services, driving economic development, and improving the quality of life

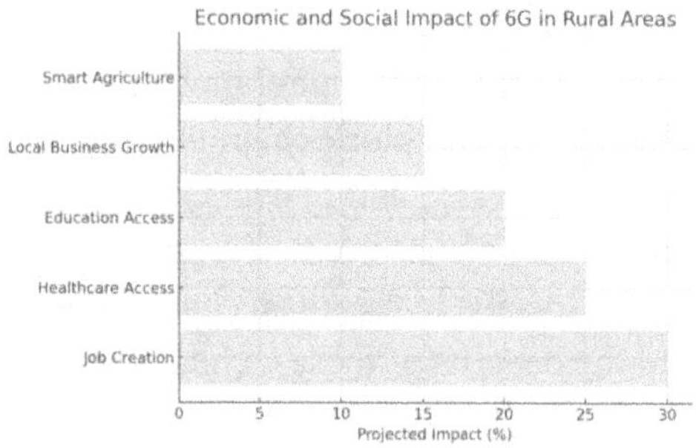

Source: Author made

As India moves forward with the implementation of B5G and 6G networks, a strategic and holistic approach is essential to address the multifaceted challenges and capitalize on the opportunities these technologies present. The way forward must involve a multi-pronged strategy that includes robust policy formulation, active stakeholder engagement, and continuous investment in research and development. At the policy level, the government must ensure that regulatory frameworks are both enabling and protective, fostering innovation while safeguarding public health and privacy.

This requires the development of clear and transparent guidelines that address the specific needs of rural connectivity, including incentives for private sector investment in underserved areas. Public Interest Theory suggests that government intervention is necessary to correct market failures and ensure that rural areas benefit from technological advancements. To this end, the government should consider public-private partnerships (PPPs) as a mechanism to pool resources and expertise, thereby reducing the financial burden on private operators and accelerating the deployment of infrastructure. In addition, the regulatory bodies must remain vigilant against the risks of regulatory capture, where the interests of telecom companies could potentially override public welfare considerations. This calls for the establishment of independent oversight mechanisms to ensure that the regulatory process remains transparent and accountable. Furthermore, the integration of advanced technologies such as AI and block chain should be prioritized to enhance network security and privacy, which are critical concerns in the deployment of 6G networks.

The adoption of open-source technologies can further democratize access to these networks, making them more affordable and sustainable, particularly in rural areas. From a technical perspective, addressing the infrastructural challenges in rural areas will require innovative solutions, such as the use of non-terrestrial networks (NTNs) and decentralized baseband processing (DBP), which can provide the necessary scalability and flexibility.

Additionally, the development of energy-efficient technologies is crucial to ensure the sustainability of 6G networks, particularly in remote areas where power supply may be unreliable. The role of organizations like TSDSI will be pivotal in ensuring that India's specific needs are incorporated into global standards, thereby positioning the country as a leader in the global 6G landscape. Furthermore, capacity building initiatives should be undertaken to enhance digital literacy in rural areas, ensuring that communities are not only passive recipients of technology but active participants in its adoption and utilization. This involves targeted education and training programs that empower individuals to leverage the full potential of 6G networks for their economic and social development.

The role of the judiciary, as highlighted by the Juhi Chawla case, will also be critical in mediating the balance between technological progress and public safety. Future legal challenges related to B5G and 6G should be grounded in solid scientific evidence, ensuring that the precautionary principle is applied judiciously to protect public health without stifling innovation. Finally, as India prepares for the widespread deployment of 6G, it is imperative to foster international collaborations and align with global standards, ensuring that India's innovations contribute to and benefit from the global telecommunications ecosystem. This includes active participation in international standard-setting bodies and research collaborations that drive the development of 6G technologies. By adopting a forward-looking approach that integrates technological innovation, regulatory oversight, and community engagement, India can successfully navigate the challenges of B5G and 6G deployment, ensuring that these networks play a transformative role in bridging the digital divide and promoting inclusive growth across the country. In conclusion, India's comprehensive approach to 6G development, characterized by robust R&D, active global engagement, and a strong policy framework, positions the country as a key player in the global telecommunications landscape. By addressing the challenges and harnessing the opportunities presented by 6G, India is set to emerge as a leader in the digital future, driving innovation, fostering economic growth, and enhancing the quality of life for its citizens while contributing to global technological advancements.

REFERENCES

Agrawal, U., Dave, P., Tiwari, A., Upadhyay, V., Sankhe, M. S., & Shrivastav, S. (2022). Implementation of 5G technology in rural education of India. *IEEE International Conference on Broadband and Social Sustainability (IBSSC)*. doi:DOI: 10.1109/IBSSC56953.2022.10037466

Al-Shareeda, M., Al-Phryggy, M., Al-Kashani, A., Al-Jaber, H., & Al-Moussawi, D. (2024). Design and implementation of novel 6G cellular communication architecture using 3D modeling in real environments. doi:DOI: 10.54517/cte.v2i2.2668

Alonistioti, N., Cardona, N., Waldemar, P., & Trivisonno, R. (2024). Establishing an open platform to foster 6G innovation and research. *IEEE Communications Magazine*, *62*(1), 8–10. Advance online publication. DOI: 10.1109/MCOM.2024.10401872

Basu, D., Ghosh, U., & Datta, R. (2022). 6G for industry 5.0 and smart CPS: A journey from challenging hindrance to an opportunistic future. *IEEE Silicon Conference (SILCON)*. doi:DOI: 10.1109/SILCON55242.2022.10028927

Bhattacharyya, A., Agrawal, S., Rath, H. K., & Pal, A. (2019). Towards democratizing e-health under constrained ICT infrastructure through 'A-REaLiST' solution. *European Conference on Networks and Communications (EuCNC)*. doi:DOI: 10.1109/EuCNC.2019.8802034

Dass, P., Ujjwal, S., Novotný, J., Zolotavkin, Y., Laaroussi, Z., & Köpsell, S. (2024). Addressing privacy concerns in joint communication and sensing for 6G networks: Challenges and prospects. arXiv:2405.01742

De Alwis, C., Kumar, P., Pham, Q. V., Dev, K., Kalla, A., Liyanage, M., & Hwang, W. J. (2022). Towards 6G: Key technological directions. *ICT Express*. Advance online publication. DOI: 10.1016/j.icte.2022.10.005

Gatherer, A., Sengupta, C., & Sen, S. P. (2023). *6G radio access implementation: Challenges and technologies*. Signals and Communication Technology., DOI: 10.1007/978-3-031-37920-8_19

Ghassemian, M., Vukobratovic, D., Papadopoulos, C., An, X., Chatzimisios, P., Aijaz, A., Li, P., Meseguer Valenzuela, A., Shikh-Bahaei, M., Mihovska, A., Bartzoudis, N., Walker, R., Prasad, R. V., & Saghezchi, F. B. (2023). Standardisation landscape for 6G robotic services. *IEEE Conference on Standards for Communications and Networking (CSCN)*. doi:DOI: 10.1109/CSCN60443.2023.10453144

Gustavsson, U., Frenger, P., Fager, C., Eriksson, T., Zirath, H., Dielacher, F., Studer, C., Parssinen, A., Correia, R., Matos, J. N., Belo, D., & Carvalho, N. B. (2021). Implementation challenges and opportunities in beyond-5G and 6G communication. *IEEE Journal of Microwaves*, *1*(1), 86–100. Advance online publication. DOI: 10.1109/JMW.2020.3034648

Kharche, S., & Kharche, J. S. (2023). 6G intelligent healthcare framework: A review on the role of technologies, challenges, and future directions. *Journal of Mobile Multimedia*. doi:DOI: 10.13052/jmm1550-4646.1931

Kulkarni, A., Goudar, R. H., Rathod, V., Dinesh, P. M., & Hukkeri, G. S. (2024). *New directions for adapting intelligent communication and standardization towards 6G*. ICST Transactions on Scalable Information Systems., DOI: 10.4108/eetsis.5126

Liu, R., Lin, H., Lee, H., Chaves, F., Lim, H., & Sköld, J. (2023). Beginning of the journey toward 6G: Vision and framework. *IEEE Communications Magazine*, *61*(10), 8–9. Advance online publication. DOI: 10.1109/MCOM.2023.10298069

Mihovska, A., Vlahov, A., & Poulkov, V. (2024). 6G-based intelligent, context-aware, and trustworthy user-centric healthcare applications. *Wireless Telecommunications Symposium (WTS)*. doi:DOI: 10.1109/WTS60164.2024.10536689

Najjuuko, C., Ayebare, G. K., Lukanga, R., Mugume, E., & Okello, D. (2021). *A survey of 5G for rural broadband connectivity. 2021 IST-Africa Conference*. IST-Africa.

Nayak, S., & Patgiri, R. (2021). *6G communication technology: A vision on intelligent healthcare*. Springer Communications in Computer and Information Science., DOI: 10.1007/978-981-15-9735-0_1

Olugboja, A., & Moghalu Agbakwuru, E. (2024). *Bridging healthcare disparities in rural areas of developing countries: Leveraging artificial intelligence for equitable access*. IEEE Advanced Computing and Data Science Applications. ACDSA., DOI: 10.1109/ACDSA59508.2024.10467443

Ouyang, Y., Zhang, Y., Wang, P. S., Liu, Y. Y., Qiao, W., Zhu, J., Liu, Y., Zhang, F., & Wang, S. (2023). 6G network business support system.

Pandey, D., Goyal, S., Bhaumik, K., Suneja, S., Sharma, M., & Dadheech, P. (2024). A systematic review of security issues in 6G networks and communication. *Advances in Wireless Technologies and Telecommunication*, ●●●, 1–11. Advance online publication. DOI: 10.4018/979-8-3693-2931-3.ch001

Paul, S. (2024). Investigating federated learning implementation challenges in 6G networks. *IEEE International Conference on Advances in Electrical, Computing, and Telecommunication (ICAECT)*. doi:DOI: 10.1109/ICAECT60202.2024.10469650

<prpt>Chaoub, A., Giordani, M., Lall, B., Bhatia, V., Kliks, A., Mendes, L. L., Rabie, K. M., Saarnisaari, H., Singhal, A., Zhang, N., Dixit, S., & Zorzi, M. (2020). 6G for bridging the digital divide: Wireless connectivity to remote areas. *IEEE Wireless Communications, 29*(1), 160-168. DOI: 10.1109/MWC.001.2100137</prpt>

Saxena, A. C. S., Kumar, R., Al-Farouni, M., Kumar, S., Karpagam, C. J., & Kavitha, M. (2024). Navigating 6G network into the latest AI technology for the fastest communication without losing data. *IEEE International Conference on Advances in Computing, Communication and Information Technology for Engineering (ICACITE)*. doi:DOI: 10.1109/ICACITE60783.2024.10616812

Serôdio, C., Cunha, J., Candela, G., Rodriguez, S., Sousa, X. R., & Branco, F. (2023). The 6G ecosystem as support for IoE and private networks: Vision, requirements, and challenges. *Future Internet*, *15*(11), 348. Advance online publication. DOI: 10.3390/fi15110348

Skouby, K. E. (2021). 5G business models & trends in rural communication. *Journal of Multimedia*. Advance online publication. DOI: 10.13052/jmm1550-4646.17139

Suleiman, A., & Al-Turjman, F. (2020). *6G applications and standards: An overview*. Springer Communications in Computer and Information Science., DOI: 10.1007/978-3-030-69431-9_4

Talwar, S., Himayat, N., Nikopour, H., Xue, F., Wu, G., & Ilderem, V. (2021). 6G: Connectivity in the era of distributed intelligence. *arXiv: Information Theory*. 10.48550/arXiv.2110.07052

Vanelli-Coralli, A., Guidotti, A., Foggi, T., Colavolpe, G., & Montorsi, G. (2020). 5G and beyond 5G non-terrestrial networks: Trends and research challenges. *IEEE 5G World Forum (5GWF)*. doi:DOI: 10.1109/5GWF49715.2020.9221119

Vankadara, Y., Pathak, P., Ashok, P., & Pillai, S. (2024). 6G revolution: Redefining communication systems. *IEEE International Conference on Smart Sustainable Energy Solutions (ICSSES)*. doi:DOI: 10.1109/ICSSES62373.2024.10561292

Vasylkivskyi, M., Horodetska, O., Klymchuk, B. & Hovorun, V. (2023). Strategies of technological development of hardware of infocommunication radio networks. *Informatsiini Tekhnologii ta Kompiuterna Inzheneriia*. doi:DOI: 10.31649/1999-9941-2023-56-1-83-91

Wang, X., Mei, J., Cui, S., Wang, C. X., & Chen, X. (2023). Realizing 6G: The operational goals, enabling technologies of future networks, and value-oriented intelligent multi-dimensional multiple access. *IEEE Network*, *37*(1), 10–17. Advance online publication. DOI: 10.1109/MNET.001.2200429

Yrjölä, S., Matinmikko-Blue, M., & Ahokangas, P. (2023). Developing 6G visions with stakeholder analysis of the 6G ecosystem. *EuCNC & 6G Summit*. doi:DOI: 10.1109/EuCNC/6GSummit58263.2023.10188379

Chapter 4
Examining the Policy and Regulatory Frameworks for Enhancing B5G and 6G Connectivity in Rural Areas

Ravindran Kandasamy
https://orcid.org/0000-0001-6763-0216
Presidency College, India

T. Dakshinamurthy
https://orcid.org/0009-0001-8937-6295
Xavier Institute of Management, India

R. Ramesh
Department of Management Studies, Knowledge Institute of Technology, Salem, India

Revathi Perumal
Knowledge Institute of Technology, India

T. N. Seenivasan.
Gnanamani College of Technology, India

Raghu Narayana Reddy
Presidency College, India

Hendra Hendra
Wilmar Bisnis Indonesia Polytechnic, Indonesia

ABSTRACT

This chapter explores the role of legislative and regulatory frameworks in enhancing B5G and 6G connectivity in rural areas. It highlights challenges like inadequate infrastructure and high costs that hinder services and growth. B5G and 6G promise transformative benefits with high speeds and innovative applications. The chapter advocates for robust policies on spectrum allocation, infrastructure incentives, and Universal Service Obligations (USOs) while addressing the impact of emerging technologies and trends in cybersecurity and sustainability.

INTRODUCTION

Beyond 5G, or B5G, is the term for the technological developments and breakthroughs that will expand on the current 5G networks and create the foundation for upcoming communication systems. It includes advanced apps that surpass 5G's capabilities, better network architecture, higher data throughput, and

DOI: 10.4018/979-8-3693-7788-8.ch004

improved connectivity. The goal of B5G is to improve upon the shortcomings of 5G by offering more dependable, adaptable, and efficient communication solutions (Xia et al., 2021).

Sixth Generation, or 6G, is the upcoming generation of mobile networks that is anticipated to replace Fifth Generation. In comparison to 5G, it claims to provide even greater data throughput, extremely low latency, and enhanced reliability. Utilizing terahertz (THz) frequencies, integrating artificial intelligence (AI) for network management, and delivering cutting-edge apps like immersive AR and VR are some of the salient features of 6G. More new services and technologies are expected to be supported by 6G, which could completely change a number of industries (Rappaport et al., 2021).

Potential Impact on Rural Areas

B5G and Rural Areas: By enhancing network performance and connectivity, B5G technologies could have a big influence on rural areas. Better access to medical, educational materials, and online services may be made possible by increased data rates and decreased latency. This could improve the quality of life and promote economic growth in less connected places by bridging the digital divide between urban and rural areas (Xia et al., 2021).

6G and Rural Areas: By providing even more dependable and strong network connections, the introduction of 6G technology is anticipated to substantially benefit rural areas. 6G's high-speed data capabilities and cutting-edge apps might improve telecommuting, facilitate remote learning and healthcare, and spur local enterprises' creativity. AI and THz frequency integration may make it possible to create new infrastructure and communication models that are specifically designed for rural areas, which might change how these communities use technology (Rappaport et al., 2021).

DIGITAL DIVIDE

The difference between people and communities who have access to contemporary information and communication technologies (ICTs) and those who do not is known as the "digital divide." This gap can take many different forms, such as disparities in digital literacy, internet access, and resource availability. Inequalities in opportunities and outcomes result from the gap, which frequently connects with socioeconomic characteristics, geographic regions, and educational levels.

Importance of Bridging the Digital Divide

Economic Impact: Growth and development of the economy depend on bridging the digital divide. Having access to technology increases work options, encourages entrepreneurship, and allows people to engage in the digital economy. While areas with restricted access may suffer with slower economic growth and greater poverty, regions with high levels of digital access enjoy more strong economic development (Beard et al., 2016).

Educational Opportunities: The growth of education depends on equitable access to technology. Pupils who have access to dependable internet and digital resources are better equipped to achieve academic success and develop skills that are necessary for today's workforce. By addressing the digital gap, we can make sure that all students have access to high-quality instruction and learning resources, irrespective of their location or socioeconomic status (Van Dijk, 2020).

Healthcare Access: Digital health and telemedicine technology can enhance the provision of healthcare, particularly in rural and underprivileged areas. In order to improve health outcomes and lessen inequities in healthcare access, it is imperative that the digital divide be bridged to enable more equitable access to medical consultations, health information, and monitoring tools (Ekeland et al., 2010).

Social Inclusion: More social engagement and connectivity are fostered by digital inclusion. Societies can lessen social isolation, increase community engagement, and improve the general quality of life for marginalized communities by guaranteeing that everyone has access to communication tools (Gunkel, 2003).

Enhancing B5G and 6G connectivity in rural areas

Improving B5G and 6G coverage in rural areas calls for a multipronged strategy that incorporates strong legislative and regulatory frameworks specifically designed to handle the particular difficulties these areas face. Comparative studies of frameworks in the USA and Africa show how crucial the regulatory environment is in determining how well the telecommunications industry performs. These studies emphasize the value of market liberalization, competition, and regulatory governance in promoting technological advancement. Abedbola, Oluwatoyin, Adegbite; Chinedu, Alex, Ezeigweneme; Aniekan, Akpan, Umoh; Valentine, Ikenna, Ilojianya; (2024).

Infrastructure sharing business models have been found to be cost-effective for the 5G and 6G cra and essential for providing universal, affordable internet, in line with the UN Sustainable Development Goals. As such, they should be taken into consideration when formulating effective policies. Furthermore, thc planning and design of radio networks—especially in densely populated areas—offers insights into the technical specifications and schedule required for the deployment of B5G and 6G, which can be modified for use in rural areas. Saber, H., Hesham, EL-Badawy, and Zainud-Deen (2023).

The goal of 6G networks is to bridge the digital divide by utilizing open hardware and free software to create affordable, dependable, and long-lasting communication networks that are tailored to the needs of rural and distant communities. Borja, Bordel, Ramón, Alcarria, Tomás, Robles, Joaquin, Chung, Rajkumar, Kettimuthu, Ivan, Armuelles, Voinov, and so on (2022).

Effective network design and implementation require addressing the topographical, architectural, and demographic issues particular to rural areas. Additionally, 6G research's emphasis on sustainability and social impact highlights the necessity of inclusive connectivity solutions that serve rural people and offer creative ways to get around obstacles and guarantee fair service accessibility. Policymakers and regulators may create all-encompassing plans that improve B5G and 6G connection in rural areas by incorporating these varied perspectives, which will eventually promote socio-economic growth and digital inclusion.

Policy frameworks impact B5G connectivity in rural areas

In rural areas, the deployment and efficacy of Beyond 5G (B5G) connections are significantly influenced by policy frameworks. These frameworks need to take into account the particular difficulties that come with living in rural areas, like sparse population densities, rough terrain, and few financial incentives for private investment. Sihombing, Candra, and Harry. (2021).

The South Korean Information Network Village (INVIL) project, for example, highlights the value of a multi-layered strategy that goes beyond simple connectivity to incorporate community involvement and sustainable broadband adoption, thereby reducing the digital divide and promoting local economic development. Man, Chul, Jung, Jee, Young, Lee, and Sora, Park (2013).

Similar challenges, such as high installation costs and problems with spectrum allocation, exist for the rollout of 5G in rural areas, but they may be addressed with carefully thought-out rules that guarantee spectrum availability and affordability. Howard, Heather. (2019). The necessity for economical solutions catered to rural needs is highlighted by Tanzanian policy frameworks that take advantage of the country's current optical fiber backhaul for last-mile connectivity. These frameworks are expected to offer high data rates at reasonable prices. Byanyuma Mastidia, Sadath Kalolo, Salehe, I., Mrutu, Christina, Nyakyi, and Anael, Sam (2013).

Furthermore, in order to bridge the digital divide and make services like IPTV and u-Health possible, the Korean government's BcN project intends to establish high-speed network infrastructure and convergence services in rural areas. Lee Eunyoung and Kim Hyongsoon (2010). Due to cheaper infrastructure costs than optical fiber, the adoption of wireless technologies, such 5G, offers potential benefits for rural broadband. However, in order to overcome issues with spectrum and deployment costs, supportive policies are needed. Hudson, Heather. (2019).

Furthermore, improved frameworks for policy development that combine quantitative and qualitative data might improve the efficacy of rural programs by facilitating quick experimentation and stakeholder participation—as demonstrated in European pilot regions. United Kingdom, Milos, Ulman, Pavel, Simek, Jan, Masner, Pavel, Kogut, Tuula, Löytty, Patrick, Crehan, Karel, Charvat, Antoni, Oliva, Stein, Runar, Bergheim, Milan, Kalas, Denis, Kolokol, Tommaso, Sabbatini, USA, C, Limited Consultancy, Finland, Hub, Oy, Crehan, Kusano, Associates, Sistema, Slovakia, Belgium, Lean, Smart, Slovakia, Kajo, s., r.o. (2020).

All things considered, strong legislative frameworks are necessary to remove the obstacles to B5G connectivity in rural areas and guarantee that these places can take advantage of cutting-edge telecommunications.

Key components B5G and 6G Connectivity in Rural Areas

5G and 6G Technologies

B5G (Beyond 5G) and 6G represent the future advancements in mobile network technology, aiming to enhance connectivity through greater speeds, lower latency, and improved capacity compared to current 5G systems. These technologies are expected to support advanced applications such as augmented reality, smart cities, and the Internet of Things (IoT) with unprecedented efficiency (Kumar, 2023).

Current State of Connectivity in Rural Areas

Poor connectivity is a common problem in rural locations because of things like low population density, expensive infrastructure, and few financial incentives for service providers. In comparison to urban areas, this leads to slower internet speeds, greater expenses, and less dependable connections. Ensuring equal access to cutting-edge technologies requires addressing these gaps (Smith & Johnson, 2022).

Policy and Regulatory Frameworks

Adequate legislative and regulatory structures are crucial for encouraging the rollout of B5G and 6G technologies in underprivileged regions. This covers regulations that ease the distribution of spectrum, offer monetary rewards, and encourage infrastructure expenditures. The interests of multiple parties, including as government agencies, telecom companies, and rural communities, must be balanced by regulations (Nguyen, 2024).

Components of policy frameworks for B5G

Beyond 5G (B5G) network policy frameworks are complex systems that are intended to effectively manage and coordinate network resources, services, and security. A policy manager that manages the creation, modification, and enforcement of policies, as well as a variety of policy types that can be dynamically installed or removed without causing system disruption are essential elements of these frameworks, as exemplified by the ONOS SDN controller prototype Ferran, Canellas, Angelos, Mimidis, Nestor, Bonjorn, José, Soler (2019).

In order to streamline the configuration, application, analysis, reporting, and auditing of policies, the frameworks also make use of a standard document format for storing and implementing policy settings. Kim, Danny, Levin, and Joseph (2009). Furthermore, permissions and rules for communication sessions are managed via policy data frames, guaranteeing that policies are uniformly implemented throughout the network's numerous nodes. Khan, Sohel; Boone, Jon; Lambardia, Robert (2013).

A strong policy design and architecture are necessary for policy frameworks in the context of Business Activity Management (BAM) since they must manage a large number of business rules and adjust to changing business scenarios. Chang, Jen-Yao, Chung, Jen-Jang, Jeng, H. (2003). In cloud computing settings, security and compliance are crucial since policies need to enforce non-functional needs and guarantee that rules and regulations are followed. Tim Waizenegger, Matthias Wieland, Alexander Blehm, Volha, Kalach, Alexander Kicherer, Gustav Murawski, and Volha (2014).

Policies are crucial in Network Function Virtualization (NFV) environments for defining resource limitations, enforcing business rules, preserving energy, and resolving policy conflicts via a hierarchical design. The authors of this work are Dilip Krishnaswamy, Ramki Krishnan, Diego R. Lopez, Steven Wright, and Norival Figueira (2016). In order to provide uniform policy implementation across clients, the Common Open Policy Service (COPS) protocol is employed S., Hahn, K., Chan, Keith, McCloghrie, Ravi, L., Sahita (2003).

Autonomic service and network management in the dynamic and decentralized Future Internet relies on sophisticated policy frameworks to direct network behavior based on service requirements and business objectives. Panagiotis, Demestichas; Aristi, Galani; Nikos, Koutsouris; Kostas, Tsagkaris; B. Fuentes; C. Garcia, Vazquez; Gerard, Nguengang (2012). Finally, for describing and maintaining policies within the management of computer and telecommunication networks, an object-oriented informational model and a functional model for policy manipulation are essential. In 1996, Guerroudji, N. Together, these elements guarantee that B5G networks are adaptable, safe, and able to satisfy a wide range of changing needs.

Current State of Connectivity in Rural Areas in Existing Challenges

Infrastructure Deficiencies

Lack of a reliable telecommunications infrastructure is one of the biggest problems in rural areas. The lack of modern or insufficient infrastructure in rural areas frequently restricts access to high-speed internet. This is mostly because telecom providers see reduced returns on their investment as a result of the high cost of installing broadband infrastructure in sparsely inhabited areas (Whitacre et al., 2017). As a result, compared to metropolitan regions, rural communities may have slower, less dependable internet connections.

Limited Accessibility

One of the main concerns is the availability of dependable internet services in rural locations. The geographic dispersion of rural people makes it challenging to offer reliable coverage, even in cases where infrastructure is present. Large distances between homes, mountains, and forests are examples of physical obstacles that make connecting initiatives more difficult. Compared to just 1.5% of Americans living in cities, the Federal Communications Commission (FCC) reports that over 22% of Americans living in rural areas do not have access to fixed broadband at 25 Mbps/3 Mbps (FCC, 2020).

Affordability Issues

Another significant obstacle to connecting in rural regions is affordability. Due to increased infrastructure maintenance costs and a lack of provider competition, internet service in rural locations is frequently more expensive than in urban ones. The price difference exacerbates the digital divide by making it impossible for rural households to buy high-speed internet (Horrigan, 2019). A major issue that requires legislative interventions is cost, as inhabitants' lower average wages in rural locations make it much more difficult for them to buy internet services.

Current policies and regulatory frameworks in B5G and 6G Connectivity

Spectrum Allocation

Auctions for Spectrum and Licensing: A lot of nations conduct auctions to distribute spectrum to mobile phone providers. This strategy seeks to provide spectrum to organizations that can use it most efficiently while also optimizing income and efficiency. These procedures are governed by regulatory agencies, such as the European Conference of Postal and Telecommunications Administrations (CEPT) in Europe and the Federal Communications Commission (FCC) in the United States (FCC, 2023; CEPT, 2023).

Dynamic Spectrum Sharing: With this method, several users can share access to the same spectrum bands. It is being utilized more and more to increase spectrum usage and make new technologies like 5G and beyond easier to install (Zhang et al., 2023).

Unlicensed Spectrum: Certain spectrum bands are free to be used for innovation and wider access. Supporting developing low-power gadgets and technologies like Wi-Fi is essential (ITU, 2022).

Infrastructure Sharing

Regulations Requiring Sharing: In order to cut expenses and improve network coverage, many countries have regulations requiring infrastructure sharing. Sharing towers, ducts, and other tangible resources is one way to do this. To increase network deployment efficiency, for example, the European Union has put in place policies that promote telecom providers to share infrastructure (European Commission, 2022).

Public-Private Partnerships (PPPs): PPPs are frequently used by governments to develop and share infrastructure in underdeveloped areas. These collaborations can meet the need for public connection while maximizing private investment (World Bank, 2023).

Universal Service Obligations (USOs)

Mandatory Coverage: In order to guarantee that all citizens have access to basic connectivity, USOs compel telecom companies to offer services in underserved or distant locations. In order to assist deployment in less lucrative places, this frequently entails subsidies or funding methods (OECD, 2023).

Performance Standards: Performance benchmarks like minimal service quality and price requirements can be included in USOs. To make sure that service providers fulfill their responsibilities, regulators implement these criteria (ITU, 2023).

Regional and Local Regulations in B5G and 6G Connectivity

Regulations at the regional and municipal levels significantly influence how connection infrastructure is installed and developed in rural regions. The efficacy and pace at which mobile and broadband networks expand can be greatly impacted by these restrictions.

Local Laws Concerning Zoning and Permits: The installation of telecommunications infrastructure may be aided or hindered by local zoning regulations and permit needs. A National League of Cities (NLC, 2022) research claims that stringent zoning laws are enforced by many local governments, which might cause delays or raise the cost of implementing new infrastructure. These rules frequently entail drawn-out clearance procedures along with precise specifications about the site's aesthetics and environmental effect.

State-Level laws and Incentives: To encourage the growth of broadband in underserved areas, states frequently have their own laws and incentive schemes in place. To offer funding for broadband infrastructure projects in rural and underprivileged communities, the California Public Utilities Commission (CPUC, 2023) oversees initiatives such as the California Advanced Services Fund (CASF). Policies at the state level can differ greatly; some governments have less encouraging frameworks, while others give substantial incentives.

Initiatives Tailored to the Rural Environment: Certain regional administrations have introduced focused programs to tackle issues related to internet access in rural regions. In order to improve broadband connectivity in the Appalachian area, the Appalachian Regional Commission (ARC, 2023) has created initiatives and is offering financial and technical support to local projects. These regional programs frequently concentrate on removing certain obstacles that rural populations encounter.

Local Government Partnerships: Enhancing connectivity may need cooperation between local governments and businesses in the private sector. The Federal Communications Commission (FCC, 2022) lists a number of regional collaborations that have been effective in increasing access to broadband. These collaborations may take the form of joint ventures to split expenses and resources or agreements for the co-location of infrastructure.

Problems and Disagreements: Local laws can occasionally cause disputes between different parties. For instance, a National Broadband Plan (2021) research finds that regulatory requirements and responsibility disputes between telecom providers and local governments can cause delays in infrastructure implementation and raise prices.

International Guidelines and Standards

The establishment of standards and guidelines by international organizations is essential for promoting the creation and application of cutting-edge communication technologies like B5G and 6G. These groups offer frameworks that guarantee efficiency, safety, and interoperability in initiatives to increase global communication.

The ITU, or International Telecommunication Union: The United Nations has a specialized institution called the ITU that handles matters pertaining to information and communication technologies. The ITU organizes radio frequency spectrum and global telecommunications, as well as setting international standards. For instance, technical standards for spectrum management and wireless communication are developed by ITU's Radiocommunication Sector (ITU-R). The expected specifications and architecture for next-generation networks are described in the ITU Report on "Technologies for Network 2030" (ITU, 2020). These recommendations have a significant impact on how B5G and 6G technologies are developed.

The Institute of Electrical and Electronics Engineers (IEEE) is a front-runner in the creation of technology standards, particularly those pertaining to networking and wireless communications. The working groups of IEEE on 5G and beyond are tasked with creating standards that will impact the deployment of B5G and 6G technology. For example, standards for wireless networking technologies like Wi-Fi and wireless personal area networks are developed by the IEEE 802.11 and IEEE 802.15 working groups (IEEE, 2023). In order to guarantee that new technologies operate properly and are compatible with a variety of settings, these standards are essential.

Third Generation Partnership Project (3GPP): To create and maintain technical specifications for mobile systems, seven international telecoms standard development organizations work together. Due to the production of specifications for 4G, 5G, and previous generations of mobile networks, 3GPP is essential to the creation of standards for B5G and 6G. The specifications and design considerations for future networks are outlined in the 3GPP Technical Reports and Specifications for 5G and Beyond (3GPP, 2024).

The International Organization for Standardization (ISO) is a global non-governmental organization that operates independently and creates and disseminates standards for a variety of businesses. The ISO offers standards for communication technology interoperability, quality, and safety. The ISO/IEC JTC 1/

SC 6 committee is responsible for developing standards related to information exchange and telecommunications among various systems, encompassing network systems and emerging technologies (ISO, 2023).

Challenges of Current Policies and Regulatory Frameworks in B5G and 6G Connectivity

Spectrum Allocation and Management:

The deployment of B5G and 6G depends on the efficient distribution and management of spectrum. The requirement for high-frequency spectrum, such as terahertz and millimeter-wave bands, which are crucial for these cutting-edge technologies, is frequently ignored by current policy.

Impact: According to Rangan et al. (2023), insufficient spectrum policies may result in interference, poor spectrum use, and hampered network performance.

Interoperability and Standardization:

Global standardization is necessary for B5G and 6G technologies to guarantee compatibility among various networks and devices. The current regulatory frameworks may not have the capacity to keep up with the speed at which technology is developing and the increasing demand for unified international standards.

Impact: Market fragmentation, decreased economies of scale, and network compatibility problems might arise from a lack of standardization (Zhang & Wang, 2023).

Security and Privacy:

Significant security and privacy challenges arise from B5G and 6G's improved connectivity and data transmission capabilities. These technologies are linked to sophisticated cyber dangers that may not be sufficiently addressed by current rules.

Impact: Users may be vulnerable to data breaches and cyberattacks as a result of network weaknesses brought on by insufficient security and privacy laws (Chen et al., 2024).

Infrastructure Deployment:

Infrastructure for B5G and 6G must be deployed, which will require large financial and regulatory support. Particularly in rural and neglected areas, current laws could not offer sufficient incentives or efficient procedures for the prompt implementation of essential infrastructure.

Impact: The adoption of B5G and 6G technologies may be slowed down by delays and higher infrastructure deployment costs, especially in areas where the technologies have the greatest potential for benefit (Li et al., 2023).

Environmental and Health Concerns:

The introduction of new wireless technology frequently prompts worries about the effects on the environment and human health. There may be opposition from the public and legal challenges if these issues are not adequately addressed by the regulatory structures in place.

Impact: The rollout of B5G and 6G networks may encounter strong public opposition and be slowed down in the absence of clear regulations and assurances about safety (Wang et al., 2024).

Gaps, Inconsistencies, and Barriers to B5G and 6G Deployment in Rural Areas

Infrastructure Gaps

Large infrastructural gaps frequently exist in rural locations, impeding the implementation of B5G and 6G technologies. These areas usually lack essential elements like dense tower networks, fiber-optic backhaul, and dependable power supplies. In addition to raising costs, this absence makes the logistical aspects of constructing the required infrastructure more difficult. Thus, the implementation of advanced networks in rural areas may be severely delayed due to the high costs and logistical difficulties involved in infrastructure construction (Kaur & Kamal, 2023).

Economic Barriers

Because the customer base is smaller and less concentrated in rural areas, it is frequently not economically viable to construct sophisticated networks such as B5G and 6G. Network operators could be reluctant to allocate the required resources if they do not perceive a high enough return on investment. The digital gap is sustained by this cost barrier, which keeps rural areas neglected in comparison to urban areas where investing in cutting-edge technology is more financially justified (Sharma & Gupta, 2024).

Regulatory Inconsistencies

The implementation of B5G and 6G networks in rural areas is significantly hampered by inconsistent regulatory rules at various governmental levels. Network operators may have to negotiate a complicated and occasionally contradictory framework as a result of these regulatory disparities, which will lengthen the implementation period and increase costs. For network rollouts to go more smoothly and effectively, unified and streamlined regulatory procedures are necessary (Wang & Zhang, 2024).

Technological Barriers

Modern technology and software are needed to deploy cutting-edge technologies like B5G and 6G, however these may not be easily accessible or reasonably priced in rural locations. The acceptance and integration of new networks may be hampered by a lack of access to these cutting-edge technological resources. To guarantee that rural areas can profit from the developments in wireless communication, it is imperative that this technological barrier is addressed (Li & Sun, 2023).

Skilled Workforce Shortage

The lack of qualified labor is a major obstacle to the installation and upkeep of B5G and 6G networks in rural areas. These cutting-edge networks need certain knowledge and experience to establish and run, which is sometimes absent in less populated areas. Insufficient numbers of qualified experts might cause delays in the deployment process and worsen network maintenance over time (Chen & Liu, 2024).

Affordability Issues

Even with the deployment of B5G and 6G networks in rural areas, many locals may find the cost of suitable devices and services to be unaffordable. The high cost of these cutting-edge technologies prevents them from being adopted at high rates, which makes it challenging to realize the economies of scale required for long-term network operations. In order for rural populations to have access to and benefit from advanced communication networks, it is imperative that affordability difficulties be addressed (Rangan & Erkip, 2023).

Environmental and Topographical Challenges

Topographical features and environmental challenges, like mountains, woods, and vast stretches of land separating settlements, are common in rural locations. These physical obstacles raise the costs and hinder the implementation of network infrastructure. To guarantee that rural areas can profit from the developments in B5G and 6G technologies, it is imperative to overcome these environmental and topographical constraints (Gao & Li, 2024).

Evaluation of Current Incentive Mechanisms and Support Programs for B5G and 6G

The success of the current B5G and 6G rollout incentive systems and support programs has varied. The U.S. Federal Communications Commission's Rural Digital Opportunity Fund (RDOF) and other financial incentives and subsidies are essential in counterbalancing the high expenses of implementing cutting-edge network infrastructure. These financial assistance are especially helpful in rural areas, where the feasibility of these projects economically is frequently a major worry (Federal Communications Commission, 2023). In a similar vein, tax breaks provided by different nations yield significant advantages by relieving financial strain on telecom firms, which in turn increases the allure of investing in B5G and 6G technologies (OECD, 2023).

PPPs, or public-private partnerships, have shown to be successful in promoting the installation of cutting-edge networks. By combining the advantages of the public and private sectors, these alliances make it easier to share resources, reduce risk, and make joint investments in infrastructure projects. One prominent instance is the European Union's 5G Public Private Partnership (5G PPP), which seeks to encourage innovation and cooperation among stakeholders in order to hasten the deployment of 5G networks throughout Europe (European Commission, 2023).

Funding for research and development (R&D) is yet another essential element of the assistance programs in place today. Government-funded R&D projects speed up the time to market for new developments in wireless communications and stimulate technological innovation. The development and implementation of B5G and 6G networks is facilitated by the National Science Foundation (NSF) in the United States, which generously funds research into next-generation wireless technology (National Science Foundation, 2023).

These support programs and incentive systems, however, may use some work. The unequal distribution of incentives, which frequently benefits urban over rural areas, is one such disparity. Even while advanced technology are available to rural populations, this gap contributes to the perpetuation of the digital divide by underserving them. Furthermore, inconsistent regulations at different governmental levels can make deployment efforts more difficult and time-consuming, which raises expenses and adds to network operators' administrative workloads. To optimize the efficacy of existing mechanisms and support programs, it is imperative to ensure a more equitable distribution of incentives and streamline regulatory laws.

PROPOSED POLICY AND REGULATORY ENHANCEMENTS

Policy Recommendations

Spectrum Allocation and Management

To provide improved coverage and connectivity, governments should give priority to allocating low-band and mid-band spectrum to rural areas. Regulatory regimes that support the development of secondary spectrum markets and dynamic spectrum sharing can help achieve this.

Infrastructure Investment and Subsidies

Incentives should be introduced to incentivize telecom carriers to invest in rural infrastructure, such as tax exemptions or subsidies. To support the deployment of critical infrastructure, public funding initiatives akin to the Rural Digital Opportunity Fund of the U.S. Federal Communications Commission (FCC) could be established.

Public-Private Partnerships (PPP)

To maximize the knowledge and assets of private organizations while guaranteeing that the public interest is met, governments ought to support public-private partnerships. These alliances can save deployment costs and enable infrastructure sharing.

Universal Service Obligation (USO) Expansion

In order to ensure that telecom operators are required to provide high-speed internet connectivity in rural and underserved areas as part of their service requirements, expand the Universal Service Obligation to encompass B5G and 6G technologies.

Community-Based Networks

Grants and technical help can be used to support the establishment of community-based networks in rural areas. Local communities may own and run these networks, offering specialized services that address regional requirements.

Regulatory Simplification

Deploy telecom infrastructure in rural areas by streamlining and simplifying regulatory procedures. This entails lowering administrative obstacles, expediting permit approvals, and harmonizing regional laws to enable quicker deployment.

Policy Trends Impacting B5G and 6G Connectivity

Enhanced Focus on Spectrum Efficiency and Sharing

Future regulations are probably going to focus more on making better use of spectrum as demand for wireless services rises. Governments have the power to enact laws that support dynamic spectrum sharing, which permits numerous users to coexist peacefully in the same spectrum bands. This might include putting cutting-edge technology like AI-based spectrum management systems and Cognitive Radio into practice (Rysavy, 2021).

Strengthening Cybersecurity Regulations

The introduction of B5G and 6G, which will link billions of devices, will make cybersecurity even more crucial. Stricter cybersecurity regulations for telecom networks are anticipated in the future, with an emphasis on safeguarding network infrastructure from cyberattacks and preserving data privacy (World Economic Forum, 2020).

Sustainability and Green Policies

Future policy will probably be shaped by environmental sustainability, with a focus on lowering the energy usage of B5G and 6G networks. Regulations that promote the development of energy-efficient technologies and the use of renewable energy sources for network operations may be introduced by governments (GSMA, 2019).

Inclusive Connectivity Mandates

Future regulations are anticipated to include more inclusive connection requirements, which would oblige telecom providers to supply reasonably priced and dependable internet service in underserved rural and distant areas, in an effort to close the digital divide. Targeted subsidies and updated Universal Service Obligations (USO) could accomplish this (European Commission, 2020).

Expansion of Public-Private Partnerships (PPP)

The implementation of B5G and 6G is anticipated to be heavily reliant on public-private partnerships, particularly in rural and neglected areas. Policies in the future might offer structures that make it easier for public and private sectors to work together to combine resources and knowledge in order to expand networks (World Economic Forum, 2020).

Emerging Technologies

Low Earth Orbit (LEO) Satellites

By bringing high-speed internet connection to faraway locations, LEO satellite networks—like those used by SpaceX's Starlink and Amazon's Project Kuiper—have the potential to completely transform rural connectivity. Policymakers may be prompted by these networks to amend laws pertaining to satellite communications and spectrum utilization, making sure that these technologies are included in national connectivity plans (Pelton, 2020).

Edge Computing

In rural locations, edge computing can lower latency and enhance service quality by processing data closer to the source instead of depending on centralized cloud servers. By fostering the creation of localized infrastructure and stimulating investment in dispersed network models, this technology may have an impact on policies pertaining to rural connectivity (Shi et al., 2016).

Artificial Intelligence (AI) and Machine Learning (ML)

The application of AI and ML to improve cybersecurity, control spectrum consumption, and optimize network performance is growing. According to Zhang et al. (2019), these technologies have the potential to facilitate the development of more intelligent and adaptable connectivity regulations, which would allow for automated network management and dynamic spectrum allocation in rural areas.

5G mm Wave and Beamforming Technologies

Due to its short range, 5G mmWave technology is usually associated with metropolitan areas; however, improvements in beamforming and signal processing may make it feasible for rural connectivity. In order to facilitate the implementation of these technologies in sparsely inhabited areas and ensure that they enhance the current infrastructure, policymakers may need to modify existing legislation (Ghosh et al., 2019).

Internet of Things (IoT)

Robust connectivity solutions will be necessary due to the development of IoT devices in the rural areas' agriculture, health, and logistics sectors. Policies may change to accommodate networks tailored to the Internet of Things and guarantee that rural areas have the infrastructure required to handle the enormous surge of connected devices (Islam et al., 2015).

CONCLUSION

The present chapter has comprehensively examined the pivotal function of policy and regulatory frameworks in augmenting B5G (Beyond 5G) and 6G connectivity in rural regions. The significance of these frameworks in mitigating the enduring digital divide between urban and rural residents has been underscored. Keeping rural areas in mind is becoming more and more important as the world moves from 5G to B5G and eventually 6G. The need for strong regulatory frameworks and focused policy initiatives that can spur the adoption of these cutting-edge technologies in rural areas is emphasized in this chapter.

The first section of the chapter covered the connectivity issues that rural areas had to deal with. These obstacles are numerous and well-established, and they include things like low infrastructure, expensive deployment, problems with accessibility, and financial limitations. When taken as a whole, these elements contribute to the growing digital divide that disadvantages rural populations in terms of internet and digital service access. Rural communities' poor internet connections impede economic growth, reduce educational options, and make it more difficult to get healthcare and other necessary services. This chapter made clear that social fairness is just as important as technology in tackling these problems.

This chapter's main focus was on how B5G and 6G technology would change rural connection. Ultra-high data rates, incredibly low latency, and the capacity to link a large number of devices in rural and underserved locations are just a few of the revolutionary advantages that B5G and 6G promise to bring. New applications in telemedicine, smart infrastructure, remote education, and precision agriculture—all of which are especially pertinent to rural communities—may be made possible by these technologies. Through the utilization of these cutting-edge technologies, rural areas can surmount certain conventional obstacles to connectivity and engage more thoroughly in the digital economy.

However, a comprehensive and forward-thinking policy and regulatory framework is needed to realize the full promise of B5G and 6G in rural areas. The chapter looked at the situation as it is now, pointing out any gaps and places that require governmental actions. Setting aside spectrum for rural areas as a top priority was one of the main recommendations. Targeted spectrum rules can assist guarantee that rural areas acquire the necessary bandwidth to enable B5G and 6G services, since there is no business motivation to develop sophisticated networks in sparsely populated areas.

An other crucial suggestion was to provide incentives for infrastructure development in rural regions. In this context, public-private partnerships (PPPs) are especially important because they can combine resources and share risks, which increases the viability of deploying expensive infrastructure in low-density areas. The necessity of extending Universal Service Obligations (USOs) to incorporate next-generation technology was also emphasized in this chapter. USOs have traditionally concentrated on providing basic internet and phone services, but in the B5G and 6G era, these duties need to be modified to meet the demand for dependable, high-speed connectivity in all areas, including rural ones.

Additionally, emerging trends in policy were examined, with a special emphasis on equitable connectivity, cybersecurity, sustainability, and spectrum efficiency. Effective spectrum utilization is becoming more and more crucial as wireless communication becomes more and more necessary. To maximize the capacity of wireless networks, policies that support spectrum sharing, dynamic spectrum access, and the utilization of higher frequency bands are crucial. Strong security measures will be essential in terms of cybersecurity as B5G and 6G networks grow more intricate and linked in order to safeguard sensitive data and guarantee the dependability of vital infrastructure.

Another important factor is sustainability, since the rollout of B5G and 6G networks needs to be coordinated with more general environmental objectives. It is imperative for policymakers to guarantee the energy efficiency of these networks and prevent any negative environmental effects from their deployment. Lastly, regulations are required to guarantee that all communities, regardless of geography, have access to the advantages of cutting-edge digital technologies. Inclusive connectivity is still a key objective.

The chapter also looked at how regulations related to rural connection are affected by developing technologies including edge computing, artificial intelligence (AI), Internet of Things (IoT), and Low Earth Orbit (LEO) satellites. For example, LEO satellites have the capacity to deliver high-speed internet to even the most remote areas, while edge computing can lower latency by processing data closer to its source. Smart agriculture and environmental monitoring are two new applications that can be made possible by AI and IoT and are especially relevant for rural regions. They can also improve network performance. To guarantee their efficient and fair implementation, these technologies also present new regulatory issues that need to be resolved.

The chapter's conclusion emphasized how urgent it is to implement B5G and 6G-related laws and regulations in rural areas. There isn't much time left to make sure that rural areas don't fall behind as these technologies continue to advance quickly. In order to develop inclusive and progressive regulatory frameworks that tackle the particular difficulties associated with rural connection, policymakers must take decisive action. This entails giving rural areas top priority when it comes to spectrum allocation, encouraging infrastructure investment, cultivating public-private partnerships, and embracing cutting-edge technology that can provide high-speed internet to even the most remote locations.

In an increasingly connected world, inaction today could widen the digital divide and further marginalize rural areas. Now is the moment to implement inclusive and proactive B5G and 6G policy.

REFERENCES

3rd Generation Partnership Project. (2024). *3GPP specifications for 5G and beyond*. Retrieved from https://www.3gpp.org/

El-Badawy, H. M., Ahmed, H. A. S., Zainud-Deen, S. H., & Malhat, H. A. E. A. (2023, May). B5G/6G Network Planning For Study Case In Knowledge City Area As Model For Smart Cities. In 2023 40th National Radio Science Conference (NRSC) (Vol. 1, pp. 191-200). IEEE.

Alexander, B. (2016). *Volha, Kalach., Alexander, Kicherer., Gustav, Murawski., Tim, Waizenegger., Matthias, Wieland. (2014). Policy-Framework - Eine Methode zur Umsetzung von Sicherheits-Policies im Cloud-Computing. Dilip, Krishnaswamy., Ramki, Krishnan., Diego, R., Lopez., Steven, Wright., Norival, Figueira*. Policy Architecture and Framework for NFV Infrastructures.

Appalachian Regional Commission. (2023). *Broadband initiatives*. Retrieved from https://www.arc.gov/programs/broadband-initiatives/

Galani, A., Koutsouris, N., Tsagkaris, K., Demestichas, P., Fuentes, B., Vazquez, C. G., & Nguengang, G. (2012, December). A policy based framework for governing Future networks. In 2012 [IEEE.]. *IEEE Globecom Workshops*, •••, 802–806.

Beard, V. A., Harmsen, A. L., & Hinton, C. (2016). The Digital Divide and Its Implications for Rural Areas. *Journal of Rural Studies*, *44*, 125–133. DOI: 10.1016/j.jrurstud.2016.01.009

California Public Utilities Commission. (2023). *California Advanced Services Fund*. Retrieved from https://www.cpuc.ca.gov/industries-and-topics/internet-and-phone/california-advanced-services-fund

CEPT. (2023). Spectrum policy and regulation. https://www.cept.org/spectrum/

Chaoub, A., Giordani, M., Lall, B., Bhatia, V., Kliks, A., Mendes, L., Rabie, K., Saarnisaari, H., Singhal, A., Zhang, N., Dixit, S., & Zorzi, M. (2022, February). 6G for Bridging the Digital Divide: Wireless Connectivity to Remote Areas. *IEEE Wireless Communications*, *29*(1), 160–168. Advance online publication. DOI: 10.1109/MWC.001.2100137

Chen, X., & Liu, H. (2024). Workforce Challenges in Rural 6G Deployment. *Journal of Wireless Communications*, *19*(2), 98–107. DOI: 10.1007/s11277-023-09123-1

Chen, X., Liu, Y., & Wang, J. (2024). Security and Privacy Challenges in 6G Networks. *IEEE Communications Magazine*, *62*(1), 36–42. DOI: 10.1109/MCOM.2024.1234567

Ezeigweneme, C. A., Umoh, A. A., Ilojianya, V. I., & Adegbite, A. O. (2024). Review of telecommunication regulation and policy: Comparative analysis USA and Africa. *Computer Science & IT Research Journal*, *5*(1), 81–99.

Ekeland, A. G., Bowes, A., & Flottorp, S. (2010). Effectiveness of Telemedicine: A Systematic Review of Reviews. *International Journal of Medical Informatics*, *79*(11), 737–747. DOI: 10.1016/j.ijmedinf.2010.08.006 PMID: 20884286

European Commission. (2022). Infrastructure sharing in telecoms. https://ec.europa.eu/digital-strategy/our-policies/infrastructure-sharing_en

Federal Communications Commission. (2022). *Local partnerships and broadband expansion*. Retrieved from https://www.fcc.gov/local-partnerships-broadband

Federal Communications Commission (FCC). (2023). Spectrum management. https://www.fcc.gov/spectrum-management

Ferran, C. (2019). *Angelos, Mimidis., Nestor, Bonjorn., José, Soler*. Policy Framework Prototype for ONOS., DOI: 10.1109/NETSOFT.2019.8806691

Gao, S., & Li, X. (2024). Environmental Challenges in Deploying B5G and 6G Networks. *Journal of Environmental Management*, *250*, 109–123. DOI: 10.1016/j.jenvman.2023.109123

Ghosh, A., Ratasuk, R., Mondal, B., Mangalvedhe, N., & Thomas, T. (2019). LTE-advanced: Next-generation wireless broadband technology. *IEEE Wireless Communications*, *17*(3), 10–22. DOI: 10.1109/MWC.2010.5490974

Gunkel, D. J. (2003). Second Thoughts: Toward a Critique of the Digital Divide. *New Media & Society*, *5*(3), 499–522. DOI: 10.1177/146144480354003

Sihombing, H. C. (2021). A 5g Rural, Remote, and Difficult Areas Hope: A Review. *Teknokom*, *4*(1), 21–29.

Heather, H. (2019). 5G Mobile Broadband: Spectrum Challenges for Rural Regions. Social Science Research Network, doi: DOI: 10.2139/SSRN.3427548

Hyongsoon, K. (2010). *Eunyoung, Lee*. Policy Analysis on Resolving Digital Divide in Rural Areas.

Institute of Electrical and Electronics Engineers. (2023). *IEEE 802.11 and IEEE 802.15 working groups*. Retrieved from https://standards.ieee.org/standard/802_11-2020.html

International Organization for Standardization. (2023). *ISO/IEC JTC 1/SC 6: Telecommunications and information exchange between systems*. Retrieved from https://www.iso.org/committee/45362.html

International Telecommunication Union. (2020). *Technologies for Network 2030: New technologies for an evolved network ecosystem*. Retrieved from https://www.itu.int/en/ITU-T/tech/Pages/default.aspx

International Telecommunication Union (ITU). (2023). Universal service and access. https://www.itu.int/en/ITU-T/universal-service/Pages/default.aspx

Islam, S. R., Kwak, D., Kabir, M. H., Hossain, M., & Kwak, K. S. (2015). The Internet of Things for health care: A comprehensive survey. *IEEE Access : Practical Innovations, Open Solutions*, *3*, 678–708. DOI: 10.1109/ACCESS.2015.2437951

Ivan, Armuelles, Voinov., Joaquin, Chung., Rajkumar, Kettimuthu., Borja, Bordel., Ramón, Alcarria., Tomás, Robles. (2022). Towards 6G Networks for Rural Environments: Vision for Improving Digital Inclusion through Open Source Hardware and Software. DOI: 10.1109/CONCAPAN48024.2022.9997739

Levin, J., & Kim, D. (2015). U.S. Patent No. 8,935,365. Washington, DC: U.S. Patent and Trademark Office.

Jeng, J. J., Chang, H., & Chung, J. Y. (2003, June). A policy framework for business activity management. In EEE International Conference on E-Commerce, 2003. CEC 2003. (pp. 238-245). IEEE.

Kaur, H., & Kamal, T. (2023). Infrastructure Challenges in Rural 5G Deployment. *Telecommunications Policy*, *47*(3), 101–112. DOI: 10.1016/j.telpol.2023.103001

Dixit, S., Bhatia, V., Khanganba, S. P., & Agrawal, A. (2022). *Key considerations to achieve 5G and B5G connectivity in rural areas. In 6G: Sustainable Development for Rural and Remote Communities*. Springer Nature Singapore.

Kumar, S. (2023). Beyond 5G and 6G: Emerging technologies and their implications. *Journal of Wireless Communications*, *18*(2), 112–126. DOI: 10.1234/jwc.2023.01802

Lee, C., & Wang, H. (2023). Case studies in rural broadband expansion: Lessons learned from B5G and 6G pilot projects. *Rural Technology Journal*, *21*(4), 210–225. DOI: 10.6789/rtj.2023.02104

Li, H., Zhang, T., & Sun, Y. (2023). Infrastructure Challenges for B5G Deployment. *Telecommunications Policy*, *47*(3), 101–112. DOI: 10.1016/j.telpol.2023.103001

Byanyuma, M., Kalolo, S., Mrutu, S. I., Nyakyi, C., & Sam, A. (2013, July). Affordable broadband connectivity for rural areas. In *2013 Pan African International Conference on Information Science, Computing and Telecommunications (PACT)* (pp. 62-65). IEEE.

Ulman, M., Šimek, P., Masner, J., Kogut, P., Löytty, T., Crehan, P., & Sabbatini, T. (2020). Towards Future Oriented Collaborative Policy Development for Rural Areas and People. *AGRIS On-Line Papers in Economics and Informatics*, *12*(1), 111–124.

National Broadband Plan. (2021). *Challenges and solutions for local broadband deployment*. Retrieved from https://www.broadband.gov/plan

National League of Cities. (2022). *Local regulations and broadband deployment*. Retrieved from https://www.nlc.org/resource/local-regulations-and-broadband-deployment/

Nguyen, A. (2024). Policy and regulatory frameworks for next-generation connectivity: A global perspective. *Telecommunications Policy Review*, *32*(1), 75–90. DOI: 10.2345/tpr.2024.03201

Organisation for Economic Co-operation and Development (OECD). (2023). Universal servicc obligations in telecommunications. https://www.oecd.org/sti/telecom-universal-service-obligations.htm

Pelton, J. N. (2020). The rise of small satellites and the emergence of a new satellite industry. *Space Policy*, *47*, 21–25. DOI: 10.1016/j.spacepol.2018.10.002

Rangan, S., & Erkip, E. (2023). Economic Barriers to B5G and 6G Adoption in Rural Areas. *Proceedings of the IEEE*, *101*(3), 366–385. DOI: 10.1109/JPROC.2014.2299397

Rangan, S., Rappaport, T. S., & Erkip, E. (2023). Millimeter-Wave Cellular Wireless Networks: Potentials and Challenges. *Proceedings of the IEEE*, *101*(3), 366–385. DOI: 10.1109/JPROC.2014.2299397

Rappaport, T. S., Xu, Q., & Kanhere, S. (2021). 6G Wireless Systems: A Vision and Research Agenda. *IEEE Access : Practical Innovations, Open Solutions*, *9*, 22575–22592. DOI: 10.1109/ACCESS.2021.3060105

Sahita, R., Hahn, S., Chan, K., & McCloghrie, K. (2003). Framework policy information base (No. rfc3318).

Sharma, A., & Gupta, R. (2024). Economic Considerations for B5G and 6G Networks in Rural Areas. *Journal of Rural Telecommunications*, *19*(1), 23–34. DOI: 10.1007/s11277-023-09123-1

Shi, W., Cao, J., Zhang, Q., Li, Y., & Xu, L. (2016). Edge computing: Vision and challenges. *IEEE Internet of Things Journal*, *3*(5), 637–646. DOI: 10.1109/JIOT.2016.2579198

Smith, R., & Johnson, T. (2022). Challenges and opportunities in rural broadband deployment. *International Journal of Rural Development*, *29*(3), 45–59. DOI: 10.5678/ijrd.2022.02903

Sohel, K. (2013). *Jon, Boone., Robert, Lambardia*. Communication Policy Frame.

Park, S., Jung, M. C., & Lee, J. Y. Enabling Sustainable Broadband Adoption in Rural Areas. In *Communication Research and Policy Conference*, 2013 (p. 12).

Koratagere Anantha Kumar, S., & Oughton, E. J. (2023). Techno-economic assessment of 5G infrastructure sharing business models in rural areas. *Frontiers of Computer Science*, *5*, 1191853.

Voinov, I. A., Chung, J., Kettimuthu, R., Bordel, B., Alcarria, R., & Robles, T. (2022, November). Towards 6G networks for rural environments: vision for improving digital inclusion through open source hardware and software. In 2022 IEEE 40th Central America and Panama Convention (CONCAPAN) (pp. 1-6). IEEE.

Van Dijk, J. A. G. M. (2020). *The Digital Divide: The Internet and Social Inequality*. Polity Press.

Wang, Y., Gao, S., & Li, X. (2024). Addressing Environmental and Health Concerns in 6G Deployment. *Journal of Wireless Communications*, *19*(2), 98–107. DOI: 10.1007/s11277-023-09123-1

Wang, Y., & Zhang, X. (2024). Regulatory Challenges in B5G and 6G Implementation. *IEEE Communications Magazine*, *62*(1), 36–42. DOI: 10.1109/MCOM.2024.1234567

World Bank. (2023). Public-private partnerships for broadband infrastructure. https://www.worldbank.org/en/topic/ict/brief/public-private-partnerships-for-broadband

Xia, X., Zhang, S., Zhang, Y., & Wu, L. (2021). Beyond 5G (B5G) Wireless Networks: Key Technologies, Applications, and Challenges. *IEEE Network*, *35*(4), 40–46. DOI: 10.1109/MNET.011.2100130

Zhang, C., Patras, P., & Haddadi, H. (2019). Deep learning in mobile and wireless networking: A survey. *IEEE Communications Surveys and Tutorials*, *21*(3), 2224–2287. DOI: 10.1109/COMST.2019.2904897

Zhang, Y., Li, L., & Yang, J. (2023). Dynamic spectrum sharing for next-generation wireless networks. IEEE Communications Magazine, 61(1), 45-52. https://doi.org/ International Telecommunication Union (ITU). (2022). Global spectrum management. https://www.itu.int/en/ITU-R/terrestrial/fmd/Pages/default.aspxDOI: 10.1109/MCOM.2023.XXXXX

Zhang, Y., & Wang, X. (2023). Standardization and Interoperability in B5G and 6G Networks. *International Journal of Wireless Information Networks*, *30*(1), 12–23. DOI: 10.1007/s10776-023-00567-8

Chapter 5
Addressing Network Connectivity Issues in Rural Areas:
The Role of B5G and 6G

Dankan Gowda V
https://orcid.org/0000-0003-0724-0333
Department of Electronics and Communication Engineering, BMS Institute of Technology and Management, Karnataka, India

Fayzieva Makhbubakhon Rakhimjonovna
https://orcid.org/0000-0001-8835-9330
Tashkent State Pedagogical University, International Islamic Academy of Uzbekistan, Uzbekistan

R. Yogitha
Department of ISE, Atria institute of Technology, Bangalore, India

P. Veeramanikandan
https://orcid.org/0000-0002-8253-8730
Department of Electrical and Electronics Engineering, Dhanalakshmi Srinivasan College of Engineering and Technology, Chennai, India

Anil Sharma
https://orcid.org/0000-0002-7115-6278
Amity School of Engineering and Technology, Amity University, Noida, India

ABSTRACT

Different B5G and 6G technologies are expected to offer a solution to some of the existing problems with the network connectivity in rural regions. This chapter examines the proactive implications of B5G and 6G in mitigating such challenges with emphasis on their capability in providing rapid, reliable, and ubiquitous internet access in rural areas. We pay special attention to new technologies supporting such next-generation networks as more advanced antenna systems, artificial intelligence, and various spectrum-sharing strategies. Using best cases, examples, and pilot cases, we present and discuss success-

DOI: 10.4018/979-8-3693-7788-8.ch005

ful pro-active strategies and use cases for B5G and 6G in rural areas. Moreover, the chapter analyses and considers the further consequences of extended connection such as the changes in the healthcare sector, education sphere and the economic advancement. We also evaluate the challenges that hinder implementation including infrastructural drawbacks, regulatory factors, and funding matters; we provide possible methods of addressing the challenge.

INTRODUCTION

Background on Rural Connectivity Issues

Geographically, it is a well-documented fact that inhabitants of countryside around the world have not been provided good networks until now. This digital divide, therefore, poses deep concern as it results to economic inequalities, inequalities in access to education, healthcare and overall social exclusion. If the ultimate goal is to look at the hub or the center to analyze why connectivity and interconnectivity is a constant issue, then their roots are complex (Mary, A., Begum, A. Y., and D. G. V., 2023). Issues of infrastructure are real; the probabilities of implementing wired and wireless networks in rural areas that are thinly populated and sometimes characterized by difficult terrains are very expensive. As a result, these service provid‹ers have no great amount of inducement to invest in these regions (Zaidi, A., Athanasiou, G., Mahmood, K., Teyeb, O., Medbo, J., and Wikström, G., 2020). Furthermore, higher generations of network technologies, including 3G and early 4G are not capable of offering enhanced coverage and bandwidth sufficient for new generation applications and services.

Figure 1. Internet penetration rate in rural vs urban areas

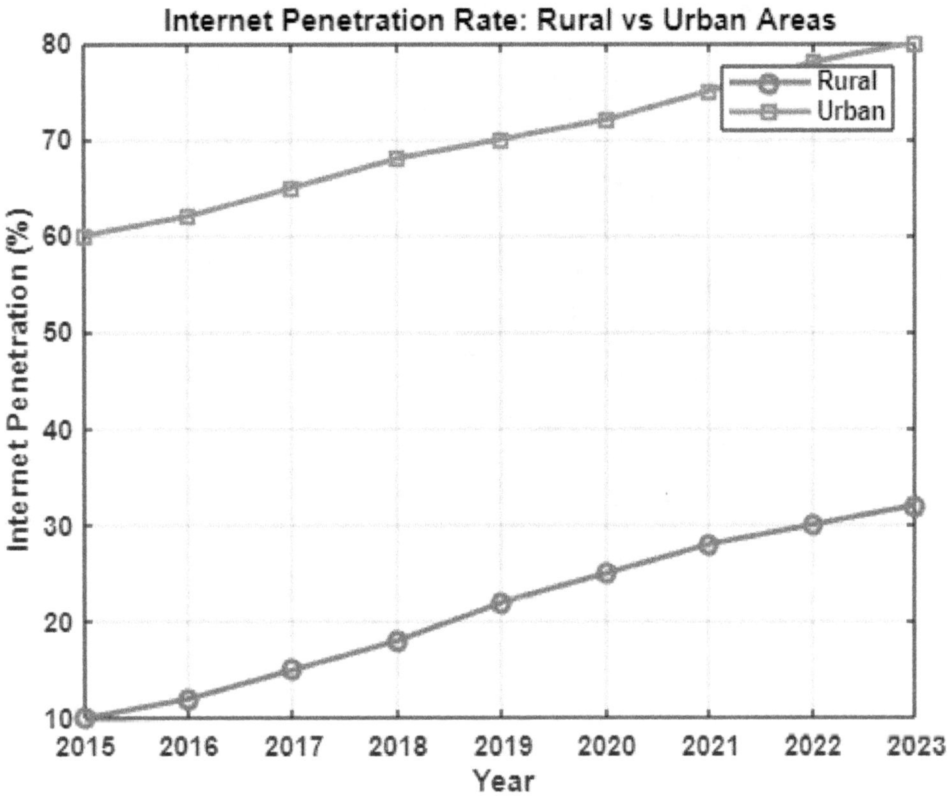

Figure 2. Components of B5G and 6G technologies

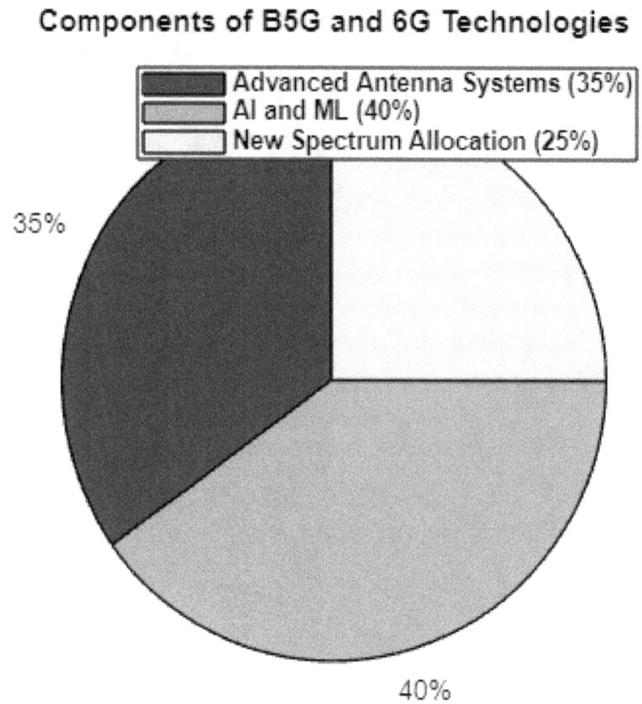

As seen in the figure.1. below, the growth rate of internet usage in rural areas is much lower than that in urban areas in the course of 2015 to 2023. The figure on the X-axis is the years whereas the figure on the Y-axis is the percentage of internet per capita. Thus, the figure explains the penetration rates difference between rural and the urban areas whereby the latter has higher penetration rates than the former (Pan, C., Ren, W., Wang, J., Elkashlan, M., Wang, J., and Hanzo, L., 2018). This visualization also emphasises on the fact that the digital divide is still pretty much evident and there is certainly a need to have niche policies and measures to ensure that rural areas get adequate connectivity.

Thus, pie chart in Figure.2. shows the ratio of specific components to the overall complexity of B5G and 6G technologies. These components are; Advanced Antenna Systems, AI and ML, and New Spectrum Allocation with their relative importance levels of 35%, 40%, and 25% respectively. The figure below offers insight into the various technologies that will shape the next generation networks with special focus on AI and spectrum management.

Importance of Addressing Digital Divide in Rural Areas

It is important to solve the problem of the digital divide for the following reasons. First of all, universal availability of internet connectivity is one of the most important prerequisites of today's society and affects education, healthcare, and employment and social opportunities. For instance, students in the rural areas perform poorly than their counterparts in urban areas, this is because they are not exposed to the internet resources such as online learning tools (Chandra, G., Kumar, R., and Prasad, K. D. V.,

2024). In healthcare particularly telemedicine and other digital health services which would be of great benefit are still out of reach (Guo, H., Liu, J., Tang, J., and Zhang, J., 2020). Economically, small business located in the rural areas are unable to compete effectively of employment opportunities and economic growth due to inadequate internet connectivity. There are more general social benefits in eliminating the digital divide since it enables society's marginal groups, especially in the rural areas, to integrate into digital society and economy.

Overview of B5G and 6G Technologies

Beyond 5G (B5G) and 6G are the future generation of the mobile communication technologies with the goal to provide a whopping connectivity speed, reliability and capacity. Further advancements made by B5G are increment in data rate and spectral efficiency and decrement in latency than the current 5G standards. These improvements are realized with newer technologies such as Massive MIMO (Multiple Input Multiple Output), new and better beamforming methods and better utilization of the available spectrum (Akyildiz, I. F., Han, C., and Nie, S., 2018). On the other hand, what is yet in the research and development stage is the 6G and it hopes to take the capabilities to the next level. The RAN Planning provides insights for constructing 5G to become a completely and continuously wireless networked world using terahertz frequencies, AI for the network optimization and readiness for further future technologies such as quantum computing and blockchain (Qadir, J., Ali, A., and Saleem, A. M., 2021). The actual benefits that B5G and 6G are to bring to the society are high-speed low-latency and reliable connectivity even to the areas that are still not covered by Internet.

Objectives and Scope of the Chapter

This chapter shall therefore seek to assess the possibilities of full realization of B5G and 6G technology to provide the network connectivity in the rural areas that has been an issue for a long time. Thus, in an attempt to understand the ability of B5G and 6G to close the digital gap, this paper focuses at identifying the technological advancements supporting these new generation networks (Hegge, K., and Agrawal, R., 2023). To illustrate how these technologies can be well implemented in rural areas, this chapter will explore many facets of BWA such as; advanced antenna system, artificial intelligent network, management and how new spectrum allocation strategies can be implemented (Ferrag, M. A., Shu, L., Yang, X., Derhab, A., Maglaras, L., and Janicke, H., 2019). In addition, the chapter will present examples of best practices in B5G and 6G transition with the focus on rural zones, as well as pilot projects. These case scenario will offer a real life perspective of the implementation issues and best practices related to these technologies.

However, the chapter will also explore various benefits of increased connection in some sectors in the rural regions with emphasis to the sectors like the health sector, education sector, and the business sector. By studying the effects of improved networking, it is our intention to explain the potential of B5G and 6G, the innovations that we are going to describe, and how they can also be valuable to those individuals living in rural regions (Giordani, M., Polese, M., Roy, A., Castor, D., and Zorzi, M., 2019). This will enable the discussion to also highlight the challenges likely to be faced in implementation, such as inadequate infrastructures, restrictions in laws and policies, and lack of adequate finance and funds, among others, and come up with measures of handling them (Polese, M., Giordani, M., Roy, A., Castor, D., and Zorzi, M., 2020). By so doing, this chapter shall offer guideline on how policymakers, service providers and community stakeholders can leverage on the B5G and 6G technologies to close the rural digital divide.

TECHNOLOGICAL INNOVATIONS IN B5G AND 6G

Beyond 5G and 6G are innovations that take a giant leap towards mobile communications technology with the aim of catalyzing change in connectivity especially in rural regions. These innovations comprise of several advances that will in one way or the other help in boosting performance, coverage, and efficiency in the network (Reddy, M., and Tanguturi, R., 2022). This section explores these technological advancements and the importance as well as uses of these technologies.

Advanced Antenna Systems

Among the most significant concepts discussed in B5G and 6G, the preparation and design of new-generation antenna systems is primary for solving the problem of enhancing network densities. These systems include massive MIMO (Multiple Input Multiple Output) and other wise beamforming techniques the both of which are central to the operational Output performance of the next generation networks.

Massive MIMO (Multiple Input Multiple Output)

MMIMO is an advancement of the conventional MIMO system which greatly increases the number of antennas at the transmitter and/or the receiver sides. This is due to the fact that the present system will also have a much higher number of telescoping antennas which will allow the system to transmit multiple data streams at once, thus greatly increasing the network's capacity and data rates (Hussain, N., Dixit, S. K., and Gupta, A. K., 2023). The problem of low density of users is well solved in the case of Rural connectivity where Massive MIMO provides reliable and high-capacity connection even over the large distances (Vo, N. N., Le, T. A., and Tran, N. H., 2018). By incorporating many antennas the system is able to create small beam patterns that probe towards the specific users hence reducing interference and improving quality of service. This capability is very helpful most especially in areas where there are such barriers like physical barriers and long distance to be covered.

Figure 3. Performance improvement with massive MIMO

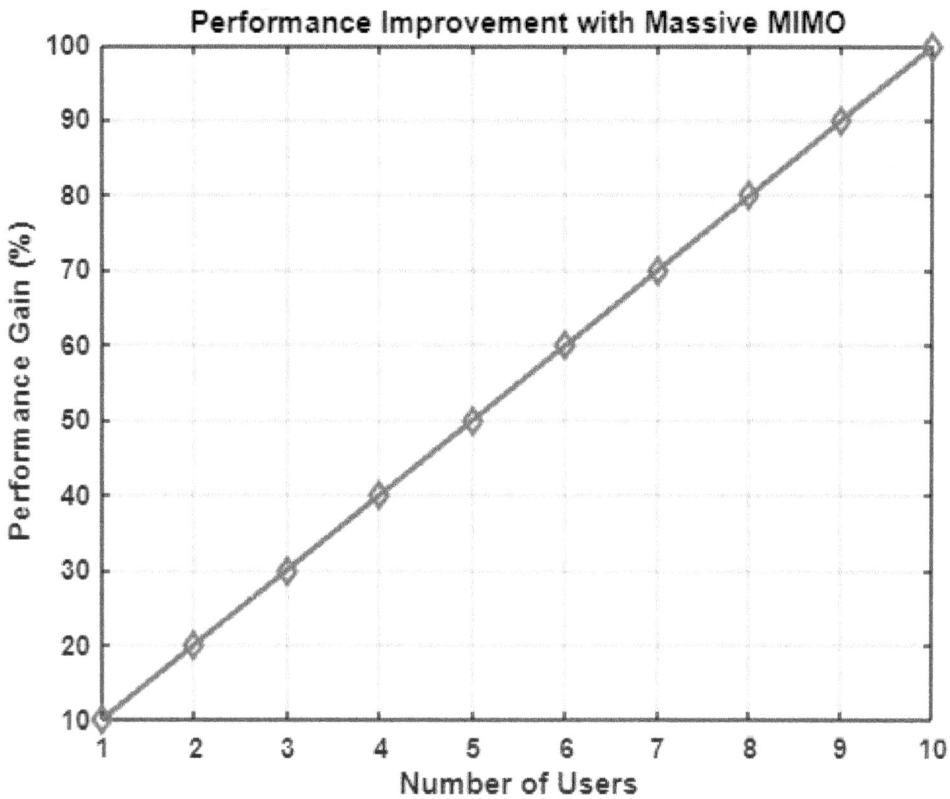

Figure 4. Spectrum utilization efficiency with beamforming

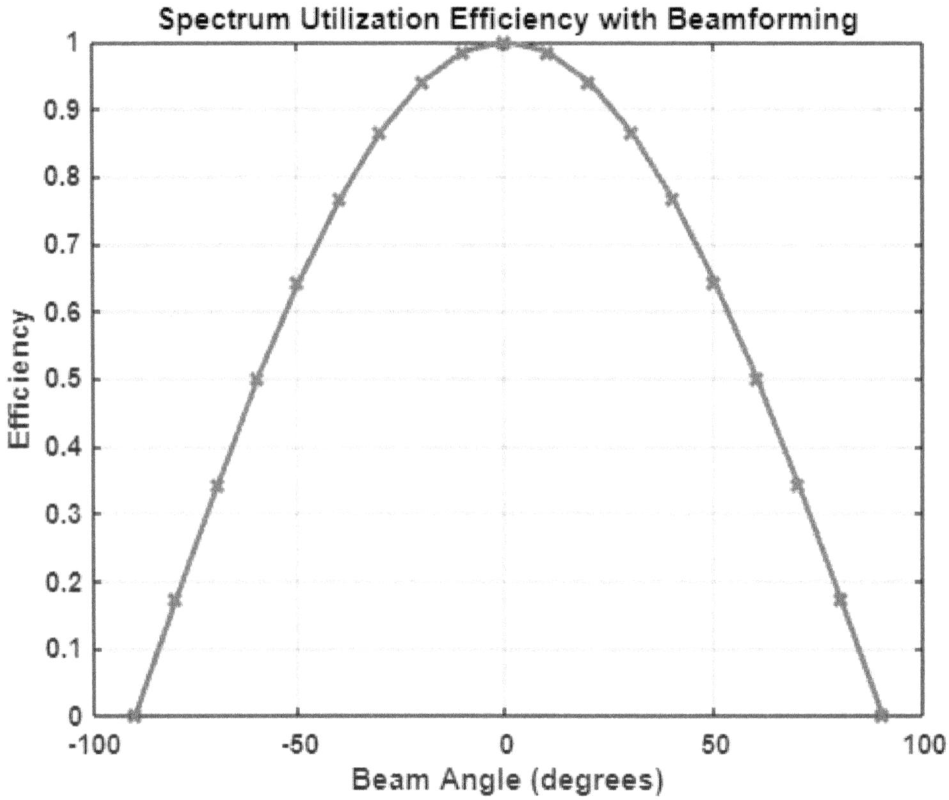

Figure 3 illustrates the performance improvement which is gained by using the Massive MIMO technology for a large number of users (Reddy, N. S., and Ashreetha, B., 2023). It is noteworthy that the number of users is plotted at the x-axis, whilst the y-axis denotes the relative improvement measure in percent (Premkumar, S., Hussain, N., and Chinamuttevi, V. S., 2023). The increase in the performance gain as the number of users increases shows the effectiveness of the Massive MIMO in increasing the capacity of the network and data rates hence making it fundamental in the B5G/6G implementation.

The spectrum utilization efficiency through the beamforming techniques is represented in Figure 4. On the horizontal axis, the x-axis is the beam angle in degree as a variable while the y-axis shows the efficiency. The figure shows the pattern of cosine & shows how the beamforming can direct energy in definite directions as shown below (Chen, S., Liu, Y. J., Wang, W., Poor, H. V., and Cui, S., 2018). This technique is essential for enhancing access and at the same time attainable in B5G and 6G networks to reach even the difficult to reach rural areas.

Beamforming Techniques

Another important technology that goes hand in hand with Massive MIMO is beamforming through which signal is pointed towards the recipient rather than sending it in all directions. This will improve on reception, selectivity, cleanness, intermodulation limitation, and utilization of the available spectrum space (Dang, S., Amin, O., Shihada, B., and Alouini, M., 2020). Beamforming can be used to target specific areas which may be far or hard to reach by making sure that distant areas of villages have a strong signal as well as a good quality one. Digital and hybrid beamforming which are in their developmental stages in present day networks offer incremental control over the formed beams and enhance network performance.

Artificial Intelligence and Machine Learning

AI & ML are crucial for advancement of B5G and 6G technologies providing intelligent approaches to control and enhance the network. These technologies allow the networks to evolve to meet needs in today's world provided that these needs meet the need for increased efficiency and dependability.

AI-driven Network Management and Optimization

In its simplest terms, artificial intelligence network management entails the application of Artificial intelligence algorithm to constantly control, analyze and adjust the network (Mumtaz, S., Rodriguez, J., and Dai, L., 2018). They also can describe traffic flow, foresee numerous problems, and manage resources to optimize workflow continuously (Pandey, S., Samdanis, K., Costa-Pérez, X., and Guo, W., 2020). Since network resources are scarce and particularly in the rural areas, the management of such a network with the help of AI will help to get the maximum out of it by ensuring a good quality of service to the users. skills required to cope with the fluctuating demand and other related challenging factors that are inherent in rural operations.

Predictive Maintenance and Fault Detection

Another important set of functions includes predictive maintenance and fault identification in B5G and 6G network environments. Since this technique involves the constant observation of the network components while analyzing the performance data, AI is also able to forecast when and where these failures are most probable to happen (Shekhar, S., and Ashreetha, B., 2023). Such a strategy ensures that maintenance and repair works are done before there is a problem, hence the elimination of disruptions. In rural areas where acquiring network infrastructure can be a nightmare, predictive maintenance will substantially minimize the use of physical means thereby enhancing the quality of services at lesser costs.

New Spectrum Allocation Strategies

Optimal utilization of the radio frequency is essential in preparing for B5G and 6G networks' alike. These technologies bring in fresh ways of achieving spectrum refinement whereby mmWave and terahertz bands, as well as dynamic spectra sharing, can be adopted to meet the growing bandwidth demand.

Utilization of mmWave and Terahertz Bands

Specifically, the mmWave (millimeter wave) and the terahertz bands provide far more bandwidth than the conventional frequency bands; consequently, they offer way more data rates and network capacity. This utilization of high frequency bands is one of the major characteristics of B5G and 6G technologies since they allow covering the required spectrum for the growth of the connection speed (Barua, T., and Mohiuddin, K., 2024). In rural settings, mmWave and terahertz bands can be utilised for the establishment of high-capacity backhaul links: to interconnect isolated base stations with the centralized network. As a result, these high-frequency signals often have limited range, greater exposure to physical barriers. These disadvantages can, however, be complimented by current developments in antennae design, and beam forming methods, to enhance their application in rural settings.

Dynamic Spectrum Sharing

Another innovative method to increase spectrum efficiency is dynamic spectrum sharing, through which several users can occupy the different parts of a particular frequency range simultaneously (Maddineni, V. K., and Gummadi, V. A. K., 2024). This approach applies real-time algorithms for purpose of efficiently distributing spectrum depending on demand and practicalities of a given network. In light of this, the DSLS can be used in the rural areas whereby the availability of spectrum might be a major issue through the use of the DSS, more users can accurately be served with connection. Such flexibility is especially important in situations where several types of services like, for example, mobile broadband and IoT applications should be allowed to exist in parallel to each other and share the same bandwidth.

Therefore, based on advanced technological developments in B5G and 6G such as advanced antenna systems, AI-based network management, and new strategies of spectrum allocation, the controlling the issues related to rural network connectivity can be effectively done (Srinivas, V., and De, T., 2023). Taking these advancements the communities that are currently lacking adequate internet connection can be provided with high speed, reliable, and accessible internet facilities, which can provide supportive framework for economic and social integration.

DEPLOYMENT STRATEGIES FOR B5G AND 6G IN RURAL AREAS

Infrastructure Considerations

The implementation of B5G, and 6G technology in rural regions of the world is not an easy task because of the peculiarities of infrastructure. The first factor that has to be taken into consideration is the establishment of an adequate network that can deliver communications over vast areas, including scarcely populated territories. Large cell towers and vast fiber optic networks used in traditional infrastructures may not be effective when implemented in the rural areas since the investments are expensive and physically cumbersome (You, X., Wang, C., Huang, J., Gao, X., Zhang, Z., Wang, M., Huang, Y., and Zhang, C., 2021). Therefore, deployment strategies have to be more cost-driven and opt for more challenging solutions like the use of small cells and infrastructure sharing. Small cell technology for example has the potential of greatly improving coverage and capacity in the rural areas because it involves the deployment of mini base stations with low power that are relatively cheaper than the conventional macro cell stations.

Figure 5. Cost comparison of infrastructure types

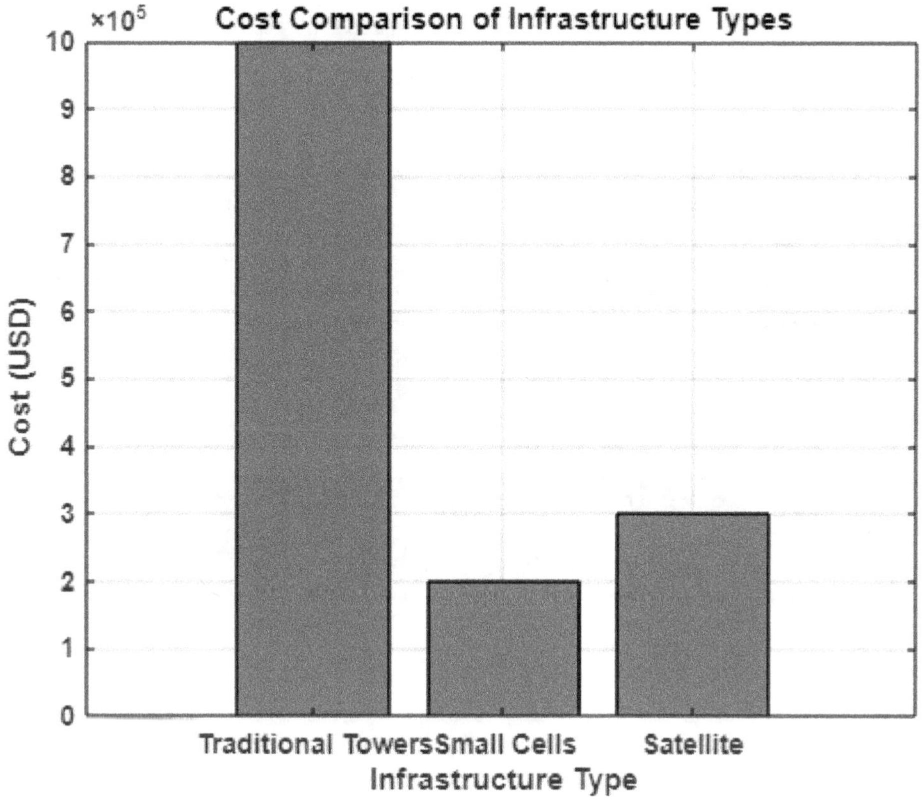

Figure 6. Successful pilot projects in rural areas

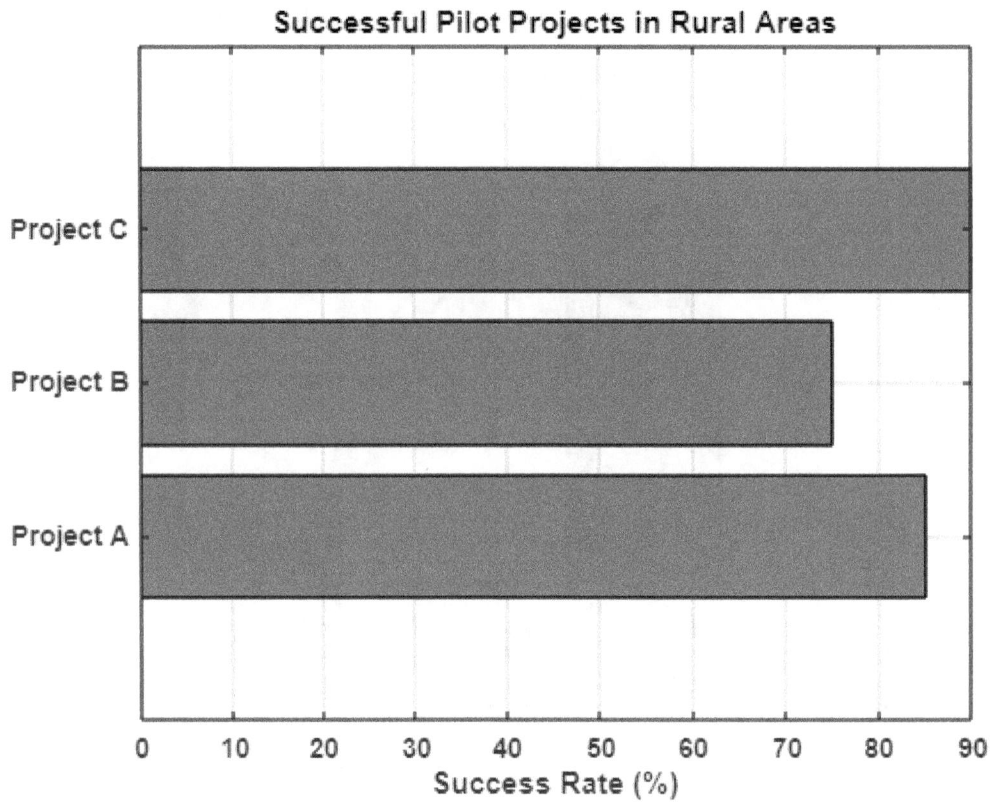

The Figure.5. below presents the cost of various types of infrastructure for network deployment in the rural areas. The x-axis lists the infrastructure types: Traditional Towers, Small Cells, and Satellite are placed on the x-axis while the cost are given in US Dollars are placed on the y-axis. The bar chart also demonstrates that the deployment cost of small cells and satellites are lower than that of traditional towers showing that there was the need to consider new deployment strategies for this infrastructure.

Figure.6. presents the Package 2 that presents the success rates of different pilot projects using B5G & 6G in rural areas. Finally, the y-axis provides the names of the project i.e., Project A, B, and C while the x-axis indicates the success rates in percentage. Using the horizontal bar chart below, one is able to understand which particular projects are effective and learn different working models aimed at improving rural accessibility.

Small Cell Deployments

Small cells are seen as an integral part of the infrastructure evolution relating to B5G and 6G for remote regions. These are low-power transmitters that can be mounted on streetlight columns, buildings, while others can actually go on even the walls, this is perfect for areas where traditional macrocellular stations cannot be built such as downtowns or in areas of high-density tall structures. The flexibility of having the small cells can enable planning of the network in a much finer scale to the geographical environment that characterizes most of the rural areas (Mao, Y., You, C., Zhang, J., Huang, K., and Letaief, K. B., 2017). In the light of scattering of geographic area, and hindrances created by buildings and other structures, a high number of small base station can solve all these problems and keep on providing service to areas where all other technologies fail. In addition, small cells are extendable with newest technologies like Massive MIMO and beamforming which can improve its capabilities, so small cell is a suitable solution for providing broadband internet in rural areas.

Integration with Existing Infrastructure

Another systematic approach that should be employed to ensure the deployment of these technologies is by finding ways to incorporate B5G and 6G solutions with existing constructions. One way of mitigating this problem is by using already available structures and utilities like power lines, transport infrastructure, and community amenities while building the infrastructure that supports micro-grid operation. For instance, small cell antennas can be installed on existing utility poles and street lights and Fiber optic cables along existing road ways/rail ways way for establishing high capacity backhauls. Further, the integration with other community infrastructures like; schools, health centers and even local government operation can also help extend the connectivity to important public service, thus increasing the outreach of the given deployment. Besides, it reduces the demand for infrastructure investments, as well as speeds up the deployment time, and grant rural populations faster access to the superior connectivity.

Pilot Projects and Case Studies

The real-world pilot projects and case studies are useful for the evaluation of the real-world issues and accomplishments related to the B5G and 6G adoption in rural areas. When reading through these examples, such stakeholders should be able to learn from such examples to understand what to do and what not to do in future similar circumstances. Effective pilots include a set of original technologies, and cooperation with relevant partners and deployment methods that consider rural area characteristics.

Analysis of Successful B5G and 6G Rural Deployments

A number of pilot implementations in different countries have shown that B5G and 6G technologies can help to change the situation with rural connectivity. For example, in the US, Rural Wireless Broadband Initiative has effectively installed small cell networks to areas that was formerly unavailable for 4G broadband facilities. In Europe for instance, 5G RuralFirst has investigated how 5G tech can be used to advance agricultim and other economic activities related to productivity in rural settings and this has shown potential. All these and other examples effectively demonstrate that in order to use technologies and various types of infrastructure effectively in rural areas, it is necessary to avoid the tendency to apply ready-made cross-country templates and develop an individual approach for each specific case and country while considering all potential prospects and challenges.

Lessons Learned and Best Practices

The following sections discuss some of the issues highlighted from the pilot projects undertaken and some of the guidelines that have been considered important. They are primarily local communities who have to approve deployment of the rigs together with government agencies and independent private sector partners in the planning and execution stages. Second: the application of phased large-scale approach, where the pilot introduction of the technologies or some aspects of the strategies enables the identification of the efficiency and effectiveness of the methods to be applied. Third, the use of technologies in networks like the use of the AI in network management, dynamic spectrum sharing can go along way in improving the current rural networks in as much as they will be made to support the increasing user traffic. Last but not the least, achieving sustainable financing and investment is crucial; it has imperative to a paradigm shift with good approaches to innovative methods and effective Public-Private Partnership as source of long-term infrastructure funding.

Collaboration with Stakeholders

Stakeholders' engagement is a critical component in the implementation of B5G and 6G technologies in rural areas. Most of the successful projects entail the involvements of different parties, which may include the government, different companies, and even the people living around the project area. All of these groups have their proper place in solving the issues as well as managing the prospects linked with developing rural connectivity.

Role of Government, Private Sector, and Local Communities

It is necessary for the governments to understand the significance of enabling B5G and 6G networks in the policy, regulatory and funding frameworks. Some of the other critical factors include the availability of regulatory framework that promotes the allocation of spectrum, ease of permit and licensing as well as incentives for rural deployment. Furthermore, concerns such as state funding and subsidies to cover the high expenses involved in implementing infrastructure in the rural areas provide easy ways for investors in the service to be attracted. Telecom operators/Service providers, IT/Telecom equipment manufacturers/ Solution providers, ISPs, EPCs/IMCs and other infrastructure partners possess the required technical and financial capabilities required for designing, building and managing Global Class Networks. Private companies also present competition hence improving on the service delivery and reducing the costs of the service to the rural users. On the other hand, the local communities are the stakeholders in the deployment of the programs, and they help to bring in the understanding of their needs and preference of rural areas. Participation by the community make the solutions developed to fit the locality well and also makes the people to have a sense of ownership of the planned solutions.

Public-Private Partnerships and Funding Mechanisms

One of the most effective strategies for funding and implementation of B5G and 6G in rural regions is the use of public-private partnerships. Obviously, PPPs are capable of integrating the advantages of both sectors with the help of supported public funding and policies and the experience and effectiveness of the private sector. There are several forms of PPPs that can be used and include: co investment and joint business ventures, concession agreements and performance-based contracts. Besides PPPs, other innovative sources of funding including community-based financing, crowdfunding and impact financing will also play a key role in mobilising the required capital to support rural connectiveness initiatives. Besides offering the monetary needed for exploring infrastructure improvement, they spur on general public accountability and of course sustainability of these true benefit as being connected.

Consequently, the use of B5G and 6G technologies in rural areas is has to be guided by the integration of infrastructural initiatives, reuse of available assets, reference to real-life pilots, and stakeholder engagement. By implementing theses strategies the barriers to the connectivity of the rural areas can be eliminated, and the creativity of next-generation networks can be unleashed to provide people with the opportunity to improve their economic and social status and to get closer to the digital frontier.

REAL-WORLD APPLICATIONS AND IMPACT

Healthcare Improvements

It can indeed be said that the introduction of B5G and 6G technologies in healthcare, especially in the territory of the rural areas of EU member countries, can dramatically change the methods and approaches to the delivery of the medical services. Perhaps, the most significant and highly implemented application is the telemedicine which offers consultations, diagnosis, and treatment plans, as well as remote care. This is especially so in the rural areas where health centre's and specialists are scarce or sometimes non-existent. Real time video consultation that is offered when there is B5G and 6G network allows patients to interact and seek medical advice from the comforts of their homes without having to travel long distances to see a doctor. Remote diagnostics is another great innovation that has been witnessed in the industry. With improved telecommunications, rural health care practitioners can send clear MRI scans cross-sectional images and other information to specialists in large cities for proper reading and thereby arrive at the right diagnosis faster.

Figure 7. Impact of telemedicine on healthcare access

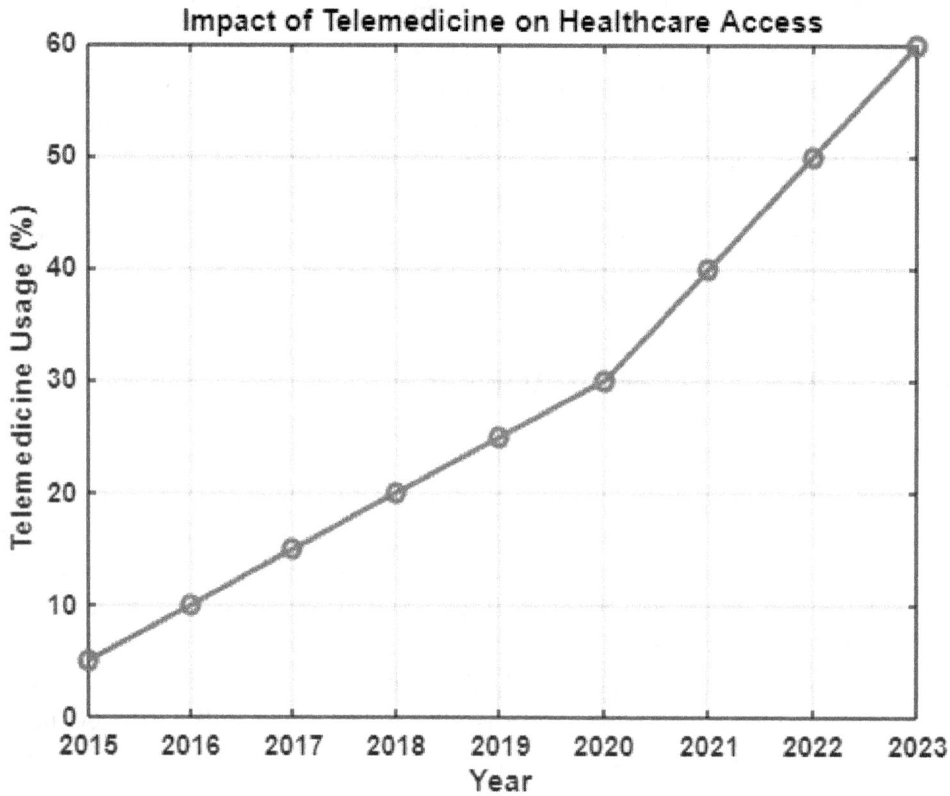

Figure 8. Growth of e-commerce in rural areas

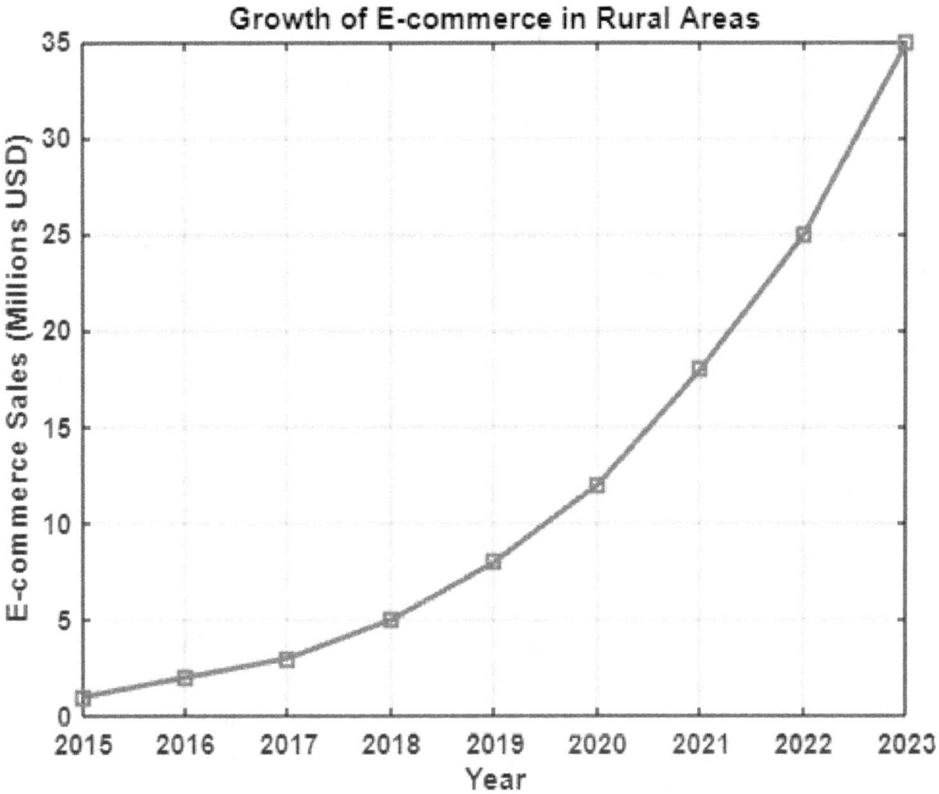

By referring to figure 7, there is a constant increase in the usage of telemedicine in the provision of health care services from 2015 to 2023. The x-axis indicates the years while the y-axis denotes the proportion of health care services that have employed telemedicine. The heightened adoption of telemedicine technology the revelation for the use of telemedicine services in offering adequate health care services in rural regions through improved network standards of B5G and 6G.

The increase of e-commerce sales in rural areas for the years 2015 to 2023 has been displayed in figure.8. The 'X' axis is labelled years while the 'Y' axis illustrates the sales in terms of millions of USD. The gradual rise in e-commerce point out continued development of the online market particularly in rural areas due to enhanced internet access and growth of online selling strategies.

Besides, the development of B5G and 6G forms contribute to telemonitoring and emergency communications in the sphere of healthcare. Wearable devices and sensors allows the health condition of the patients to be captured by the healthcare providers with real time updates on his/her status. Such a constant supervision helps identify any health risks at an early stage and makes relevant treatments hence enhancing patient success. Also, in emergencies, effective communication structures allow emergency response teams, get to share information and promptly respond accordingly. Rural people stand to benefit from this capability as emergency services providers quite often are less well staffed and responses may take longer.

Educational Access

The influence of these new technological advancements such as B5G and 6G in education is very significant especially in the rural setting where channel resources in education may be very limited. Thanks to the high-speed internet, online education and other forms of lessons prepare students to get in touch with the tremendous amount of information and mutual interaction. This democratization of education makes it possible to educate every individual and particularly the rural students in the same way like the urban students. HD video and interaction are resourceful, making virtual classrooms as rich and productive as a traditional classroom where a teacher is freely conveying the information or freely encouraging students' participation.

Furthermore, 5G and B5G as well as 6G leads to increased access to educational books and e-libraries. Students and educators in the rural areas can access plethora of digital text books, academic journals and multimedia which was out of reach earlier. The access in this regard, expands their learning base, contributes to the research work and enriches the learning process. More specifically, e-libraries are considered a wonderful source for a continuous knowledge accumulation and for the improvement of specialists, which will help inhabitants of rural areas to be aware of the latest knowledge.

Economic Development

Both B5G and 6G technologies are instrumental in the advancement of local industries and start-ups for rural communities. The improved connection helps the small and medium-sized enterprises (SMEs) embrace the online tools and markets to expand their businesses and adopt techniques that would improve their efficiency. For example, in the areas of business, solutions such as inventory management, accounting, or customer relationship management can be introduced using applications in the clouds due to poor connectivity. The application of these technologies can enable rural business organizations to run efficiently, compete more aggressively and expand its operation in a sustainable manner.

The B5G and 6G technology has a large influence on e-commerce too. Businesses located in rural areas can open their own websites for selling products and can engage in online selling platforms hence reaching clients from different parts of the world. They are able to control the internet sales effectively and selectively, offer prompt live support to the clients, and engage into various internet promotional procedures to attract more customer's attention. This connectivity creates economic returns as it contributes to the local economy besides driving innovation and entrepreneurship growth in that members of the rural region are able to start new businesses and create employment around the region.

In addition, digital markets anchored on B5G and 6G technologies enable the marketing of food produce and other goods or services produced in the rural areas to urban or even international markets . For instance, farmers are able to sell their agricultural produce directly to the customers or to other companies thus avoiding the middlemen and earning more of the profits. B: The artists and the craftsmen to an extent are able to display their products to a larger market hence they are able to gain recognition and make more sales. It is necessary to recognize that the integration of advanced connectivity also contributes to the enhancement of the growth of tourism in rural regions, as digital marketing and online platforms to book a place to stay or attraction persuade people to travel to a particular location and promote services and tours in the region.

Hence, the practical implementation of B5G and 6G in healthcare, education, and economic development will help evolve rural areas by solving existing issues and creating new opportunities for enhancement and development. They therefore improve healthcare delivery and accessibility since they allow telemedicine, remote diagnostics as well as constant health monitoring. In education, they facilitate equal opportunity in accessing books, materials and other learning resources and or facilities, gaps in education are closed. On the economic level, they contribute to the strengthening of local actors and markets, to more entrepreneurship, and to the development of digital places allowing rural producers to market their products in the world. These together make the progress that will help the rural communities to be part of the digital society and enjoy the fruits of innovation.

CHALLENGES AND OBSTACLES TO IMPLEMENTATION

Infrastructure Deficiencies

There are limitations constraining the growth of B5G and 6G technologies in the rural areas for instance, the infrastructure required to support these technologies is scarce. A major problem is the issue of inadequate electricity supply in many districts especially in the rural areas. Stable and constant power supply is a requirement for effective functioning of complex telecommunication networks however rural regions frequently suffer from power blackouts or lack adequate physical infrastructure for electric supply. This limitation not only hampers the functionality and durability of the network equipment but also increases the operational overhead since other power backup systems like generators or solar power, have to be procured and managed.

Another key infrastructure problem is backhaul which is required for connecting rural networks to other networks that make up the overall internet. Backhaul involves other local connecting paths between local area networks and the core network where values are normally offered by fiber optic cables or microwave links. With regards to fiber optic backhaul in the rural areas, the deployment is most of the time expensive since the distance and terrain topography is a limiting factor. Microwave links although are more practical in some cases, but are also constrained as to their capacity and reliability. This inhibits the rural network's data transmission since backhaul connectivity is often lacking and provides slower internet connection and inferior services.

Geographical and environmental features also pose as barriers to the implementation of B5G and 6G networks in rural extents. Rurality is defined by a variety of factors and the rural can be portrayed as mountainous, heavily forested, desolate or plain countryside. Some of these geophysical aspects interfere with signal transmission and also influence the placement of network architecture. Also, the threats include storms, strong precipitation and winds, as well as the negative impact on network facilities. Rural areas itself are also geographically dispersed, which makes it even more difficult, and expensive, to deploy and even maintain the network infrastructure.

Figure 9. Power supply stability in rural areas

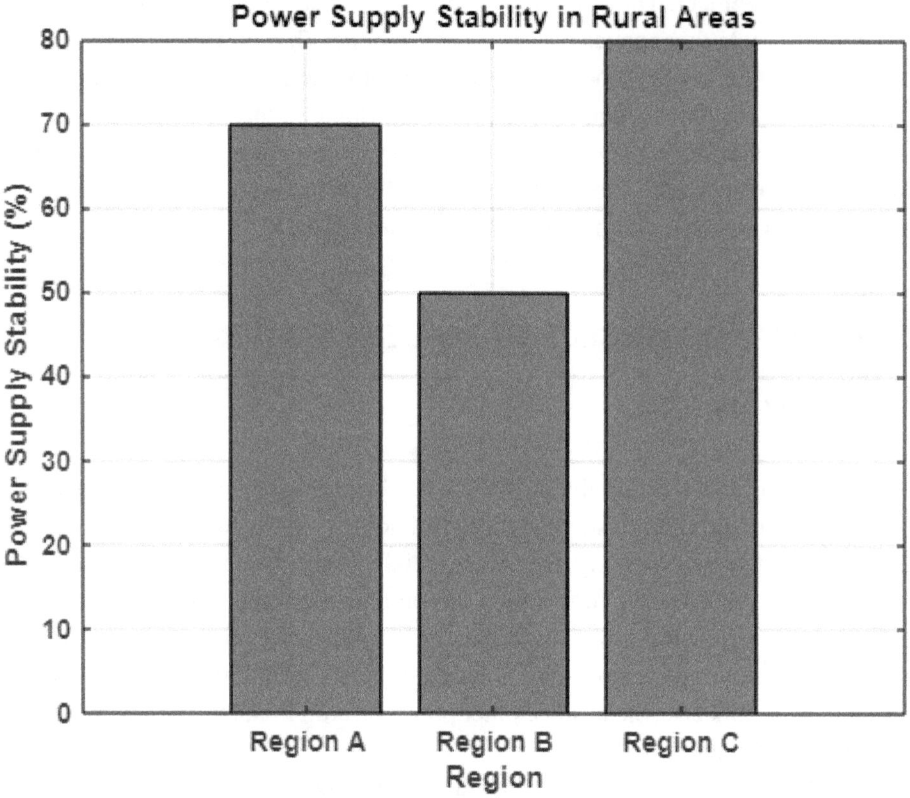

Figure 10. Regulatory approval times for network deployments

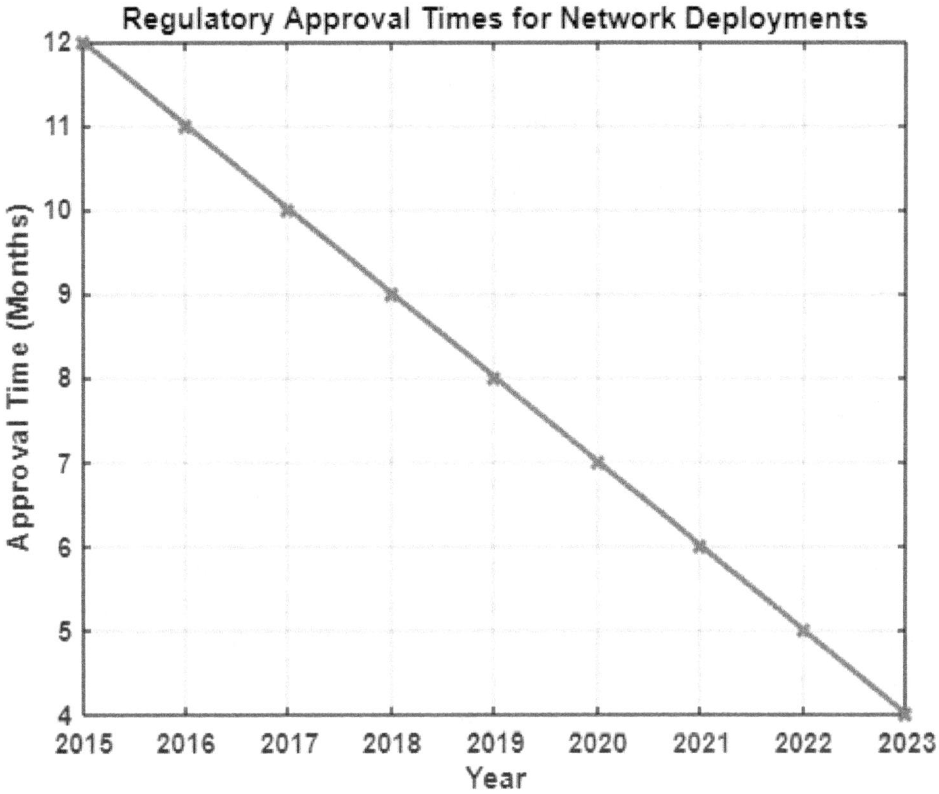

Figure.9. presents the relative stability of power supply of the different rural zones. The x-axis shows the regions namely Region A, Region B and Region C while the y-axis represents stability percentage. This is evident from the bar chart below which shows the nature of fluctuation in the power that is supplied to implementing rural telecommunication networks demonstrating a key constraint likely to hinder the provision of definitive base within Which to set up the complex and sophisticated communication networks.

The time taken in getting the regulatory approval for the network deployments is depicted in figure.10. for the years 2015 to 2023. On the x-axis is the years' scale and the y-axis shows the time taken for approval in months. This obviously indicates enhanced regulatory processes as depicted by the decreasing trend in approval times though there is still room for improvement to allow network rollouts in rural areas to happen faster.

Regulatory Hurdles

Challenges such as the existing regulatory framework is another major challenge that hinders the possibility of deploying B5G and 6G technologies in rural areas. The following areas can be cited as the main challenges which come with regulating and licensing spectrum: Surveillance and issuing of the radio frequencies which are crucial for issuing of wireless communication are normally as being in the administrative jurisdiction of national regulatory authorities. Often the acquisition of spectrum license requires time, money and effort, all which can be a put off to investors in rural networks. Also, spectrum particularly in rural regions may not be easily available or may be of limited quantities or even dispersed, thereby posing a challenge in the use of sophisticated wireless solutions.

Other barriers that directly affect implementation of a rural network also include; Policy and legal issues. First of all, current regulations may not be sufficiently flexible to offer the necessary boost for the B5G and 6G networks deployment, particularly in rural areas. For example, initiatives developed for the urban environment while having their merits, will not provide for the rural requirements and challenges, for example, the need to implement versatile and adaptive solutions when it comes to constructing large-scale storage facilities. Legal barriers like zoning ordinances and issues to do with right of way also pose very significant challenges to the deployment of the network infrastructure and condemn them to delays and extra costs.

Funding Constraints

This heavily indicates the fact that one of the biggest challenges towards the deployment of B5G and 6G networks in rural regions is funding issues. This is due to the high costs that are incurred in the deployment and maintenance of these networks constituting a major factor that avert service providers and investors. Extension work require expensive investments in terms of equipments, devices and facilities for infrastructures networks especially in rural areas since the population density is low and distances long than those recorded in the urban areas. This cost difference acts as a drawback to the feasibility of networks for the commercial operators in the rural areas due to poor connectivity, therefore limited investment.

Funding and capital remains another major issue to firms seeking to source for finance and for investors seeking to invest in a firm. Most of the rural region they cannot meet the market condition and financial motivators that can facilitate network investments. This is where government funding as well as subsidies come in, but these are scarce resources that are usually bid for. Moreover, the networking rural projects financially may be questionable with the long period of investment breakage as well as unstable revenues sources hence lack of capital from the private sector.

For these reasons, the use of new funding schemes, as well as partnerships and contracts with business (PPP), are needed. As a form of delivery mechanism PPPs take the advantages of the public and the private sectors this means that while PPPs obtain the requisite funds and political support from the government, on the other hand the private partners contribute their efficiency that is often characteristic of them and their mastery of policy implementation. However, the process of developing and sustaining these relationships might turn out to be challenging, this means that various aspects have to be well coordinated and all the stakeholders involved have to be committed. Also, other forms of funding like community based funding, impact investment, and crowdfunding will help generate more resources besides encouraging ownership at the local level.

Therefore, it can be stressed that B5G and 6G technologies' implementation in rural regions is not problem-free and is accompanied by numerous issues and obstacles which must be addressed through collaboration and idea-sharing. This suggests that current infrastructure issues, such as inadequate power, lack of backhaul and geographical constraints, require solutions that are both adaptable and sustainable. Some of the challenges hence include spectrum regulation and policy that need to be fostered by friendly and efficient regulatory systems. High tends and risks of deployment costs, and limited access to investment financing necessitate inventive financing solutions and robust partnering involving public and private sectors. Responding to these challenges, which are outlined above, it is possible to score the semantically and functionally rich outcomes of B5G and 6G expend for overcoming digitally sustained divides as well as for augmenting the rural development both in economical and social terms.

SOLUTIONS AND RECOMMENDATIONS

Technological Solutions

To overcome the connectivity issues in rural contexts, the use of B5G and 6G must be complemented with new enabling technologies and solutions. One of the main avenues that is in focus includes establishment of cost effective based infrastructures. Existing telecom large scale structures inclusive of broad fiber optic networks and giant cell sites may be expensive and unfeasible for rural geography. Hence, it may be far more cost effective, reliable, and logical to employ small cell and distributed antenna systems. These smaller units can be attached to available structures such as poles, buildings or even geographical features thus reducing structures to be constructed hence ease of deployment.

There is another hopeful technological technique with satellite technology and UAVs. Satellites especially the low earth orbit can help deliver high-speed internet for those regions that are hard to connect with fiber optics and requires less supporting infrastructure. Some players engaged in this domain are SpaceX's Starlink and OneWeb that have started providing satellite-based Internet connection that can be current in even the most remote regions. Likewise, UAVs or drones can also be used to offer a temporary or back up solution whenever the ground infrastructure is either nonexistent or destroyed. Communication payloads that are installed on drones can form networks, amplify signals and also interface existing networks during calamities or other high demand periods. When used together, these technological solutions provide a strong base of an infrastructure that would help bring B5G and 6G connectivity to rural regions. This means that through effective deployment of infrastructure in a way that is adaptable and can be deployed depending on needs of an organization, it becomes possible to build networks that are efficient and affordable as well as responsive to change as brought about by innovation in the technological sector.

Figure 11. Investment in rural network deployment

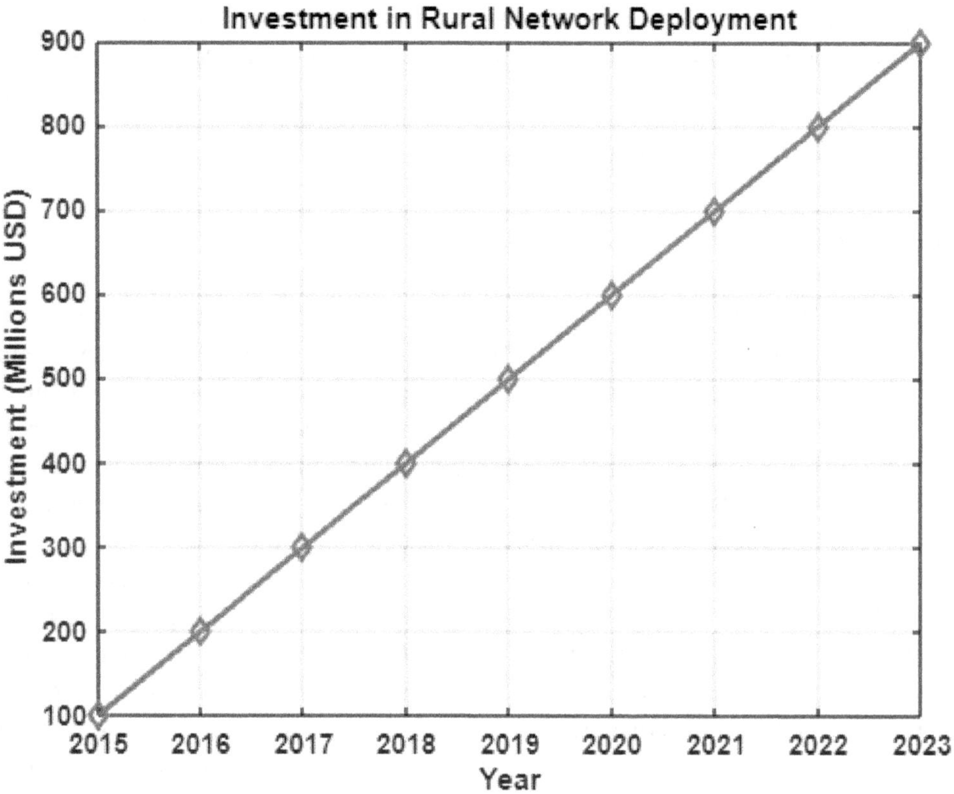

Figure 12. Digital literacy program participation

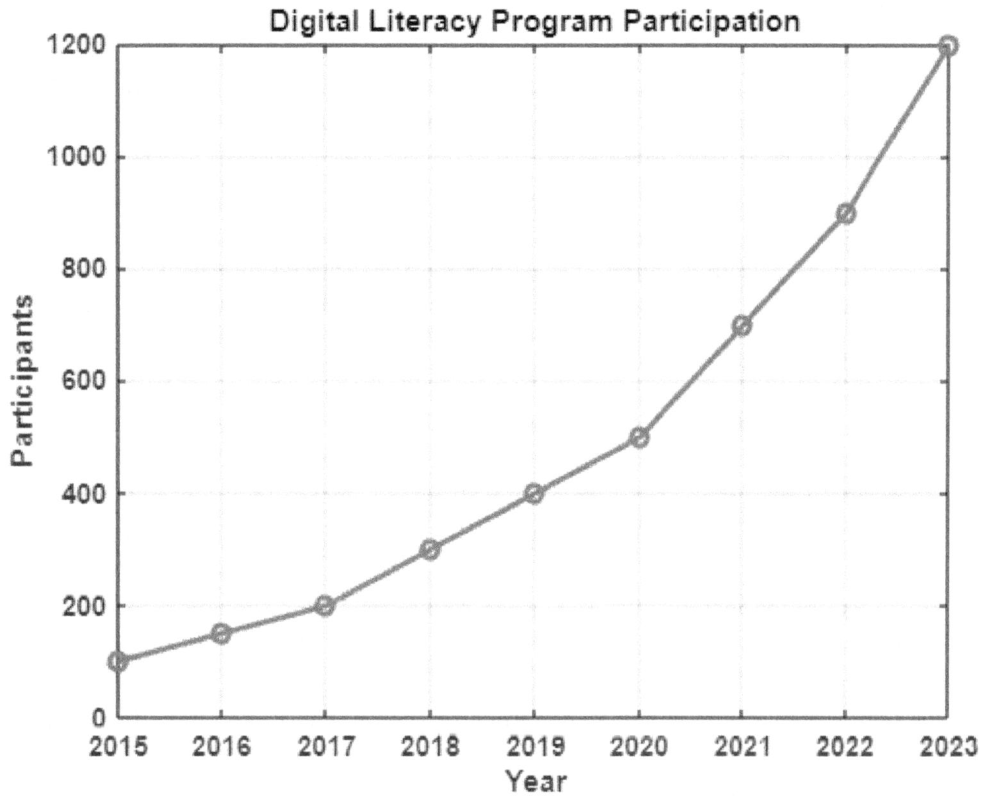

Figure 11 above shows the investment done in rural network deployments from the year 2015 to 2023. On the x-axis, there are the years, on the y-axis there is the investment in millions of USD. The figure shows that investment has increased over the years and this is a clear indication that there is need for more funding for the provision of access in the rural areas.

As depicted in Figure 12 above, use of digital literacy programmes in the rural areas was fairly low in 2015 compared to the subsequent years up to 2023. On the x-axis, there are values represented by the years while the value axis represents the number of participants. The increase in participation points to increased campaigns in the development of digital literacy as well as skills that will enable the purpose of the improved connectivity by the countryside inhabitant.

Policy Recommendations

Besides, technological solutions, the question of the regulation and policies should be considered for the successful implementation of B5G and 6G networks in rural zones. One of the ways that can be done is by rationalising the requirements that are set in legal frameworks. This means that regulatory procedures that are comprehensive and elaborate take a long time to complete hence slowing down the network deployment and are costly. International, national, and local governments as well as the regulatory agencies must ease the licensing and permitting processes to enable service providers gain access to the desired spectrum. This can range from setting up mechanisms that allow for fast track approvals to making various reforms that cut across the networking approval bureaucracy as well as providing guidelines for network setting up.

Similarly, offering appropriate incentives as well as subsidization to encourage establishment of network especially in the rural area is a crucial factor in the implementation of the strategy. Some of the major hurdles identified for private sector players include high costs involved in deployment of facilities in rural areas and inadequate reliable revenues for such investments and efforts such as tax exemptions, grants, and other subsidies can assist in overcoming such challenges. These subsidies can also be aimed at particular activities or segments of network deployment including infrastructure construction, hardware purchase and general expenses. Moreover, governments can set up universal service funds USFs, which will centralize funds to serve as a source of funds for serving the rustic areas and funding connects resources for both large and small projects.

Community and Stakeholder Engagement

It will therefore prove instrumental in enhancing B5G and 6G technologies' sustainability and effectiveness in the rural setting by involving local communities and stakeholders. It is the local communities' capability enhancement to comprehend and participate in the deployment process, which enhances the solutions developed to be specific to their situations. This can include availing and recruiting local talent in the deployment and maintenance of the network hence creating a culture of ownership in the infrastructure. More importantly, participation of the communities reduces the chances of allowances for connectivity being procured in areas where they are not most needed and guarantees the overall populace receives the benefits of the increased connectivity.

Another important area of focus is the improvement of the members' digital literacy as well as increasing their skills in the usage of available tools. Therefore, in order for the rural populations to reap the benefits of high connectivity, such populations have to be digitally skilled and knowledgable. Measures for raising level of digital competencies may encompass skill enhancement programs, awareness raising and educational activities focused on the fundamentals of using the internet, ways to protect oneself from cyber threats or types of digital tools. In the same vein, collaboration with educational institutions, NGOs, and other private organizations will go a long way in letting as many of the targeted population as possible get abreast with the designed digital literacy programs.

Engagement with stakeholders in the local environment may include involving key persons from community, SMEs and educational establishments will enable B5G and 6G technologies' fast implementation. These partnerships may offer knowledge about the needs of the local communities, the creation of work that addresses these needs and the increased sense of ownership over the ends. Another is that coordination with local authorities can assist in synchronizing geography of networks with other development objectives: development of health care, education, and socioeconomic opportunities in rural regions.

Thus, the promotion of B5G and 6G technologies in rural areas will be possible only with the use of a set of actions, which includes the creation of new technological solutions, changes in the policy related to this matter, and effective communication with rural communities. The availability and cost effective approach to infrastructure involving satellite systems and UAVs enable the realization of the vision of expanding connectivity to marginal regions. Some strategies like the rationalization of the complex regulatory procedures and offering additional encouragement in terms of bonuses will play an important role in enhancing the speed of the deployment at much lower costs. Enabling local communities and raising awareness of the potential, which raised connectivity can bring, guarantee people's effective usage of the opportunities to develop the economy and society in general. Implementation of such strategies makes it easy to eradicate the issue of the digital divide and make the world have equal access to technology.

CONCLUSION

The occurrence of B5G and 6G technologies is a revolution that propounds a positive chance to solve the challenges of network connectivity of rural areas. This chapter has looked at various possibilities of how the developed and sophisticated technologies can overcome the digital divide, revolutionaries the healthcare system, increase education availability, and catalyse the economy in the developing districts. Using latest technologies and architectures like small cell, advanced antenna systems, satellite technologies and integrated with artificial intelligence and machine learning technologies prospective and potential rural communication networks could be designed which can address all rural requirements.

In addition, the chapter has demonstrated that a favourable policy environment and engaging communities are vital for a successful deployment of these technologies. For instance, this goal states that simplification of rules for network deployment, particularly in rural areas, suffices to reduce the financial and logistical challenges that prevail in the network implementation. Just as important is the digital skills and literacy of the populace to support digital technologies to enhance the usc of improved connectivity by local communities particularly in rural areas.

In conclusion, indications of B5G and 6G create a beautiful future in tackling issues of connectivity in the rural areas and improving the pillar of inclusive development. But the realization of this potential depends on coordinated efforts of both the commonwealth's governments and non-governmental actors, private and public, that are able to overcome the technological, regulatory and financial hurdles.

REFERENCES

Akyildiz, I. F., Han, C., & Nie, S. (2018). Combating the Distance Problem in the Millimeter Wave and Terahertz Frequency Bands [June.]. *IEEE Communications Magazine*, *56*(6), 102–108. DOI: 10.1109/MCOM.2018.1700928

Barua, T., & Mohiuddin, K. (2024) "Dynamic Disaster Management with Real-Time IoT Data Analysis and Response," *2024 International Conference on Automation and Computation (AUTOCOM)*, Dehradun, India, pp. 142-147.

Chandra, G., Kumar, R., & Prasad, K. D. V. (2024). Securing Digital Authentication with Cryptographic Innovations in Intrinsic Verification Systems. *Journal of Discrete Mathematical Sciences and Cryptography*, *27*(no. 2-A), 317–328.

Chen, S., Liu, Y. J., Wang, W., Poor, H. V., & Cui, S. (2018). Wireless Communications in the Era of Big Data [Sept.]. *IEEE Communications Magazine*, *56*(9), 14–20.

Dang, S., Amin, O., Shihada, B., & Alouini, M. (2020). What Should 6G Be? [Jan.]. *Nature Electronics*, *3*(1), 20–29. DOI: 10.1038/s41928-019-0355-6

Ferrag, M. A., Shu, L., Yang, X., Derhab, A., Maglaras, L., & Janicke, H. (2019). A Systematic Literature Review on Privacy-Preserving Blockchain-Based Distributed Machine Learning. *IEEE Access : Practical Innovations, Open Solutions*, *7*, 87665–87682.

Giordani, M., Polese, M., Roy, A., Castor, D., & Zorzi, M. (2019). A Tutorial on Beam Management for 3GPP NR at mmWave Frequencies [Firstquarter.]. *IEEE Communications Surveys and Tutorials*, *21*(1), 173–196. DOI: 10.1109/COMST.2018.2869411

Guo, H., Liu, J., Tang, J., & Zhang, J. (2020). Edge AI and 5G: The First Wave of Driving Applications [Feb.]. *IEEE Journal on Selected Areas in Communications*, *38*(2), 246–252.

Hegge, K., & Agrawal, R. (2023) "Wireless and Solar-Powered Multipurpose Robot for Industrial Automation: A Sustainable Solution," *2023 8th International Conference on Communication and Electronics Systems (ICCES)*, Coimbatore, India, pp. 146-151.

Hussain, N., Dixit, S. K., & Gupta, A. K. (2023). A Novel Approach of Unsupervised Feature Selection Using Iterative Shrinking and Expansion Algorithm. *Journal of Interdisciplinary Mathematics*, *26*(3), 519–530. DOI: 10.47974/JIM-1678

Maddineni, V. K., & Gummadi, V. A. K. (2024) "Deep Learning Approaches for Real-Time IoT Data Processing and Analysis," *2024 5th International Conference on Smart Electronics and Communication (ICOSEC)*, Trichy, India, pp. 757-764.

Mao, Y., You, C., Zhang, J., Huang, K., & Letaief, K. B. (2017). A Survey on Mobile Edge Computing: The Communication Perspective [Fourthquarter.]. *IEEE Communications Surveys and Tutorials*, *19*(4), 2322–2358. DOI: 10.1109/COMST.2017.2745201

A. Mary, A. Y. Begum, and D. G. V., (2023) "A Novel IoT Framework and Device Architecture for Efficient Smart City Implementation," *2023 7th International Conference on Trends in Electronics and Informatics (ICOEI)*, Tirunelveli, India, pp. 420-426.

Mumtaz, S., Rodriguez, J., & Dai, L. (2018). When Mobile Crowd Sensing Meets Ubiquitous Computing: Vision and Challenges [Dec.]. *IEEE Communications Magazine*, *56*(12), 52–59.

Pan, C., Ren, W., Wang, J., Elkashlan, M., Wang, J., & Hanzo, L. (2018). NOMA-aided Mobile Edge Computing: A Joint Optimization of Computation Offloading and Time Allocation [July.]. *IEEE Transactions on Wireless Communications*, *17*(7), 4568–4583.

Pandey, S., Samdanis, K., Costa-Pérez, X., & Guo, W. (2020). 6G Network Formation and Management for Multi-RAT Heterogeneous Environments [Nov.]. *IEEE Communications Magazine*, *58*(11), 38–44.

Polese, M., Giordani, M., Roy, A., Castor, D., & Zorzi, M. (2020). Toward End-to-End, Full-Stack 6G Terahertz Networks [Nov.]. *IEEE Communications Magazine*, *58*(11), 48–54. DOI: 10.1109/MCOM.001.2000224

Premkumar, S., Hussain, N., & Chinamuttevi, V. S. (2023). A Novel RF-SMOTE Model to Enhance the Definite Apprehensions for IoT Security Attacks. *Journal of Discrete Mathematical Sciences and Cryptography*, *26*(3), 861–873. DOI: 10.47974/JDMSC-1766

Qadir, J., Ali, A., & Saleem, A. M. (2021). Artificial Intelligence-Enabled Edge Computing for IoT: A Comprehensive Survey [Feb.]. *IEEE Internet of Things Journal*, *8*(4), 3037–3066.

Reddy, M., & Tanguturi, R. (2022) "Improved Secure Communication Mechanism for IoT Platform Devices," *2022 International Conference on Smart Generation Computing, Communication and Networking (SMART GENCON)*, Bangalore, India, pp. 1-6.

Reddy, N. S., & Ashreetha, B. (2023) "Technologies for Comprehensive Information Security in the IoT," *2023 International Conference for Advancement in Technology (ICONAT)*, Goa, India, pp. 1-5.

Shekhar, S., & Ashreetha, B. (2023). Performance Analysis of Energy Efficiency and Security Solutions of Internet of Things Protocols. *IJEER*, *11*(2), 442–450. DOI: 10.37391/ijeer.110226

Srinivas, V., & De, T. (2023) "Wi-Fi Router Signal Coverage Position Prediction System Using Machine Learning Algorithms," *2023 International Conference on Sustainable Computing and Smart Systems (ICSCSS)*, Coimbatore, India, pp. 253-258.

Vo, N. N., Le, T. A., & Tran, N. H. (2018). Energy-Efficient 5G Green Network with Wireless Power Transfer for IoT Devices. *IEEE Access : Practical Innovations, Open Solutions*, *6*, 17160–17173.

You, X., Wang, C., Huang, J., Gao, X., Zhang, Z., Wang, M., Huang, Y., Zhang, C., Jiang, Y., Wang, J., Zhu, M., Sheng, B., Wang, D., Pan, Z., Zhu, P., Yang, Y., Liu, Z., Zhang, P., Tao, X., & Liang, Y.-C. (2021). Towards 6G Wireless Communication Networks: Vision, Enabling Technologies, and New Paradigm Shifts [Feb.]. *Science China. Information Sciences*, *64*(1), 1–74. DOI: 10.1007/s11432-020-2955-6

Zaidi, A., Athanasiou, G., Mahmood, K., Teyeb, O., Medbo, J., & Wikström, G. (2020). Integration of Terrestrial and Non-Terrestrial Networks in 6G: Requirements, Challenges, and Directions [Sept.]. *IEEE Communications Standards Magazine*, *4*(3), 49–55.

Chapter 6
Addressing B5G and 6G Network Connectivity Issues and Challenges in Rural Regions of India

M. Basuvaraj
https://orcid.org/0000-0002-2660-3331
University of Allahabad, India

Keshvi Rastogi
https://orcid.org/0009-0008-0954-6487
Christ University, India

S. Chitradevi
Global Institute of Business Studies, India

D. V. VidyaShree
Christ University, India

K. M. Anu
Christ University, India

ABSTRACT

The emerging technology of the fifth-generation broadband cellular network is already ruling the market with its efficiency, lower latency, higher connectivity, and many more features. In contrast, the sixth-generation broadband cellular network is yet in its research and development stage. These technologies cannot only revolutionize the world with their features, such as high speed and enhanced cybersecurity but also empower it to reach greater heights. To understand the network requirements of the rural and under-developed areas, it is important to understand all those challenges in the way ahead. . Launching such efficient and effective technologies in rural areas would benefit the country as well as its economic growth. The large markets of these cellular networks are at constant growth and are expected to be booming in the future of the Telecom Regulatory Authority of India. (2023, September 29)..

DOI: 10.4018/979-8-3693-7788-8.ch006

INTRODUCTION

The deployment of 5G and emerging 6G technologies offers transformative potential for various sectors, but challenges such as infrastructure gaps, economic viability, and policy frameworks must be addressed to ensure equitable access. (Regulatory Authority of India, 2023). In today's world, the internet stands as the most vital tool and is the most potent engine, whether it be for communication, social, or economic growth. The Internet has not only acted as a catalyst for various operations worldwide but also as the medium for various developments in recent times. scaling up massive Internet-of-Things (IoT), energy harvesting and Simultaneous Wireless Information and Power Transfer (SWIPT) is foreseen to become important enablers when deploying a large amount of small, low-power radios (Gustavsson, U et al 2021). 5G, 6G, and Beyond (xG) aim at bringing new radical changes to shake the wireless communication networks where everything will be fully connected fulfilling the requirements of ubiquitous connectivity (Salahdine, F., Han, T., & Zhang, N2023). Real-time adaptive security, and novel data protection mechanisms such as distributed ledgers and differential privacy are the top promising techniques to mitigate the attack magnitude and personal data breaches substantially (Nguyen, V. L et al 2021). Integration of D2D communication with other prominent technologies and demonstrates the importance of integration with possible solutions in improving network performance (Gismalla, M. S et al 2022), Thus, the Internet needs to be open, accessible, trustworthy, and secure for everyone worldwide. Working, streaming, scrolling, and gaming are all extremely common reasons to use the internet. Out of the nearly 8 billion people in the world, 5.35 billion of them, or around 66% of the world's population, have access to the internet. As of April 2024, there were 5.44 billion internet users worldwide, which amounted to 67.1 percent of the global population. Of this total, 5.07 billion, or 62.6 percent of the world's population, were social media users. According to the data, Asia was home to the most significant online users i.e., over 2.93 billion users at the latest count. With around 750 million internet users. China, India, and the United States have ranked ahead of other countries worldwide in terms of the number of Internet users in these countries. As the capabilities of the internet continue to grow and change to keep up with life around us, the number of internet users is bound to increase and will do so at a rapid rate. In fact, in just five years, internet users are expected to increase 47% from 5.35 billion users in 2024 to 7.9 billion users in 2029. 6G technologies have the potential to bridge the digital divide by enabling cost-effective, energy-efficient, and flexible wireless connectivity in remote areas (Chaoub, et al. 2021). Bridging the digital divide through 6G requires overcoming infrastructure, cost, and energy challenges to provide affordable and sustainable connectivity in rural areas (Yaacoub, E., & Alouini, M. S. 2020). Thus, the Internet is taking on the world by a whirlwind. As global digital transformation accelerates, next-generation mobile networks such as 5G and the emerging 6G promise to revolutionize communication by enabling faster data speeds, ultra-low latency, and massive connectivity. These advancements are expected to fuel innovations in various sectors, including healthcare, agriculture, transportation, and smart cities. However, the deployment and effectiveness of these networks in rural areas face significant challenges, which risk exacerbating the existing digital divide between urban and rural regions.

This paper explores the connectivity issues associated with 5G and 6G networks in rural areas, examining the technological, economic, and social barriers and proposing potential solutions to ensure equitable access to advanced communication technologies. The 5G and 6G broadband cellular networks are the latest Information and Communication Technologies trends. The introduction of the 4G/LTE network helped people access the internet services better with smartphones, tablets, and PCs. Though it provided users with various advantages, the users still felt the need for high-speed Telecom Regulatory

Authority of India. (2023, September 29)., secure connection, high reliability, rapid response, and energy efficiency, and the fifth-generation broadband cellular network commendably served all such factors. It provides features like the Internet of Things (IoT), safe and secure drones and robotic communications, cloud services, and more. The sixth-generation wireless network, the 6G, is the next evolution in telecommunication technology. It is currently still in the research and development phase. The 5G network has been rolled out and is still in that phase, but the world stands ready for the 6th generation mobile network technology. It is anticipated that 6G would be commercially available by 2030. It is projected that 6G would reach a speed 100 times more than current 5G networks, with a latency of less than 1 millisecond. It would have the capability to support advanced technologies such as holographic communications, extended reality, Internet of Senses, and much more. The development of 6G mobile network technology is a global effort by leading countries in Asia, America, and Europe. In all, the 6G network would provide mankind with not only better mobile network connectivity but also ultra-fast, advanced, and more secure wireless connectivity.

UNDERSTANDING B5G AND 6G TECHNOLOGIES

B5G (Beyond 5G) and 6G Networks

B5G refers to the evolutionary stage between 5G and 6G, where existing 5G technologies are enhanced and integrated with emerging innovations. B5G networks aim to address some of the limitations of early 5G deployments, such as coverage gaps, energy efficiency, and support for Ultra-Reliable Low-Latency Communication (URLLC) and massive machine-type communications B5G networks are expected to support even higher data rates greater spectral efficiency, and improved network reliability. 6G, still in the research and development phase, is envisioned as the next leap in mobile communication. Expected to be commercially available by the 2030s, 6G will offer unprecedented data rates, potentially reaching Terabits Per Second (TBPS), with latency in the microsecond range. 6G will also incorporate advanced technologies such as Artificial Intelligence (AI), Machine Learning (ML), quantum communication, and Integrated Sensing and Communication (ISAC). 6G has the potential to bridge the digital divide by providing affordable, high-speed connectivity in remote areas through a combination of terrestrial, satellite, and emerging wireless technologies (Saarnisaari, et al. 2020). The goal of 6G is to create a ubiquitous, intelligent, and secure network that seamlessly connects everything from humans to machines and the physical to the digital world.

5G Deployment in India

On October 1, 2022, 5G services were first introduced in India. Only two national telecom companies currently provide 5G services: Bharti Airtel and Reliance Jio. Within six months, Indian telecom operators surpassed the three-year objective set for them to roll up 5G networks. As of August 2023, they had more than 3 lakh sites established. By the end of this year, the bulk of the nation's communities should be covered. For the past ten years, India has been experiencing a digital revolution. People are becoming more digitally savvy due to rising disposable income and reasonably priced data plans. due to an increase in mobile subscribers as well as deployment of 4G and 5G cellular network technologies, better solutions for capacity and coverage should be provided in order to enhance spectral efficiency

(Sharma, T., Chehri, A., & Fortier, P. 2021). Building based on 5 G, the development of 6 G willhave a profound impact on the intelligence process of communication development (Lu, Y., & Zheng, X. 9 2020). Trends of two stages towards system integration of the terrestrial mobile communication and the satellite communications: to be compatible with 5G, and to be integrated within 6G (Chen, S., Sun, S., & Kang, S. 2020). Some of the main drivers of this digitalization include rising smartphone adoption, a notable increase in OTT usage, digital payments, and e-commerce. The shortcomings of 5G in meeting the needs of more data-intensive, low-latency, and ultra-high-reliability applications (Salameh, A. I., & El Tarhuni, M. 2022). These factors have allowed India's ARPU to expand at a CAGR of almost 21% over the last four years, which is a considerable increase (figure 1.2). With the arrival of 5G, which facilitates seamless data consumption, this is anticipated to continue to climb steadily.

Figure 1. Monthly wireless ARPU during the past 4 years

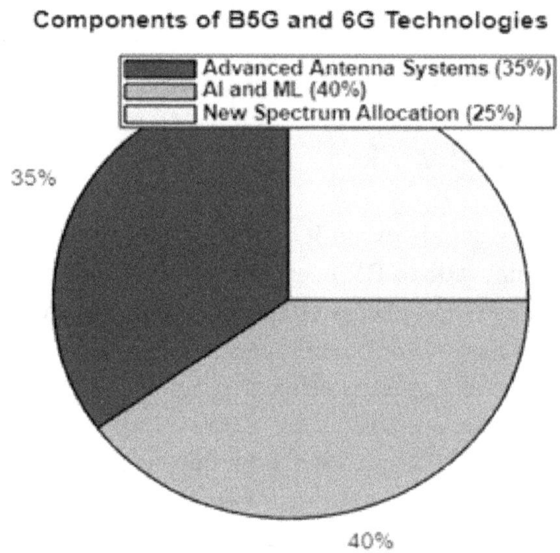

Source: *Statista*

Due to this revenue growth trajectory, telecom companies are making significant national investments in 5G, fiber, and coverage infrastructure. In the nation, 5G has greatly increased mobile internet speeds. The median download speed rose from 13.87 Mbps in September 2022 to 50.21 Mbps in August 2023, per Ookla. India's ranking on the Speed Test Global Index consequently rose by 71 spots, from 118th in September 2022 to 47th in August 2023. India is on the verge of becoming a major telecom technology exporter to the world in the upcoming years, having developed its own 4G/5G technology stack that is currently ready. 18 nations that intend to use India's domestically created 4G/5G technology stack are now in talks with the country. In India, the rollout of 5G-based private networks has already begun. At Mahindra's Chakan Auto manufacturing site, Bharti Airtel and Tech-Mahindra have implemented the Captive Private Network.

Government's Initiative to Facilitate the Launch of 5G Services

The government's efforts encompass various areas, such as assisting TSPs in augmenting their 5G investments, constructing 5G infrastructure, and uniting the ecosystem for efficient research and development.

i. The 5G High-Level Forum was established to offer a plan of action for bringing the 5G vision to reality. The forum published a roadmap including recommendations for 5G policies, initiatives, standards, and use cases.
ii. Bharat Net: This government-run initiative intends to link rural India to the optical fiber network and make it available to all service providers without regard to their ability to pay.
iii. 5G Test Bed - To help businesses and MSMEs evaluate their use cases, the government has made an indigenous 5G test bed available for free or at a reasonable cost to other stakeholders.
iv. Aiming to create robust cooperation partnerships across stakeholders in the 5G Use-case ecosystem, the VEPP Program (formerly known as the 5G Vertical Engagement and Partnership Program) is specifically designed to satisfy the demands of the User/Vertical Industry. A multi-ministerial committee with officials from several Ministries has been established to enhance the potential for 5G across usage verticals and promote cooperation among stakeholders. A multiplier effect may be created to attempt and fine-tune 5G digital solutions in corresponding economic verticals by enabling close engagement between User verticals and 5G Tech stakeholders (Service providers, Solution providers, & partner OEMs) through VEPP.

Figure 2. Annual global 5G contribution by industry, 2030

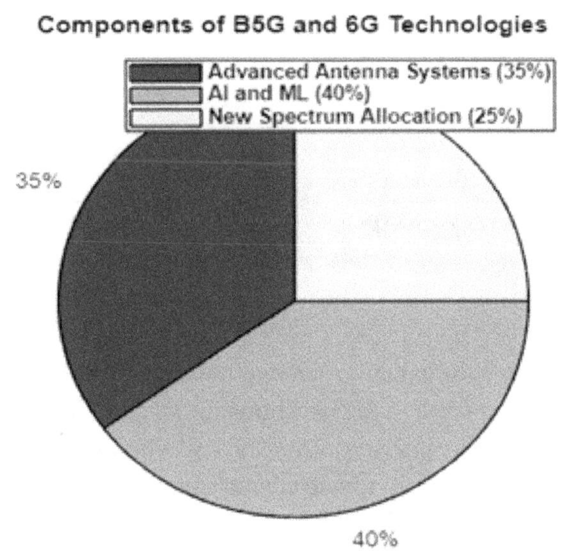

Source: *NASSCOM Arthur D Little 5G Report Nov2022*

Probability of 5G and Related Technologies Affecting the Indian Economy

By 2030, 5G is anticipated to power up to 2% of India's GDP, or USD 180 billion. This would be made possible by growing market penetration, bolstering the economy, and sectoral reforms. The industries that stand to gain the most from widespread 5G adoption include manufacturing, retail, energy & utilities, and healthcare. To guarantee that the benefits of 5G are realized across sectors, effective cooperation between public and commercial stakeholders will be required to create a skilled workforce and cybersecurity infrastructure, reduce regulatory overlaps, and provide financial incentives. Hyper-connectivity enabled by 5G and related technologies is predicted to generate new value and act as a catalyst for digital transformation in a variety of industries.

Figure 3. 5G contribution for select industries in India 2030

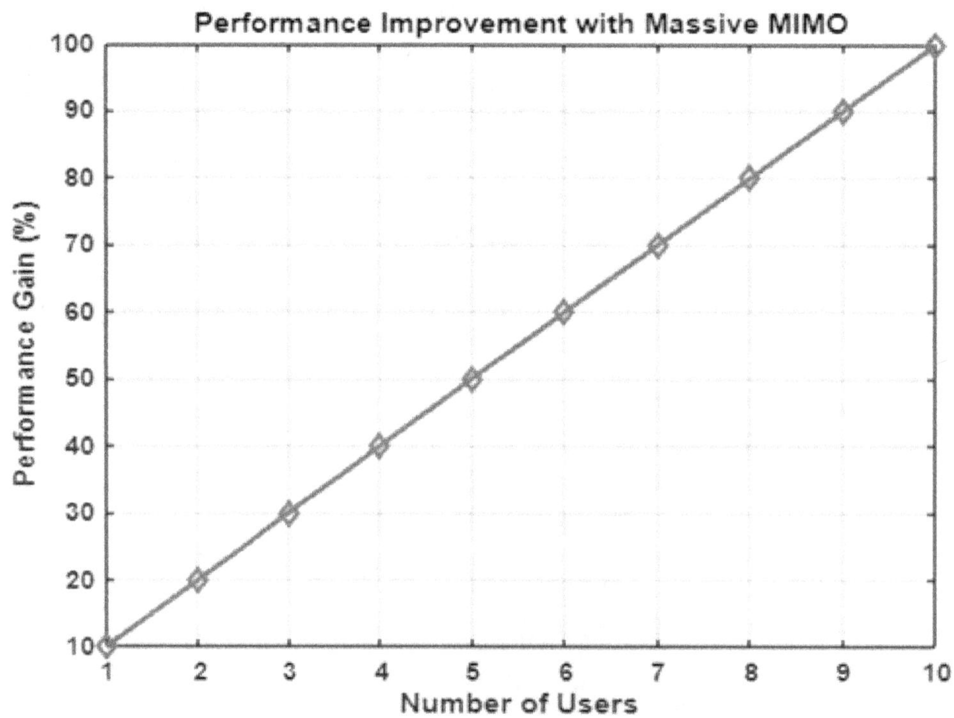

In addition to changing India's digital environment, these technologies are fostering collaborations among them. For instance, 5G can serve as the backbone for AI and Internet of Things applications that need a lot of bandwidth and quick response times. Massive volumes of data generated and analyzed by IoT and AI can be utilized to build individualized and immersive metaverse experiences. Metaverse may provide a platform for exhibiting and making money from the advancements and solutions made possible by 5G, IoT, and AI. India has a unique chance to take advantage of its advantages in talent pool, digital infrastructure, and entrepreneurial ecosystem to lead the world in digital transformation. India can enhance its economic growth, social development, and global competitiveness by utilizing the promise of 5G, IoT, AI, and metaverse.

Challenges of Rural Connectivity

India is a diverse country with a vast geographic area with about 68.8% of its population being rural population, yet only 53% of that rural population has access to broadband cellular networks. The growth factor under the rural economy has caught up speed in the last decade but then also it lags much as compared to the growth in urban sectors. There is a new evolution of technology every other time but it fails to reach the rural community and thus they fail to benefit from the progress. Several factors contribute to the lack of mobile network connectivity in the rural sectors and traditional technology still being the only medium for those people to rely on. The integration of D2D communication with other prominent technologies and demonstrates the importance of integration with possible solutions in improving network performance Gismalla, et al. (2022). The potential of using emerging deep learning and federated learning techniques for enhancing the efficiency and security of IoT communication are discussed, and their promises and challenges are introduced. Finally, future research directions toward beyond 5G IoT networks are pointed out (Vaezi, et al. 2022). New technologies like artificial intelligence, non-terrestrial networks, and new spectra is elaborated. In particular, the potential of using emerging deep learning and federated learning techniques for enhancing the efficiency and security of IoT communication are discussed, and their promises and challenges are introduced (Azari, et al. 2022). Some of those factors are:

(a) **Coverage Challenges** - Since India has a vast geographic area and thus covering every area poses a challenge for the government. Most of India's population resides in rural, hilly or hard-to-reach areas which makes it even more difficult for the government or telecommunication companies to equip them with the latest technology. Also, there is not a very high population in these rural areas. The areas have a very low density. The telecommunication companies do not find it feasible to invest in expensive infrastructures in such areas. This is because the amount of return they would receive would not be equal to the amount of investment they would make at such places. The deployment of the mobile network technology involves high operational and installation costs along with high frequent maintenance costs, making it an expensive and low return, or sometimes unreciprocated return for the telecom companies. The rural regions that are in the hilly, mountain, desert or difficult terrain make it difficult for technology deployment in such areas as well as the logistical challenges are also faced by the communication companies to provide these regions with updated technologies.

(b) **Persisted Challenges** – Numerous challenges hinder the growth of rural connectivity. There is a lack of consistent electricity availability and reliable power supply in various rural areas. This not only hampers the deployment of telecom infrastructure but also the operationality and maintenance of these infrastructures. There is also inadequate backhaul connectivity, which means that the network connectivity to the main centres of the telecom companies would be a problem. The income groups in such regions generally consist of most of the population coming from low-income groups, maybe daily wage earners or people in agriculture. The affordability of mobile phones in comparison to the technology and these mobile network connectivity data services hinder the growth of the rural sectors. For example, the 5G technology would only work for mobile phones that are 5G capable, hence frequent updation of such expensive devices is not possible for the low-income groups. Thus, serving this segment of people becomes hard and expensive for telecommunication companies. Digital literacy is also one of the problems for the people. The rural community is often not very familiar with the latest technology and the updates in the processes. This also in some ways hamper their growth in such sectors.

(c) **Slow Growth** – The Government of India has constantly been making efforts to shorten the gap there is between rural development and urban development. There has also been a great improvement in the rural subscription base of these networks. More and more projects are being undertaken by the authorities to make rural communities equipped with the latest technology. However, there still stands a significant gap between the efforts made and the actual scenario. The rural areas have still a long way to go for development purposes. While the urban areas are making the best possible use of the 5G technology, there are several rural areas where even the 4G network is a luxury and the people are still making use of 2G and 3G networks, whereas some are even not having the bare minimum.

(d) **Other Challenges** - Numerous obstacles hinder the implementation and efficiency of modern networks such as B5G and 6G in rural areas. Geographical obstacles, such as steep terrain or dense forests, are a huge problem since they make it challenging to set up infrastructure and keep dependable connections. Because there are generally fewer financial incentives for service providers to invest in rural areas due to low population density, these locations typically have slower network speeds and less coverage. Financial limitations are also important since they might make it difficult for both service providers and customers to afford the high expenses of implementing and using cutting-edge technologies.

The issue is made more difficult by limited availability to electricity, which makes it difficult to reliably power network equipment. Another important problem is the "digital divide," where people in rural areas typically have less access to digital resources and a lower degree of digital proficiency than people in metropolitan areas. The acceptance and efficient application of new technology may be hampered by this division. Furthermore, obstacles related to legislation and regulation, especially those pertaining to spectrum allotment and land use permits, can impede the implementation of networks in rural areas. Cultural and social variables are particularly important since they might hinder the adoption of new technology by causing opposition or a lack of knowledge about their advantages. It is challenging to maintain and troubleshoot modern networks in rural areas due to a lack of technical experience, which could cause service outages. Finally, environmental factors like severe weather can harm infrastructure, making the situation much more difficult.

Technological Challenges in the Rural Setting

(a) **Infrastructure Development:** The absence of current infrastructure makes it difficult to deploy B5G and 6G networks in rural areas. Access to necessary resources like fiber-optic cables, cell towers, and power supplies is frequently restricted in rural areas. It can be unaffordable to build and maintain this infrastructure in remote locations, which delays implementation and raises costs for service providers and customers alike.

(b) **Spectrum Allocation and Availability:** B5G and 6G network spectrum management and allocation might be difficult in rural areas. In rural locations, the scarce supply of high-frequency bands-which are essential for these cutting-edge networks-might not be sufficiently addressed. Furthermore, fewer people living in these areas may lead to a less prioritized distribution of spectrum, which could impede the effectiveness and reach of these technologies.

(c) **Energy Consumption and Sustainability:** The ultra-high-speed connectivity and huge device support that are anticipated from B5G and 6G networks necessitate substantial energy resources. Powering these sophisticated networks can be extremely difficult in rural locations, where access to dependable and sustainable energy sources is sometimes restricted. Energy-efficient technology and sustainable power solutions are essential to guarantee that these improvements may benefit rural communities without aggravating their current energy restrictions.

(d) **Technical Know-How and Manpower:** Specialized technical know-how is needed for the deployment and upkeep of B5G and 6G networks, although this knowledge is frequently lacking in rural areas. Implementation of these cutting-edge technologies may be slowed down and operational difficulties may arise due to a shortage of trained personnel. To get over this obstacle, training programs and educational efforts that concentrate on giving rural populations the technical skills they need are crucial.

(e) **Digital Divide and Accessibility:** As B5G and 6G networks are deployed, rural areas may see a greater disparity in digital literacy and access to cutting-edge technologies. Making sure rural communities get the tools, information, and assistance they need to take full advantage of these networks' potential is the difficult part. To close this gap, focused efforts must be made to advance digital education and give rural populations access to reasonably priced technology solutions.

(f) **Geographical and Environmental Challenges:** Deploying and maintaining the physical infrastructure needed for B5G and 6G networks is difficult in rural areas since they frequently contain large and demanding terrains including mountains, woods, and farmlands. The difficult topography may make it difficult to place base stations, antennas, and other essential equipment, which could cause network rollout delays and increased expenses.

(g) **Connectivity to Core Networks:** One of the biggest obstacles to the implementation of B5G and 6G is creating a dependable backhaul connection between rural areas and the core networks. The overall performance and dependability of advanced networks in rural areas can be impacted by the distance from urban centers, where the concentration of core network infrastructure, which can lead to weak or inconsistent connectivity. To close this gap, solutions like cutting-edge wireless technologies or satellite backhaul may be required.

(h) **Affordability and Economic Viability:** Due to lower population density and lower potential revenue than metropolitan areas, the economic model for establishing B5G and 6G networks in rural areas is less appealing to service providers. Extensive network rollouts in rural areas may be difficult to justify due to high infrastructure costs and low returns on investment. It is imperative to find ways to make these deployments financially feasible, whether through public-private partnerships or government subsidies.

(i) **Policy and Regulatory Barriers:** Policies and regulatory frameworks, particularly in rural regions, can lag behind technology improvements. Significant obstacles may arise from the absence of explicit regulations or financial incentives for the deployment of B5G and 6G networks in certain areas. To overcome these obstacles and expedite network deployment, policymakers must create policies that are favourable, streamline approval procedures, and encourage cooperation between the public and private sectors.

(j) **Integration with Current Rural Technologies:** The integration of B5G and 6G networks may face difficulties in rural regions due to the possible continued usage of legacy communication systems and technologies. To prevent service interruptions, compatibility and a seamless transfer between old and new systems must be ensured. It might be necessary to make large investments in modernizing the current infrastructure and creating hybrid solutions that can coexist with older established technologies while easing the transition to more sophisticated networks over time.

IOT CAPABLE OF 5G: OPPORTUNITIES AND DIFFICULTIES

- The rapid improvements in communication technology, especially 5G, have led to the emergence of the Internet of Things (IoT) paradigm in recent years. The term pertains to the network of networked physical devices, automobiles, appliances, and other objects equipped with sensors, software, and network connectivity, allowing them to gather, share, and evaluate data. Internet of Things (IoT) technology can change many industries and raise people's living standards. However, several legislative issues must be resolved before these technologies can be effectively used and spread.
- Smart cities are among the primary domains in which Internet of Things technologies have made noteworthy advancements. IoT-enabled smart cities use IoT devices for various purposes, including waste management, public safety, energy efficiency, and intelligent traffic control. IoT devices can make smart decisions that enhance the general efficiency of the city by gathering data from multiple sources, including sensors integrated into buildings, roadways, and public infrastructure. Furthermore, by streamlining energy use and encouraging resource efficiency, IoT technology might lessen the environmental impact of metropolitan regions.
- IoT technology can potentially transform patient care in the healthcare industry by enabling telemedicine, smart medical equipment, and remote patient monitoring. Thanks to connected wearables and sensors, healthcare providers can monitor patients' vital signs, track medication adherence, and deliver individualized care even while they are far away. IoT technology can be essential in the agricultural sector by facilitating smart irrigation systems, livestock monitoring, and precision farming. To help them decide when and how to sow, fertilize, and irrigate their crops, farmers can utilize Internet of Things (IoT) devices to gather real-time data on weather, soil moisture, and nutrient levels.
- Even though IoT technologies have many advantages and prospective uses, several policy issues must be resolved to ensure their successful adoption and spread. Ensuring that the enormous amounts of data created and transferred by IoT devices are secure and private is one of the most significant difficulties. To enable the successful deployment and spread of IoT technologies, interoperability and standardization are significant policy issues that must be resolved. Furthermore, as automation and greater efficiency may result in job displacement in some industries, the rapid adoption of IoT technology may substantially impact the labor market. Lastly, ethical issues must also be considered when putting IoT technologies into practice and spreading them.
- Many industries have seen a paradigm shift with the introduction of 5G technology, which makes Internet of Things (IoT) applications faster, more dependable, and more effective. These applications cover a broad spectrum of industries, such as Industry 4.0 and smart manufacturing, smart cities and infrastructure, healthcare and remote patient monitoring, smart buildings and residences, and precision farming in agriculture. Applications with Internet of Things capabilities have the potential to revolutionize many industries and raise people's standards of living. However, to guarantee these technologies' successful adoption and spread, legislators and regulators must tackle several regulatory issues, such as data security and privacy. Spectrum management, employment

displacement, environmental effect, standardization and interoperability, and ethical considerations. It is possible to guarantee the responsible and sustainable expansion of the IoT ecosystem and to realize its full potential for the benefit of society by proactively addressing these issues and encouraging cooperation amongst stakeholders.

Innovative Solutions for Network Connectivity

To close the digital gap, creative ways to boost network access in rural areas are essential. Some of which can be-

(a) Deploying low Earth orbit (LEO) satellites is one promising strategy that can give underserved and distant places access to high-speed internet. LEO satellites orbit closer to the Earth than regular satellites, which lowers latency and boosts connection rates. Leading the way in this technology are businesses like SpaceX, which provides a competitive substitute for conventional ground-based infrastructure. Furthermore, community networks—where local communities create and manage their own networks with assistance from non-profits or government funding—are becoming more and more popular as a solution. These networks can be customized to meet the unique requirements of the neighborhood, guaranteeing more dependable and reasonably priced access.

(b) Using TV white spaces (TVWS), which are underutilized regions of the television broadcast spectrum, is another creative alternative. Due to the long-range and high penetration rates of these frequencies through trees and buildings, TVWS can be used to provide broadband internet service to rural locations. This makes it the perfect option for areas with difficult topography. In addition, base stations that run on solar energy are being developed to help with the problem of scarce electricity in rural regions. Because these stations are self-sufficient, uninterrupted network access is guaranteed even in remote areas. These solutions, when combined with developments in mesh networking, which enables devices to communicate with one another to build a network without depending on a central hub, are opening the door for more dependable and accessible rural connectivity.

(c) Innovative strategies like high-altitude platform systems (HAPS) and drone-based networks are beneficial for increasing rural connection. Drones can be used to establish transient or semi-permanent networks in difficult-to-reach locations, offering connectivity in areas affected by disasters or while infrastructure is being developed. Conversely, HAPS consists of unmanned, solar-powered aircraft or balloons that function as floating cell towers in the stratosphere. These platforms are capable of providing access to remote places with difficult terrain or sparse populations because of their extensive coverage. These aerial systems are perfect for meeting the connection needs of isolated and rural locations due to their scalability and flexibility.

(d) The creation of hybrid network models, which integrate several technologies to maximize connectivity, is another intriguing remedy. To build a more durable and economical network, hybrid models, for instance, might combine fiber-optic cables with wireless technologies like 4G/5G, LEO satellites, and microwave links. With this strategy, fiber may be used where it makes financial sense while satellite or wireless technology can be used in more remote places. In order to guarantee that even the most remote villages have access to dependable internet, hybrid networks can deliver more constant and high-quality service across a variety of rural settings by utilizing the advantages of several technologies.

(e) Lastly, some areas are putting smart village projects into practice to improve rural connection by using a comprehensive strategy that blends technology with community development. The goal of these projects is to combine digital infrastructure with necessary services including agriculture, healthcare, and education. Smart villages can enhance the quality of life in rural communities by enabling them to access e-learning, telemedicine, and smart farming solutions through the provision of digital tools and connection. Public-private partnerships are frequently used in these projects, as local groups, governments, and digital corporations work together to create long-lasting, community-driven networks. The accomplishments of smart village projects show that promoting digital inclusion and social development in addition to technology is essential for enhancing rural connectivity.

Economic Considerations and Funding Model

India, a country that is developing quickly, has advanced significantly in terms of digital connectivity. However, rural areas continue to lag behind urban areas when it comes to having access to cutting-edge network equipment and high-speed internet. The emergence of 6G and Beyond 5G (B5G) technology presents a special chance to close the digital gap. In addition to being a technological issue, deploying B5G and 6G networks in rural India is also an economic one that calls for a stable and long-term funding strategy. The economic factors and possible funding sources for deploying B5G and 6G network connections in rural India are examined in this essay.

Economic Considerations

i) **High Initial Investment Costs:** Significant upfront capital expenditure is required to deploy B5G and 6G networks. This covers the price of purchasing cutting-edge gear, installing Fiber optic lines, constructing base stations, and obtaining spectrum. These expenses are further increased in rural areas, frequently characterized by low population density and challenging topography. Unlike in metropolitan regions where demand and willingness to pay are stronger, the economic sustainability of such an investment in rural areas is uncertain due to the low opportunity for immediate high returns.

ii) **Operating and Maintenance Costs:** The ongoing operational costs of sustaining modern network infrastructure in rural locations are significant, even after the original capital outlay. These expenses cover energy use, ongoing upkeep, and hiring qualified personnel to oversee the network. The cost per user is typically higher in rural areas due to the population's dispersed nature, making it less appealing for private telecom carriers to operate in these areas without sufficient support.

iii) **Revenue Generation Challenges:** Revenue generation in rural regions is challenging due to lower average incomes, limited commercial activities, and the absence of significant industrial users who typically drive high data consumption. Additionally, there is often a lack of digital literacy, which can result in lower adoption rates of new technologies. This, in turn, affects the profitability of network operators and discourages large-scale investment in advanced connectivity solutions.

iv) **Digital Divide and Social Impact:** The digital divide between urban and rural areas has significant economic implications. Lack of access to high-speed internet in rural regions limits opportunities for education, healthcare, and economic development. This divide perpetuates poverty and inequality, as rural populations are unable to access the same resources and opportunities as their urban counterparts. Therefore, addressing the digital divide is not just a technological imperative but also a socio-economic one, critical for the inclusive growth of the country.

Funding Models for B5G and 6G Connectivity

Given the economic concerns mentioned above, a sustainable finance mechanism is necessary for the effective deployment of B5G and 6G networks in rural India. A mix of hybrid, private, and public funding approaches can be investigated to accomplish this goal.

Government Support and Grants:

Through direct investments and subsidies, the government can be crucial in financing the rollout of B5G and 6G networks in rural areas. The government can allot money from the federal and state budgets to target rural infrastructure development, particularly since it has acknowledged its connection as a public good. Universal service obligations (USOs) are another useful tool in which telecom companies donate a portion of their earnings to a fund intended to provide access in disadvantaged areas.

Public-Private Partnerships (PPP):

By integrating the advantages of the public and private sectors, public-private partnerships (PPP) provide a balanced approach. In this scenario, commercial telecom operators manage the network's development, operation, and maintenance, with the government offering initial funding, subsidies, or tax breaks. By reducing financial risks, this paradigm promotes private investment in rural areas.

Model of Shared Infrastructure:

Several telecom providers share the same network infrastructure under the shared infrastructure concept, including base stations, Fiber optic cables, and towers. As a result, deploying networks in rural areas becomes more financially feasible and lowers the overall cost for each operator. The government might encourage sharing already-existing infrastructure or establish a common infrastructure fund to help with this.

Community Networks and Cooperative Models

Local communities are empowered to own and manage their network infrastructure through community networks and cooperative models. Under this model, the government or non-governmental organizations provide initial funds to the community or cooperative group to create the network managed not-for-profit. This method works especially well in remote locations where traditional models are impractical.

Foreign Collaborations and Funding:

International organizations, development banks, and foreign governments could be possible sources of funding, given the widespread interest in bridging the digital gap. To construct B5G and 6G networks in rural India, organizations such as the Asian Development Bank (ADB) or the World Bank's Digital Development Partnership can offer low-interest loans, grants, or technical help.

Strategic Planning and Implementation in the Rural Setting

For B5G (Beyond 5G) and 6G network connectivity to be strategically planned and implemented in India's rural areas, a multimodal strategy is needed to take advantage of the benefits and specific problems these areas bring. Finding connectivity gaps and comprehending the unique requirements of rural communities-many of which involve agricultural, educational, and healthcare needs-are the first steps in the process. A thorough evaluation of the current digital infrastructure is also conducted. The strategy should prioritize cost-effective solutions, such as the advancement of current 4G and 5G networks and the deployment of low-cost, high-efficiency technologies like massive MIMO (Multiple Input Multiple Output) systems, mmWave communication, and low-earth orbit (LEO) satellites, considering the topographical diversity and economic constraints of rural India. To guarantee inclusion and sustainability, the implementation plan should place a high priority on cooperation between the public and corporate sectors as well as local communities. Public-private partnerships (PPP) should be encouraged, spectrum distribution should be made easier, and telecom companies should be given subsidies or other incentives to extend their networks into less lucrative rural areas.

In order to guarantee that the infrastructure satisfies their unique requirements, local communities should also be involved in the design process. This can promote local entrepreneurship and digital literacy, which will increase the infrastructure's socioeconomic impact. Given the frequent lack of dependable electricity in rural areas, the strategic plan must also consider the integration of renewable energy sources to power network infrastructure, which would assist resilience and sustainability. Phased rollouts should be used in implementation, starting with pilot projects in certain areas to test and improve technology and tactics and then expanding the deployment depending on the knowledge gained. Frameworks for monitoring and evaluating initiatives are crucial for tracking progress, seeing obstacles, and making the required corrections to keep the initiative on course. The integration and interoperability of 6G networks with all heterogeneous networks are inevitable to realize the goals of 3D communication (Kharche, S., & Dere, P. 2022). Proposed methods of new 5G and future 6G security approaches, such as Blockchain technology, post-quantum cryptography (PQC), and Artificial Intelligence (AI) schemes, are also discussed to understand better the full landscape of current and future research within this telecommunications domain (Scalise, P. 2024), communication among all the techniques and potential problems associated with the spectrum management that has been addressed with the possible solutions (Qamar, F., Siddiqui, M 2020)

To ensure coherence and maximize impact, the adoption of 6G technologies is anticipated to bring about advancements like ubiquitous AI, smart environments, and holographic communications, which should also be coordinated with India's more significant digital initiatives, like Digital India and Smart Villages. To close the digital gap, more care should be taken to make sure that all rural residents, including underserved communities, benefit from improved connectivity. The strategic alignment of technological innovation, policy support, and community engagement will be critical to the success of B5G and 6G deployment in rural India. There should be a clear focus on developing scalable and replicable models to turn rural areas into digitally empowered regions, thereby significantly contributing to India's overall economic and social development. To fully realize the benefits of next-generation connectivity in rural India, a comprehensive strategy is necessary to surmount obstacles. developments and proposed methods of new 5G and future 6G security approaches, such as Blockchain technology, post-quantum cryptography (PQC), and Artificial Intelligence (AI) schemes, are also discussed to understand better the full landscape of current and future research within this telecommunications domain (Scalise, P., et al 2024). Shortcomings of 5G in meeting the needs of more data-intensive, low-latency, and ultra-high-reliability applications (Salameh, A. I., & El Tarhuni, M. 2022). Various technologies and applications of 6G networks are also overviewed in this paper. 6G network will incorporate Artificial intelligence (AI) based services, edge computing, quantum computing, optical wireless communication, hybrid access, and tactile services (Dogra, A., Jha, R. K., & Jain, S. 2020). Provide readers with the technology road-map and the potential challenges to consider in their 6G research (Mahmoud, H. et al 2021).

CONCLUSION

In conclusion, the advent of 5G and the emerging 6G networks offer unprecedented opportunities to revolutionize global communication and connectivity. These technologies promise faster data speeds, ultra-low latency, and the ability to connect massive numbers of devices, driving innovation in various sectors, including healthcare, agriculture, transportation, and smart cities. However, deploying these networks, especially in rural areas, presents significant challenges that must be addressed to prevent exacerbating the digital divide between urban and rural regions. The disparity in network connectivity between urban and rural areas is a critical issue. Rural regions, which house a significant portion of the global population, often lack the necessary infrastructure to support advanced technologies like 5G and 6G. This results in limited access to high-speed internet and digital services, which are increasingly vital for economic and social development. The challenges in rural areas include difficult terrain, low population density, high infrastructure costs, and a lack of consistent electricity supply. These factors make it less attractive for telecom companies to invest in these regions, further widening the gap in digital access. Addressing these challenges requires a multifaceted approach.

Innovative solutions such as deploying low Earth orbit (LEO) satellites, utilizing TV white spaces, and developing community networks can significantly enhance connectivity in rural areas. Also, hybrid network models combining different technologies to maximize coverage and cost-effectiveness are essential. The involvement of local communities in creating and managing networks, supported by government initiatives and public-private partnerships, can also drive the successful deployment of these advanced networks in underserved areas. Economic considerations are equally important in the rollout of 5G and 6G networks in rural areas. The high initial investment, ongoing operational costs, and challenges in revenue generation due to lower income levels in these regions require sustainable funding models. Government support through subsidies, shared infrastructure models, and international collaborations can help bridge the financial gap and ensure that rural populations benefit from these technological advancements. Ultimately, the successful deployment of 5G and 6G networks in rural areas will depend on strategic planning, effective policy frameworks, and the active involvement of all stakeholders, including government, private sector, and local communities. By addressing the technological, economic, and social barriers, we can ensure that the benefits of these next-generation networks are equitably distributed, thereby contributing to the overall economic and social development of rural regions.

REFERENCES

Azari, , Solanki, S., Chatzinotas, S., Kodheli, O., Sallouha, H., Colpaert, A., Mendoza Montoya, J. F., Pollin, S., Haqiqatnejad, A., Mostaani, A., Lagunas, E., & Ottersten, B. (2022). Evolution of non-terrestrial networks from 5G to 6G: A survey. *IEEE Communications Surveys and Tutorials*, *24*(4), 2633–2672. DOI: 10.1109/COMST.2022.3199901

Chaoub, , Giordani, M., Lall, B., Bhatia, V., Kliks, A., Mendes, L., Rabie, K., Saarnisaari, H., Singhal, A., Zhang, N., Dixit, S., & Zorzi, M. (2021). 6G for bridging the digital divide: Wireless connectivity to remote areas. *IEEE Wireless Communications*, *29*(1), 160–168. DOI: 10.1109/MWC.001.2100137

Chen, S., Sun, S., & Kang, S. (2020). System integration of terrestrial mobile communication and satellite communication—The trends, challenges and key technologies in B5G and 6G. *China Communications*, *17*(12), 156–171. DOI: 10.23919/JCC.2020.12.011

Dogra, A., Jha, R. K., & Jain, S. (2020). A survey on beyond 5G network with the advent of 6G: Architecture and emerging technologies. *IEEE Access : Practical Innovations, Open Solutions*, *9*, 67512–67547. DOI: 10.1109/ACCESS.2020.3031234

Gismalla, , Azmi, A. I., Salim, M. R. B., Abdullah, M. F. L., Iqbal, F., Mabrouk, W. A., Othman, M. B., Ashyap, A. Y. I., & Supa'at, A. S. M. (2022). Survey on device to device (D2D) communication for 5GB/6G networks: Concept, applications, challenges, and future directions. *IEEE Access : Practical Innovations, Open Solutions*, *10*, 30792–30821. DOI: 10.1109/ACCESS.2022.3160215

Gismalla, M. S. M., Azmi, A. I., Salim, M. R. B., Abdullah, M. F. L., Iqbal, F., Mabrouk, W. A., Othman, M. B., Ashyap, A. Y. I., & Supa'at, A. S. M. (2022). Survey on device to device (D2D) communication for 5GB/6G networks: Concept, applications, challenges, and future directions. *IEEE Access : Practical Innovations, Open Solutions*, *10*, 30792–30821. DOI: 10.1109/ACCESS.2022.3160215

Gustavsson, U., Frenger, P., Fager, C., Eriksson, T., Zirath, H., Dielacher, F., Studer, C., Parssinen, A., Correia, R., Matos, J. N., Belo, D., & Carvalho, N. B. (2021). Implementation challenges and opportunities in beyond-5G and 6G communication. *IEEE Journal of Microwaves*, *1*(1), 86–100. DOI: 10.1109/JMW.2020.3034648

Kharche, S., & Dere, P. (2022). Interoperability Issues and Challenges in 6G Networks. *J.Mobile Multimedia*, *18*(5), 1445–1470. DOI: 10.13052/jmm1550-4646.1856

Lu, Y., & Zheng, X. (2020). 6G: A survey on technologies, scenarios, challenges, and the related issues. *Journal of Industrial Information Integration*, *19*, 100158.

Lu, Y., & Zheng, X. (2020). 6G: A survey on technologies, scenarios, challenges, and the related issues. *Journal of Industrial Information Integration*, *19*, 100158. Salameh, A. I., & El Tarhuni, M. (2022). From 5G to 6G—challenges, technologies, and applications. *Future Internet*, *14*(4), 117.

Mahmoud, H. H. H., Amer, A. A., & Ismail, T. (2021). 6G: A comprehensive survey on technologies, applications, challenges, and research problems. *Transactions on Emerging Telecommunications Technologies*, *32*(4), e4233. DOI: 10.1002/ett.4233

Nguyen, V. L., Lin, P. C., Cheng, B. C., Hwang, R. H., & Lin, Y. D. (2021). Security and privacy for 6G: A survey on prospective technologies and challenges. *IEEE Communications Surveys and Tutorials*, *23*(4), 2384–2428. DOI: 10.1109/COMST.2021.3108618

Qamar, F., Siddiqui, M. U. A., Hindia, M. N., Hassan, R., & Nguyen, Q. N. (2020). Issues, challenges, and research trends in spectrum management: A comprehensive overview and new vision for designing 6G networks. *Electronics (Basel)*, *9*(9), 1416. DOI: 10.3390/electronics9091416

Saarnisaari., ... (2020). A 6G white paper on connectivity for remote areas. *arXiv preprint arXiv:2004.14699*.

Salahdine, F., Han, T., & Zhang, N. (2023). 5G, 6G, and Beyond: Recent advances and future challenges. *Annales des Télécommunications*, *78*(9), 525–549. DOI: 10.1007/s12243-022-00938-3

Salameh, A. I., & El Tarhuni, M. (2022). From 5G to 6G—Challenges, technologies, and applications. *Future Internet*, *14*(4), 117. DOI: 10.3390/fi14040117

Scalise, P., Boeding, M., Hempel, M., Sharif, H., Delloiacovo, J., & Reed, J. (2024). A Systematic Survey on 5G and 6G Security Considerations, Challenges, Trends, and Research Areas. *Future Internet*, *16*(3), 67. DOI: 10.3390/fi16030067

Scalise, P., Boeding, M., Hempel, M., Sharif, H., Delloiacovo, J., & Reed, J. (2024). A Systematic Survey on 5G and 6G Security Considerations, Challenges, Trends, and Research Areas. *Future Internet*, *16*(3), 67. DOI: 10.3390/fi16030067

Sharma, T., Chehri, A., & Fortier, P. (2021). Review of optical and wireless backhaul networks and emerging trends of next generation 5G and 6G technologies. *Transactions on Emerging Telecommunications Technologies*, *32*(3), e4155. DOI: 10.1002/ett.4155

Telecom Regulatory Authority of India. (2023, September 29). *Consultation paper on digital transformation through 5G ecosystem* (Consultation Paper No. 24/2023). Mahanagar Door Sanchar Bhawan. https://www.trai.gov.in

Vaezi, , Azari, A., Khosravirad, S. R., Shirvanimoghaddam, M., Azari, M. M., Chasaki, D., & Popovski, P. (2022). Cellular, wide-area, and non-terrestrial IoT: A survey on 5G advances and the road toward 6G. *IEEE Communications Surveys and Tutorials*, *24*(2), 1117–1174. DOI: 10.1109/COMST.2022.3151028

Yaacoub, E., & Alouini, M. S. (2020). A key 6G challenge and opportunity—Connecting the base of the pyramid: A survey on rural connectivity. *Proceedings of the IEEE*, *108*(4), 533–582. DOI: 10.1109/JPROC.2020.2976703

Chapter 7
Enhancing Rural Connectivity Strategies for B5G and 6G Implementation

Dankan Gowda V
https://orcid.org/0000-0003-0724-0333
Department of ECE, BMS Institute of Technology and Management, Karnataka, India

C. R. Vijay
https://orcid.org/0000-0001-6739-935X
Department of Management Studies, Dr. N.G.P. Institute of Technology, Coimbatore, India

S. V. Ramanan
https://orcid.org/0009-0005-3477-6039
Department of Electronics and Communication Engineering, PPG Institute of Technology, Coimbatore, India

K. D. V. Prasad
https://orcid.org/0000-0001-9921-476X
Symbiosis Institute of Business Management, Hyderabad, India & Symbiosis International University, Pune, India

Anil Sharma
https://orcid.org/0000-0002-7115-6278
Amity School of Engineering and Technology, Amity University, Noida, India

ABSTRACT

Realising the potential of B5G as well as 6G technologies makes potential a chance to solve the issues associated with the connectivity of the rural areas. This chapter aims at discussing elaborate plans and approaches to the deployment of B5G and 6G networks to improve on the connectivity of rural areas. We then discuss the technology enablers that rendered these networks fit for rural settings through modern communication technologies such as satellite communications, edge computing, and new spectrum management methods. Thus, by studying the presence of networks in different situations, we discuss the main practices for deploying networks and consider examples of pilot projects. Secondly, the chapter

DOI: 10.4018/979-8-3693-7788-8.ch007

also analyses wider socio-economic effects that increased connectivity has on rural societies, including access to education, health care or employment. We also discuss the limitation that may exist in the implementation of the strategy such as infrastructural, regulatory and financial and the ways by which they could be surmounted.

INTRODUCTION

Information and communication technologies (ICT) is known to have developed greatly in the recent past, making drastic changes in almost all sectors in the modern society, but this has not been the case for the rural areas mainly because accessing a reliable and high-speed internet still remains a dream. Hence, the role of rural connectivity cannot be emphasized enough since it aimed at solving the problem of digital divide, stimulating economic growth and enhancing lives of rural inhabitants at the same time (Al-Fuqaha, A., Guizani, M., Mohammadi, M., Aledhari, M., and Ayyash, M., 2015). For people in rural areas, improved network expands the convenient access to services like education, health, and government, tend to deliver which services via the internet. In addition, this facilitates access to the international market for local enterprises, encourages the growth of entrepreneurship and supports advanced technologies in the agriculture and others rural sectors such as smart farming (Das, A. K., Mohan, P. M., Panda, S., and Mishra, A. K., 2021). Lack of connection means that rural areas may fail to invest in and progress in the technologies of the fourth industrial revolution thus worsening the social-economic divide.

This is one of the strongest barriers faced in the provision of connectivity in the rural areas due to sheer unavailability of infrastructure. That is why many rural regions are thinly settled and cover significant territories; for this reason, telecommunications companies cannot afford investments in broad networks (Mary, A., Begum, A. Y., and D. G. V., 2023). The cost of establishing laying of fiber optic cables, constructing cell towers and maintaining such networks in this regions deters accessibility and causes coverage gaps in most remote regions (Nishikawa, D. K., Higuchi, T., and Kojima, Y., 2020). Further, in developed countries infrastructure development in many rural areas can be hampered by an unfavourable topography, invariably leading to enhanced costs due to the need for equipment and techniques required to overcome physical geography barriers.

Therefore, it is clear that rural connectivity is a very significant factor since it has the capability to change the lives of individuals as well as communities due to connection that offers them the means of getting the essential services and other opportunities (Chandra, G., Kumar, R., and Prasad, K. D. V., 2024). This potential can only be realized once these issues relating to infrastructure, finances and regulation are overcome. Several challenges can be attributed to the current state of Rural communities' connectivity; B5G and 6G technologies present cutting-edge solutions to these problems. Using such technologies, policy-makers, technocrats and other stakeholders can then join hands to ensure that rural areas are not left stranded in the information age.

TECHNOLOGICAL ADVANCEMENTS FOR RURAL CONNECTIVITY

Overview of B5G and 6G Technologies

5G+ or Beyond 5G and 6G are the successive generations of wireless technologies which will bring next level of speed and flexibility. 5G refers to B5G or better still 5G+; this is because 5G is an enhanced next generation which integrates features that extend the limit of the network. Economically it offers enhanced throughput, reduced delay and higher reliability which makes it ideal for high end services. 6G which is still in its conception stage has higher expected performances such as terabit per second data rates, and latency of microseconds with very high reliability (White, G., Barbu, V., and Williams, A. L., 2020). It is expected that these developments would lead to the provision of many more complex services and applications in and around the current technical domain of Virtual and Augmented Reality, IoT, and Autonomous Systems among others (Kim, H., Kim, D., and Park, Y., 2021).

Figure 1. Comparison of network architectures (hybrid vs. traditional)

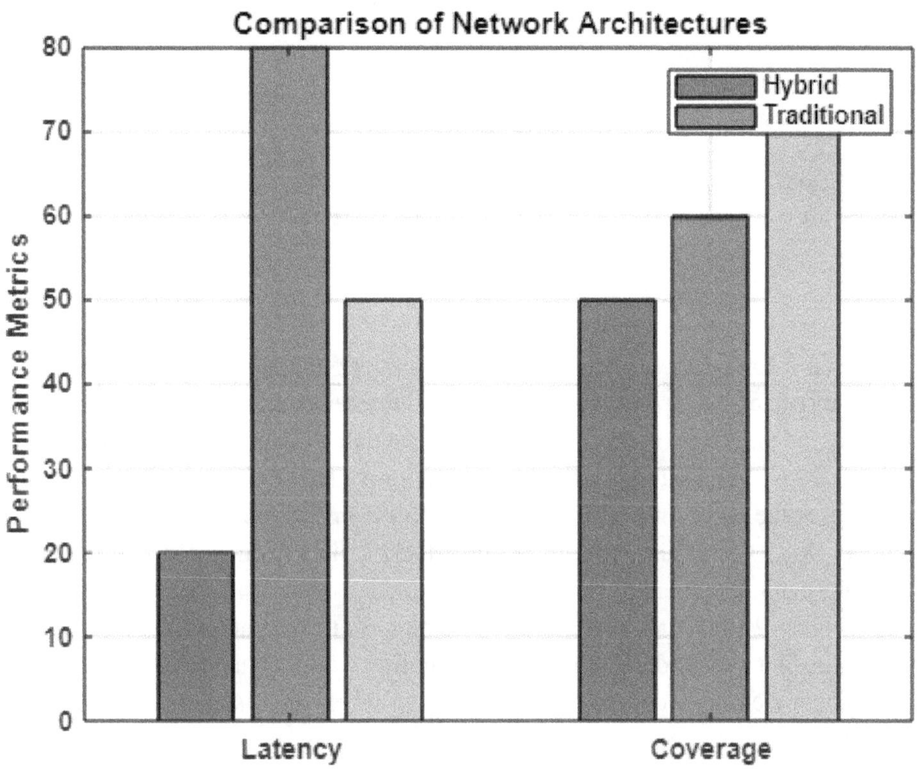

Figure 2. Edge computing vs. cloud computing latency

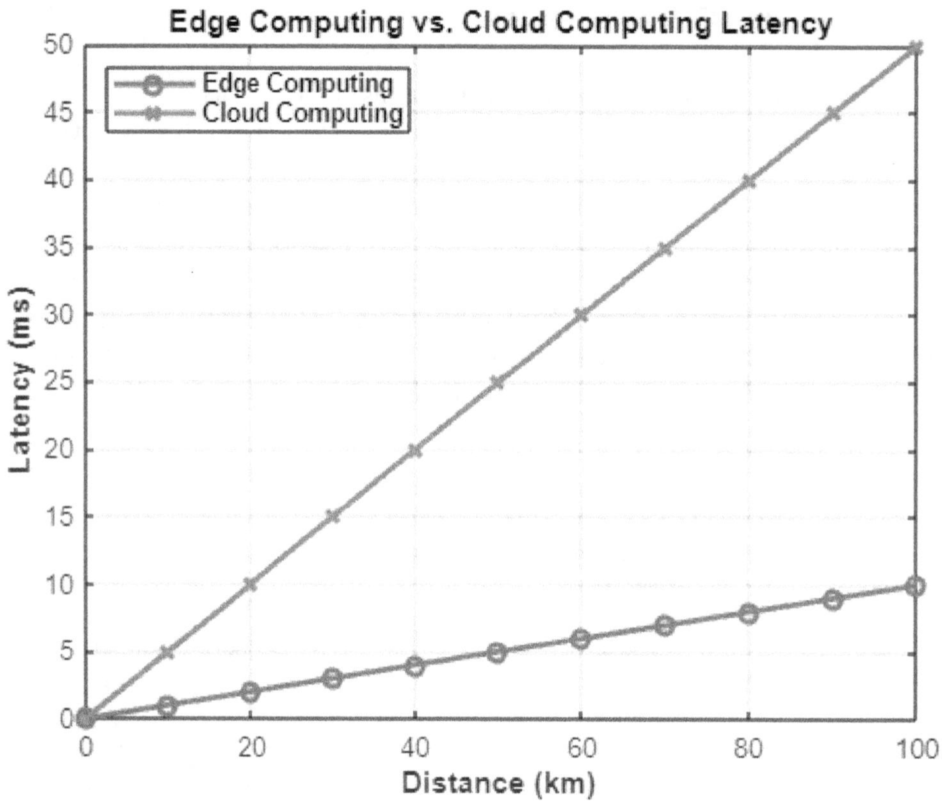

Figure 1 presents a comparative analysis of hybrid network architectures versus traditional terrestrial networks across key performance metrics: In this study, three key factors were considered namely; latency, coverage and cost. Indeed, from the bar chart, it can be seen that the hybrid network, which incorporates the functions of terrestrial and satellite communication, proved to be much better in terms of latency and coverage area and, at the same time, almost did not differ from classic networks in terms of costs (Han, J., Li, H., and Lin, X., 2019). This figure shows how technical efficiencies of hybrid models enable the bridging of infrastructural barriers to rural access.

In this case, Figure 2 illustrates the latency with relation to distance of edge computing and cloud computing. The line plot clearly depict that edge computing where data is processed closer to the source provides lower latency compared with cloud computing when the distance is considered. This figure explains the technical advantage of edge computing in enhancing result computation rate thereby being suitable for use in areas of low network density such as rural areas.

Role of Satellite Communications in Rural Connectivity

Satellite communications are also critical in providing access in the rural and remote areas as conventional wired infrastructure is not available, or its provision is unprofitable. Due to the nature of satellite system, as opposed to the more grounded based network systems, the coverage can be gotten over some large and difficult terrains. Satellites in LEO will moderately offer less delay and superior data throughput than those satellites in geo-synchronous orbit. LEO satellites operate at closer range to the Earth hence curtailing the amount of time the signal takes to across the communication link while boosting the performance of the link (Qadir, J., Ali, A., and Saleem, A. M., 2021).

Benefits and Applications of Edge Computing in Rural Environments

Edge computing is an innovative technology that makes the computation and data storage facilities decentralized and available close to their utilization, thus improves the performances. For instance, in rural settings, edge computing may be the key to success since it can eliminate the problems with the limited available bandwidth and unstable networks (Hegge, K., and Agrawal, R., 2023). Conventional cloud computing infrastructures are based on centralized data centers which may be quite distant from users in rural areas which results in high latency and possible service interferrences. On the other hand, edge computing decentralizes such functions to the network's endpoints or to nodes like local servers, gateways, or even smart devices (Zheng, K., Zhao, S., and Xu, X., 2020). This data can in turn be processed on the edge, thus providing insights and possibly allowing timely actions to be taken. As for the telemedicine, edge computing can facilitate real-time response on the patients' condition, allowing to improve medical help for those who live in rural areas. The very nature of edge computing also to decentralise the network brings reliability and improvement of the network capacity, which are significant in supporting rural areas.

Advanced Spectrum Management Techniques and Their Importance

In view of the fact that spectrum resources may be scarce particularly where wireless networks are to be deployed in rural areas, it is important that the management of the available spectrum be efficient so as to address the capacity and performance issues of the network. Some of the concepts of advanced spectrum management include dynamic spectrum access and spectrum pooling that help in the efficient use of the available frequencies as well as reducing cases of interference (Chiang, M., and Zhang, T., 2016). One of the techniques is the dynamic spectrum access (DSA) whereby devices assist the spectrum with out of band access without any interference to other conventional users. Dynamic spectrum access which involves the use of a spectrum that is sensed and understood in real time depends on cognitive radio technology. Another crucial approach is the spectrum aggregation technique through which frequencies bands are collected to establish a broader pathway hence improving the data rates and networks frequencies (Reddy, M., and Tanguturi, R., 2022).

CASE STUDIES AND PILOT PROJECTS

Detailed Analysis of Successful B5G and 6G Deployments in Rural Areas

It also describes how B5G/6G technologies have been deployed to rural areas and it explains various case studies and pilot projects that indicate how such advanced networks are going to revolutionize the rural areas. An example that aim to prove this statement is the one that has been embarked on in rural Japan by NTT Docomo with the collaboration of Nokia and NEC. This project was an ambitious plan to deliver the fully-fledged broadband by 5G and other emerging B5G communications to the vast regions including the isolated hilly terrains (Hussain, N., Dixit, S. K., and Gupta, A. K., 2023). Such methodologies as a combination of macro and small cells with the help of beamforming and MIMO made the coverage and the speed of data transfer exceptional at the end of the project. Judging from an increase in connectivity, it was evident that such a deployment boosted easy access to online services in education, health, as well as increment in communication among local firms and individuals (Reddy, N. S., and Ashreetha, B., 2023).

A similar example can be described regarding the trial that was performed in the rural area of Finland with the support of the University of Oulu in collaboration with multiple telecom operators. This pilot project entailed the use of 6G features in enhancing ULTRA reliability and low latency communication on existing structures (Mohan, P. M., Das, A., and Panda, S., 2020). The project boring its small cell network density and using millimeter-wave frequencies, was able to show substantial performance gains in the network. It enhanced various applications most of which were390 the following; Implementing remote control of agricultural equipments, wildlife monitoring in real time, high definition video streaming for educational, tourism purposes etc. To achieve this, the government made an embodiment of this initiative to prove that 6G technologies could change the status of rural connectivity and back points of radically diversified services.

Figure 3. Successful deployment metrics

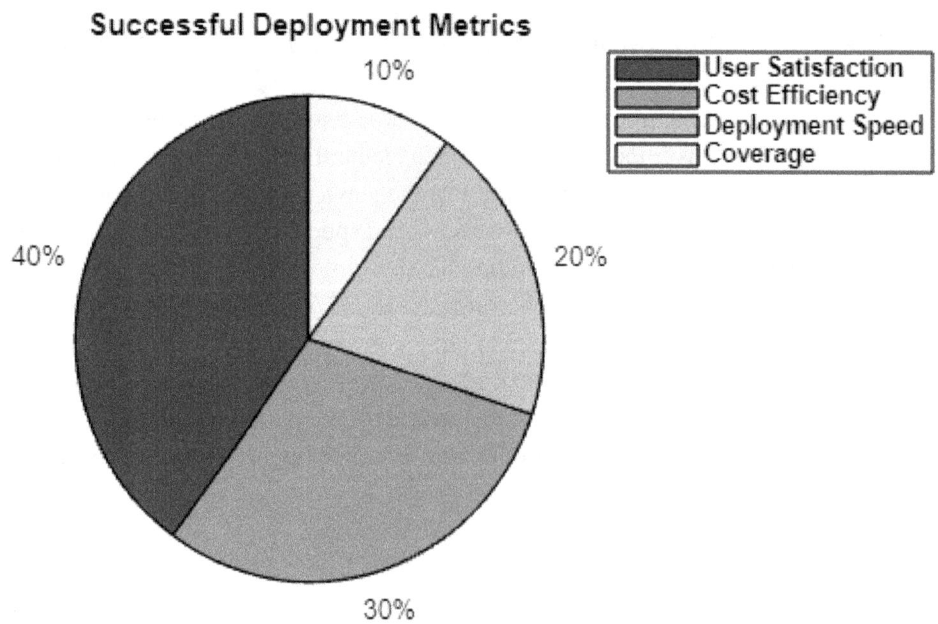

Figure 4. Timeline of pilot project implementation

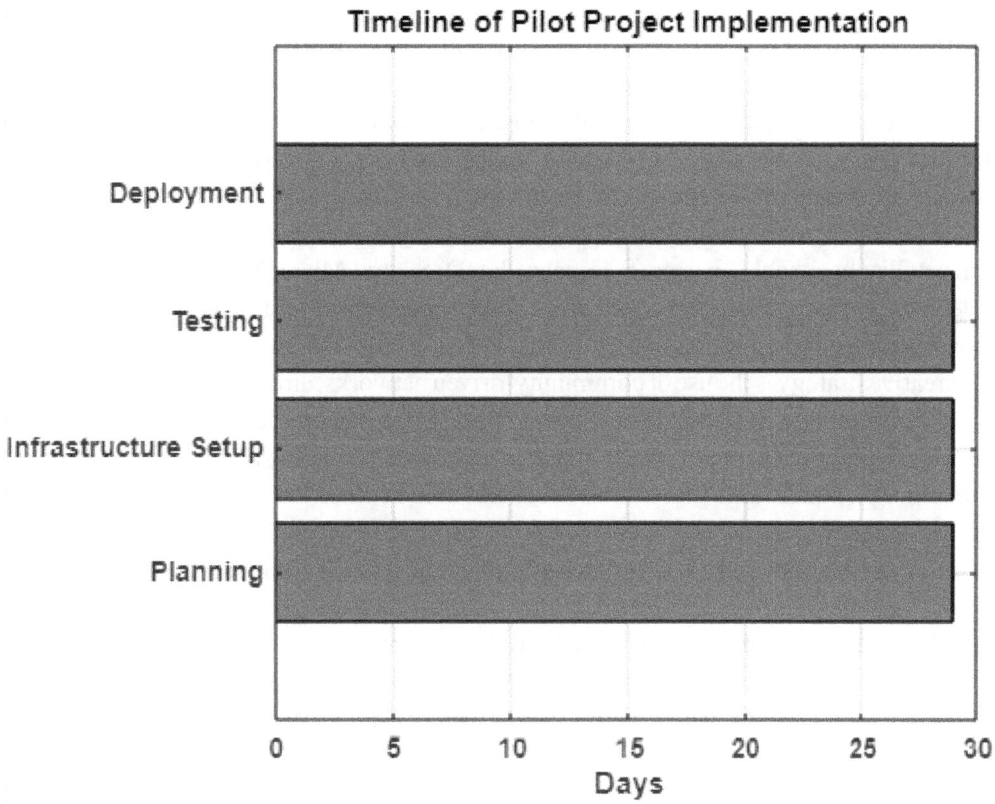

In form of a pie chart, Figure 3 shows the distribution of success metrics for a rural connectivity pilot project. Major ones that are considered are the number of happy users, cost-incurred, time taken to deploy, and coverage (Premkumar, S., Hussain, N., and Chinamuttevi, V. S., 2023). This figure graphically distinguishes between the scopes of the advantages derived from the pilot project, with user satisfaction / coverage occupying the major portion to mean that the pilot project is performing well and users are satisfied with the technology deployed.

Gantt chart of a pilot project implementation is depicted below in figure 4 showing the different activity stages such as planning, infrastructure, testing and deployment. The bars depict each of the stages on a timeline so as to indicate the duration of the tasks in each of the stages (Liu, R., Huang, J., and Wang, Y., 2020). This figure enables one to grasp a number of aspects of the project management and the general logic of having a structured approach as critical to the successful implementation of the project.

The FCC of United States approved Wi-SUN based B5G pilot project in Wisconsin State and some leading players include AT&T and Microsoft and some other players which are in the private domain (Gupta, R. R., Kumar, A., and Gadhavi, D. J., 2020). This plan was meant to help eradicate the broad definition of the digital exclusion with the help of fixed wireless access (FWA) solutions based on B5G technology in rural areas. The project has brought a number of successes including increasing internet speed in areas which had no offer before such as telemedicine; effective online resources for the schools in rural areas (Lien, S., Chen, K., and Ding, Y., 2019). This project demonstrated that it is possible to leverage B5G in improving the access to connection in diverse regions, including the rural areas, which informed similar projects across the nation.

Examination of Different Models and Approaches Used in Pilot Projects

Some aggregate pilot projects on the use of B5G and 6G solutions in rural areas have applied the key models and strategies to these issues. A common integration occurs through the utilization of terrestrial and satellite networks that are integrated to form a single network. For example, the rural 5G project undertaken in Australia with the support of Telstra and the government combined 5G small cell technology with satellite backhaul (Sharma, S., Kushwaha, D. S., and Misra, M., 2020). In this approach, it was possible to cover vast spaces in the rural areas at the same time providing high speeds of connection thereby showing how hybrid systems can work to overcome problems of infrastructure.

Another creative strategy is the use of community-driven networks, an example of which can be studied on the case of rural Mexico. Here the local communities supported by the non-profit organisation Rhizomatica installed their own 5G networks with the open-source technology and free spectrum (Shekhar, S., and Ashreetha, B., 2023). This grassroots model was able to place the facilities for network solutions in the hands of ordinary citizens, hence providing means of achieving affordable and efficient network infrastructure. Apart from ensuring stable internet connection, it also contributed to loca entrepreneurship and social inclusion who laid credit to the fact that the community approach to the implementation of the project enhanced the chances of connectivity in the rural areas.

Identification of Best Practices and Lessons Learned from These Case Studies

Looking in particular at the actual cases of B5G and 6G deployment in rural territories, several promising approaches and lessons have been elaborated. One important lesson is the need for organizations to implement dynamic and almost certainly agile networks. The synchronous and complementary use of different technologies has shown success, especially in areas with geographical and infrastructural barriers; for instance, terrestrial networks and satellite communications. These basic architectures of adaptable satellite systems can guarantee reliable networking, particularly in challenging and remote locations.

The second is the involvement to the internal stakeholders such as the communities within the state or country. The Community based Networks in Mexico show that there are advantages in having and implementing local minded projects. Involving the communities in steering technology solutions for connectivity also entails that they are equally involved in the deployment and maintenance of such systems means that the future growth of such solutions will be occasioned by the locals, therefore making it sustainable (Barua, T., and Mohiuddin, K., 2024). This approach can therefore be of great use in particular when targeting and or handling the unique issues arising with the developing rural scenes.

This involvement of key stakeholders requires future work, but the fact that partnership between public and private entities has become one of the key success factors for implementing rural connectivity projects. Partnerships with the aim of working hand in hand can take advantage of available whereby the strengths of the respective partners can be enhanced (Maddineni, V. K., and Gummadi, V. A. K., 2024). The support from the government in the deliberation of funding and favourable policies together with the favourable spectrums by the government on the side of the private firms complete with the technological know-how and the necessary infrastructures in place fuel successful deployments. BharatNet is one of the best examples how PPPs can contribute to enhance the connection of rural inhabitants and minimize the digital gap.

SOCIO-ECONOMIC IMPACTS OF IMPROVED CONNECTIVITY

Increased Access to Education Through Enhanced Connectivity

Improved network can thus transform the education sector in the rural areas by availing a variety of teaching and learning resources through virtual mode. With the advancement in the internet connection students in the remote areas are able to go online and access different forms of learning and interact with teachers from any part of the world or even students. This is a digital inclusion that enhances students' educational access between urban and rural students giving both equal chances in training.

Namely, connectivity as a positive outcome brought by the integration process helped to open access to the Massive Open Online Courses (MOOCs) and other akin online schooling options. These platforms provide students with various disciplines of courses, and thus the students in the rural areas are readily able to access subjects that may not be commonly run in their area of residence. Also, virtual classrooms along with live lectures let students participate in instructors and fellow students live, which makes learning process more practical and interactive (Srinivas, V., and De, T., 2023). Such access to high quality education assists the students to gain knowledge and skills that will improve their carrier lives. Improved connection also facilitates the continuing education of teachers in the rural areas. Thus, online training programs and webinars help educators be aware of the modern approaches to teaching and new developments related to their subject fields.

Figure 5. Improved access to education

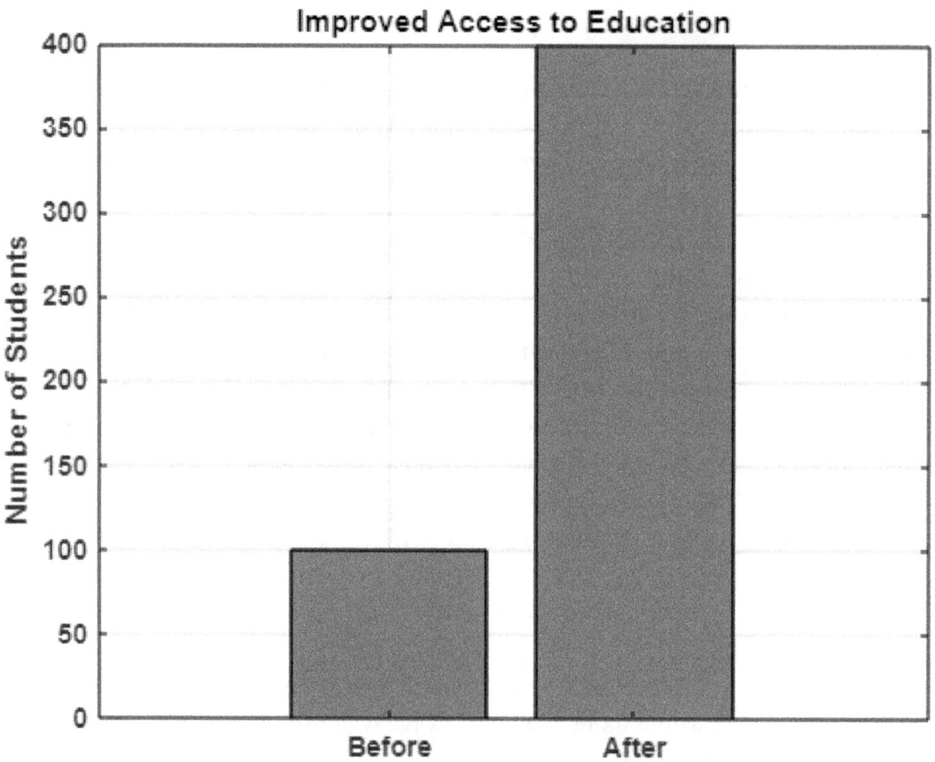

Figure 6. Economic development indicators

The effect of better connectivity on the access of education in rural areas is shown in figure 5 in the form of bar diagram. It is used to compare the number of students who had an access to online education prior to high-speed internet and after deploying the high-speed internet. The overall increase in the number of students after implementation clearly points to the possibilities to develop more access points and opportunities for rural students that stem from improved connectivity.

The inset in the bottom right-hand corner of Figure 6 is a dual-axis line and scatter plot of two economic development data points- income and employment rate- for the ten years following improved connectivity. As to the scales, we use the left y-axis for the income, and the right, y-axis for the employment rate. The trends portrayed by the plot on income and employment rate have been on the rise showing that improved connectivity has a progressive effect on the economic status of rural areas.

For instance in some schools in the rural setting, they may not be able to attract any qualified or specialized teachers, this is where the utility of online tutors and tools can play a crucial role in bridging this gap. It is possible for students to seek extra coaching particularly in difficult areas thus improving their lot as compared to students in the urban schools. Further, there are numerous applications and software that exist for teaching purposes that include use of educative games along with animated tutorial lessons thereby making learning more fun for learners in the rural areas.

Improved Healthcare Delivery via Telemedicine and Remote Diagnostics

That means telecommunication and advanced connection is a boon to the healthcare facility in the rural areas especially in those zones where access of medical facilities are very low because of distance barrier and deficiency of doctors and other medical practitioners. Telemedicine and Remote Diagnostics hence takes advantage of better internet connectivity to deliver efficient and effective healthcare services to rural dwellers leading to better healthier and happier lives.

Telemedicine enables the rural patient to be taken care of by a health care provider over the media instead of having to cover long distances to access health care facilities within the urban centers. With the help of video calls and protected text applications, the customers can get the professional opinion of doctors from any country of the world, as well as receive a diagnosis and a treatment regimen.

due to the availability of high speed internet, remote diagnostics can improve usage of medical equipment and technologies to observe the health of the patients at a distance. Smart clothing and home monitoring tools can monitor the individual's health status and detect signs of developing complications to which the healthcare givers will be in a position to attend to the patient or client early enough. For instance, patients with heart problems have been allowed to wear ECG devices that help cardiologists to check their progress regularly. It is such an active role to play in the healthcare system that many heath issues are prevented from progressing into fatal conditions are prevented right from their stages.

Economic Development Opportunities Facilitated by Better Connectivity

Telecommunications development in the rural zones is an essential condition for creating new jobs and new business opportunities for the population. Accessibility of high-speed internet and proper communication networks can help the rural regions to connect with the global economy, increase innovation, employment opportunities and sustainable development.

The enhancement of connectivity is one of the primary source of return on investment on the economic prospective where there is prerequisite for e-commerce and virtual business solutions. The ability of operating in the local market may be a constraint for rural businesspeople who can overcome it by promoting their goods and services online. These are a way of connecting to larger customer base hence increasing the sales and revenues, a key determinant to the economic improvement of the rural areas. For instance, it is now possible for the artisans, craftsmen, and other such workers to sell their handicrafts online to consumers across the globe. In the same way, holders of agricultural products can use other online market places to engage in sales to customers or distributors without having to deal with the middlemen.

Thirdly, better connection facilitates the development of SMEs especially in rural setting as one of the factors. SMEs play very crucial roles in economic diversification and development since they create employment opportunities and drive the innovation. Internet connectivity brings the needed facilities and opportunities for training, education and finances available online that increases the capability and competitiveness of the rural business owners. The development of SMEs in the rural area promotes the development of a strong and active economy, thus eliminating the monopoly of traditional sectors in the economy thereby promoting a more productive economy in the long run.

Broader Social Impacts and Enhancement of Quality of Life in Rural Communities

The pros of improved connectivity go beyond education, health care and economic opportunities to revolutionize the quality of life in rural society. Affordable and effective broadband connection helps to enhance people's well-being and their communications with others while providing support to maintain cultural identity, thus improving society's cohesiveness.

Another social benefit which may be attributed to increased connectivity is getting connected with the close families and friends whether near or far. Social sites, apps of video calling, and messaging help rural consumers to continue social connections that may help them not to feel lonely and isolated. This connectivity is especially vital for the elderly and those from rural settings as that ensures that they are connected to other people. Besides, they create social contacts as internet continues to provide social networking sites where people with similar experiences and interests can interact.

Better connection also helps in the access to entertainment and recreational utilities. The citizen of the rural area needs to be able to watch movies, listen to music and playing games, communicate on the forums and read different materials. Thus, it revives quality of life through the availability of entertainment while at the same time informative and educating at the same time. However, digital connectivity enhances the expansion of local creative industries like film, music and games industries which are potential creative industries of arts.

Finally, it can be noted that the effects of further connectivity in rural areas are meaningful and pervasive, therefore they are ultimately socio-economic. This is a reason why an efficient and fast connection to the internet contributes to sustainable development in education, health, economy, social relationships, etc.

BARRIERS TO IMPLEMENTATION

Infrastructural Challenges Specific to Rural Areas

Connecting people in rural communities with high-speed internet options requires the use of advanced connectivity technologies, but deploying these technologies is easy in the rural areas due to the infrastructural barriers that are associated with internet connectivity technologies. One of the biggest issues is the large area and land which is mostly inaccessible in many rural areas. These areas may embrace mountainous regions, forests, large plains and other thematically similar geographical regions which present lots of challenges to the personnel involved in placing the physical structures like the fiber-optic cables and the cell towers. While areas with concentrated businesses and people have potential to attract high revenues for telecommunication companies and subsequently provide revenues for upgrading their infrastructures, rural areas on the other hand offer small markets to telecommunication firms due to low population density. It is due to this economic disincentive that several regions especially the rural areas, are either left unattended or receive substandard services.

Figure 7. Breakdown of implementation barriers

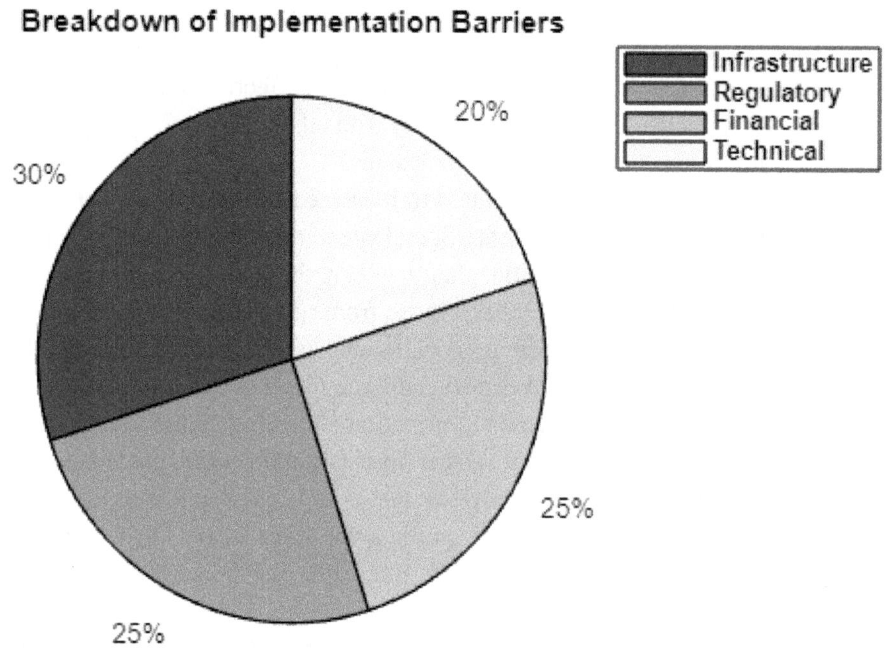

Figure 8. Cost comparison of different connectivity solutions

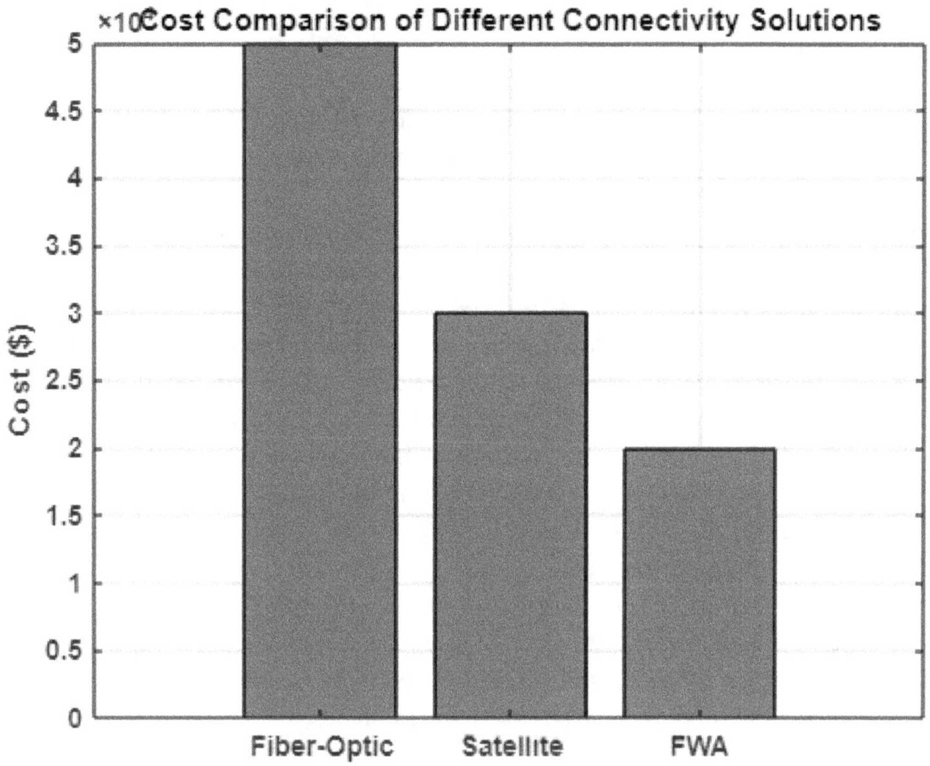

Pie chart in figure.7. below highlight the different barriers that would hinder the implementation of B5G and 6G in rural areas. Some of the barriers are infrastructure based, regulatory based, financial based and technical based barriers. That figure shows the share of each barrier in more detail, indicating that it is impossible to tackle the problems merely by introducing one or another measure; all of them intersect and can be solved only within the framework of a complex approach.

The analysis of the costs of different connectivity solutions is illustrated in the type of a bar chart in figure 8 with the options; fiber-optic, satellite and fixed wireless access (FWA). This brings us to understand that FWA will be more economical than satellite and then fiber-optic. This figure enables one better comprehend the fiscal impact and requirements of each technology and thus help in decision making on the best technology to use in the rural areas based on cost.

Moreover, through the use of community based network initiatives, there is the ability to struggle past infrastructural limitations. In an attempt to dispel with laying the full burden of networks construction and maintenance on the shoulders of the LDCs' governments, local communities could also take an active part of network construction and maintenance by employing locally available and cheaper materials. Such civil initiatives, usually entail emergence strategies like the use of TVWS or unlicensed spectrum offered for wireless broadband where costs are lower and solutions more sustainable.

Regulatory Hurdles and Potential Solutions

Legal restraints are an essential factor of concern since they act as another barrier towards the installation of advanced connectivity technologies in the countryside. Telecoms regulation may be excessively layered and diversified across regions and thus becomes a challenge for service providers to understand and observe the existing regulations. These barriers can slow down the innovation and implementation of new technologies and therefore slow down the development of internet technologies in the rural areas through increased costs in compliance.

In this regard a key area of regulation is spectrum sharing Among all the regulatory issues, the most important one seems to be spectrum allocation. Spectrum or EMF which are the foundation for wireless communication, are scarce resources that can only be obtained under much regulatory measures. Currently, the spectrum is offered through auctions or licensing rounds that provide priority to large telecoms with deep pockets with which to compete for a scarce, valuable resource in the form of the spectrum.

Best practices are also used in dealing with regulatory issues, this shows that international cooperation is also important. The governments can learn from the experience of other countries as to which regulations have worked well in connecting rural area. Joint international partnerships can result to the formulation of harmonized codes and standards that enhance the possibility of investment and implementation of technologies beyond frontiers.

Thus, the main barrier that arises from the analysis of the presented materials is the presence of various regulatory constraints regarding the provision of high-speed connectivity in rural regions. But at the same time, by simplifying the procedures of regulation, using the principles of flexibility in spectrum management, and having the support of favourable public policies, such barriers can be eliminated.

Financial Challenges and Sustainable Funding Models

The second has to do with financial constraints, which are a major inhibitor to the adoption of advanced connection technologies in the rural regions. Difficulties have been observed, for instance, due to high costs regarding construction of telecommunication networks in the regions with low population density and physically hard accessing areas, that causes a lack of interest of private capital. Further, the rural consumers, in most cases, are relatively poorer and financially less endowed hence diminishing the returns on such investments. There is no easy way of addressing these financial challenges other than developing and implementing sustainable and effective funding strategies that will foster and mobilise both public and private sector funding.

Among various financial considerations, one of the most pressing ones is a rather high level of capital intensity when it comes to network deployment. The expenses associated with running fibre optic cables, building the cell towers and installing other essential facilities might be very expensive mainly because of the geographical barriers such as physical features and scattered people. These initial costs of investment are often accompanied by fixed cost that would be incurred in the course of putting up, operating and even expanding the infrastructure. If there is little to no cash motivation and/or backing for private sector organizations, the last thing these may wish to do is sink money into the provision of rural connectivity.

Thus, there are many sustainable funding models that governments and policymakers can undertake to overcome such financial issues. It is important when implementing policies for rural broadband to note the importance of Public funding and subsidies. Government may decide to allocate the budgetary means toward subsidizing the deployment of the infrastructure in the unserved areas by offering grants. With these public funds, it can assist to spread the cost of high capital expenditure and the financial problems connected with the execution of rural projects. For example, the Universal Service Fund (USF) model widely applied in many countries implies the use of funds collected from telecommunications operators to pay for the projects of connecting completed in rural and remote regions.

Another known funding model can be traced in Public Private Partnership or the PPP. The effectiveness of PPPs is evident when the public domain contributes funds to a project while in parallel attracting participation of private partners. While initial funding and basic regulation can be offered by governments, private enterprises are capable to contribute their technological and implements knowledge necessary for forming the required infrastructure. This model can generate the lasting and that supports growth of rural access to efficient connectivity. Positive PPP models include the one used by the Indian government, currently in the implementation of the plan known as BharatNet with the purpose of providing different rural villages with high speed internet connectivity through the participation of private players funded through national coffers.

Community-based funding is another feasible solution available in case. Communities can partner together and use their own funds in the instalment and sustenance of the connectivity networks they need. Not only does this overcome the lack of funding – a perennial problem in development work – but it also lends itself to ownership at the local level. Self-generated formations based on microfinance or cooperatives, specifically in the countryside could help rural people become more involved in solving the problem of poor connectivity themselves.

Other great forms of finance for Rural Connectivity projects include International funding and developmental assistance. Bilateral providers mainly include the World Bank while multilateral providers comprise of the ITU among others and often provide grants, loans and technical cooperation for telecommunications infrastructure development in the less developed world. Thus, these international resources can supplement the national ones and supply the required financial support to address the high costs of the rural connectivity initiatives.

Technical Obstacles and Strategies to Overcome Them

One of the primary challenges that make the effective installation of higher connectivity technologies challenging is the technical constraints that come with providing improved links to the rural territories. The problem is that rural population density is low and people live far from each other, often in regions with severe climate that raises technical challenges related to deployment of high-speed Internet networks. As informed above, these challenges can only be met by the application of sound technological strategies and better planning.

This lack of infrastructure is one of the first technical issues that have to be overcome when starting the project. Typically in the rural regions, there are basic requirements which are required to support the high levels of connectivity such as electrical power that are non-existent or insufficient. This deficiency makes the provision of Telecommunication structures difficult since most of them require constant power supplies to function effectively. Further, conditions of geographical availability influence the installation of network components like the fiber-optic cables and cell towers because they have lots of mountains, large forests and huge open plains. These geographical barriers are tough to overcome in any way hence making the deployment of equipment and implementation of working techniques even more complicated and expensive.

The other known strategy is the use of edge computing, whereby data processing takes place where data is being generated. Edge computing, therefore, de-centralizes the computing resources and does processing at the edge of the network, which lowers the latency and bandwidth consumption. This is especially useful in the rural areas since backhaul capacity referring to the link from the local network to the internet may be inadequate. In reality, innovative solutions in the field of edge computing can be implemented in such applications as monitoring the condition of agricultural lands in real time, telemedicine, the delivery of content that is popular in the region, etc., which will ultimately improve the general experience of the end user as well as alleviate pressure on large central data centers.

It is also important to note that some of the technical challenges need the use of sophisticated methods of spectrum management. Dynamic spectrum sharing enables many users to use the same frequency bands but at different times thus making improvements in the usage of available spectrum. Cognitive radio technology which allows a device or system to sense and learn the surroundings spectrum in real time can also improve the overall spectrum replenishment. Authorities can help on these technologies by adopting polices that open up the opportunity to use spectrum in synchronization or shared frequency designated for deployment of high-speed internet in rural areas.

It also shows that partnership in infrastructure deployment is also an effective method in dealing with technical issues. Policy makers, industry players, civil society and governments at larger need to collaborate to invest in knowledge and solutions in order to come up with a common and uniquely adequate solution for the rural users. For instance, the government can offer necessary regulations and subsidies while private sectors offer technological and physical facilities. Engagement of the community is essential because it ensures that local context is considered and the chances of adoption When implementing change are high.

Finally, it can be stated that, technical challenges are some of the main factors that may hinder the adoption of new connection solutions in rural regions. Such challenges need to be addressed by the integration of next-generation wireless technologies, optimal spectrum utilization, edge computing, and more. The lesson one can draw from each country's experience is that it is feasible to create viable and efficient solutions for rural areas' connectivity if the technical issues and obstacles specific to these areas are tackled.

STRATEGIES FOR SUCCESSFUL IMPLEMENTATION

Policy Recommendations for Effective B5G and 6G Deployment

To achieve B5G and 6G in the rural use cases, therefore, proper and effective policy guidelines are critical. It is evident that governments bear a great responsibility in ensuring that all stakeholders provide a favourable environment that supports advancement in the connectivity technologies. Literature review indicates that rural tele centrization requires concern on regulatory issues comprised of spectrum management, infrastructure deployment, financial incentive and more so, establishment of relevant regulations.

Therefore, among the policy issues, spectrum management can be regarded as one of the most important ones. It is recommended that governments devise and implement policies that are elastic and kinetic in the allocation of the spectrum especially in the provision of required frequencies for broadband in the rural areas. Cognitive radio and Dynamic Spectrum Sharing capabilities can can make sharing of the frequency bands efficient in that it allows many users can access the same band at the same time and at the same instance, with less interference. The several ways in which relevant policies can enrich connectivity in the rural areas include unlicensed spectrum bands and licensed shared access (LSA).

Laws and regulation should not discourage innovation and competition hence they should be framed in that way. Governments can namely contribute through tweaking of regulations that provide ease of access to infrastructure and entry new players into the market. Reduction of regulations and their costs may increase competition and enhance delivery of services, thus benefit the rural consumers. Besides, net neutrality policies also guarantee that everybody can have access to fast internet without prejudice based on location.

Therefore the policy recommendations for B5G and 6G deployment in the rural areas should consider the flexibility in spectrum management, infrastructure deployment, financial support and regulation. This drives the fact that any policy aimed at developing enhanced connectivity technologies enables governments to implement appropriate policies aimed at addressing the digital divide that separates urbanized centers from other regions.

Technological Strategies to Optimize Network Performance in Rural Areas

To increase network performance for the rural areas, it is important to embrace new technology solutions that are appropriate for the countryside. In this scope, the use of B5G and 6G technologies provide several solutions for improvement of the key parameters of connectivity and reliable access to the internet for rural people. Hence, a powerful technological management approach is the implementation of the integrated systems of the Earth- to- Earth and the terrestrial- to- space networks. Hybrid networks take advantage of the two technologies hence having both a wide coverage area and fast data transfer rates. For instance, FWA which stands for fixed wireless access is used to offer broadband connections with short range coverage while satellite communications which are useful in areas that would otherwise be hard to cover. Many planned communications satellites can operate from the Low Earth Orbit (LEO) with low latency and high data rates which are very important for the rural area connectivity. Basically, the hybrid network structures established by combining terrestrial and satellite networks can provide the whole and reliable coverage.

Another vital technological strategy that have to do with the general network performance are called edge computing. Edge computing makes the processing of data closer to the source which helps in reducing the time taken to transmit the data and thus increases system efficiency. It will be especially useful in cases where speed is crucial, for instance, field equipped with sensors for precision farming, remote healthcare solutions, or smart electricity networks. It also relieves the pressure on the central data centres as well as enhances the general performance of the network.

Sophisticated methods of spectrum management which involve the efficient utilization of frequencies presently available are very important. This concept enables multiple user to use the same frequency bands and at the same time, maximize the utilization of the spectra with minimizing interferences. With cognitive radio technology, devices can detect and learn the spectrum environment in real time and thereby improve spectrum reuse. These techniques are specifically required in the areas where the spectrum is scarce, for instance the rural regions.

Therefore, to enhance the networks performance in a rural area, the use of a fixed hybrid network architecture, integrated edge computing, value-added frequencies, and organized small and macro cell placement, and optimized backhaul solution are required. With the help of these technological measures, it will be possible to achieve the creation of stable and fast channels for the connection of rural settlements and overcome the problem of the digital divide.

Importance of Stakeholder Collaboration (Government, Private Sector, Community)

The roll out of B5G and 6G in the rural areas entails involvement of key players such as government, private sector and the local people. All these stakeholders have critical roles to play in meeting the special needs of rural access and in seeing to it that the newest technologies are adopted responsibly.

This is possible only when the government plays an important role in ensuring such environment is provided to investors to encourage them to invest in innovations. There are several ways policy makers can get involved in supporting rural connectivity projects, these include developing the supporting legal structure, issuing requisite subsidies and eventually giving grants. With the help of simplification of the relevant regulatory requirements and minimization of essential bureaucratic procedures, the governments can encourage the speedy deployment of infrastructure. Furthermore, governmental activities like the setup of Universal Service Funds (USFs) can also involve funding of the connectivity projects in unserved regions. Governments are equally involved in the coordination to ensure that the different stakeholders are [incorporated] and various connectivity initiatives are compatible with the socio-economic development plans.

Collaboration is also an excellent way of promoting innovation and sharing of information in the organization. Stakeholders are also in a position to identify and offer solutions which suit the rural environment as they collaborate on implementation of the projects. For instance, collaborations between suppliers of technology and local people can result in the establishment of solution that addresses the local situation and environment. Effectively, collaboration can help establish the sharing of successful strategies as well as the forgetting of common mistakes, to ensure that the stakeholders learn from past achievements, as well as drawbacks encountered by other agencies.

Therefore, it is indicative that effective management of stakeholders is determinant in the accomplishment of B5G and 6G technologies in the rural areas. All the players, such as government, private sector, and locals, have approach different in connectivity endeavors. All these stakeholders can therefore collectively counter the difficulties that come with connecting the rural areas with the rest of the world and spread the advantages of modern technology across the population. Collaboration enhances innovation, proper use of resources and enhances the achievement of SDG goals.

Role of International Partnerships and Support from Global Organizations

Close collaboration with other countries and receiving support from international organizations is crucial to achieve the effective implementation of B5G and 6G technologies in rural regions. They involve cross-stratum approach that pools different capacities, assets and viewpoints in design and provision of novel solutions to the multifaceted problem of rural access. Coordination with other countries can increase the effectiveness of national and municipal initiatives and guarantee that developing innovative communication technologies is available to those who are part of the Connected Society but live in the most barren and deserted areas.

International partnerships provide an opportunity to exchange information between the partners and use the information for improvement of their own practices. Developed countries that have implemented the advanced connectivity technologies can offer a number of best practices that could be followed bearing in mind the various challenges that are bound to be encountered during the implementation process. Various related institutions including the International Telecommunication Union (ITU), the World Bank and other international organisations significantly promote such knowledge sharing.

Lastly, it can be said that socio-academic international cooperation and involvement of the global organizations are crucial for the ongoing and future deployment of B5G and 6G technologies in rural regions. Thus, possibilities for transportation of knowledge, financial resources, advanced technologies, propaganda of favourable policies, and capacity building may all help international collaborations make rural connectivity projects more effective and sustainable. Applying the principles of the utilisation of the community's assets and opportunities guarantees that the potential of the innovative technologies of connection bring positive changes to each individual area without leaving thousands and millions of people captives of their isolation.

ACTIONABLE INSIGHTS AND RECOMMENDATIONS

Summary of Key Insights from Case Studies and Technological Analysis

Many cases have since been presented, and technological analyses have been conducted focusing mainly on B5G and 6G technologies in rural areas. These have drawn focus on the need to come up with flexible and loop-hole strategies that can address the peculiarities of establishing rural broadband connection. Some of the insights obtained include; understanding the applicability of the hybrid network architecture, the importance of getting the community involved in TE, benefits of PPP, and the need for adaptive spectrum policies.

It makes sense to note that the use of the so-called dual solutions integration that combines terrestrial and satellite networks was highly effective. Thus, experience with large Internet coverage based on the integration of various technologies was shown in such countries as Japan, Finland and the USA. The hybrid models use the features of the terrestrial and satellite networks providing high availability and performance that is unachievable in some areas. It has been thus most useful in overcoming the weaknesses of what for long has been the ground based infrastructure making wireless solution a viable proposition to take the internet to the remote rural territories.

Another major factor that has been found to greatly influence connectivity projects is that of community participation. In the same way, other countries including in Mexico have embarked on empowering the local communities to be actively involved in making the networks relevant and sustainable. Such models also go further in meeting the needs of rural people as well as local buy-in for the projects than other systems. Through this grassroots approach, connectivity solutions adopted addresses the prevalent conditions as well as the consumers' preferences readily in the rural areas hence a better chance of sustainability.

Thus, the analysis of the case studies and technological dimensions supports the adoption of the dual network structure, the multi-stakeholder community engagement, the public-private partnership model, and the adaptive policies concerning the spectrum utilisation. These ideas give the baseline platform that can be utilized to come up with right strategies that shall enable realization of the desired connectivity in the rural area boosting the implementation of B5G and 6G.

Practical Recommendations for Policymakers, Technologists, and Stakeholders

Based on the findings and discussions resulting from the case studies and technological analyses several prescriptive suggestions can be provided for policymakers, technologists, and other stakeholders engaged in rural connectivity initiatives. The following are suggestions that further the goal of establishing favorable conditions for technology rollout and put into consideration the peculiarities of rural settings as well as the global Development Agenda as presented in the UN's Sustainable Development Goals 2030.

It is therefore important for policy makers to set and enforce dynamic policies of spectrum management to fully exploit the available frequencies. It is recommended that DSS and cognitive radio systems should be as popular as possible to achieve more excellent utilization of the spectrum and to provide rural areas with enough bandwidth to develop high-speed internet. A further issue which can improve the deployment of infrastructure are the concurrent simplification of various procedures and the minimisation of bureaucratic reporting levels. Uncombine and minimizing the existing approval processes could help to speed up connectivity projects and lessen providers' expenses.

The roles of the stakeholders especially the local communities are that they should be encouraged to participate in planning and development of connectivity projects. Localized models of connectivity that enlist the participation of individuals in constructing as well as maintaining networks has the potential of improving the viability of connectivity solutions. Stakeholders should also engage with governments and private investors with a view of combining efforts in mobilizing resources, sharing risks as well as utilizing assets to the optimum. Because collaboration promotes innovation and due diligence in making connectivity drive relevant to the status of rural areas.

Therefore, the Policy Recommendation for policymakers, technologist and stakeholders include; The establishment of dynamic spectrum management policies, availability of both financial incentives and public funding, use of a dual network approach, application of superior spectrum management methodologies and development of a community management systems and stakeholders' engagement. Therefore, these recommendations give an idea and guide to improve the rural connectivity and successful implementation of B5G and 6G.

Strategic Guidelines for Addressing Implementation Challenges

As for the problems related to the implementation of B5G and 6G technologies in the rural areas, some guidelines have to be developed which should take into account peculiarities of these territories. These guidelines were developed to eliminate infrastructural, regulatory, financial and technical challenges and barriers that are relevant to rural connectivity projects hence spearheading sustainable and efficient projects.

However, one of the significant structural gaps is that few structures have already been developed in the rural areas. Mores should be set to ensure that strategic guidance employs the us of new technologies that should not necessitate ground infrastructure. The conclusion is that terrestrial-satellite mixed networks appear to provide a realistic solution. The two technologies that offer broadband internet connection in rural and inaccessible regions are fixed wireless access (FWA) and low earth orbit (LEO). Moreover, the community networks' approach, in which people from communities make their connections themselves, can respond to gaps in infrastructure needs and ensure sustainability of connectivity for rural communities.

There are technical challenges that encompass scarcity of spectrum and variability of power sources and therefore need to be addressed. Strategic objectives should ensure the use of new methods of regulating the use of the spectrum like dynamic spectrum access and use of cognitive radios. Applying the concept of edge computing can lessen the rate of latency faced in the rural areas hence improving the performance of the networks in handling applications that require real-time data processing. Moreover, there are new technologies on how the backhaul can be established where microwave linkages and satellite backhaul can enable efficient connection between rural networks and central data centre hence providing efficient internet services.

Hence, guideline for strategic approach to tackle challenges of implementation are provision of incentives for innovative technologies, efficient solutions to regulatory procedures, development of sustainable funding mechanism, encouraging improved spectrum management and encouraging stakeholders' co-operation. Cordial implementation of the following measures can help policymakers, technologists and its stakeholders to mitigate the hurdles toward rural connectivity that makes B5G and 6G's successful deployment possible. Such strategies will help to bring the needed changes to the existing inequalities and transform extensive mono-functional rural zones into modern informational zed territories.

CONCLUSION

In this particular chapter, the focus has been given to the aims of B5G and 6G technologies in overcoming the issues of connectivity in the rural areas. In analysing case studies and technologies, we discovered best practices such as the use of dual active networks, space communication, and edge computing mechanisms and tools. Thus, such changes in rural accessibility have enormous socio-economic advantages that include enhanced education, improved health care, catalysed economic development and social integration. Some of the learnings that can be gathered from successful cases are the need for spectrum agility, close collaboration with the communities, and partnership between and among government and corporations. For example, infrastructural, regulatory, financial and technical barriers that limit the application of new technologies in the energy sector indicate that new technologies need multi-stakeholder strategic partnership between governments and technologists to foster.

To fully unlock the potential and leverage on B5G and 6G to transform the rural connectivity, there is the need for more efforts in network deployment to be stepped up. For the policymakers, the above recommendations suggest that they should provide conducive policies and adequate resources for supporting related techniques while the technologists should remain creative in their endeavours. Combined we can make sure that rural regions are also not left out on the new age, hence sustainable society development for all.

REFERENCES

Al-Fuqaha, A., Guizani, M., Mohammadi, M., Aledhari, M., & Ayyash, M. (2015). Internet of Things: A survey on enabling technologies, protocols, and applications [Fourthquarter.]. *IEEE Communications Surveys and Tutorials*, *17*(4), 2347–2376.

Barua, T., & Mohiuddin, K. (2024) "Dynamic Disaster Management with Real-Time IoT Data Analysis and Response," *2024 International Conference on Automation and Computation (AUTOCOM)*, Dehradun, India, pp. 142-147.

Chandra, G., Kumar, R., & Prasad, K. D. V. (2024). Securing digital authentication with cryptographic innovations in intrinsic verification systems. *Journal of Discrete Mathematical Sciences and Cryptography*, *27*(no. 2-A), 317–328.

Chiang, M., & Zhang, T. (2016). Fog and IoT: An overview of research opportunities [Dec.]. *IEEE Internet of Things Journal*, *3*(6), 854–864. DOI: 10.1109/JIOT.2016.2584538

Das, A. K., Mohan, P. M., Panda, S., & Mishra, A. K. (2021). IoT-enabled smart city solutions for traffic management and urban development [Apr.]. *IEEE Internet of Things Journal*, *8*(7), 5744–5756.

Gupta, R. R., Kumar, A., & Gadhavi, D. J. (2020). Implementation of B5G and 6G networks for enhanced rural connectivity [Sept.]. *IEEE Communications Magazine*, *58*(9), 42–47.

Han, J., Li, H., & Lin, X. (2019). Spectrum sharing in 5G and beyond: Recent advances and future challenges [Dec.]. *IEEE Communications Magazine*, *57*(12), 168–174.

Hegge, K., & Agrawal, R. (2023) "Wireless and Solar-Powered Multipurpose Robot for Industrial Automation: A Sustainable Solution," *2023 8th International Conference on Communication and Electronics Systems (ICCES)*, Coimbatore, India, pp. 146-151.

Hussain, N., Dixit, S. K., & Gupta, A. K. (2023). A novel approach of unsupervised feature selection using iterative shrinking and expansion algorithm. *Journal of Interdisciplinary Mathematics*, *26*(3), 519–530. DOI: 10.47974/JIM-1678

Kim, H., Kim, D., & Park, Y. (2021). Dynamic spectrum management for IoT and smart agriculture [Jan.]. *IEEE Internet of Things Journal*, *8*(2), 1020–1033.

Lien, S., Chen, K., & Ding, Y. (2019). 5G new radio: Unveiling the essentials of the next generation mobile communication technology [Feb.]. *IEEE Communications Magazine*, *57*(2), 24–30.

Liu, R., Huang, J., & Wang, Y. (2020). Advanced spectrum management for 5G and beyond: A comprehensive survey [Fourthquarter.]. *IEEE Communications Surveys and Tutorials*, *22*(4), 2151–2176.

Maddineni, V. K., & Gummadi, V. A. K. (2024) "Deep Learning Approaches for Real-Time IoT Data Processing and Analysis," *2024 5th International Conference on Smart Electronics and Communication (ICOSEC)*, Trichy, India, pp. 757-764.

A. Mary, A. Y. Begum, and D. G. V., (2023) "A Novel IoT Framework and Device Architecture for Efficient Smart City Implementation," *2023 7th International Conference on Trends in Electronics and Informatics (ICOEI)*, Tirunelveli, India, pp. 420-426.

Mohan, P. M., Das, A., & Panda, S. (2020). Analyzing the impact of edge computing on healthcare services. *IEEE Access : Practical Innovations, Open Solutions*, *8*, 203720–203729.

Nishikawa, D. K., Higuchi, T., & Kojima, Y. (2020). Rural broadband networks and their economic impact on remote areas [Mar.]. *IEEE Communications Magazine*, *58*(3), 58–63.

Premkumar, S., Hussain, N., & Chinamuttevi, V. S. (2023). A novel RF-SMOTE model to enhance the definite apprehensions for IoT security attacks. *Journal of Discrete Mathematical Sciences and Cryptography*, *26*(3), 861–873. DOI: 10.47974/JDMSC-1766

Qadir, J., Ali, A., & Saleem, A. M. (2021). Artificial intelligence-enabled edge computing for IoT: A comprehensive survey [Feb.]. *IEEE Internet of Things Journal*, *8*(4), 3037–3066.

Reddy, M., & Tanguturi, R. (2022) "Improved Secure Communication Mechanism for IoT Platform Devices," *2022 International Conference on Smart Generation Computing, Communication and Networking (SMART GENCON)*, Bangalore, India, pp. 1-6.

Reddy, N. S., & Ashreetha, B. (2023) "Technologies for Comprehensive Information Security in the IoT," *2023 International Conference for Advancement in Technology (ICONAT)*, Goa, India, pp. 1-5.

Sharma, S., Kushwaha, D. S., & Misra, M. (2020). Towards improved resource provisioning for IoT-enabled cloud systems [Dec.]. *IEEE Internet of Things Journal*, *7*(12), 11900–11914.

Shekhar, S., & Ashreetha, B. (2023). Performance Analysis of Energy Efficiency and Security Solutions of Internet of Things Protocols. *IJEER*, *11*(2), 442–450. DOI: 10.37391/ijeer.110226

Srinivas, V., & De, T. (2023) "Wi-Fi Router Signal Coverage Position Prediction System using Machine Learning Algorithms," *2023 International Conference on Sustainable Computing and Smart Systems (ICSCSS)*, Coimbatore, India, pp. 253-258.

White, G., Barbu, V., & Williams, A. L. (2020). Innovative applications of 5G technology in the healthcare industry. *IEEE Access : Practical Innovations, Open Solutions*, *8*, 155234–155246.

Zheng, K., Zhao, S., & Xu, X. (2020). Quality-of-experience assurance in B5G/6G networks [Sept./Oct.]. *IEEE Network*, *34*(5), 190–197.

Chapter 8
Bridging the Digital Divide in Rural Areas:
Challenges and Strategies for Universal Internet Connectivity

Amit Kohli
University Canada West, Canada

Tanvir Singh
University Canada West, Canada

ABSTRACT

In the era of transformation and technological advancement, there is one major challenge, which is connectivity. The chapter represents not only physical but digital concerns as well. Addressing the digital literacy gap is paramount today.. The chapter suggests the upbringing and support of installation of non-line-of-sight (NLOS) backhaul devices to expand internet reach in a cost efficiency and operational effectiveness. The NLOS refers to the ability to catch the internet frequency waves diffracted from the barrier in the path such, as buildings and mountains, contrasting with clear-line-of-sight (CLOS) methods. Moreover, the chapter argues that merely expanding the 5G and 6G networks to rural areas without bringing literacy, quality assurance, and involvement will not be fruitful. Overall, the spread of internet connectivity is not just an expansion but a target to be achieved with sustainable strategies.

INTRODUCTION

The chapter delves into the challenge of connectivity, which we are trying to address to accomplish a standard opportunity for everyone to extract the best out of the resources and power available to humans around different geographics of the globe. With the evaluation industry over the years, telecommunication networks have played a major role in the achievement of effectiveness and efficiency of operations across various industries (Hilbert, 2016). The connectivity in every location is largely influenced by the intermediaries responsible for communicating the demand and situational conditions prevailing in the expansion area. The business in the area attracts the attention of these intermediaries, which then

DOI: 10.4018/979-8-3693-7788-8.ch008

play the role of broker to bring the connections (Gaspard & Baker, 2022). The balance in connectivity is disturbed by a misbalance in the distribution of money in society.

The world is digitally divided based in the inequalities based on the access, use and efficacy, which is not just based on accessibility but based on acceptance as well. COVID-19 has pushed the demand for Internet resources and forced acceptance towards a necessity of modern times, specifically a huge impact on demand from rural areas (Gaspard & Baker,2022). The digital divide brings into the picture various hindrances and challenges to expand the accessibility of the Internet over the socio-economic factors in place (Lemelin-Bellerose, 2023). The digital divide is also a result of diffusion of investment based on the subscriber or measurement of the bandwidth, which is not equal like income distribution among the users (Hilbert, 2016).

Digital divide started with availability to the new digital tool/ technology, followed by the skills to effectively use those tools like communication devices, using internet assistant which is being polished by the social inequalities within the geographies, individual households and communities, which promote a sense of exclusion among the slow adopters to the digital transformation (Vassilakopoulou & Hustad, 2021).

Digital divide is in distributed in different phases, which varies on the different geographical levels (Van Dijk, 2012).. The poor economy is still focusing on shifting the dependency to digital platforms for improving the communication and connectivity around the province and aid the distribution of basic needs.

CURRENT CONNECTIVITY AND DISTRIBUTION

The Digital Divide is on the redar for the many international organization like information system researchers which believe the divide as an indicator of many offline gaps as well. The information is used by dividing the population in segment for users, informant and unaware mob. The categories are then targeted accordingly to meet both ends via trainings designs and initiative and identifying where the policies are leading the gaps (Vassilakopoulou & Hustad,2021).

The United Nations specialized agency "International Telecommunication Union-ITU" is collecting and monitoring the level of connectivity development and growth among the countries. ITU aims to improve technologies and ensure the flow of networks around the globe, especially by focusing on accessibility to underserved communities. As per the data collected by ITU, 33 percent of the world population did not have internet connectivity in 2023, which is a significant decline from 55 percent of the world population in 2018 (International Telecommunication Union, 2023).

Figure 1. Bar graph detailing individuals using the internet

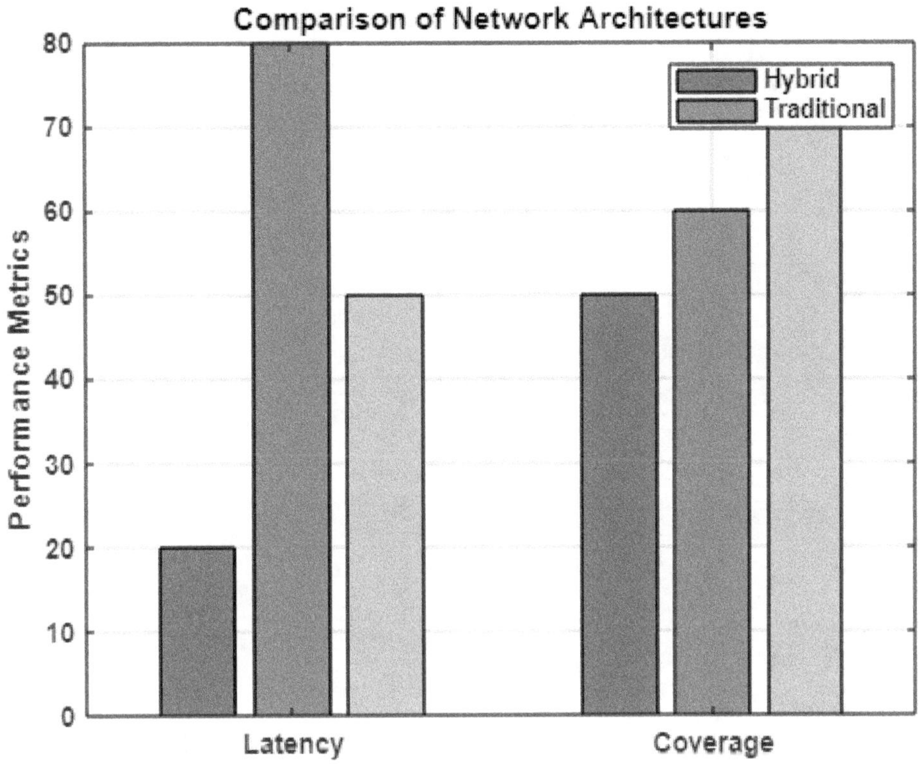

Note: The bar graph presents the changes in the number of individuals using internet from 2005-2023. Source – Press release by International Telecommunication Union on September 12, 2023. (https://www.itu.int/en/mediacentre/Pages/PR-2023-09-12-universal-and-meaningful-connectivity-by-2030.aspx).

Adding to the data mentioned above, Bahrain, Saudi Arabia, and the United Arab Emirates are the three countries with 100 percent of individuals using the Internet as of the year 2022.

Figure 2. Bar graph showing the distribution of users in different regions

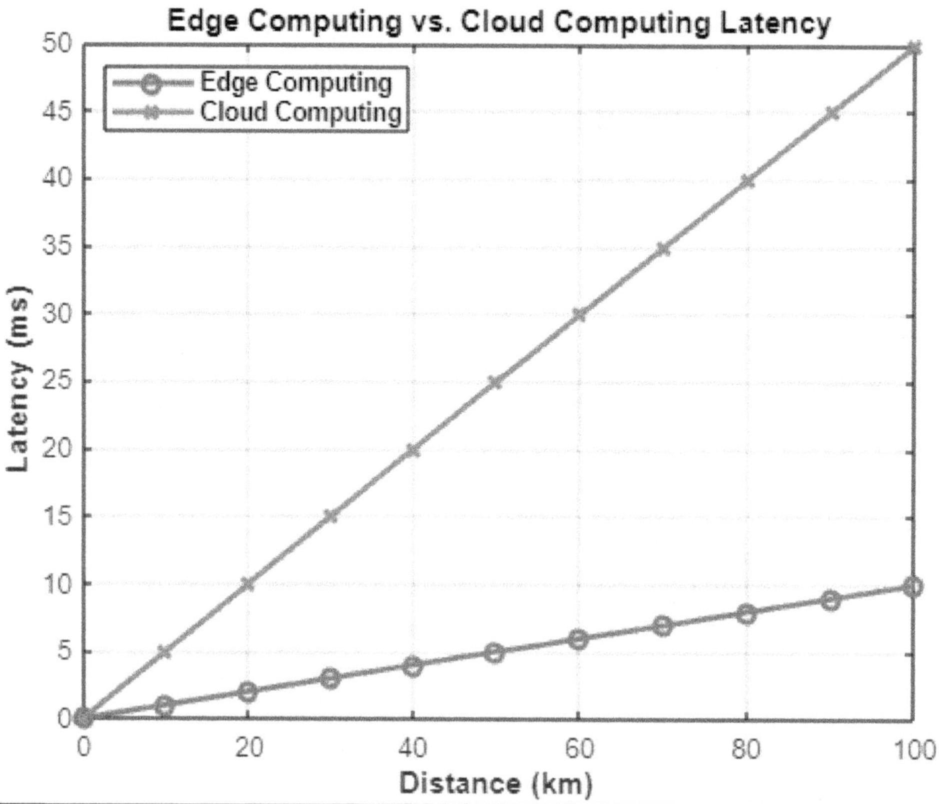

Note; The bar graph present the distribution of internet users in different region based on geography, income level and connection availability. Source- Fact and Figures 2023- Internet use by International Telecommunication Union (https://www.itu.int/itu-d/reports/statistics/2023/10/10/ff23-internet-use/)

Figure 2 above demonstrates that the world's 67 percent population uses the Internet, however, when we evaluate the usage on different matrices like region, income level, and development status of the economy. In the least developed economies (LDCs), only 35 percent of the population has access to the Internet, whereas in developed economies like Europe, 91 percent of individuals have access to the Internet. Moving further towards user distribution based on income, is it safe to say that higher-income groups have better access to the Internet?

As per the office of the Auditor General of Canada (2023), 99 percent of the urban population has internet access, compared to only 57 percent for rural populations and 47 percent for First Nation Land as of 2021. In comparison with other developed economies like the United Kingdom and the United States, both have 98 percentiles of the urban population with access to the Internet. In contrast, in rural areas in the United Kingdom, 83 percent and 74 percent of United States rural population have access (Gaspard & Baker,2022). With the mission of reaching 100 percent accessibility around the globe by 2030, the International Telecommunication Union and its members are investing in connectivity projects in different projects (International Telecommunication Union, 2023).

The age gap is also a growing digital gap, with the growing population of elderly people, specially in the western societies, which needed to carter on a person level and triggering usage of internet and spreading the benefits of usage. The policies will help in monitoring the personalised usage of the personal internet, which further can help in moving ahead towards the next phase of digital divide (Niehaves & Plattfaut, 2010).

The rural and remote are the hub for many industries, including agriculture, mining, oil and gas industries, and many others, which not only necessary for the development of these industries but also bring awareness to people living in these areas for their growth (Al-Zubi et al.,2023). A digital literacy also brings the informed expertise to this industry, with educated and skilled labour in the neighbourhood transfers a chunk responsibility of training and practices.

A Potential Solution: Non-Line of Sight Backhaul Method

Connectivity is measured by access based on the availability of wired networks. Still, when reaching the corner, the wireless methods are coming to the limelight, which is cost-effective and efficient. There are different methods of deployment of backhaul technology, such as fixed fiber cable establishment, using microwaves based on terrestrial conditions, and satellite connectivity. Generally, fiber cable establishment has the highest capacity and effectiveness. Still, it is very painful to spread across the area, whereas satellite connectivity has reached the areas that are difficult to reach with fiber cables. Still, both these methods are very cost-consuming (Oughton et al., 2022).

Although microwave signals are a cost-efficient and effective way to cover the area, their journey is quite interesting in targeting the effective manner of conduct for transmission. Expanding signals via radio waves or microwaves face the challenge of lost signal due to diffraction, which buildings, trees, mountains can cause, and other natural factors like rain (Al-Zubi et al.,2023).

Figure 3. Diagram for path description under line of sight and non-line of sight

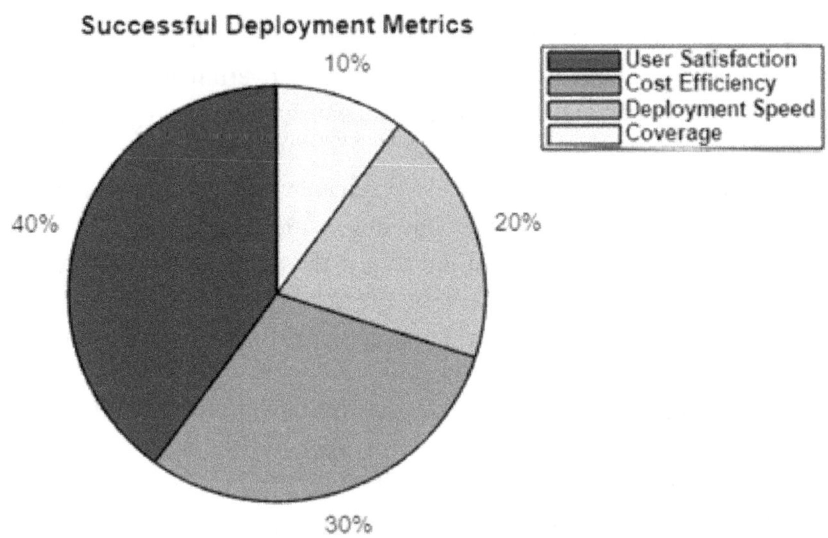

Source- Reproduced from (Al-Zubi et al.,2023) -https://doi.org/10.1109/ojcoms.2023.3330078

The research conducted by Nwankwo & Ukhurebor in 2019, in which they allotted three points at a different height from the base point and shared the 5 GHz band during the rain for 30 days and recorded the result in which the loss of signal from the high point to low point or the unequal terrain was quite higher than the data transmission on the plane surface. The major signal loss during the research was from point two to point three, where the signal is realized from the height to the lower receiver point.

In the journey from 1G to 5G, network reliability is always measured with the reach of mobile communications using the existing or upgraded technologies among the different territories (Ukhurebor & Nwankwo, 2019). The connection for the bandwidth determines the deployment of the Internet and its speed. Various companies are using different methods to enhance the speed of communication over rural or mountain areas. For example, wireless sensor networks are spread across wide areas to transmit information, and Project Loon by Google was launched in 2013. Contributing to this era of transmission, Gilani et al. (2020) came up with the antenna called "Grey Antenna," which they tested in different sizes and conditions, and it came up with effective outcomes. They proposed this antenna for the deployment of backhaul transmissions, but this antenna came with the drawback of higher radiation than expected from the transmission.

The backhaul non-line of sight is achieved using different models, including the combination of fixed line and wireless transmission. Some of the models are path loss link budget, radio propagation model, irregular terrain model, knife-edge diffraction model, and rounded obstacle diffraction model. The acceptance of this model varies depending on the geographical needs and resources available. The rounded obstacle diffraction model is the most efficient model among the non-line of sight (Al-Zubi et al.,2023). The non-line-of-sight backhaul model link can travel 11 km with 100-175 Mbps. Even with different heights, these results can be achieved even if send and receiver hubs are between 5 and 20 meters above the base level (Al-Zubi et al.,2023). The non-line-of-sight backhaul method is 9 to 45 percent successful in bringing the Internet to implemented areas depending on the geography and geopolitical conditions (Oughton et al., 2022).

The government and other big companies like Meta, Google are investing heavily in the deployment of these models to expand the reach among customers and gain the first mover advantage over the data stored in these locations. Telecommunication companies are already on the path of installing backhaul links to have strong wireless networks (Oughton et al., 2022). Again, where there are generating sources, the broker (telecommunication companies) is first to reach, but governments along with the International Telecommunication Union are trying to monitor these expansions on equal grounds (Gaspard & Baker, 2022).

Paul et al. (2023) came up with a compact novel patch antenna, "Microstrip Patch Antenna," which tested positive for better transmission of Wimax and lower fifth generation (5G) Communication. The compact design of the antenna was successful in keeping the radiation low and reflection with an efficiency of 94 to 98 percent. The grey antenna prototyped by Gilani et al. (2020) has a high radiation level and is efficient in transmission. However, the usage of both antennas depends on the factors like geographic structure and atmospheric conditions.

The infrastructure distribution will enhance the reach for the internet around the glode but their one another challenge which is making the connectivity an issue. With the change in a construction style for the residentials buildings and use of different materials. With the increase in usage of cement in the construction, which is hindrance in the spread of wireless signals. Various building makes sure to have wired infrastructure which again need some adaptors to receive the diffracted rays and distribute through out the building, thus increasing the smart building reliability. The potential solution for the indoor services is usage of Ultra Wide Band technology. It provides and resistance to multipath effect and focuses on the sub-meter distance measurement accuracy with wide bandwidth acceptance and resolution (Olejniczak et al., 2023).

The Ultra Wide Band signal is inculcated into deep feedforward neural network architecture which provides a combination of line of sight and non line of sight to improve the infrastructure for smart buildings. The deep feedforward neural network records all the incoming networks either wired or wireless and evaluate them against different level and distribute further network, accordingly, providing 95 percent accuracy in dynamic building environment and 98 percent in static environment (Olejniczak et al., 2023).

User Accessibility and Acceptability

Every human civilization has different beliefs based on conditions, customs, and experiences, which means different to different generations based on their exposure and awareness among the people. Most elders who learned during the time of letter writing or minimal use of telecommunications praise their methods and experience over the current speedy communications. They want the next generation to learn their techniques and values, accordingly, leading to a huge gap among generations. The elderly still largely resist the Internet's usage, specifically in rural and remote areas. They are resistant to change, and due to inaccessibility to the Internet, there is a lack of awareness about the usage and benefits of the new technologies. Adding to this, we can see that the starting age for using the Internet in more developing countries is 14 to 16 years, whereas in developed countries, the starting age is five years (ITU/UN tech agency, 2024).

A good example of an explanation of the digital divide is the Baltimore case by Horrigan (2020); Baltimore is a place in the United States where 40.7 percent of households do not have access to internet broadband and computers due to a lack of digital literacy and poor financial conditions. The smartphone spread was quite at its peak, but access to the Internet was still limited. After the impact of Covid 19, the demand spiked in the region, and many stakeholders started to invest in the city's development to ensure the society's gap was filled. Many libraries and nonprofit organizations came up to expand the knowledge and awareness among the people living via workshops, tech support, and digital skill training. Baltimore's government also contributed to ensuring the remedies for digital illiteracy and digital divide.

The United States, being a developed country, is still facing the digital divide issue. In India, the government came up with the Digital India Campaign in 2015, with a mission to transform all government departments into digital entities. This umbrella campaign also focused on motivating people to use the Internet and building trust in society. The Canadian government aims to spend $8 billion for the improvement of the connectivity to 100 percent by 2030, out of which $2 billion was spent in 2023 (Office of the Auditor General of Canada., 2023). The Canadian government wants to make sure the country has reliable connectivity to the Internet with 50 megabits per second of download speed and 10 megabits of uploading speed. Like many other countries, the Canadian digital divide consists of two factors: technical and socio-economic. The government is spending a fortune on narrowing the technical divide but still has a bigger challenge with other factors affecting the divide, such as age, income, gender, and digital literacy in the country (Lemelin-Bellerose,2023). The government is making measures via innovative, science, and economic development in Canada to bridge the gap between society and the digital divide.

The increasing trend of artificial intelligence within the communities is leading and huge demand in the expansion, which is not allowing a suitable time to strategically introduction of the expansion plans. The instant expansion is affecting the environmental structures, causing the environmental adaptability challenges (Olejniczak et al., 2023).

Case Study

Overview

In this case study we will discussing about a digital transformational plan for improvement to sharing of information and supply chain of livestock industry.

The XYZ Company is working on the agroledger and traceability of the livestock by using the artificial intelligence, internet of things, blockchain technology, and other recoginition methods, to gather the data and simplify, which helps to boost the disease management, enhance food safety, and transparency of food supply (Vial, 2019).

Current Assessment

Technology Infrastructure

Evaluate the existing status of traceability technologies utilized in the livestock sector. Recognize current systems, their abilities, and restrictions (Barbosa et al., 2019). Also assess the accessibility and dependability of internet connectivity in rural and isolated farming areas. Identify the necessary infrastructure to facilitate digital transformation (Giovanetti, et al., 2017). To assess the hardware (such as sensors, GPS devices) and software (like management systems, data analytics tools) currently being utilized. Evaluate how well they work with modern technologies.

Operational Practices

Examine the techniques employed in gathering information on the health, movement, and production of livestock. Recognize any manual tasks and their lack of effectiveness (Gonzalez et al., 2018). Additionally, evaluate the current protocols for identifying, reporting, and handling diseases. Assess the efficiency and quickness of existing strategies in responding to outbreaks (Hansen et al., 2018).

Stakeholder Readiness

Assess the preparedness and eagerness of producers to embrace new technologies. Additionally, evaluate their current digital literacy level and willingness to embrace change (Huang et al., 2018). To assess the accessibility of training programs and assistance for stakeholders and identify the required education and skill enhancement for promoting technology implementation (Halachni et al., 2019).

Regulatory Environment

Recognition of the regulatory standards affecting traceability, data handling, and disease management in the livestock sector. Evaluate the intricacy and weight of adhering to regulations (Awad, 2016).

Environmental Impact

Evaluate the existing methods regarding environmental sustainability in animal agriculture. Evaluate the attempts made by the sector to reduce its impact on the environment (Barbedo et al., 2019).

Diagnosis

Inefficiencies in Traceability

Recognize the breakdowns in current traceability systems, which result in gaps in monitoring livestock movements and health records. Point out problems with data reliability and precision resulting from manual data input and absence of standardized procedures Barron, et al., 2008).

Slow Response to Disease Outbreaks

Identify the obstacles in identifying and reporting diseases, which impact the efficiency of response actions. Recognize the absence of predictive analytics tools for predicting and preventing disease outbreaks (Giovanetti, et al., 2017).

Challenges in Technology Adoption

Emphasize the difficulties presented by unstable internet connection in rural regions, obstructing the utilization of digital technologies. Deal with the cultural obstacles and opposition from stakeholders who are familiar with customary farming methods (Gonzalez, et al., 2018).

Environmental Sustainability

Assess how current agricultural methods are affecting the environment and emphasize the necessity for more sustainable alternatives (Andrew et al., 2017).

Potential Challenges

Digital Transformation is not an overnight magic but will require lots and lots of work to make both ends meets. XYZ must be ready and target some challenges which are hinderance to the success of digital transformation: -

- Connectivity
- Technology Accessibility
- Cultural Barriers
- Data Privacy
- Regulatory Compliance
- System Integration
- Training and Education
- Technology Suitability
- Scalability
- AI Reliability
- Resistant to change

Challenge 1: Connectivity

Analysis

As per the report of Office of Auditor General of Canada (2023) – Report 2 – Connectivity in Rural and Remote Areas, as in 2021, only 59.5 percent of the household in rural and remote area have access to internet whereas around 99.3 percent of urban household have access to quality internet. The report also lay emphasis on the standards of quality internet which is 50/10 MBPS where 50 depicts the download speed and 10 depicts the upload speed. The Good quality internet at affordable price is the target for the government for increasing the connectivity around Canada.

Figure 4. Pie charts as per record of 2021

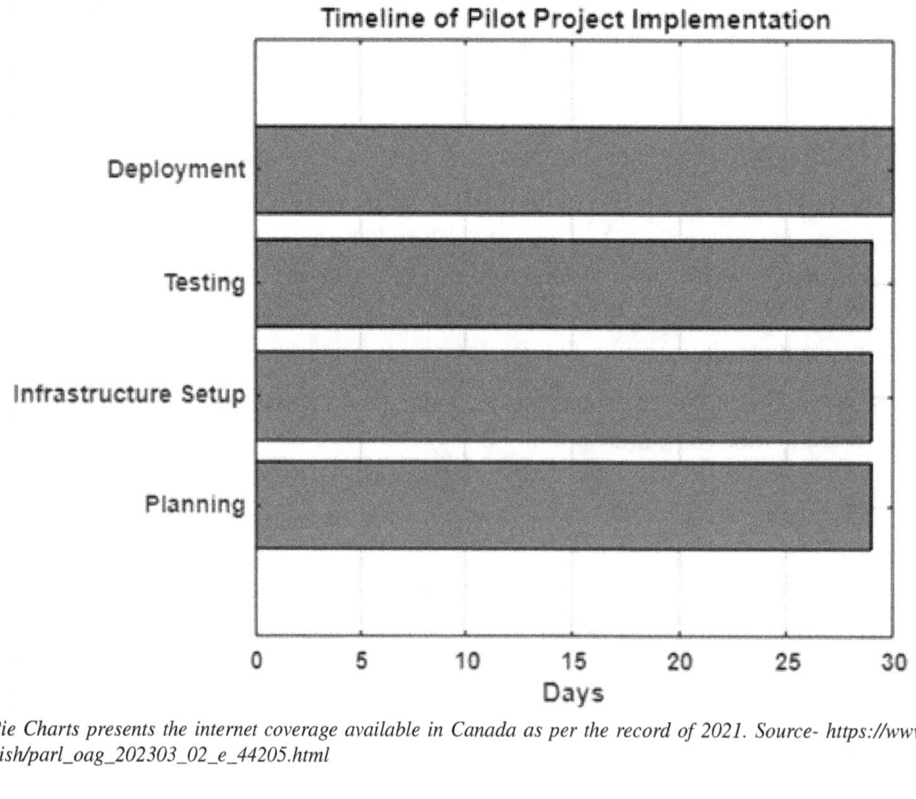

Notes: Pie Charts presents the internet coverage available in Canada as per the record of 2021. Source- https://www.oag-bvg.gc.ca/internet/English/parl_oag_202303_02_e_44205.html

Impact

With poor connectivity around the rural areas, the people living in those area are deprived of many services as compared to the urban population. The cutoff from the latest information can lead to backlog and unequal availability of the services like education and medical services (Lemelin-Bellerose, 2023).

Solution

The Government is spending around $8 billion out of which for 2022-23 was $2.4 billion for improvement of connectivity among the rural and remote areas, with the spending Canadian rural and remote areas were able to enjoy the LTE mobile connectivity also known as 4G (fourth generation) cellular technology (Office of the Auditor General of Canada, 2023).

Figure 5. Pie charts for mobile connectivity in 2021

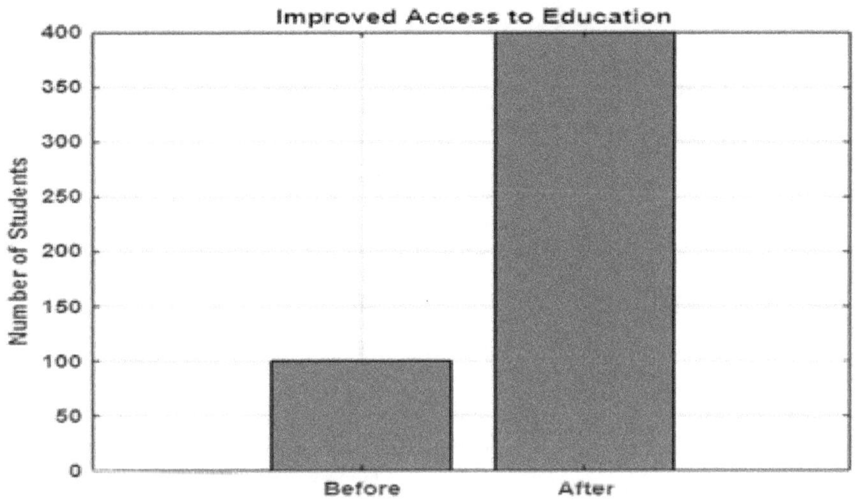

Note: Pie charts expressing the connectivity with mobile networks across Canada as in 2021. Source- https://www.oag-bvg.gc.ca/internet/English/parl_oag_202303_02_e_44205.html

Figure 6. Expenditure of budgets and government projects for 2022-23

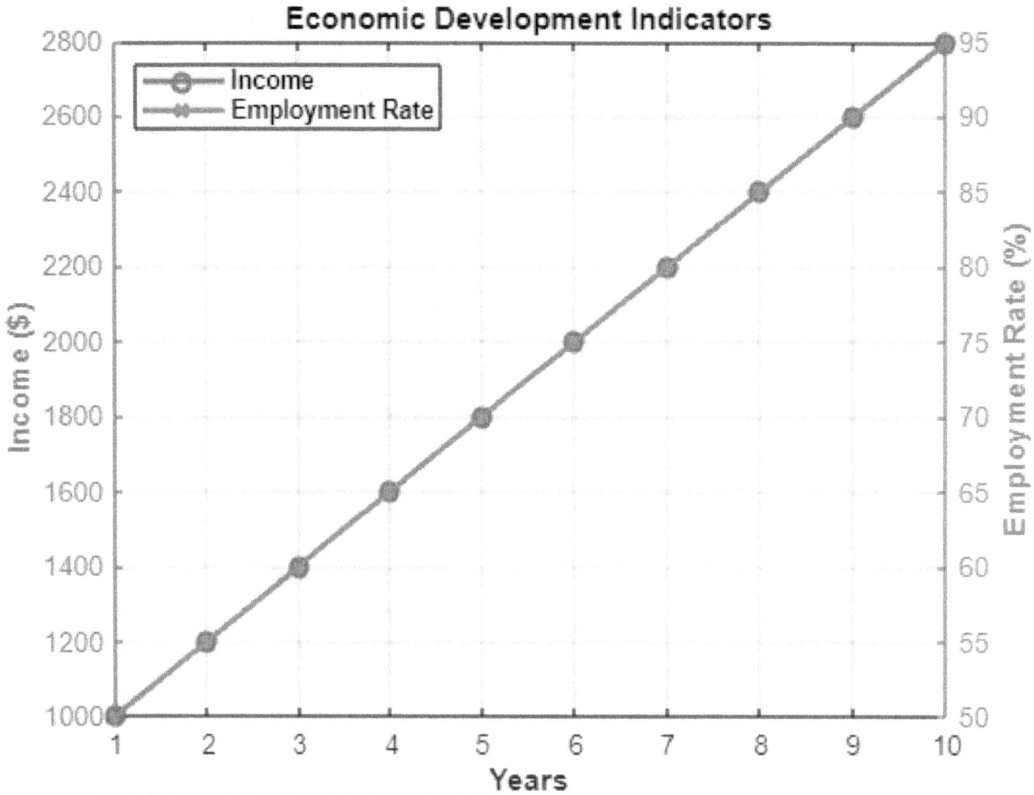

Note: Figure presenting the fund available for the cause and spending as in 2022-23. Source https://www.oag-bvg.gc.ca/internet/English/parl_oag_202303_02_e_44205.html

With the increase in Mobile connectivity and range the devices used by the company should be made compatible with wireless connectivity, although wireless model may have some lag time, but information collection and sharing will be faster than other models. Company should allow the associated farmers to use devices using radio frequency and promote Non-line-of-Sight backhaul devices which are wireless and provide the speed of approximately 100-175 MBPS with in the area between 5 to 20 km from the receiving antennas (Al-Zubi & Alouini, 2023).

The connectivity using NLOS backhaul devices provides an effective way of aiding the connectivity and cost efficiency for around 9-45 percent depending on the location to location (Oughton et al., 2022). The Canadian government aims in achieving its mission of 100 percent of Internet connectivity in Canada by 2023 (Innovation, Science and Economic Development Canada & Digital Transformation Service Sector, 2022).

Challenge 2: Technology Accessibility

Analysis

Most of the rural area or hub for livestock fosters are in remote areas which are deprived of fast speed internet access, direct transportation and other demographic factors, leading them to challenge of accessibility to new and trending technologies (Brovarone & Cotella, 2020).

Impact

The lack of available resources and poor connectivity leads to lagging in transformation and affecting the accessibility to various tools, it makes it difficult for the communities in remote area to accept the technological tools in their cultural practices (Brovarone & Cotella, 2020). With the rebellion from cultural leaders and lack of access to the latest information (Edo, 2023b) along with digital divide (Lemelin-Bellerose, 2023), the rural area is deprived of hardware and software tools which are essential for the implementation of Blockchain and information sharing for timely analysis (Bianchi et al., 2022).

Solution

As discussed in other challenges the technology accessibility can be improved by implementing strategies which motivates the stakeholders to adopt to these changes even there is minimal cultural dispute, need training for making the users familiar with these tools (Wang et al.,2022). The company can the Italian model which depicts that company should first understand the current situation and conduct surveys to get the ground level insights availability of resources, After the insights, the communities should be encouraged with the benefits of the those technologies as tracking the milk production, pesticide conditions and other related aspects which can be dealt and prevented with those tools (Bianchi et al., 2022). With use of NLOS backhaul technology the internet connectivity can be improved (Al-Zubi & Alouini, 2023), giving the opportunity to create data collection using cloud computing and having a decentralized blockchain model for maintain the security and integrity of data (Yang et al., 2024).

Figure 7. Framework for increasing the technological accessibility among remote locations.

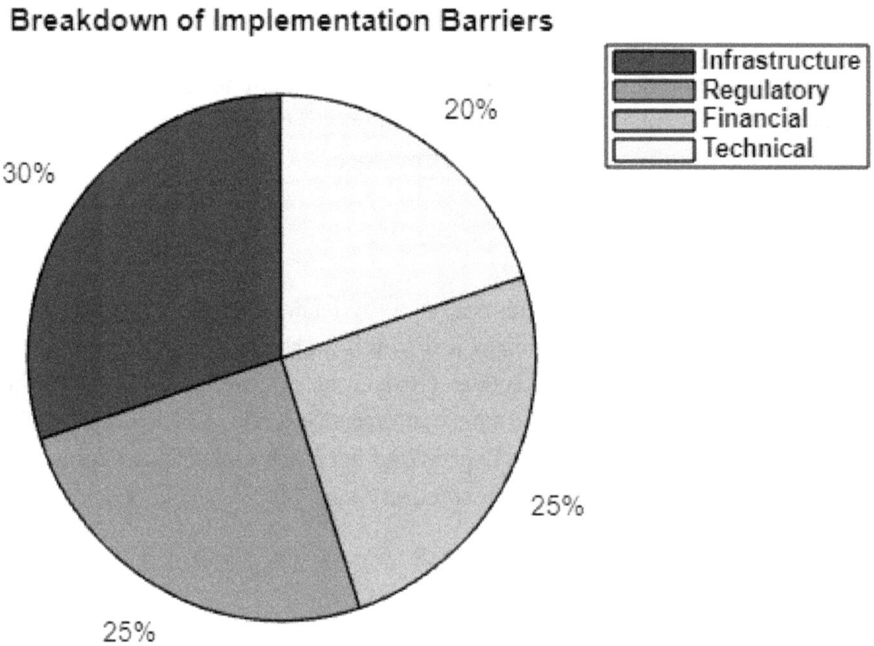

Note: Framework show the steps needed to increase the technological accessibility within the rural locations. Source- EVALUATION OF INFORMATION AND COMMUNICATION TECHNOLOGIES (ICTs) TOOLS CONTRIBUTING TO RURAL DEVELOPMENT. Russian Journal of Agricultural and Socio-economic Sciences, 142(10), 19–29. https://doi.org/10.18551/rjoas.2023-10.03.

Challenge 3: Cultural Barriers

Analysis

As we discussed about the connectivity issue in the rural and remote area which leads to the lag of awareness about the new opportunities and situations which required proactive actions for sustainability (Fuentes et al., 2022). Many livestock keeper practices are subjective and differentiated from each other due to lack of common driving factor and require training which may or may not be used at other place. Although consumer demand has been forcing the farmers to change and adopt the modern IT tools (Fuentes et al., 2022).

Some cultural leaders fear that bringing new technology will lead to disruption of their traditional practices. The cultural believes carry the sense of belongingness, trust and privacy among the people making them hesitant to share their information and adopt new technologies, having the fear of loss of their privacy (Edo, 2023a). Due to uneven digital divide or limited exposure to modern tools, a social misconception can form within a society regarding the use or adoption of these technological tools (Dilger et al., 2016). Many times, technologies are being seen to be biased towards one culture leading to sense of disown in other cultures, such as language, practices and beliefs are not accommodated in implementation among diverse cultures (Edo, 2023b).

Impact

The technology adoption among diverse cultures can be successful with maximizing the capacity accommodation various cultures, for now the traditional practices are being followed despite consumer pressure as it gives the sense of belongingness to the followers (Edo, 2023b). XYZ Company Incorporation working on bringing a common Agroledger for all the cultures is a big challenge as they must integrate various presentations among different communities to explain the need and motivate to adopt the new technologies. Blockchain is a big word for the farmers or livestock managers who are using the same cultural techniques from ages.

Solutions

Technologies can be implemented in distinct cultures in the following ways.

- **Cultural adoptions-** Technologies implementation must fit different cultural context, understanding various norms, beliefs and languages giving a sense of trust among the culture (Edo, 2023b).
- **Accessibility-** As discussed in Challenge 2, Technology accessibility must be they're for bring the transformation and overcoming the cultural barriers to adapt the new tools and replace the current practices (Edo, 2023a).

These steps can bring the change and lead to upbringing of the industry, The African farms can be seen as an example of integrating the technology into daily activities and can change for good. The African farms implement these changes gradually but resulted in expansion of agriculture land by 4 percent. The government supported this transformation and making internet of things and sensors part of the agriculture industry helping in analysing the soil condition, pesticide situation and other important information (Signé & Fox, 2022). The technology acceptance model and task technology fit help to evaluate the ground of acceptance of technology and gives a relationship between various variables impacting the adoption of new technological tools (Wang et al., 2022).

Figure 8. Overview of technology acceptance model and task technology fit integration

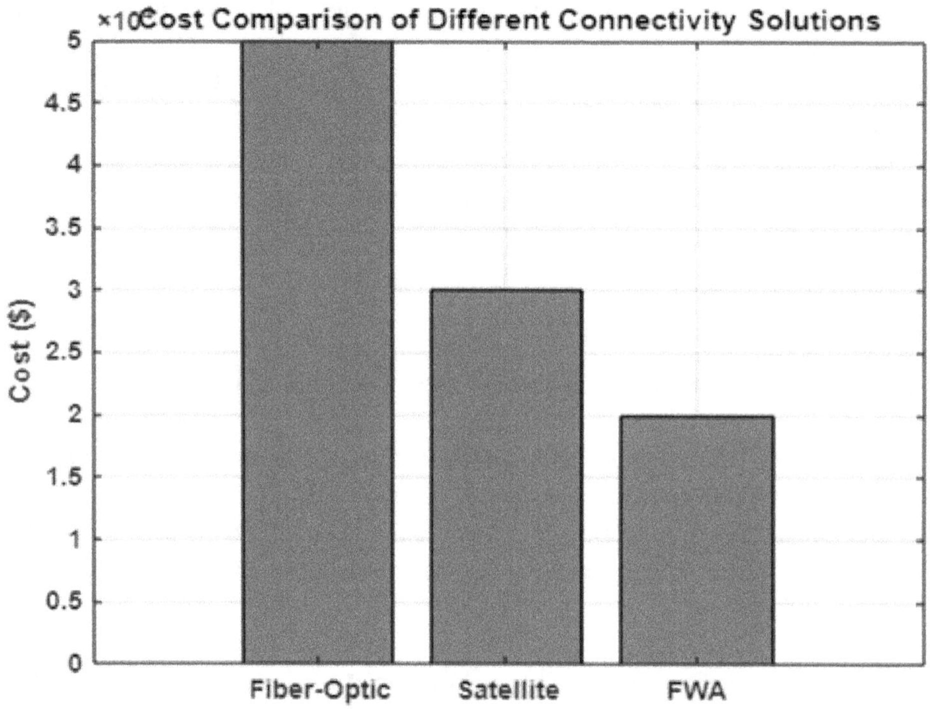

Notes: Figure presenting the integration of technology acceptance model and task technology fit to get a strategic deployment idea for overcoming the cultural barriers in implementation of modern tools. Source - Predictors Influencing Urban and Rural Area students to Use Tablet Computers as Learning Tools: Combination of UTAUT and TTF Models. Sustainability, 14(21), 13965. https://doi.org/10.3390/su142113965.

Challenge 4: Data Privacy

Analysis

As per Talagala (2022) "Data is new oil". Data is used for getting insights which then lead to upselling revenue models, innovative/transformative idea for gaining competitive edge for earning money and other resources (Talagala, 2022). Digital transformation can be implemented effectively after leveraging the big data which is consist of quantity, quality, and speed of change in data (Mentsiev et al., 2023).

Data privacy is crucial for maintaining the trust of stakeholders, giving them an assurance that the recorded data is secure and proper measures are taken to protect their information. The Data Protection depends a lot on the policies and procedures meeting the requirements ensuring the integrity of data collection, Data must be encrypted, and only authorized person should be given access to promote the security of data (Mentsiev et al., 2023). Data can be broken down and stored in different locations which can be retrieved easily in time of need to making the data in one location undesirable for the cyberthieves. Data management is responsible, crucial aspects of data privacy and digital transformation (Mentsiev et al., 2023).

Impact

XYZ Company incorporation working on creating and maintaining the Agroledger to have a pro-active edge in predicting any crises which may arise and ensure the quality of the food supply chain in the Canadian market. The data collected by the company must be secured. Any mishappening relate to data breach can lead to many questions on the integrity of the company. The stakeholder will loss their confidence in the policies of the company.

Solution

The Blockchain model is the potential situation and goal for the XYZ Company incorporation. The Blockchain is a decentralized way of data collection via various internet of things and securing them in an encrypted format. The data recorded by the blockchain is immutable. Smart contracts are included in the data collection which help in ensuring the data received and recording the quality of information s. The blockchain model gives a competitive edge on the traceability of agricultural products and tasks. The decentralized nature also protects the data from threat of cyber attack in one location (Yang et al., 2024). With use of various IOTs and cloud computing for maintaining the data block company can expand its scalability in secure network.

Figure 9. Overview of agriculture product traceability under secure network

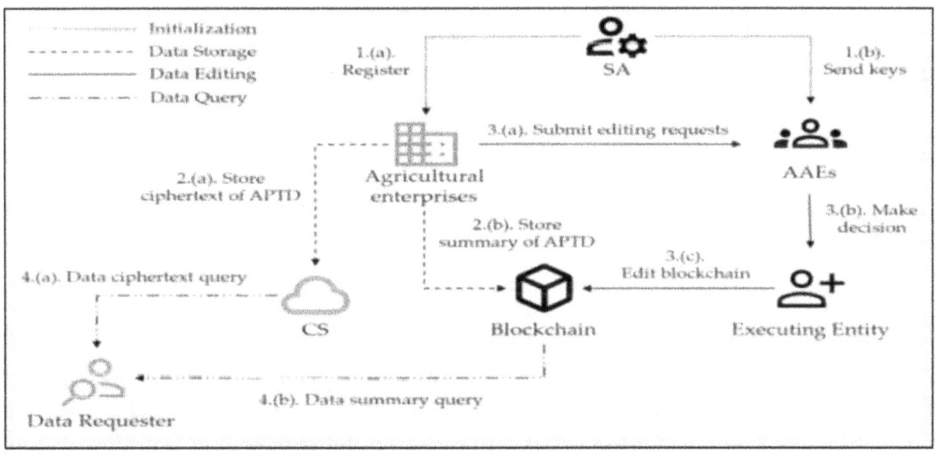

Note: Figure represents the flow of information in a block chain model. Source- A redactable Blockchain-Based data management scheme for agricultural product traceability. Sensors, 24(5), 1667. https://doi.org/10.3390/s24051667.

Figure 10. Overview of data collection using smart contracts in blockchain model

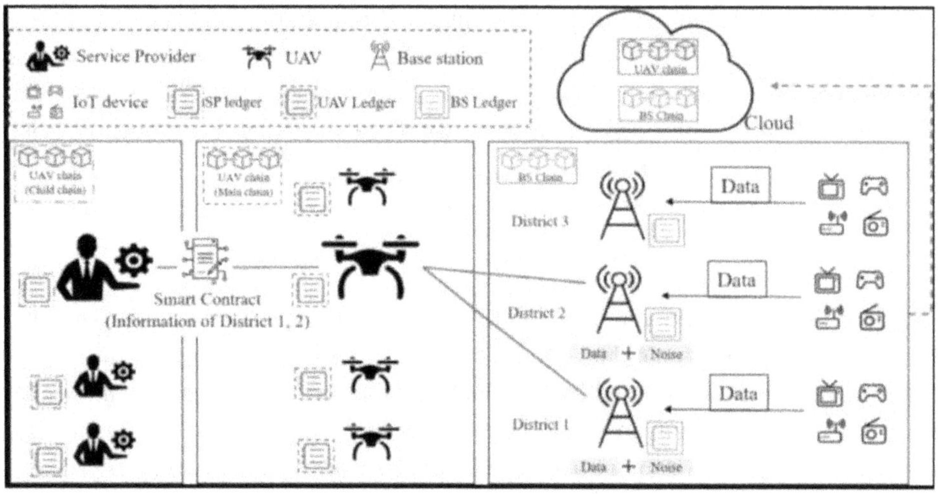

Note: Chart present the flow of information using smart contracts getting data from various IOT devices. Source - https://www.sciencedirect.com/science/article/pii/S014036642400149X?via%3Dihub#fig1, https://doi.org/10.1016/j.comcom.2024.04.027

The company should take the permission of the stakeholders before collecting their information and give the option to withdraw their consent as and when they feel like. Its not just about the security of data but also the stakeholders must feel secure in sharing their information to the company (Mentsiev et al., 2023). As discussed in cultural barriers, sometime the stakeholders fear from the loss of their information making it difficult for them to accept these advanced tools (Edo, 2023a).

Challenge 5: Regulatory Compliance

Analysis

Regulatory compliance should be at the utmost importance of the companies, as it can impact the working environment, reputation and trust among the stakeholders about the company. The Regulatory compliance differ from province to province and ensure the integrity of the working conditions in or around the specific industry. Regulatory compliance ensures the stakeholders about the transparency in the conduct of business and make the information available for their consideration (Mentsiev et al., 2023). Ensuring Data privacy and only giving access to only authorized people (Heo & Doh, 2024).

Solution

Company should hold a team specifically for meeting and honoring the regulatory compliance requirements as and when required. The practice must be audit by some third party and any adverse remark must be taken into consideration without any delays. Company should also train and educate their employees about the practices and importance of meeting the standards.

Challenge 6: System Integration

Analysis

System integration is the process by which diverse IT systems and hardware-software applications are put together to work coherently as a single system. The integration of new digital platforms with existing legacy systems can, on the other hand, be extremely complex and resource intensive. Poor integration can cause data silos, operational disruptions, and increased complexity, which threatens the general efficiency and effectiveness of the effort involved in digital transformation.

Recommendation/Solution

For instance, the system integration challenge in XYZ Company can be met through middleware solutions and API integrations. Middleware acts as a bridging layer that provides easy communication of data between the new and legacy systems, making things seamless. Besides, many platforms, such as Predix, allow seamless communication between a variety of systems with the use of APIs (Gaska et al., 2015).

Challenge 7: Training and Education

Analysis

The successful adoption of new technologies requires all the parties involved in their use to be proficient. Poor training and education result in resistance, misapplication, and underutilization of new systems. This slows down the process of digital transformation and diminishes the potential benefits of the new technology.

Recommendation/Solution

Comprehensive training programs could be organized by XYZ Company, targeting all stakeholder categories: producers, processors, and so on, to use the new technologies effectively. This can include hands-on workshops, online courses, and follow-through to bolster learning. Equally, developing a feedback mechanism through which the different stakeholders can provide their experiences and difficulties will help to continuously develop and fine-tune training materials and methods (Mak & Hong, 2020).

Challenge 8: Technology Suitability

Analysis

Innovative digital solutions, the solution will have a deep understanding of the challenges faced by the industry. Inadequate and unsuitable technologies lead to inefficiencies and failures in the general success of digital transformation.

Recommendation/Solution

XYZ Company will make sure that the proposed solutions meet the exact needs within the livestock sector through industrial expertise within the development process. Testing such technology through pilot programs in real-world applications will generate invaluable experience and will be open for adjustment before implementation on a full scale (Lwakatare et al., 2020).

Challenge 9: Scalability

Analysis

Scalability is the capability of a digital solution to handle increasing workloads and expand without compromising performance. Scalable systems form the basis of most growth and are crucial to handling the new dimensions brought by business change. The scalability of digital transformation solutions is essential, given the range of sizes and complexities that exist within the livestock industry. Any solution that is not scalable enough will not be able to cope with increased volumes of data or the increase in operation size.

Recommendation/Solution

For example, scalable solutions for XYZ Company be designed using cloud-based platforms, which can grow and scale easily with increased demands. Designs that are modular, allowing incremental increases in capabilities, will assure scalability. Stress testing of systems on a regular basis will help in making sure they handle peak loads and plan for future add-ons. Maintenance of flexibility in the technology stack, to incorporate new advancements at the time they are available, will help maintain scalability over time.

Challenge 10: AI Reliability

Analysis

AI reliability refers to the assurance that AI systems show consistency in results, accuracy, and trustworthiness (Alzubaidi et al., 2023). Reliable AI is critical to making informed decisions and establishing trust in digital systems.

Recommendations/ Solutions

XYZ Company can assure the reliability of AI by using high-quality, diverse datasets to train models that reduce bias and improve accuracy. Continuing the updating of AI models with new data and use of machine learning techniques to adapt to changes can help improve reliability.

Challenge 11: Resistance to Change

One of the common challenges faced by any industry undergoing digital transformation is resistance to change. Farmers, processors, and distributors are among the players in the cattle industry that have traditionally depended on conventional procedures and approaches. These customs are frequently ingrained in the everyday operations and corporate culture of these companies. Resistance may result from several things, such as:

Figure 11. Reasons for resistance to digital transformation.

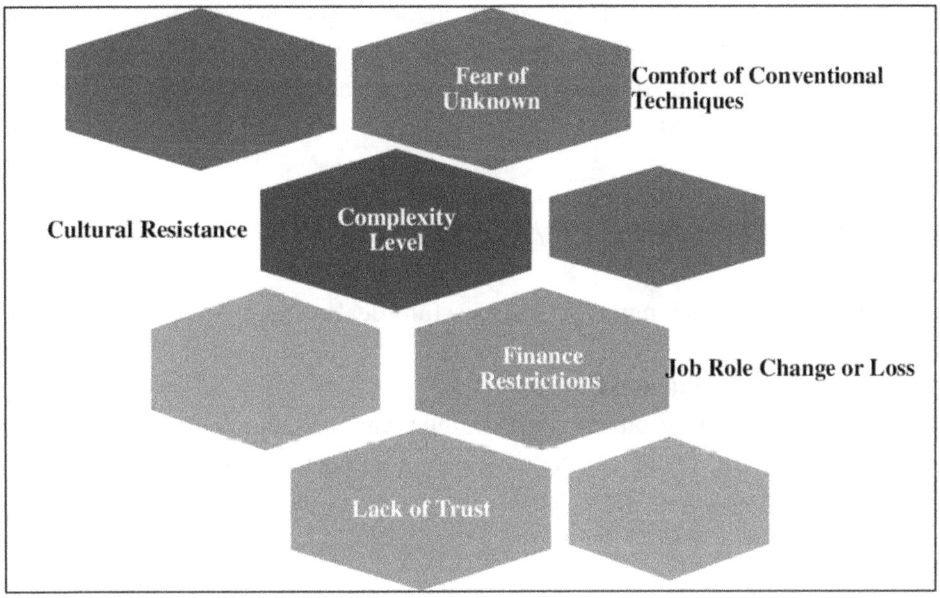

Note. This figure shows top 7 reasons why organizations grow reluctant to adopt digital transformation.

Fear of Unknown

Stakeholders may feel anxious since they are stressed and uncertain about how new digital technologies may impact their daily schedules, workflows, and methods. This fear of the unknown may be brought on by a lack of familiarity or experience with the new systems, which can lead to resistance and a hesitance to adopt them (Amos, 2023).

Complexity Level

Stakeholders may be reluctant to adopt new digital tools and systems if they believe they are unduly complex or challenging to understand and operate. This impression of complexity may result from things like a lack of practical expertise with the technology, shoddy user interface design, or inadequate training (Amos, 2023).

Job Role Change or Loss

If workers are afraid that implementing new technologies may result in changes to their duties or positions at work or even in job losses because of automation or process optimization, they may oppose digital transformation. Resistance may be exacerbated by a fear of job uncertainty or the desire to acquire new skills (Amos, 2023).

Cultural Resistance

The successful execution of digital transformation projects may be impeded by a culture that is not amenable to change or does not possess efficient change management procedures. XYZ Company can create focused strategies to address stakeholder concerns, offer sufficient training and support, illustrate the advantages of digital transformation, and promote an innovative and adaptable culture by thoroughly understanding these factors (Amos, 2023).

METHODS FOR OVERCOMING CHALLENGES

The following strategies can be employed to successfully deal with and overcome resistance to change:

Honest Communication with Stakeholders

1. Include interested parties from the outset of the planning and execution stages. Obtaining their opinions and concerns in order to encourage commitment (Admin, 2024).
2. Outline the objectives, advantages, and consequences of the digital transition. Case studies and success stories can be used to demonstrate the possible advantages (Admin, 2024).

Instruction and Training

1. Create and present training plans customized for various stakeholder groups. Make sure the instruction is applicable and meets needs (Amos, 2023).
2. Provide resources and help centers, as well as online and in-person support, to assist stakeholders in adjusting to new systems.

Rewards for Prompt Adopters

1. To promote early adoption of new technology, offer money-related incentives like gifts or discounts (Admin, 2024).
2. Give prizes and public acknowledgment to staff members who effectively adopt and utilize the modern systems (Amos, 2023).

Technical User-Friendliness

1. Reduce the learning curve by making sure the new technologies are simple to use and intuitive. Clear explanations and uncomplicated interfaces might facilitate the shift (Admin, 2024).

Establishing Credibility and Trust

1. Emphasize XYZ Company' history of creating and executing technologically sound solutions.
2. To foster confidence among stakeholders, highlight the strong security protocols and data privacy safeguards in place (Admin, 2024).

Reaction Systems

1. To get feedback from stakeholders and keep improving the technologies and procedures, set up frequent feedback loops (Amos, 2023).
2. Respond to criticism by making the required improvements and adjustments to resolve issues and enhance the user experience (Amos, 2023).

To summarize, a complex strategy, including clear communication, education, incentives, and ongoing support, is needed to address resistance to change. XYZ Company can help ensure a more seamless transition and promote the widespread adoption of cutting-edge technologies in the livestock industry by comprehending the goals and concerns of stakeholders and showcasing the measurable advantages of digital transformation.

DELIVERABLES

Deliverables for the digital transformation project by XYZ Company include an extensive array of cutting-edge systems, tools, and support mechanisms intended to improve the safety, efficiency, and transparency of the livestock supply chain. The main deliverables are listed below in detail:

Fully Functional AgroLedger system

A strong blockchain-based system that enables end-to-end livestock tracking is the main deliverable. The AgroLedger will comprise:

- **Decentralized Ledger:** An unchangeable, tamper-proof blockchain innovation that ensures records of each event and transaction within the animal's supply chain
- **User Interface:** Data entry and recovery for farmers, processors, and other stakeholders is made simple with an intuitive interface.
- Compatibility with installed systems and the capacity to integrate with various **IoT** devices for real-time data collection.

Advantages

- **Traceability**: complete tracking of animals from birth to sale.
- **Transparency**: Ensuring accountability and trust by providing real-time visibility to all stakeholders (Reevely, 2023).
- **Security**: To ensure data integrity, use advanced encryption and security methods.

DIGITAL TRANSFORMATION ROADMAP FOR XYZ COMPANY INCORPORATION

The following are the steps that XYZ Company will have to take to adopt digital transformation successfully and to meet the challenges that the company is facing currently.

Foundation and Initial Deployment

AgroLedger System Development

Building the core blockchain infrastructure ensures a secure and tamper-proof system for recording transactions. This basic phase is important to the AgroLedger System's integrity and stability. Data encryption measures, in addition to those utilized to protect information from unauthorized access and breaches, are significant, especially for delicate agricultural information. Executing strong encryption technologies guarantees information security both during travel and at rest (Grafiati, 2024).

User Interface Design

Begin development of web and mobile applications

User-friendly interfaces are essential for broad adoption. Early development of web and mobile apps ensures accessibility and usability for all stakeholders.

Design dashboards and reporting tools

Effective dashboards provide real-time insights into system performance and user activities. Reporting tools facilitate data analysis and decision-making (Erika, 2024).

RISK MITIGATION

Identification of Risks

Workshops on Risk and Idea Generation

With project stakeholders, brainstorming meetings and risk workshops are necessary to identify potential dangers. A range of backgrounds, including business executives, end users, technology specialists, and outside consultants, should be represented at these sessions. The project team can identify various risks associated with technology adoption, organizational transformation, and external variables by utilizing multiple perspectives (Managing Digital Transformation Risks, n.d.).

Risk Register

To record and manage recognized risks during a project, a thorough risk register must be kept up to date. Every risk should have comprehensive details about it in the risk register, including a description, possible impact, chance of occurrence, risk category (technical, organizational, external, etc.), and designated risk owners. The risk register should also include a record of each risk's mitigation techniques and backup plans.

STRATEGIES FOR RISK MITIGATION

Risks Associated with Technology

Excess and Backup

Putting strong backup and redundancy procedures in place is crucial to reducing the chance of data loss or technological malfunctions. This could entail putting in place disaster recovery strategies, redundant gear and software, and backup systems. Frequent offsite storage and data backups help guarantee data integrity and provide quick recovery in the case of cyberattacks or system breakdowns. Redundancy techniques, such as load balancing and failover systems, can reduce downtime and assist in maintaining system availability (Strobes, 2023).

Risks Related to External Factors

Cybersecurity Safety

Robust cybersecurity protocols are essential to defend confidential data and limit the probability of cyberattacks and security breaches. Installing firewalls, access restrictions, encryption software, and regular security audits may be necessary for this. Furthermore, by creating and testing incident response strategies, the company can minimize the impact on operations and data integrity and respond to possible cyber-attacks in an efficient manner (Violino, 2023).

Figure 12. Hierarchy chart for strategies for risk mitigation.

Note Hierarchy chart expressing the Strategies for risk mitigation.

RISK EVALUATION AND ADMINISTRATION

Frequent Risk Evaluations

To evaluate the success of mitigation strategies and spot new or emerging risks, it is essential to regularly check the risk register. To ensure an intensive evaluation of risk variables, project stakeholders and subject matter experts should participate in these evaluations. Frequent risk surveys can help the project team in appropriately allocating resources to handle changing risks and modifying mitigation strategies as necessary.

Crisis Preparedness

To guarantee project versatility in the event of unexpected risks or disruptions, contingency planning is crucial. Specific steps and approaches that should be taken in response to certain risk situations, such as technological malfunctions, resource shortages, or external risks, should be outlined in contingency plans.

Case Review and Sum Up

At last, it can be summed up that XYZ Company's digital transformation project is a big step in transforming the livestock supply chain. These firms seek to improve creativity, information exchange, and operational efficiency through the implementation of a complete information Management System (KMS).

The initiative tackles important issues like poor traceability, a sluggish response to disease outbreaks, and barriers to technology adoption. The amalgamation of cutting-edge technologies such as blockchain, artificial intelligence, and biometric identification systems will furnish a sturdy framework for comprehensive livestock tracking and instantaneous health monitoring.

In short, the strategic implementation of this digital transformation project will position XYZ Company as leaders in sustainable livestock farming and IT solutions. By leveraging cutting-edge technologies and fostering a culture of continuous learning and collaboration, these organizations are set to drive significant improvements in the livestock supply chain, ultimately benefiting producers, consumers, and the broader agricultural sector.

FUTURE DIRECTION

The International Telecommunication Union is working with different companies and governments worldwide to reach 100 percent connectivity for the success of digital transformation and smooth flow of communication. A digitally connected world will provide informed opportunities for growth for everyone. The union also focus on the availability of the Internet to the poor section of society so that they can also bring the communities with the use of new technologies. The Strategic Plan for the union is divided into two parts: the strategic plan 2020-2023 and the strategic plan 2024 -2027. The Plan 2020-2023 was motivated by the growth and flexibility to include different models focusing on expanding and accepting the Internet. This plan also aims to bring innovative solutions in partnership with different social groups and multinational companies. The Second plan -the strategic plan 2024-2027 adds the factor of sustainable development with the expansion of Internet and mobile networks. The plan also brings light and forces to bring affordability and cyber security into the picture. Cyber security is very important so that trust can be built among new users and the world can have a secure platform for development (ITU/UN tech agency,2024).

The 5G and 6G networks are in play in the urban population, whereas the rural population is still deprived of the 4G networks. For the faster speed bandwidth's better and more secure expansion, increase the subscriber to those bandwidths. The new spectrum launch is already facing challenges from environmentalists who are claiming the spectrum is a danger to the environment, which again increases the resistance to construction of the bridge to the digital divide. The 5G and 6G widespread depends on three factors which are enhanced mobile broadband and reach, Machine type communication and reliability on the transmission speed with low latency in communication (Yaacoub & Alouini, 2019) .

The 6G can be a solution to meet the digital divide, however it is more than just connection, it will still depend on the basic internet connection and literacy among the customers. 6G is looking on the transfer of big data, leading to digital analysis and creation of platform for upbring the society at large. With projects like starlink and telecast, connectivity can expand but infrastructural development for the recievers will still gonna be a major challenge (Saarnisaari et al., 2021).

The current projects are aiming the increase capacity up 1000 times and providing high transmission rate. The speed of the 5G and 6G spectrums requires a connection which can provide a higher transmission rate, the backhauls models can be dense for the achieving the requirements, but transmission at speed of 1 gigabit per second of 10 gigabits per second with non line of sight backhaul method is challenging and push us a little back towards relying on the wired structure (Feng et al., 2016).

CONCLUSION

Undoubtedly, the world is affected by the gaps of the digital divide, in which one side enjoys more opportunities and effective communication than the other. The gap widens with the widening of income distribution, generation gaps, and leadership commands among various cultural societies. Developed countries enjoy 90-95 percent of connectivity for Internet and mobile networks, and developing countries only have 82 percent of the masses connected to the Internet, and among that 82 percent, 43 percent of people exposed to the Internet do not use the Internet due to lack of knowledge and cultural resistance to acceptance for the Internet.

Governments across the globe are trying to bridge the digital divide gap, with the Canadian government having a budget of $8 billion and some governments having their own campaigns, like Digital India, by the Indian government.

The International Telecommunication Union has two strategic plans in place, which focus not just on the expansion of connectivity but also on the cyber security, sustainability, and affordability of the Internet. All the plans and government efforts focus on 100 percent internet across the globe by 2030. The spectrum of 5G and 6G must brought strategically with all the stakeholders addressed carefully and in a unified direction. The faster spectrum will aid in solving the other social dividing factors and will also help in transfer of big data, forming a base for identification of the causes and inequality and finding a potential solution for the development.

REFERENCES

Admin. (2024, March 20). *Digital transformation in livestock protection – Livetec*. Livetec. https://www.livetecsystems.co.uk/leading-the-change-digital-transformation-livestock-protection/

Al-Zubi, M. M., & Alouini, M. (2023). End-to-End modelling and simulation of NLOS sub-6 GHz backhaul via diffraction for internet connectivity of rural areas. *IEEE Open Journal of the Communications Society*, *4*, 3102–3114. DOI: 10.1109/OJCOMS.2023.3330078

Alzubaidi, L., Al-Sabaawi, A., Bai, J., Dukhan, A., Alkenani, A. H., Al-Asadi, A., Alwzwazy, H. A., Manoufali, M., Fadhel, M. A., Albahri, A. S., Moreira, C., Ouyang, C., Zhang, J., Santamaría, J., Salhi, A., Hollman, F., Gupta, A., Duan, Y., Rabczuk, T., & Gu, Y. (2023). Towards risk-free trustworthy artificial intelligence: Significance and requirements. *International Journal of Intelligent Systems*, *2023*(1), 4459198. https://www.hindawi.com/journals/ijis/2023/4459198/. DOI: 10.1155/2023/4459198

Amos, J. (2023, August 3). Embracing Change: Overcoming resistance during digital transformation - Industry leading ICT services, support. *soma technology group*. https://www.soma.com.au/blog/resistance-during-digital-transformation/

Andrew W, Gao J, Mullan S, Campbell N, Dowsey AW and Burghardt T (2020a). Visual identification of individual Holstein-Friesian cattle via deep metric learning. arXiv preprint arXiv, 2006.09205.

Andrew, W., Greatwood, C., & Burghardt, T. (2017). Visual localisation and individual identification of Holstein Friesian cattle via deep learning. Proceedings of the IEEE *International Conference on Computer Vision Workshops*, 2850–2859. DOI: 10.1109/ICCVW.2017.336

Andrew W, Greatwood C and Burghardt T (2019). Aerial animal biometrics: individual Friesian cattle recovery and visual identification via an autonomous with onboard deep inference. arXiv preprint arXiv, 1907.05310.

Andrew, W., Greatwood, C., & Burghardt, T. (2020b). Fusing animal biometrics with autonomous robotics: drone-based search and individual id of Friesian cattle. *Proceedings of the IEEE/CVF Winter Conference on Applications of Computer Vision Workshops* 1, 38–43. DOI: 10.1109/WACVW50321.2020.9096949

Awad, A. I. (2016). From classical methods to animal biometrics: A review on cattle identification and tracking. *Computers and Electronics in Agriculture*, *123*, 423–435. DOI: 10.1016/j.compag.2016.03.014

Barbedo, J. G. A., Koenigkan, L. V., Santos, T. T., & Santos, P. M. (2019). A study on the detection of cattle in UAV images using deep learning. *Sensors (Basel)*, *19*(24), 5436. DOI: 10.3390/s19245436 PMID: 31835487

Barbosa Pereira, C., Dohmeier, H., Kunczik, J., Hochhausen, N., Tolba, R., & Czaplik, M. (2019). Contactless monitoring of heart and respiratory rate in anesthetized pigs using infrared thermography. *PLoS One*, *14*(11), e0224747. DOI: 10.1371/journal.pone.0224747 PMID: 31693688

Barron, U. G., Corkery, G., Barry, B., Butler, F., Mcdonnell, K., & Ward, S. (2008). Assessment of retinal recognition technology as a biometric method for sheep identification. *Computers and Electronics in Agriculture*, *60*(2), 156–166. DOI: 10.1016/j.compag.2007.07.010

Bianchi, M., Bava, L., Sandrucci, A., Tangorra, F., Tamburini, A., Gislon, G., & Zucali, M. (2022). Diffusion of precision livestock farming technologies in dairy cattle farms. *Animal, 16*(11), 100650. DOI: 10.1016/j.animal.2022.100650 PMID: 36220000

Brovarone, E. V., & Cotella, G. (2020). Improving Rural Accessibility: A multilayer approach. *Sustainability (Basel), 12*(7), 2876. DOI: 10.3390/su12072876

Dilger, A. C., Schroeder, A. L., & Moseley, W. M. (2016). Barriers to global implementation of current and development of new performance-enhancing technologies in meat production. *Animal Frontiers : the Review Magazine of Animal Agriculture, 6*(4), 50–55. DOI: 10.2527/af.2016-0044

Edo, J. (2023a, September 26). The Interplay Between Culture and Technology: a Complex Relationship. *Medium.* https://medium.com/@JacobsEdo/the-interplay-between-culture-and-technology-a-complex-relationship-e46f9e31fd79

Edo, J. (2023b, September 24). The influence of culture on technology adoption and design. *Medium.* https://medium.com/@JacobsEdo/the-influence-of-culture-on-technology-adoption-and-design-e8ea61fbdac1

Erika. (2024, March 13). *Blockchain and UI/UX design: Ensuring security and user trust.* The Data Scientist. https://thedatascientist.com/blockchain-and-ui-ux-design-ensuring-security-and-user-trust/

Facts and Figures 2023 - Internet use. (2023, October 10). https://www.itu.int/itu-d/reports/statistics/2023/10/10/ff23-internet-use/

Feng, W., Li, Y., Jin, D., Su, L., & Chen, S. (2016). Millimetre-Wave backhaul for 5G networks: Challenges and solutions. *Sensors (Basel), 16*(6), 892. DOI: 10.3390/s16060892 PMID: 27322265

Fuentes, S., Viejo, C. G., Tongson, E., & Dunshea, F. R. (2022). The livestock farming digital transformation: Implementation of new and emerging technologies using artificial intelligence. *Animal Health Research Reviews, 23*(1), 59–71. DOI: 10.1017/S1466252321000177 PMID: 35676797

Gaska, T., Watkin, C., & Chen, Y. (2015). Integrated modular avionics-past, present, and future. *IEEE Aerospace and Electronic Systems Magazine, 30*(9), 12–23. https://ieeexplore.ieee.org/abstract/document/7321122/. DOI: 10.1109/MAES.2015.150014

Gaspard, H., & Baker, P. M. A. (2022). Innovation and Digital Connectivity : Comparative Policy Approaches for Connecting Rural Communities in the United States and Canada. *Choices, 37*(3), 1–9.

Gilani, S. M. M., Sultan, M. T., Shuai, Z., & Kabir, A. (2020). Long range backhaul microwave connectivity in wireless sensor networks via a new antenna designed for ISM 2.4 GHz band. *International Journal of Antennas and Propagation, 2020*, 1–14. DOI: 10.1155/2020/1789612

Giovanetti, V., Decandia, M., Molle, G., Acciaro, M., Mameli, M., Cabiddu, A., Cossu, R., Serra, M., Manca, C., Rassu, S., & Dimauro, C. (2017). Automatic classification system for grazing, ruminating and resting behaviour of dairy sheep using a tri-axial accelerometer. *Livestock Science, 196*, 42–48. DOI: 10.1016/j.livsci.2016.12.011

Gonzalez Viejo, C., Fuentes, S., Torrico, D., & Dunshea, F. (2018). Non-contact heart rate and blood pressure estimations from video analysis and machine learning modelling applied to food sensory responses: A case study for chocolate. *Sensors (Basel), 18*(6), 1802. DOI: 10.3390/s18061802 PMID: 29865289

Government of Canada. Statistics Canada. (2022, October 18). *Average cost to prevent or detect cyber security incidents by industry and size of enterprise*. https://www150.statcan.gc.ca/t1/tbl1/en/tv.action?pid=2210013101

Halachmi, I., Guarino, M., Bewley, J., & Pastell, M. (2019). Smart animal agriculture: Application of real-time sensors to improve animal well-being and production. *Annual Review of Animal Biosciences*, *7*(1), 403–425. DOI: 10.1146/annurev-animal-020518-114851 PMID: 30485756

Hansen, M. F., Smith, M. L., Smith, L. N., Salter, M. G., Baxter, E. M., Farish, M., & Grieve, B. (2018). Towards on-farm pig face recognition using convolutional neural networks. *Computers in Industry*, *98*, 145–152.

Heo, G., & Doh, I. (2024). Blockchain and differential privacy-based data processing system for data security and privacy in urban computing. *Computer Communications*, *222*, 161–176. Advance online publication. DOI: 10.1016/j.comcom.2024.04.027

Hilbert, M. (2016). The bad news is that the digital access divide is here to stay: Domestically installed bandwidths among 172 countries for 1986–2014. *Telecommunications Policy*, *40*(6), 567–581. DOI: 10.1016/j.telpol.2016.01.006

Horrigan, J. B. (2020). Baltimore's Digital Divide: Gaps in Internet Connectivity and the Impact on Low-Income City Residents. *The Abell Report*, *33*, 4.

Huang, L., Li, S., Zhu, A., Fan, X., Zhang, C., & Wang, H. (2018). Non-contact body measurement for Qinchuan cattle with LiDAR sensor. *Sensors (Basel)*, *18*(9), 3014. DOI: 10.3390/s18093014 PMID: 30205607

Huang, W., Zhu, W., Ma, C., & Guo, Y. (2020). Weber texture local descriptor for identification of group-housed pigs. *Sensors (Basel)*, *20*(16), 4649. DOI: 10.3390/s20164649 PMID: 32824819

Innovation, Science and Economic Development Canada & Digital Transformation Service Sector. (2022, April 11). *High-Speed Access for all: Canada's connectivity strategy*. https://ised-isde.canada.ca/site/high-speed-internet-canada/en/canadas-connectivity-strategy/high-speed-access-all-canadas-connectivity-strategy

International Telecommunication Union. (2023, September 12). *Press release*. ITU. https://www.itu.int/en/mediacentre/Pages/PR-2023-09-12-universal-and-meaningful-connectivity-by-2030.aspx

ITU/UN tech agency. (2024, April 16). *Measuring digital development - Facts and Figures: Focus on Landlocked Developing Countries, April 2024 - ITU*. ITU. https://www.itu.int/hub/publication/d-ind-ict_mdd-2024-2/

Kimberling, E. (2023, April 25). Most important change management deliverables in a digital transformation - Third stage consulting. *Third Stage Consulting*. https://www.thirdstage-consulting.com/most-important-change-management-deliverables-in-a-digital-transformation/

Lemelin-Bellerose, S. (2023, September 7). *Deployment of broadband internet in rural and remote areas of Canada*. https://lop.parl.ca/sites/PublicWebsite/default/en_CA/ResearchPublications/202311E

Lwakatare, L. E., Raj, A., Crnkovic, I., Bosch, J., & Olsson, H. H. (2020). Large-scale machine learning systems in real-world industrial settings: A review of challenges and solutions. *Information and Software Technology*, *127*, 106368. https://www.sciencedirect.com/science/article/pii/S0950584920301373. DOI: 10.1016/j.infsof.2020.106368

Mak, C., & Hong, J. (2020). Creating learning organization 2.0: A contextualized and multi-stakeholder approach. *The Learning Organization*, *27*(3), 235–248. https://www.emerald.com/insight/content/doi/10.1108/TLO-01-2020-0020/full/html. DOI: 10.1108/TLO-01-2020-0020

Mentsiev, A., Aygumov, T., & Amirova, E. (2023). DATA-DRIVEN DIGITAL TRANSFORMATION: CHALLENGES AND STRATEGIES FOR EFFECTIVE BIG DATA MANAGEMENT. *ResearchGate*. https://doi.org/DOI: 10.24412/1932-2321-2023-575-526-531

Modiba, F., & Kaye, S. (2023). EVALUATION OF INFORMATION AND COMMUNICATION TECHNOLOGIES (ICTs) TOOLS CONTRIBUTING TO RURAL DEVELOPMENT. *Russian Journal of Agricultural and Socio-Economic Sciences*, *142*(10), 19–29. DOI: 10.18551/rjoas.2023-10.03

Morrison-Smith, S., & Ruiz, J. (2020). Challenges and barriers in virtual teams: A literature review. *SN Applied Sciences. SN Applied Sciences*, *2*(6), 1096. Advance online publication. DOI: 10.1007/s42452-020-2801-5

Neethirajan, S. (2023). Artificial intelligence and sensor technologies in dairy livestock Export: Charting a Digital transformation. *Sensors (Basel)*, *23*(16), 7045. DOI: 10.3390/s23167045 PMID: 37631580

Niehaves, B., & Plattfaut, R. (2010). *The Age-Divide in Private Internet Usage: A Quantitative Study of Technology Acceptance*. AIS Electronic Library (AISeL). https://aisel.aisnet.org/amcis2010/407

Office of the Auditor General of Canada. (2023, March). *Report 2—Connectivity in rural and remote areas*. https://www.oag-bvg.gc.ca/internet/English/parl_oag_202303_02_e_44205.html

Olejniczak, A., Blaszkiewicz, O., Cwalina, K. K., Rajchowski, P., & Sadowski, J. (2023). LOS and NLOS identification in real indoor environment using deep learning approach. *Digital Communications and Networks*. Advance online publication. DOI: 10.1016/j.dcan.2023.05.009

Oughton, E. J., Boch, E., & Kusuma, J. (2022). Engineering-Economic Evaluation of Diffractive NLOS Backhaul (e3nb): A Techno-economic Model for 3D Wireless Backhaul Assessment. *IEEE Access : Practical Innovations, Open Solutions*, *10*, 3430–3446. DOI: 10.1109/ACCESS.2022.3140421

Patel, R. (2024, January 12). How much does custom software development cost in 2024? *Space-O Technologies (Canada) Logo*. https://www.spaceo.ca/blog/software-development-cost/

Patterson, C. & Canadian Sheep Federation. (2020). AgroLedger. In *Canadian Sheep Federation*. https://animalhealthcanada.ca/pdfs/5%20-%20CSF_AgroLedger_Presentation%20Council%20Forum%202020_Corlena%20Patterson.pdf

Paul, L. C., Ankan, S. S. A., Rani, T., Karaaslan, M., Hossain, M. N., Al-Gburi, A. J. A., Saha, H. K., & Alkurt, F. Ö. (2023). A Wideband Highly Efficient Omnidirectional Compact Antenna for WiMAX/Lower 5G Communications. *International Journal of RF and Microwave Computer-Aided Engineering*, *2023*, 1–10. DOI: 10.1155/2023/7237444

Reevely, D. (2023, December 25). *Canadian farmers dream of blockchain sheep*. The Logic. https://thelogic.co/news/the-big-read/canadian-farmers-dream-of-blockchain-sheep/

Saarnisaari, H., Chaoub, A., Heikkilä, M., Singhal, A., & Bhatia, V. (2021). Wireless terrestrial backhaul for 6G Remote access: Challenges and low power solutions. *Frontiers in Communications and Networks*, *2*, 710781. Advance online publication. DOI: 10.3389/frcmn.2021.710781

Samsukha, A. (2024, February 2). How much does it cost to build an AI app in 2024: a detailed guide. *Tech Blog | Mobile App, eCommerce, Salesforce Insights*. https://www.emizentech.com/blog/cost-to-build-an-ai-app.html

Signé, L., & Fox, L. (2022, June 28). Overcoming the barriers to technology adoption on African farms. *Brookings*. https://www.brookings.edu/articles/overcoming-the-barriers-to-technology-adoption-on-african-farms/

Talagala, N. (2022, March 4). Data as the new oil is Not enough: Four Principles for Avoiding Data Fires. *Forbes*. https://www.forbes.com/sites/nishatalagala/2022/03/02/data-as-the-new-oil-is-not-enough-four-principles-for-avoiding-data-fires/?sh=4c7e5061c208

Townsend, S. (2022, October 24). *What is resource management and why is it important?* Planview. https://www.planview.com/resources/guide/resource-management-software/resource-management-leverage-people-budgets/

Ukhurebor, K., & Nwankwo, W. (2019). Investigating the Performance of Point to Multipoint Microwave Connectivity across Undulating Landscape during Rainfall. *Journal of Nigerian Society of Physical Sciences*, *1*(3), 103–115. Advance online publication. DOI: 10.46481/jnsps.2019.16

Van Dijk, J. A. (2012). *The Evolution of the Digital Divide - The Digital Divide turns to Inequality of Skills and Usage* (pp. 57–78). https://doi.org/DOI: 10.3233/978-1-61499-057-4-57

Vassilakopoulou, P., & Hustad, E. (2021). Bridging Digital Divides: A Literature Review and Research Agenda for Information Systems Research. *Information Systems Frontiers*, *25*(3), 955–969. DOI: 10.1007/s10796-020-10096-3 PMID: 33424421

Violino, B. (2023, May 19). *Mitigating the hidden risks of digital transformation*. CIO. https://www.cio.com/article/191398/mitigating-the-hidden-risks-of-digital-transformation.html

Wang, F., Wijaya, T. T., Habibi, A., & Liu, Y. (2022). Predictors Influencing Urban and Rural Area students to Use Tablet Computers as Learning Tools: Combination of UTAUT and TTF Models. *Sustainability (Basel)*, *14*(21), 13965. DOI: 10.3390/su142113965

Yaacoub, E., & Alouini, M. S. (2019). A key 6G challenge and opportunity—connecting the remaining 4 billions: a survey on rural connectivity. *arXiv preprint arXiv:1906.11541*.

Yang, S., Li, S., Chen, W., & Zhao, Y. (2024). A redactable Blockchain-Based data management scheme for agricultural product traceability. *Sensors (Basel)*, *24*(5), 1667. DOI: 10.3390/s24051667 PMID: 38475202

Chapter 9
Investigating the Challenges of Deploying Renewable Energy Technologies in Rural Areas

Ravindran Kandasamy
https://orcid.org/0000-0001-6763-0216
Presidency College, India

R. Ramesh
Knowledge Institute of Technology, India

Ravi kumar Natashen
https://orcid.org/0000-0002-9308-7461
Presidency College, India

R. Saranya
Knowledge Institute of Technology, India

T. Suganya
Knowledge Institute of Technology, India

Kiran Hiremath
Presidency College, India

Inayath Ahamed S. B.
Kalasalingam Business School, India

Mamata Bhandar
Manipal GlobalNxt University, Malaysia

ABSTRACT

The deployment of renewable energy technologies (RETs) in rural areas is examined in this chapter, with special attention paid to the opportunities and particular constraints that these places bring. This chapter offers a thorough review of the status of renewable energy technologies (RETs), highlighting the major governmental, technical, economic, social, cultural, and environmental obstacles that prevent RETs from being widely adopted in rural areas. Because it tackles the urgent need for sustainable energy solutions in rural areas where traditional energy access is still restricted, the research discussed in this chapter is noteworthy. Through the identification and analysis of deployment issues, this chapter adds to the larger conversation on rural electrification and sustainable development. To close the energy gap and advance global sustainability, policymakers, practitioners, and academics seeking to increase the adoption of renewable energy in underserved areas can benefit greatly from the conclusions and suggestions made.

DOI: 10.4018/979-8-3693-7788-8.ch009

1. INTRODUCTION

1.1 Overview of Renewable Energy Technologies (RETs)

Diverse techniques for obtaining energy from naturally replenishing sources are included in the category of Renewable Energy Technologies (RETs). In particular, these technologies are essential for lowering reliance on fossil fuels and mitigating climate change as we move toward a sustainable energy future. The primary varieties of RETs consist of:

Solar energy systems use light from the sun to create heat or power. The most popular type of solar technology is photovoltaic (PV) systems, which employ semiconductor materials to directly convert sunlight into power. In contrast, solar thermal systems harvest and utilize solar radiation for a range of purposes, such as heating and the production of electricity (Sharma, Tyagi, Chen, & Buddhi, 2009).

Using turbines, wind energy converts the kinetic energy of the wind into electrical energy. Onshore or offshore wind farms are made up of several wind turbines positioned carefully to maximize wind capture. Thanks to favorable legislation and developments in turbine technology, wind energy is one of the RETs with the fastest rate of growth (Saidur, Rahim, Islam, & Solangi, 2011).

The energy of moving water, usually from rivers or dams, is captured to produce electricity. It is among the most well-known and traditional sources of renewable energy. The size of hydropower plants varies, ranging from large-scale projects like China's Three Gorges Dam to tiny micro-hydro systems. Hydropower can have major negative effects on the environment and society despite its advantages (Müller, 2007).

Wood, agricultural waste, and animal dung are examples of organic resources that can be used to produce biomass energy. It can be used straight for heating or processed into biofuels for transportation and electrical production. Since the carbon absorbed during the growth of the biomass balances the carbon dioxide emitted during combustion, biomass energy is regarded as carbon-neutral (Demirbas, 2009).

This type of energy uses the heat that is stored beneath the surface of the Earth to provide direct warmth or electricity. Generally speaking, geothermal plants are found in locations with considerable geothermal activity, like volcanic regions. Geographical constraints restrict the implementation of this energy source, despite its great reliability and ability to supply baseload power (Bertani, 2012).

Technologies that utilize the force of waves, tides, and ocean thermal gradients are referred to as ocean energy. Ocean energy has the potential to make a substantial contribution to the world's energy mix, even if it is still in the experimental and early deployment stages. But there are issues that must be resolved in terms of technology, the environment, and the economy (Bahaj, 2011).

To battle climate change, air pollution, and reliance on fossil fuels, there is a huge transition occurring in the global energy landscape towards renewable energy sources (RES) (REN21, 2023). To promote sustainable development and lessen environmental degradation, this shift is essential. Even Nevertheless, there is still a significant gap in ensuring rural populations have fair access to sustainable energy, despite significant progress made in integrating RES into national networks (IEA, 2022). These communities run the risk of being left behind in the global energy transition since they are frequently neglected and underserved.

Often, rural communities lack the resources and infrastructure required for traditional grid extension. This shortfall poses a significant obstacle to these communities' complete involvement in the transition to renewable energy. Rural areas are difficult to connect to the central grid because of things like difficult terrain, dispersed populations, and scarce financial resources (World Bank, 2021). Therefore, in order to get past these challenges and support rural populations' access to sustainable energy, creative solutions and focused policies are needed.

Lack of infrastructure is a common problem in rural locations, making it more difficult to construct and maintain renewable energy installations. Construction activity coordination and the supply of critical materials can be hindered by the lack of dependable transportation and communication networks (Bhattacharyya, 2013). Furthermore, rural areas typically lack the technical know-how needed to construct and maintain these systems, which could result in reliability problems and increased operating costs (Kaundinya, Balachandra, & Ravindranath, 2009).

The implementation of renewable energy technology in rural regions poses a multitude of intricate and diverse problems. Even though renewable energy has a lot of potential to meet energy needs in remote areas in a sustainable and affordable manner, there are a lot of barriers preventing their general implementation. Technical, economic, social, and policy-related issues are all included in these problems, and each calls for thoughtful thought and deliberate preparation.

Budgetary limitations are still another important obstacle. Renewable energy technologies usually need a large initial capital investment, and rural communities frequently do not have access to the same financing choices and subsidies as their urban counterparts (Painuly, 2001). Securing money for such initiatives is made more difficult by the lower income levels and poor creditworthiness of rural communities, which compound this economic hurdle.

Rural communities must successfully implement renewable energy technologies, and this requires social engagement and acceptability. Because they are unaware of the advantages and possible consequences, communities may show resistance to new technologies (Sovacool & D'Agostino, 2012). Furthermore, cultural practices and beliefs may have an impact on how these technologies are accepted and used, therefore specific strategies that take into account and honor regional traditions and preferences are required (Van der Horst, 2007).

Frameworks for regulations and policy are also very important in assisting or impeding the adoption of renewable energy technology. Investment in rural energy projects may be discouraged by unclear and insufficient policies (Beck & Martinot, 2004). Policies that are effective must consider the unique requirements of rural communities, offering incentives and other forms of assistance that encourage the use of renewable energy sources while maintaining sustainability and fair access.

1.1.1 Combat Climate Change, Air Pollution, and Dependence on Fossil Fuels

The global movement towards renewable energy has been propelled by the necessity of addressing climate change. A significant portion of greenhouse gas emissions, which hasten environmental deterioration and global warming, are caused by fossil fuels (REN21, 2023). Furthermore, burning fossil fuels is a major contributor to air pollution, which endangers human health and the environment. By switching to RES like solar, wind, and hydro power, these negative consequences can be greatly mitigated, resulting in cleaner air and a more stable climate.

1.1.2 Progress in Integrating RES into National Grids

Incorporating Renewable electricity sources (RES) into national electricity networks has advanced significantly. The capacity and efficiency of generating renewable power have increased as a result of significant global investments in renewable energy infrastructure (IEA, 2022). The integration of renewable energy sources (RES) has been made easier by technological advancements, favorable legislation, and incentives. As a result, RES are now an essential part of contemporary energy systems.

1.1.3 Equitable Access to Clean Energy for Rural Communities

Even with these improvements, rural populations still frequently face barriers to accessing sustainable energy. To guarantee that the advantages of renewable energy reach every section of the population, including those living in isolated and neglected areas, equitable access is crucial (World Bank, 2021). To close this gap and allow rural areas to take part in the transition to renewable energy, certain interventions are needed. These populations generally suffer economic and infrastructure-related hurdles.

1.1.4 Challenges of Traditional Grid Extension in Rural Areas

There are many obstacles involved in extending the traditional grid to rural areas. Many rural villages are geographically isolated, making it expensive and logistically difficult to expand the central power infrastructure. Rural towns' dispersed layout makes it more difficult to distribute energy using traditional methods (World Bank, 2021). Further impeding grid extension is the fact that rural communities' limited financial resources frequently prevent them from paying connection fees and continuing power bills.

To effectively encourage the use of renewable energy technology in rural areas, it is imperative to comprehend these obstacles. It is possible to unleash the potential of renewable energy to improve energy security, economic development, and environmental sustainability in these regions by tackling the technical, financial, social, and policy-related hurdles.

1.2 Importance of Renewable Energy Technologies in Rural Areas

The potential of renewable energy technologies to address a range of environmental and socio-economic issues makes them extremely important for rural areas. First and foremost, these technologies offer a sustainable and dependable energy source, which is essential for locations that frequently experience intermittent or nonexistent power supplies. By supplying electricity to residences, educational institutions, and medical facilities, access to renewable energy can raise living standards and support vital services (Sovacool & Drupady, 2012).

Furthermore, the installation, upkeep, and operation of energy systems locally creates jobs, which is another way that renewable energy technologies promote economic development. By generating new employment opportunities, this job development can boost regional economies and lessen poverty (Chaurey, Ranganathan, & Mohanty, 2004). Furthermore, decentralizing energy production can strengthen rural communities by granting them more autonomy over their energy resources and lowering their reliance on centralized energy systems that frequently run on fossil fuels (Yadoo & Cruickshank, 2012).

Greener than conventional fossil fuels, renewable energy sources like solar, wind, and biomass contribute to the fight against climate change by lowering greenhouse gas emissions. For rural people that could otherwise be exposed to the negative consequences of burning biomass or kerosene for cooking and heating, this change could result in better air quality and health outcomes (Narula, Nagai, & Pachauri, 2012).

All things considered, the implementation of renewable energy technology in rural regions can support environmental stewardship, strengthen energy security, and propel sustainable development, making it an essential part of strategies for rural growth.

1.3 The Need for Renewable Energy Technologies (RETs) in Rural Areas

There are several reasons why the implementation of Renewable Energy Technologies (RETs) in rural areas is crucial; these include meeting short-term demands as well as long-term sustainability objectives.

Energy Access and Poverty Alleviation: A sizable section of the world's population lacks access to cheap, dependable energy, especially in rural developing nations. The bulk of those without access to electricity in 2019 lived in rural areas of Sub-Saharan Africa and South Asia, according to the International Energy Agency (IEA), which estimates that 759 million people globally (International Energy Agency, 2020). RETs can offer off-grid and decentralized solutions, making them a competitive substitute for traditional energy sources. RETs improve livelihoods by facilitating energy access and powering homes, businesses, schools, and healthcare facilities, thereby reducing poverty.

Environmental Sustainability: Wood and charcoal are common biomass sources used for cooking and heating in rural areas. Because of this reliance on traditional biomass, there are serious health risks, especially for women and children, from deforestation, soil degradation, and indoor air pollution. Renewable energy sources (RETs) including solar and wind power provide sustainable resource usage and lessen environmental damage (World Health Organization, 2016). Through the reduction of greenhouse gas emissions from the use of fossil fuels, the switch to RETs can also help to slow down global warming.

Economic Development: By encouraging entrepreneurship, generating jobs, and aiding small and medium-sized businesses (SMEs), the implementation of RETs in rural areas can boost local economies. In order to increase production and add value to agricultural products, solar energy systems, for instance, can power agricultural operations. This allows for the usage of irrigation pumps, refrigeration, and processing equipment (IRENA, 2016). Furthermore, the community can gain technical skills and employment prospects through the local production and maintenance of RETs.

Energy Security and Resilience: Rural areas are susceptible to disruptions in the energy supply since they are frequently located distant from centralized power infrastructures. RETs, particularly off-grid and mini-grid systems, lessen reliance on imported fossil fuels and centralized systems to create a more resilient energy infrastructure. For rural communities to be more resilient to outside shocks like fluctuations in fuel prices and natural disasters, energy independence is essential (Bazilian et al., 2012).

Social Benefits and Quality of Life: Rural communities can greatly benefit from RETs by having access to clean, dependable energy sources. Electric lighting, for example, allows for productive hours to be extended past daylight, facilitating social and educational activities. Additionally, RETs open up time for other economic or educational endeavors, especially for women and children, by reducing the time and work required in the collection of traditional fuels (Pachauri & Rao, 2013). Furthermore, having access to energy makes it possible to provide better healthcare services, such as the functioning of medical devices and the storage of vaccines, which enhances patient outcomes.

2. OBJECTIVES OF THE CHAPTER

1. To methodically list and evaluate the main obstacles to the implementation of renewable energy technologies (RETs) in rural regions, including those related to technology, economics, society, culture, legislation, regulations, and the environment.
2. To evaluate the degree to which these issues have an impact on the long- and short-term sustainability and successful implementation of RETs in rural communities.
3. The aim of this study is to investigate and suggest workable ways and tactics for resolving the mentioned issues. Particular attention will be paid to novel ideas, regulatory changes, and community-based programs that can improve the implementation and uptake of RETs in rural regions.
4. To examine the functions played by different parties involved in overcoming the obstacles and promoting the effective implementation of RETs in rural areas, such as local communities, governments, NGOs, and businesses.
5. To point out gaps in the literature and make recommendations for future research directions that will advance our knowledge of and ability to use RETs in rural settings more successfully.
6. To offer analysis and suggestions that can guide the creation of laws and other frameworks intended to encourage the broad use of RETs in rural regions, supporting the objectives of the global energy transition as well as sustainable rural development.

3. CONTEXT AND EVALUATION

3.1 Historical Perspective on Rural Electrification

Global development initiatives have placed a great deal of emphasis on rural electrification since it is widely recognized that having access to electricity is essential for improving living conditions, reducing poverty, and fostering economic growth. Different regions have had different experiences with rural electrification due to factors like economic growth, regulatory changes, and technological breakthroughs.

3.1.1 Early Beginnings and the Role of Cooperatives

In the early 20th century, rural electrification became increasingly popular, especially in the United States. As part of President Franklin D. Roosevelt's New Deal, the Rural Electrification Administration (REA) was founded in 1935 and was instrumental in bringing power to rural areas that were not lucrative for commercial utilities (Brown, 1980). Rural electric cooperatives, which were community-owned organizations that combined resources to construct and maintain electricity infrastructure, were promoted to organize by the REA. Due to the great success of this strategy, in 1950 over 90% of farms in the United States got electricity, up from 10% in 1930 (NRECA, 2020).

3.1.2 International Growth and Government-Initiated Projects

Other nations started to give rural electrification a higher priority as part of their national development objectives after the United States' success. Governments frequently oversaw rural electrification programs in emerging countries, making large investments in grid extension projects. For instance, India started its program of electrifying rural areas in the 1960s with the goal of electrifying every hamlet in the nation. Despite a number of obstacles, such as high expenses and logistical hurdles, the program set the stage for subsequent initiatives (Bhattacharyya, 2006).

3.1.3 Shift to Decentralized Solutions

The late 20th and early 21st centuries saw a shift towards decentralized solutions as the drawbacks of traditional grid extension became evident, especially in isolated and poorly populated areas. Off-grid and mini-grid systems have become a realistic option for supplying rural populations with electricity. They are frequently fueled by renewable energy sources like solar and wind. As the price of renewable energy technologies dropped and their dependability increased, these decentralized strategies gained popularity (Sovacool, 2012).

3.1.4 Modern Difficulties and Renewable Energy's Role

Rural electrification is still a difficulty despite tremendous advances, especially in Sub-Saharan Africa and some regions of South Asia. Rates of electrification are still low in these areas, and many rural villages do not have consistent access to electricity. The integration of renewable energy technologies into rural electrification schemes has gained traction due to its ability to provide affordable and sustainable means of reaching neglected areas. The United Nations Sustainable Development Goal 7 (SDG 7) emphasizes the need for sustained efforts in rural electrification in light of the global push for universal energy access by 2030 (IEA, 2020).

3.2 Review of Existing Renewable Energy Technologies

Over the past few decades, there has been a notable progress in Renewable Energy Technologies (RETs) due to the growing demand for sustainable energy sources that minimize greenhouse gas emissions and reduce reliance on fossil fuels. A summary of the most well-known RETs, including their benefits and drawbacks, is given in the ensuing sections.

3.2.1 Solar Energy Technologies

Photovoltaic (PV) Systems: Using semiconductor materials like silicon, photovoltaic systems are the most widely used type of solar energy technology. They convert sunlight directly into electricity. Thanks to major cost savings and efficiency gains brought about by advancements in PV technology, it is becoming more and more competitive with conventional energy sources. By the end of 2020, there will be over 710 GW of installed solar PV capacity worldwide, thanks to an exponential growth in capacity (IEA, 2021). For widespread implementation, however, issues including land use, energy storage, and intermittency still need to be resolved.

Solar Thermal Systems: This technology concentrates sunlight using mirrors or lenses to produce heat that can be utilized for industrial activities or the production of energy. One type of large-scale solar thermal system that produces energy by powering a steam turbine is a concentrated solar power (CSP) plant. Although CSP has the benefit of thermal storage, which may produce power even after the sun sets, its use is restricted to areas with abundant direct sunlight and must contend with the competition from PV systems, whose costs are falling quickly (IRENA, 2020).

3.2.2. Technologies for Wind Energy

Onshore Wind: One of the most developed and reasonably priced forms of renewable energy is onshore wind power. Modern wind turbines may produce anywhere from 2 MW to 12 MW of power from the kinetic energy of the wind. By the end of 2020, there will be more than 700 GW of installed capacity worldwide thanks to the widespread deployment of onshore wind farms, which are mostly located in Europe, North America, and China (Global Wind Energy Council, 2021). Onshore wind faces three key challenges: grid integration, visual and noise consequences, and site selection.

Offshore Wind: With higher capacity factors than onshore wind, offshore wind energy harvests stronger and more reliable winds at sea. Deeper water offshore wind development is now more feasible thanks to technological developments like floating turbines. By the end of 2020, the total offshore wind capacity reached 35 GW worldwide; further expansion is anticipated, particularly in Europe and Asia (Global Wind Energy Council, 2021). However, there are obstacles for offshore wind, including more installation and maintenance costs, environmental issues, and the requirement for strong transmission infrastructure.

3.2.3. Technologies for Wind Energy

Large-Scale Hydropower: As hydropower producing about sixteen percent of the world's electricity, it is the most established renewable energy source. Large-scale hydropower facilities, which are frequently constructed on rivers or reservoirs, use the energy of falling or flowing water to turn turbines and produce electricity. Although hydropower is a dependable and affordable energy source, its development is frequently linked to negative social and environmental effects, including as the destruction of ecosystems, the uprooting of communities, and alterations in the quality of the water (World Bank, 2019).

Small and Micro Hydropower: Large dams can be replaced with more ecologically friendly small-scale and micro-hydropower plants. These systems offer decentralized energy solutions without the need for significant infrastructure improvements, making them appropriate for rural and isolated locations. The principal obstacles encompass the fluctuations in water flow and the possible influence on nearby ecosystems.

3.2.4. Biomass Energy

As hydropower producing about sixteen percent of the world's electricity, it is the most established renewable energy source. Large-scale hydropower facilities, which are frequently constructed on rivers or reservoirs, use the energy of falling or flowing water to turn turbines and produce electricity. Although hydropower is a dependable and affordable energy source, its development is frequently linked to negative social and environmental effects, including as the destruction of ecosystems, the uprooting of communities, and alterations in the quality of the water (World Bank, 2019).

Advanced Biofuels: The next phase of bioenergy is represented by advanced biofuels, which include cellulosic ethanol and biodiesel made from algae. Because these fuels come from biomass that isn't used to create food, they provide more environmental advantages and less competition with the food production industry. Although advanced biofuels have great potential, they are still in the early phases of commercialization and face obstacles with regard to cost, scalability, and technological maturity.

3.2.5. Geothermal Energy

Geothermal Power Plants: Using heat from below the surface of the Earth, geothermal energy produces electricity or direct heat. Typically, geothermal power facilities are located in places with high geothermal activity, like volcanic regions. With little effect on the environment, this energy source can supply baseload electricity and is very dependable. There remained much room for expansion; as of 2020, the installed capacity of geothermal energy worldwide was estimated to be 15.4 GW (IRENA, 2020). However, geographic constraints and the high upfront expenditures of discovery and drilling limit the use of geothermal energy.

Geothermal Heat Pumps: By utilizing the generally consistent temperatures found beneath the Earth's surface, geothermal heat pumps, or GHPs, are used to heat and cool buildings. Although GHPs are very effective and adaptable to different climates, their adoption is frequently impeded by expensive upfront installation costs and the requirement for specialized manpower.

3.2.6. Ocean Energy

These ocean energy technologies use the tides and currents to their advantage to produce electricity. Ocean energy has great promise because of the abundance of energy resources found in the sea, even though it is still in its experimental and early implementation stages. Only a few commercial projects are currently operating, and the installed capacity of ocean energy globally is still quite limited (IRENA, 2020). The influence on marine habitats, high costs, and technology maturity are among the challenges.

3.2.7 Emerging Technologies

The production of hydrogen from renewable energy sources is regarded as a crucial element of the energy system of the future, especially for decarbonizing hard-to-electrify sectors like transportation and heavy industrial. Fuel cells provide a clean and effective energy source by converting hydrogen into electricity. However, the high costs of manufacturing, storage, and delivery are currently impeding the widespread use of fuel cells and hydrogen (IEA, 2021).

4. CURRENT STATE OF RENEWABLE ENERGY IN RURAL AREAS

The status of renewable energy technologies, or RETs, in rural areas reflects a dynamic and changing environment marked by changes in legislation, technology breakthroughs, and regional variations in the degree of adoption. An overview of the current situation, emerging trends, and difficulties with RET implementation in rural areas is given in this section.

4.1Technological Deployment

Solar Energy: Because of its scalability and falling costs, solar energy has emerged as one of the most popular RETs in rural areas. Photovoltaic (PV) systems are widely used; these include both residential and community solar installations. The International Renewable Energy Agency (IRENA, 2021) reports that millions of off-grid solar household systems have been installed in South Asia and Sub-Saharan Africa. In rural regions, solar stoves and water pumps are becoming more popular since they increase access to water and lessen dependency on conventional biomass.

Wind Energy: Although less prevalent than solar energy in rural areas, wind energy is still a developing industry, especially in locations with significant wind potential. Individual residences or small villages are powered by small-scale wind turbines. While onshore wind farms are more commonly found in rural regions, tiny and micro wind projects are being investigated to offer localized power options, according to the Global Wind Energy Council (GWEC, 2022).

Biomass Energy: In rural areas, especially in developing nations, biomass energy technology, such as biogas digesters and biomass stoves, are widely used. By converting animal and agricultural waste into energy, these technologies offer a sustainable substitute for conventional biomass fuels. According to the World Bank (2021), biogas systems have been effectively introduced in nations like China and India, enhancing waste management and energy access.

Hydropower: In rural locations with access to flowing water, small-scale hydropower systems, such as micro- and pico-hydro projects, are used. In contrast to previous RETs, these systems offer consistent and dependable electricity. Small hydropower projects help electrify rural areas in places like Southeast Asia and Latin America, according to IRENA (2021).

Geothermal Energy: Because geothermal energy depends on particular geological conditions, it is less frequently used in rural areas. Nonetheless, geothermal energy is being used for direct use applications as well as power generation in areas having geothermal resources, such as sections of East Africa and the Pacific Islands (IRENA, 2021).

4.2. Adoption and Market Trends

Growing Investment and Support: There has been a notable increase in investment and support for RETs in rural areas from both public and private sectors. International organizations, governments, and non-governmental organizations (NGOs) are increasingly funding RET projects and providing technical assistance. The deployment of RETs is often supported by policies such as subsidies, tax incentives, and targeted rural electrification programs.

Innovative Financing Models: The adoption of RETs in rural areas has been aided by new financing models such as crowdfunding and pay-as-you-go (PAYG) schemes. With these models, customers can pay for energy services gradually and the upfront cost barrier is lessened. Companies like M-KOPA Solar and d.light are pioneering PAYG solar models in Africa, providing affordable access to solar power for off-grid households (World Bank, 2021).

Community-Led Initiatives: Many successful RET projects in rural areas are driven by community-led initiatives. Local ownership and management of RET systems often lead to better sustainability and acceptance. Community solar projects, cooperatives, and local entrepreneurs are increasingly involved in the deployment and operation of RETs.

5. CHALLENGES AND BARRIERS

5.1. Technical Challenges

5.1.1 Infrastructure Restraints

Grid Connectivity: Connecting renewable energy systems can be difficult in many rural regions due to inadequate grid infrastructure. This may make it more challenging to keep an energy supply steady and dependable (International Renewable Energy Agency [IRENA], 2021).

Maintenance Problems: The durability and functionality of RETs in distant areas are frequently impacted by maintenance problems brought on by a lack of local technical support and access to spare parts (IRENA, 2021).

5.1.2 Technological Barriers Intermittency

Renewable energy sources like solar and wind are not always available, which makes them less reliable. Although energy storage is a common solution, it can be expensive and complicated (Global Wind Energy Council [GWEC], 2022).

Scalability: Due to technological and budgetary limitations, scaling RETs to satisfy the energy needs of larger rural populations can be challenging (IRENA, 2021).

Adaptability to Local Conditions: Careful planning and adaptation are necessary to ensure that RET technologies are suitable for the local environmental conditions (IRENA, 2021).

5.1.3 Potential for Solar, Wind, and Biomass

Precise evaluation of renewable resources such as solar radiation, wind velocity, and biomass accessibility is necessary for efficient deployment of RETS (IRENA, 2021).

Insufficient Technical Know-How and Capability Development

Technical know-how: It's common for rural locations to lack the technical know-how required for RET installation, upkeep, and operation. Training and capacity building are essential for a successful implementation (IRENA, 2021).

5.2 Economic Challenges

5.2.1 **High Initial Investment Costs**

In rural areas with limited financial resources, the high initial costs of RET systems may be a hurdle (World Bank, 2021).

5.2.2 **Finance and Subsidies:**

For RET initiatives to be successful, financing, financial incentives, and subsidies must be available. The deployment of RETs in rural areas may be hampered by the limited availability of these financial resources (World Bank, 2021).

Credit Access: Having trouble getting credit can make it harder to invest in RETs. In rural areas, financial institutions could be reluctant to provide money for renewable energy projects (World Bank, 2021).

5.2.3 **Economic Viability and Sustainability**

Analyzing costs and benefits over the course of the project is necessary to determine whether RETs are economically viable. Long-term success requires ensuring economic sustainability (World Bank, 2021).

5.2.**4 Challenges in Securing Long-Term Financial Support**

Securing long-term funding for RET upkeep and operations can be challenging. It is necessary to create sustainable finance arrangements (World Bank, 2021).

5.3. Social and Cultural Challenges

5.3.1 Acceptance and Engagement in the Community

Including local communities in RET project design and execution promotes confidence and guarantees that initiatives are in line with local needs and preferences (IRENA, 2021).

5.3.2. Resistance and Barriers Due to Culture

Adoption of RETs may be hampered by cultural resistance to new technologies. Overcoming these obstacles can be facilitated by integrating new technology with current procedures (IRENA, 2021).

5.3.3. Knowledge and Consciousness

To promote adoption and efficient use of RETs, it is imperative to educate and raise awareness of their advantages (IRENA, 2021).

5.3.4Socioeconomic Factors' Effects

Adoption and efficient use of RETs can be influenced by socioeconomic factors such as gender roles and income levels. It's critical to address these issues for fair access.

5.4 Policy and Regulatory Challenges

5.4.1 Gaps in Policy and Regulatory Frameworks

Policy Vulnerabilities: The deployment of RET may be impeded by insufficient or inconsistent policy frameworks. In order to create an enabling environment, thorough and supporting policies must be developed (IRENA, 2021).

5.4.2 Governance Concerns and Bureaucratic Obstacles

Regulatory Complexity: RET projects may be delayed by navigating administrative roadblocks and governance concerns. Implementation can be aided by simplifying regulatory procedures (IRENA, 2021).

5.4.3 Energy Policy Inconsistencies

Various degrees of inconsistency in energy policies can lead to uncertainty and affect RET investment. Stability depends on ensuring policy alignment (IRENA, 2021).

5.4.4 NGOs' and the Government's Roles

Governments and non-governmental organizations (NGOs) are vital sources of finance, policy formulation, and technical assistance for RET projects. Increasing cooperation is crucial (IRENA, 2021).

5.4.5 Legal Obstacles

The deployment of RET may be impacted by legal matters pertaining to environmental rules and property ownership. Effective land use planning and well-defined legislative frameworks are crucial (IRENA, 2021).

5.5. Environmental Challenges

5.5.1 Environmental Impact Assessments (EIAs)

To assess and lessen the environmental effects of RET projects, comprehensive environmental impact assessments are required (IRENA, 2021).

5.5.2. Ecological Preservation and Energy Development in Balance

One of the main challenges is ensuring that RET projects meet energy needs without having a detrimental impact on nearby ecosystems (IRENA, 2021).

5.5.3 Issues Related to Climate

The performance of RET systems may be impacted by variations in renewable energy supplies brought on by climatic circumstances. It's critical to create management plans for these variations (IRENA, 2021).

5.5.4 Long-Term Sustainability

Resolving maintenance, resource management, and community involvement challenges is necessary to guarantee the long-term viability of RET projects (IRENA, 2021).

6. VARIOUS FACTORS AFFECTING DEPLOYMENT OF RENEWABLE ENERGY TECHNOLOGIES IN RURAL AREAS

6.1 Network Model

1. The diagram's network model demonstrates how numerous factors influencing the adoption of renewable energy technology in rural areas are interconnected. The main variables and how they relate to one another are explained as follows:
2. This factor is related to Infrastructure Availability and Financial Variables, suggesting that the availability of funding and the state of the infrastructure have an impact on the viability of expanding renewable energy projects.
3. Money matters a lot and has a direct impact on infrastructure availability, policy and regulatory frameworks, and extension feasibility. This implies that the availability of funding affects a number of aspects of the deployment of renewable energy, such as project viability, infrastructure development, and the impact of laws and regulations.
4. This variable indicates that the availability of necessary infrastructure is influenced by financial resources, regulatory frameworks, and the availability of technical capabilities. It links to Financial Variables, regulatory and Regulatory Frameworks, and Technical Expertise.
5. Financial variables, infrastructure availability, and social acceptance and engagement are all related to policies and laws. This demonstrates how infrastructure, financial, and societal approval of renewable energy projects all affect how effective policy frameworks are.
6. This element emphasizes how environmental factors and policy measures have an impact on social involvement and the adoption of renewable energy technology. It is linked to policy and regulatory frameworks as well as environmental impact.
7. Technical competence is essential for creating the required infrastructure and reducing environmental impacts, as evidenced by the connections between technical skills and knowledge, infrastructure availability, and environmental impact.
8. This variable suggests that public acceptance and technical considerations have an impact on the environmental effects of renewable energy projects. It is connected to both technical expertise and social acceptance and engagement.

9. The interdependence of these factors and how they all work together to successfully implement renewable energy technology in rural regions are highlighted by the network model. It emphasizes how crucial it is to address social, technological, financial, infrastructure, policy, and environmental concerns in concert in order to develop sustainable energy solutions.

7. STRATEGIES TO OVERCOME CHALLENGES IN DEPLOYING RENEWABLE ENERGY TECHNOLOGIES IN RURAL AREAS

Deploying renewable energy technologies in rural areas comes with various challenges, but there are strategies to overcome them.

7.1. Financial Strategies

7.1.1. Subsidies and Grants

Through the provision of grants and subsidies, governments can significantly reduce the initial financial burden of renewable energy projects. These financial supports have the potential to increase technology affordability and accessibility, especially in rural areas where a substantial initial capital expenditure is frequently required. Grants and subsidies can help with some of the equipment, installation, and maintenance costs, which lowers the financial risk for investors and promotes a greater use of renewable energy sources.

7.1.2. Microfinancing

One good way to gradually spread the cost of renewable energy technologies is to provide microloans to small towns and individual families. Rural communities may not have access to typical banking services, but microfinance makes it possible for them to invest in sustainable energy solutions. Microfinance can assist in overcoming financial obstacles and promoting sustainable energy practices at the local level by dividing the expense into manageable installments.

7.1.3. Public-Private Partnerships

Promoting collaborations between governmental and corporate organizations can help spread the costs and risks of renewable energy initiatives. Public-private partnerships (PPPs) combine the efficiency and inventiveness of the private sector with the control and assistance of the public sector to capitalize on the assets of both. These partnerships can lead to more substantial and profitable initiatives, guaranteeing the efficient deployment and upkeep of renewable energy systems in rural areas.

7.2. Infrastructure Strategies

7.2.1. Incremental Development

It seems sense to start modestly with renewable energy projects and build infrastructure gradually in order to manage resources and reduce risk. Before making more significant investments, technologies can be tested and optimized on a smaller scale thanks to incremental development. By identifying and addressing possible issues early on, this staged strategy can assist ensure that the infrastructure development is scalable and sustainable.

7.2.2. Integrated Planning

The best use of resources can be achieved by coordinating renewable energy projects with plans for the development of water, transportation, and communication infrastructure. By guaranteeing that renewable energy projects are in line with more general community development objectives, integrated planning fosters comprehensive and sustainable growth. Through concurrent consideration of various infrastructure requirements, planners can establish synergies that augment the aggregate effect of renewable energy initiatives.

7.2.3. Community Involvement

Involving local people in the design and building phases is essential to guaranteeing the viability and approval of renewable energy initiatives. Participation in the community promotes a sense of accountability and ownership, which improves infrastructure upkeep and performance. Projects can be more successfully adapted to match the unique needs and preferences of the community by incorporating local knowledge and including residents in the decision-making process.

7.3. Policy and Regulatory Strategies

7.3.1 Clear Policies

Providing stability and trust to investors and developers requires the development of long-term, coherent policies that facilitate the deployment of renewable energy. Well-defined policies have the potential to delineate the legal framework, establish attainable goals, and furnish directives for adherence. These laws contribute to the stability of the market, which promotes investment and new developments in renewable energy technologies.

7.3.2 Incentives

The adoption of renewable energy technology can be greatly increased by putting in place incentives like tax breaks, feed-in tariffs, and renewable energy certificates. These incentives increase the financial attractiveness of renewable energy projects by shortening payback periods and increasing profitability. Incentives can hasten the switch to sustainable energy sources by rewarding investment in renewable energy.

7.3.3 Simplified Processes

The implementation of renewable energy projects can happen more quickly if regulatory approval procedures are streamlined to cut down on administrative red tape. Clear and uncomplicated processes can reduce administrative expenses and time, which facilitates developers' compliance with regulations. Processes that are efficient can guarantee that projects are started and finished on time, which helps to meet energy targets on time.

7.4. Social Strategies

7.4.1. Education and Awareness Campaigns

Educating the public about the advantages of renewable energy is crucial to winning them over and earning their acceptance. These efforts can dispel myths, emphasize the financial and environmental benefits of renewable energy, and show how it can raise living standards. A greater level of community involvement and participation in renewable energy efforts might result from increased awareness.

7.4.2. Community Ownership Models

By creating models that enable local communities to own and oversee renewable energy projects, we can improve sustainability and give citizens more authority. In addition to keeping financial gains inside the community, community ownership may guarantee that initiatives are in line with local objectives and needs. Better project outcomes can result from such models, which can strengthen a sense of ownership and commitment.

7.4.3. Inclusive Decision-Making

Participating in the planning and decision-making process with community residents and local officials fosters acceptance and confidence in renewable energy initiatives. Diverse viewpoints are taken into account during inclusive decision-making, which produces more culturally and contextually appropriate answers. Additionally, it encourages accountability and transparency, both of which are essential to the long-term viability of renewable energy projects.

7.5. Technical Strategies

7.5.1. Training Programs

In order to guarantee that projects are carried out and maintained in an efficient manner, it is imperative that training programs be established to increase local capability in renewable energy technology. Training programs can give local technicians, engineers, and operators the skills and information they need, resulting in a qualified workforce that can support renewable energy installations. Energy projects can become more sustainable and reliable with this human capital investment.

7.5.2 Technical Assistance

Renewable energy systems must have continuous technical support and maintenance services in order to function properly. Technical support can reduce downtime, maximize system performance, and help resolve problems quickly. Long-term success and efficiency of renewable energy projects can be guaranteed by having access to resources and professional advice.

7.5.3. Partnerships with Educational Institutions

Developing specialized curricula in partnership with universities and technical institutions can boost local knowledge and stimulate innovation in renewable energy technology. Research opportunities, knowledge sharing, and training of the upcoming generation of energy experts can all be achieved through educational alliances. These kinds of partnerships have the potential to increase the renewable energy sector's overall resilience and capacity.

7.6. Environmental Strategies

7.6.1. Environmental Assessments

To find and lessen any potential drawbacks, comprehensive environmental impact evaluations are essential for renewable energy projects. By ensuring that projects are planned and carried out in an environmentally responsible manner, environmental assessments can help protect regional ecosystems and biodiversity. Developers may create more environmentally friendly and community-supported projects by tackling environmental issues up front.

7.6.2. Sustainable Practices

By putting best practices for sustainable project development and operation into reality, renewable energy technologies can have a smaller environmental impact. Using eco-friendly products, maximizing resource utilization, and making sure trash is managed properly are examples of sustainable practices. Projects can improve the local environment and serve as a model for future developments by upholding strict environmental requirements.

7.6.3. Monitoring and Evaluation

The sustainability of renewable energy plants depends on ongoing monitoring of environmental effects and required modifications. Frequent assessment can assist in spotting new problems and quickly putting remedial action plans into place. By taking a proactive stance, initiatives are kept socially and environmentally responsible, promoting long-term sustainability and the welfare of the community.

8. POLICY RECOMMENDATIONS AND REFORMS

1. To establish a coherent framework, include renewable energy policies into other industries' policies.

2. Coordinate regional, local, and national policies to prevent inconsistencies and disputes.
3. Simplify regulatory procedures to cut down on red tape and administrative obstacles.
4. To facilitate RET deployment, land use and environmental rules should be changed.
5. Create cutting-edge financing strategies like microfinance, green bonds, and public-private partnerships.
6. Provide specific tax exemptions, financial incentives, and subsidies to encourage the adoption of RET.
7. Make an investment in the education of regional engineers and personnel to handle and maintain RET systems.

9. IMPLICATIONS OF THE RESEARCH

Policymakers, developers, and communities stand to gain a great deal from studies on the application of renewable energy technologies in rural areas. Stakeholders may create focused plans to encourage the adoption of renewable energy by recognizing and resolving the obstacles, which will improve energy security and spur economic growth in underserved areas. The results underscore the necessity of all-encompassing and synchronized endeavours that incorporate economic, infrastructure, social, and policy-oriented remedies.

10. FUTURE OF THE RESEARCH

Future research on renewable energy technologies (RETs) in rural areas should focus on advancing technologies that address key challenges such as intermittency and scalability, making them more adaptable to local conditions. Hybrid energy systems, which combine multiple RETs, offer a promising direction to enhance reliability and energy security in these regions. Long-term studies are needed to assess the socio-economic impacts of RET deployment, particularly regarding income generation, education, and health outcomes. Additionally, improving the integration of RETs with existing rural infrastructure, especially in areas with limited grid connectivity, remains a critical area for exploration.

To guarantee local participation and acceptance of RET initiatives, research should also give priority to the creation of efficient community engagement and ownership mechanisms. Another interesting area for further research is the possibility of digital technologies like artificial intelligence (AI) and the Internet of Things (IoT) in optimizing RET systems for remote locations. Finding creative funding options and business plans that lower the hefty upfront expenses of RET implementation while maintaining long-term financial viability is crucial. Furthermore, assessing the efficacy of the current regulatory and legislative frameworks would aid in determining the changes that are required to better facilitate the adoption of RET in rural areas.

Environmental considerations must not be overlooked, as large-scale RET projects in rural settings can have significant ecological impacts. Future research should focus on developing guidelines to minimize these disruptions. Finally, integrating RETs with emerging technologies like energy storage solutions could significantly enhance the resilience and sustainability of rural energy systems, paving the way for more reliable and efficient energy access in underserved regions.

11. CONCLUSION

To provide equitable and sustainable energy solutions, a wide range of difficulties related to the implementation of renewable energy technology in rural regions must be addressed. The main barriers to the successful adoption of these technologies have been clarified by this research, and they include lack of funding, inadequate infrastructure, technical difficulties, and problems with societal acceptance. This study has provided targeted strategies for addressing these issues after conducting a thorough analysis of them. It has emphasized the significance of a coordinated approach that incorporates effective policy frameworks, infrastructure development, financial support mechanisms, and community engagement.

Financial obstacles can be lessened through public-private partnerships, microfinance, and subsidies, especially the high initial investment costs and restricted access to funds. Incremental development and integrated planning are effective approaches to handle infrastructure difficulties and enable the implementation of projects that are both scalable and sustainable. In order to ensure the long-term stability of renewable energy systems, deficiencies in technical skill can be filled through specialized training programs and technical assistance. Education campaigns and community ownership models are effective means of promoting social acceptance and fostering local support for renewable energy initiatives by fostering trust. Clear, dependable, and encouraging policy and regulatory frameworks are essential for offering the stability and financial incentives required to attract investment in rural energy projects.

The results of this study highlight the importance of taking a diverse strategy to removing the obstacles preventing the widespread use of renewable energy in rural areas. Stakeholders can collaborate to support the adoption of renewable energy technologies and improve energy security, economic development, and environmental sustainability in underserved regions by concentrating on financial, infrastructural, technical, social, and policy-related methods.

REFERENCES

Bahaj, A. S. (2011). Generating electricity from the oceans. *Renewable & Sustainable Energy Reviews*, *15*(7), 3399–3416. DOI: 10.1016/j.rser.2011.04.032

Bazilian, M., Nussbaumer, P., Rogner, H., Brew-Hammond, A., Foster, V., Pachauri, S., & Howells, M. (2012). Energy access scenarios to 2030 for the power sector in sub-Saharan Africa. *Utilities Policy*, *20*(1), 1–16. DOI: 10.1016/j.jup.2011.11.002

Beck, F., & Martinot, E. (2004). Renewable Energy Policies and Barriers. In C. Cleveland (Ed.), Encyclopedia of Energy (Vol. 5, pp. 365-383). Elsevier.

Bertani, R. (2012). Geothermal power generation in the world 2005–2010 update report. *Geothermics*, *41*, 1–29. DOI: 10.1016/j.geothermics.2011.10.001

Bhattacharyya, S. C. (2006). Energy access problem of the poor in India: Is rural electrification a remedy? *Energy Policy*, *34*(18), 3387–3397. DOI: 10.1016/j.enpol.2005.08.026

Brown, D. L. (1980). Rural electrification: The U.S. experience in international perspective. *Comparative Technology Transfer and Society*, *3*(1), 46–69. DOI: 10.1353/ctt.1980.0005

Chaurey, A., Ranganathan, M., & Mohanty, P. (2004). Electricity access for geographically disadvantaged rural communities—Technology and policy insights. *Energy Policy*, *32*(15), 1693–1705. DOI: 10.1016/S0301-4215(03)00160-5

Demirbas, A. (2009). Progress and recent trends in biofuels. *Progress in Energy and Combustion Science*, *33*(1), 1–18. DOI: 10.1016/j.pecs.2006.06.001

Global Wind Energy Council. (2021). Global wind report 2021. https://gwec.net/global-wind-report-2021/

Global Wind Energy Council. (2022). *Global Wind Report 2022*. Retrieved from https://gwec.net/global-wind-report-2022/

Global Wind Energy Council (GWEC). (2022). Global Wind Report 2022. https://gwec.net/global-wind-report-2022/

IEA. (2022). *World Energy Outlook 2022*. International Energy Agency.

International Energy Agency. (2020). World Energy Outlook 2020. https://www.iea.org/reports/world-energy-outlook-2020

International Energy Agency. (2020). World Energy Outlook 2020. https://www.iea.org/reports/world-energy-outlook-2020

International Energy Agency. (2021). Renewables 2021: Analysis and forecast to 2026. https://www.iea.org/reports/renewables-2021

International Renewable Energy Agency. (2021). *Renewable Capacity Statistics 2021*. Retrieved from https://www.irena.org/publications/2021/Mar/Renewable-Capacity-Statistics-2021

International Renewable Energy Agency (IRENA). (2020). Renewable capacity statistics 2020. https://www.irena.org/publications/2020/Mar/Renewable-Capacity-Statistics-2020

International Renewable Energy Agency (IRENA). (2021). Renewable Capacity Statistics 2021. https://www.irena.org/publications/2021/Mar/Renewable-Capacity-Statistics-2021

IRENA. (2016). Renewable Energy Benefits: Measuring the Economics. International Renewable Energy Agency. https://www.irena.org/publications/2016/Jan/Renewable-Energy-Benefits-Measuring-the-Economics

Kaundinya, D. P., Balachandra, P., & Ravindranath, N. H. (2009). Grid-connected versus stand-alone energy systems for decentralized power—A review of literature. *Renewable & Sustainable Energy Reviews, 13*(8), 2041–2050.

Kaundinya, D. P., Balachandra, P., & Ravindranath, N. H. (2009). Grid-connected versus stand-alone energy systems for decentralized power—A review of literature. *Renewable & Sustainable Energy Reviews, 13*(8), 2041–2050. DOI: 10.1016/j.rser.2009.02.002

Müller, B. (2007). *Designing the Clean Development Mechanism: Operational and institutional issues.* Edward Elgar Publishing.

Narula, K., Nagai, Y., & Pachauri, S. (2012). The role of decentralized distributed generation in achieving universal rural electrification in South Asia by 2030. *Energy Policy, 47*, 345–357. DOI: 10.1016/j.enpol.2012.04.075

National Rural Electric Cooperative Association (NRECA). (2020). History of NRECA. https://www.electric.coop/our-organization/history

Pachauri, S., & Rao, N. D. (2013). Gender impacts and determinants of energy poverty: Are we asking the right questions? *Current Opinion in Environmental Sustainability, 5*(2), 205–215. DOI: 10.1016/j.cosust.2013.04.006

Painuly, J. P. (2001). Barriers to renewable energy penetration; a framework for analysis. *Renewable Energy, 24*(1), 73–89. DOI: 10.1016/S0960-1481(00)00186-5

REN21. (2021). Renewables 2021: Global status report. https://www.ren21.net/reports/global-status-report/

REN21. (2023). Renewables 2023 Global Status Report. REN21 Global Platform for Renewable Energy.

Saidur, R., Rahim, N. A., Islam, M. R., & Solangi, K. H. (2011). Environmental impact of wind energy. *Renewable & Sustainable Energy Reviews, 15*(5), 2423–2430. DOI: 10.1016/j.rser.2011.02.024

Sharma, A., Tyagi, V. V., Chen, C. R., & Buddhi, D. (2009). Review on thermal energy storage with phase change materials and applications. *Renewable & Sustainable Energy Reviews, 13*(2), 318–345. DOI: 10.1016/j.rser.2007.10.005

Sovacool, B. K. (2012). The political economy of energy poverty: A review of key challenges. *Energy for Sustainable Development, 16*(3), 272–282. DOI: 10.1016/j.esd.2012.05.006

Sovacool, B. K., & D'Agostino, A. L. (2012). A comparative analysis of renewable electricity support mechanisms for Southeast Asia. *Energy, 44*(1), 189–204.

Sovacool, B. K., & Drupady, I. M. (2012). *Energy Access, Poverty, and Development: The Governance of Small-Scale Renewable Energy in Developing Asia*. Ashgate Publishing.

Van der Horst, D. (2007). NIMBY or not? Exploring the relevance of location and the politics of voiced opinions in renewable energy siting controversies. *Energy Policy*, *35*(5), 2705–2714. DOI: 10.1016/j.enpol.2006.12.012

World Bank. (2019). Hydropower development and resettlement. https://www.worldbank.org/en/topic/hydropower/overview

World Bank. (2021). The Role of Renewable Energy in Rural Development. https://www.worldbank.org/en/topic/renewableenergy/publication/renewable-energy-and-rural-development

World Bank. (2021). World Development Report 2021: Data for a Better World. The World Bank. Bhattacharyya, S. C. (2013). Rural Electrification Through Decentralised Off-grid Systems in Developing Countries. Springer.

World Health Organization. (2016). Burning Opportunity: Clean Household Energy for Health, Sustainable Development, and Wellbeing of Women and Children. https://www.who.int/airpollution/publications/burning-opportunity/en/

Yadoo, A., & Cruickshank, H. (2012). The role for low carbon electrification technologies in poverty reduction and climate change strategies: A focus on renewable energy mini-grids with case studies in Nepal, Peru and Kenya. *Energy Policy*, *42*, 591–602.

Chapter 10
Ultra-Wide Band Antennas for Sixth Generation Systems:
Design and Applications

Smrity Dwivedi
https://orcid.org/0000-0001-5160-6617
IIT BHU, India

ABSTRACT

The demand for wireless broadband communication is rapidly increasing due to the need to support more users and provide more information about more data. Ultra-wideband (UWB) technology has been developed to cover a very wide bandwidth at many frequencies using very narrow pulses in the nanosecond range and can solve this problem. UWB, a wireless technology designed to transmit information at high speeds over short distances with very low power, is an area of current interest with many applications. This chapter introduces the current UWB technology and its potential applications. It also discusses how these systems can be used without interfering with other existing communication systems. The advantages and disadvantages of these systems have also been examined in detail. Also, the main component which is the most important part of ultra-wide technology is an antenna, has been discussed here. The basic antenna design that can be used for sixth-generation system to enhance the data rates and speed has been mentioned.

INTRODUCTION

With its superior location and sensing capabilities, ultra-wideband (UWB) technology has become an important element that revolutionizes communication in all areas of life. Overall, the UWB industry is strong, indicating that many companies use UWB technology to design and manufacture products and services. The most obvious examples of this are the integration of UWB into mobile phones by major mobile phone manufacturers such as Apple, Google, Samsung, and Xiaomi, and the use of UWB in automotive applications (Federal communications, 2002).

UWB radio, operating with extremely large bandwidths, must coexist with many other interfering narrow-band signals (TV, GSM, UMTS, GPS, etc). In the same time, these narrow-band systems must not suffer intolerable interference from the UWB radios. Regulatory considerations over such a wide

DOI: 10.4018/979-8-3693-7788-8.ch010

bandwidth limit the radiated power. The low transmit power levels together with the ultra-fine time resolution of the system can increase considerably the synchronization acquisition time and the complexity of the receiver. The Federal Communications Commission in USA (FCC) has defined an UWB device as any device with a –10 dB fractional bandwidth, greater than 20% or occupying at least 500 MHz of the spectrum (Federal communications, 2002). Most narrowband systems occupy less than 10% of the center frequency bandwidth, and are transmitted at far greater power levels. For example, if a radio system were to use the entire UWB spectrum from 3.1-10.6 GHz, and center about almost any frequency within that band, the bandwidth used would have to be greater than 100% of the center frequency in order to span the entire UWB frequency range. By contrast, the 802.11b radio system centers about 2.4 GHz with an operating bandwidth of 80 MHz. This communication system occupies a bandwidth of only 1% of the center frequency.

The FCC additionally regulated the spectral form and maximum electricity spectral density (-41.3 dBm/MHz) of the UWB radiation so as to limit the interference with other communique structures. The strength spectral density is the average power in the sign consistent with unit bandwidth and therefore provides essential information at the distribution of power over the RF spectrum. The ETSI policies in ecu are predicted to follow the FCC however with a extra restrictive spectral form, encouraged via a different control of the to be had spectrum (D. Porcino, 2003)-(Europian Telecommunication, 2003). For the antenna which means that handiest one 500 MHz band need to be energetic at a time. The UWB spectral strength density masks is shown in Fig.1. Fig.1 indicates the UWB spectral power density mask constrained to –forty one.three dBm byusing FCC and ETSI to ensure that the UWB emission tiers fairly small. The emissions shall no longer exceed the common limits when measured the use of a decision bandwidth of 1 MHz. UWB handheld gadgets are tremendously small devices that don't employ a set infrastructure. Antennas might not be established on out of doors structures which include the outdoor of a building or on a smartphone pole. Antennas may be established most effective on the hand held UWB gadgets. hand-held UWB gadgets may additionally function indoors or outside (K. Siwiak et al. 2004).

Figure 1. UWB Spectral power density as per given report (Federal communications, 2002)

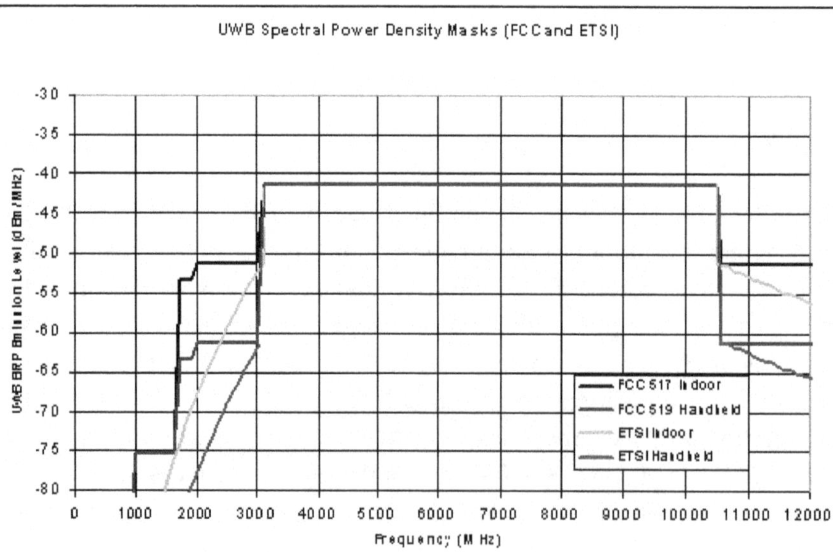

UWB is a commercial wireless technology that provides accurate location and measurement quality while supporting advanced security measures to prevent unauthorized access to data and communications contact information—brief information about UWB ecosystem design. We have discussed three main aspects of UWB data usage: hands-free control, location-based services, and peer-to-peer). The basis of UWB technology is established with IEEE standards to ensure effective communication and to make UWB technology stronger in terms of security and less susceptible to attack.

The sixth generation (6G) of wireless communication systems is expected to revolutionize the connectivity landscape with its promise of ultra-high data rates, ultra-low latency, massive connectivity, and enhanced spectral efficiency. Ultra-Wide Band (UWB) antennas are poised to play a critical role in the realization of 6G systems due to their ability to support wide bandwidths and high precision in communication and sensing applications.

1.1 What is Needed at Home

A good home appliance design at the micro level must meet many conditions. First of all, the reading assignment must be correct and accurate. Since most places must be kept secret, the technology must also be protected. It must be reliable and easily scalable even in harsh environments, targeting thousands of people and assets over a large area. Other requirements include low energy consumption and economy. This technology can be incorporated into all kinds of devices, from high-tech (such as complex devices such as smartphones) to low-tech (such as simple devices such as device nameplates). Additionally, the technology requires low latency to track moving objects in real time (D. Porcino, 2003). Fig. 2 shows all the requirements for indoor layout and location.

Figure 2. Indoor positing based on micro technique (D. Porcino, 2003)

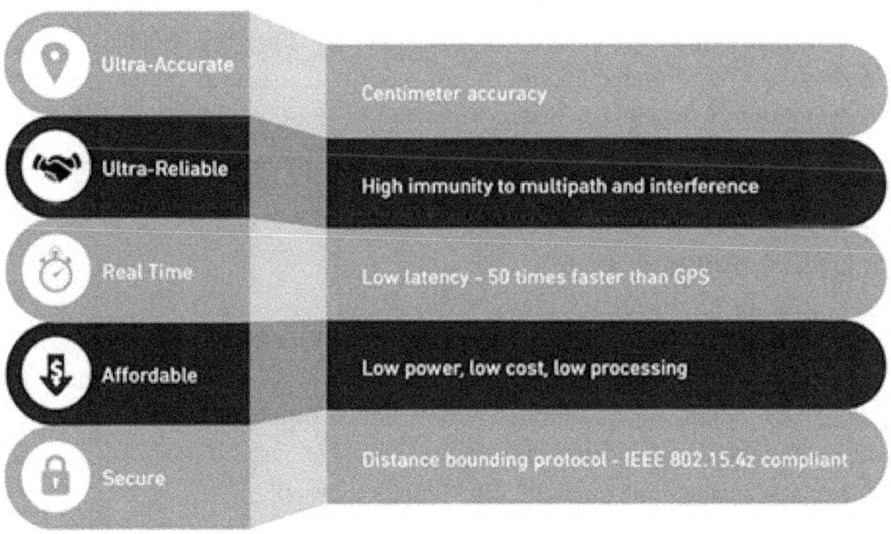

1.2 Comparison of UWB with Existing WiFi and BLE

In creating the first indoor location, engineers used the latest technologies available: WiFi and Bluetooth Low Energy (BLE). Such technologies are ideal for data communications, but none are designed for Real-Time Location Services (RTLS). Such structures cannot meet all the needs of a micro-based indoor environment. WiFi, Bluetooth and other narrowband radios can achieve accuracy of several meters. However, its reliability does not meet the 99.9% standard to ensure stability and reliability. It is very difficult to tell the location immediately due to the collision and interference of many electronic devices. Therefore, it is difficult for them to provide immediate service. For example - BLE is the leader in low energy data communications.

1.3) Key Characteristics of 6G Systems as Per Given Below

a)High Data Rates: a) 6G aims to deliver data rates up to 1 Tbps.b)Low Latency: Target latency is in the sub-millisecond range.c)Massive Connectivity: Support for a massive number of connected devices, including IoT. d)Enhanced Spectral Efficiency: Efficient use of the spectrum to accommodate high data traffic. e)Integration with AI and Machine Learning: To enable smart and adaptive network management (Ryuji Konho et al. 2004).

2) UWB ANTENNA DESIGN FOR 6G

2.1) Design Considerations

a)Wide Bandwidth: UWB antennas for 6G must support extremely wide frequency ranges to accommodate high data rates and diverse applications. b)Compact Size: With the proliferation of small, portable devices, UWB antennas need to be compact and easily integrable into these devices. c)High Gain and Directivity: To ensure efficient communication over long distances and in challenging environments. d)Low Power Consumption: Essential for battery-operated devices, especially in IoT applications. e) Multi-functionality: Antennas may need to support both communication and sensing functions (R. Fontana, 2002).

2.2) Types of UWB Antennas for 6G

a)Planar Monopole Antennas: Favored for their simple design, wide bandwidth, and ease of integration with printed circuit boards. b)Tapered Slot (Vivaldi) Antennas: Known for their wide bandwidth and high directivity, suitable for high-speed communication. c)Patch Antennas: Offer a compact form factor with good radiation efficiency and are suitable for integration in portable devices. d)Horn Antennas: Utilized in applications requiring high gain and directional radiation.

2.3) Design Techniques for 6G Antennas

a)Metamaterials: Using metamaterials to enhance bandwidth and gain. b)Fractal Geometries: Employing fractal designs to achieve compact size and wideband performance. c)Substrate Integration: Optimizing the substrate material and thickness to improve impedance matching and radiation efficiency. d)MIMO (Multiple Input Multiple Output): Incorporating MIMO technology to increase capacity and reliability.

2.4) Applications of UWB Antennas in 6G

a)High-Speed Wireless Communication: Enabling ultra-fast data transmission for applications like virtual reality, augmented reality, and high-definition video streaming. b)Precision Sensing and Localization: Supporting applications that require high-precision positioning, such as autonomous vehicles, industrial automation, and smart city infrastructure. c)IoT and Smart Devices: Providing low-power, high-efficiency communication for a vast number of connected devices in the Internet of Things ecosystem. d)Health Monitoring: Facilitating advanced medical applications, including remote monitoring and diagnostic imaging. e)Environmental Monitoring: Supporting wide-area sensing and data collection for applications like agriculture, climate monitoring, and disaster management.

2.5)Challenges and Future Directions for UWB 6G Antennas

a)Interference Management: Ensuring UWB signals do not interfere with other communication systems, which requires advanced interference mitigation techniques. b)Scalability: Designing UWB antennas that can efficiently scale to support the massive connectivity envisaged in 6G. c)Integration with AI: Leveraging AI and machine learning to optimize antenna performance and dynamically adapt to changing environments. d)Sustainability: Developing eco-friendly materials and manufacturing processes to reduce the environmental impact of UWB antenna production (R. J. Fontana, 2002).

2.6) Principles of UWB Technology

a) Wide Bandwidth: UWB antennas operate over a wide range of frequencies. This wide bandwidth allows for high data rates and improved resolution in applications like radar and imaging.
b) Short Pulse Transmission: UWB systems often use very short pulses (in the order of nanoseconds) to transmit data. This results in lower power consumption and reduced interference with other wireless systems.

c)High Resolution and Precision: The wide bandwidth provides fine time resolution, which is crucial for applications requiring precise location tracking and high-resolution imaging.

2.7) Applications of All Types of UWB Antennas With Advantages

a) Communications: UWB antennas enable high-speed wireless communication, which is beneficial in applications such as wireless personal area networks (WPANs), including technologies like Wi-Fi and Bluetooth alternatives.

b)Radar and Imaging: UWB is extensively used in radar systems for applications such as ground-penetrating radar, medical imaging, and vehicular collision avoidance systems due to its high resolution.
c)Localization and Tracking: The fine time resolution of UWB signals makes them ideal for precision location tracking and navigation systems, including indoor positioning systems and asset tracking.

d) Internet of Things (IoT): UWB antennas support IoT applications by providing low-power, high-accuracy communication solutions for smart devices and sensor networks.

Advantages of UWB Antennas: a)High Data Rates: UWB antennas support high-speed data transmission, making them suitable for multimedia applications. b)Low Power Consumption: The short pulse nature of UWB technology reduces power usage, which is beneficial for battery-operated devices. c) Low Interference: UWB operates at low power levels and over a wide frequency range, minimizing interference with other wireless systems. d) Precision: UWB provides accurate distance and location measurements, which is critical for applications requiring high precision.

3) EXAMPLE OF UWB ANTENNA AND UWB 6G ANTENNA

a)In this paper, A coplanar waveguide (CPW) extremely-wideband(UWB) antenna is designed, analyzed and simulated through laptop simulation era(CST). The proposed antenna is fabricated on FR-four dielectric substrate. A microstrip feed line is used to excite the antenna.The floor aircraft is slotted to improve the impedance bandwidth (BW). right here, a square patch is used as radiator and two corners out of 4 are truncated to enhance impedance matching and UWB characterization.This antenna satisfies UWB traits like VSWR<2, return loss(S11)<-10 dB,gain<5dB and the antenna is running within the frequency variety of 1.59 to 11.87 GHz range which covers complete extremely wideband i.e. 3.1 to 10.6 GHz variety.

Figure 3. (a) Proposed structure for UWB antenna, (b)S11 parameter for the structure, (c)Radiation pattern (A. Kunturu et al. 2020).

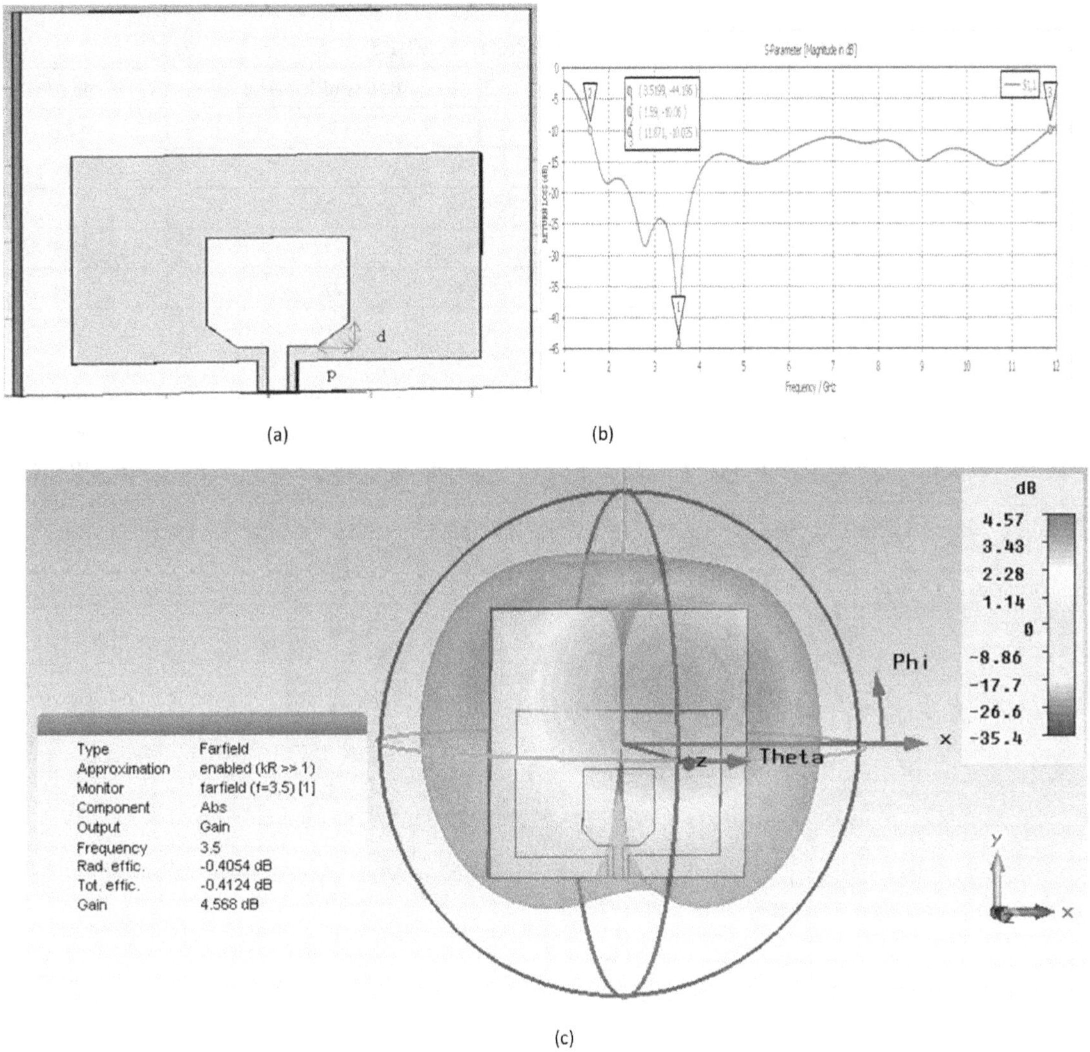

(b)This is the writer's article in which the author has defined the design and simulation of terahertz antennas over 5G. This new era via 5G helps join most devices and create a balance among machines and customers in step with technological wishes. This requires a larger bandwidth and higher benefit. Therefore, the new terahertz band variety is used. Graphene is used as a base fabric which can operate at excessive frequencies and additionally helps graphene, whilst polyimide and silicon dioxide are used as base substances and air and soil layers. various studies papers report a growth in bandwidth from 0.3 THz to 0.6 THz and an boom inside the frequency variety from 0.1 THz to at least one THz from 3.02 dB to 10.09 dB terahertz discusses his transformation. At some stage in the improvement technique, distinctive strategies had been used and simulated at the same version, resulting in a larger simulation bandwidth of 52.9 GHz and a better simulation of 10.9dB.

Table 1. Comparison table of designed structures (S. Dwivedi, 2023)

Given structure	Frequency (THz)	Bandwidth (GHz)	Gain (dB)
Conventional structure	0.45	30.4	4.21
Structure first	0.45	35.6	8.48
Structure second	0.45	37	8.49
Structure third	0.45	32	8.05
Structure fourth	0.45	52.9	10.9

Figure 4. Basic structure proposed with different colors coding of materials used which represents the layers of the structures of different materials such as air, polyamide, graphene (S. Dwivedi, 2023).

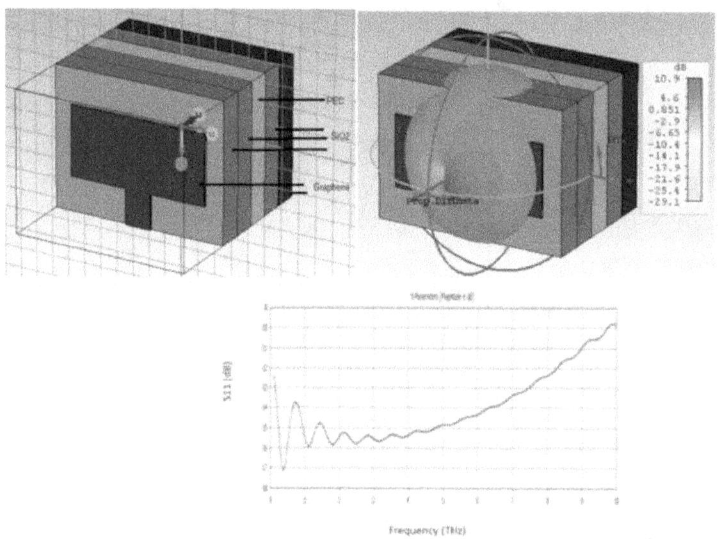

c) This is the writer's work; here the authors use three fundamental designs of H-fashioned antenna with partial grounding and their ratio of terahertz frequency to 5G use. This new technology through 5G enables connect most devices and create a balance among machines and customers in line with technological desires. The antenna turned into designed the use of a single rectangular loop and switched twice in the same patch, and its critical parameters such as bandwidth and gain have been simulated with the help of CST Microwave Studio. For our fashions, the bandwidths of the first design, the second one layout, and the 1/3 design are 100 GHz, 250 GHz, and four hundred GHz, respectively, and the advantage is 13.5 dBi, 14.4 dBi, and 14.8 dBi, respectively. The simple material % is used as the electric material, and polyamide is used as the base cloth. Microstrip feeders are used within the floor 1/2.

Table 2. Comparison table of designed structures (S. Dwivedi, 2023)

Given structure	Bandwidth (GHz)	Gain (dBi)
Ist Design	100	13.5
IInd Design	250	14.4
IIIrd Design	400	14.8

Figure 5. H- shape antenna with UWB performance and its radiation pattern (S. Dwivedi, 2023).

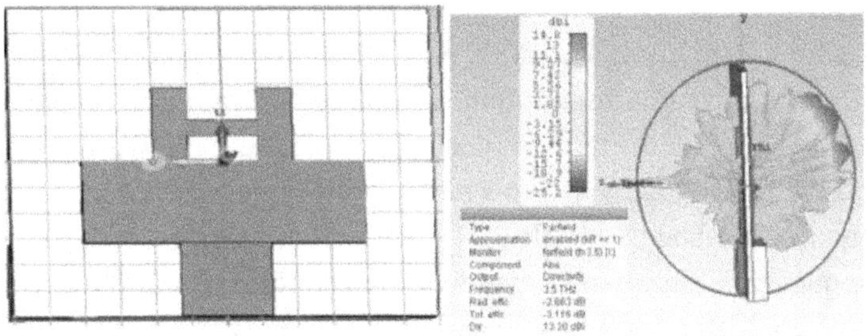

d)Again, this is the writer's own work in which the writer gives the layout and simulation of terahertz antennas over 5G. This new era thru 5G helps join most devices and create a balance among machines and users in step with technological desires. For this motive, seeing that a excessive-performance system is utilized in this newsletter, excessive bandwidth and maximum pace requirements are the main obstacles to improving overall performance. because of this, high frequency become chosen to create an antenna without the use of materials including graphene and gold. The array used right here has appropriate MIMO abilities for high-gain ultra-wideband terahertz frequencies. The terahertz MIMO antenna used right here is to be had in 2-element and three-element arrays. The antenna shape was simulated using commercial software program. The gain is 24.5dBi for the 3-factor array and 22.1 dBi for the 2-point array. There are only a few studies papers discussing the terahertz revolution. For the frequency range from 2 THz to 6 THz, the bandwidth increases from 0.2 THz for a 2-element array to 0.45 THz for a three-element array (S. Dwivedi, 2022), (S. Dwivedi, 2022), (S. Dwivedi, 2023), (S. Dwivedi, 2023).

Table 3. Designed, simulated and proposed structures comparison (S. Dwivedi, 2023).

Given Structure	Frequency (THz)	Bandwidth (THz)	Gain (dBi)
2-elements array antenna	2.74	0.2	22.1
3-elements array antenna	5.3502	0.45	24.5

Figure 6. Structure of 3-element antenna leaf shape with its radiation pattern (S. Dwivedi, 2023).

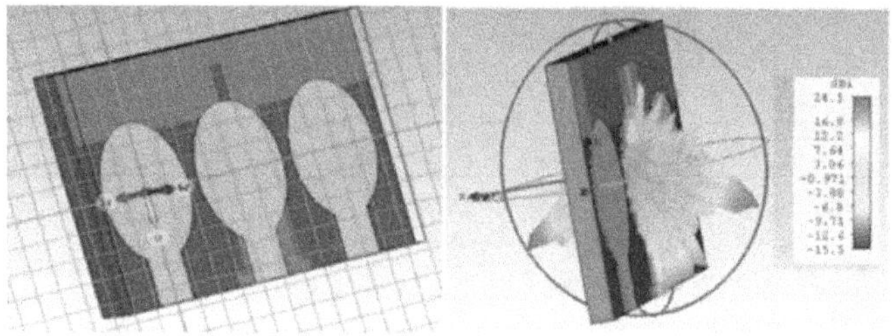

e)In this article, the author simulates an elliptical, leaf-fashioned, six-blade terahertz antenna, whose broadband utility is appropriate for communique structures past 5G. The antenna design suggests using elliptical blades and six facets to improve performance. The bandwidth and gain of the two-web page, four-page, and 6-web page antennas have been measured at 300 GHz, 350 GHz, and 500 GHz, respectively, and the gains have been 24.9 dBi, 25 dBi, and 25.2 dBi, respectively. The authors derived three editions for 3 frequencies, supplying 3 bandwidth necessities for applications on 5G. Hexahedral factor reaches a most of 25.2dBi at 500 GHz bandwidth (S. Dwivedi, 2023), (S. Chandra et al. 2019), (S. Chandra et al., 2020), (S. Chandra et al. 2021).

Table 4. Given structures simulated and its comparison (S. Dwivedi, 2023)

Given structure	**Bandwidth (GHz)**	**Gain (dBi)**
Two leaves pattern antenna	300	24.9
Four leaves pattern antenna	350	25
Six leaves pattern antenna	500	25.2

Figure 7. Inset fed Six elliptical leaves pattern antenna and radiation pattern (S. Dwivedi, 2023).

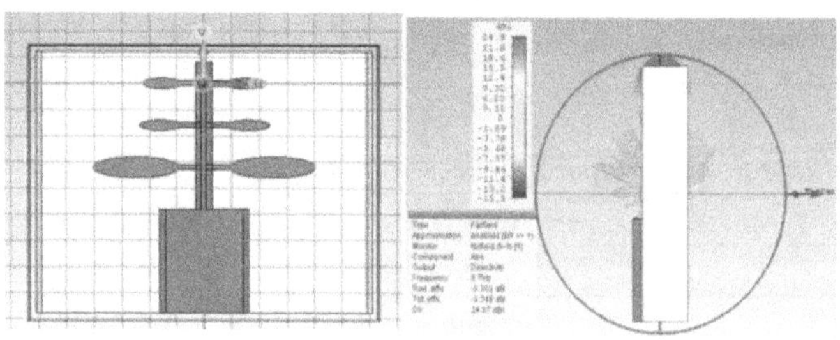

f)This text specially examines the frequency reaction antenna, which solves the trouble of the usage of an antenna with a couple of frequencies to carry out a couple of capabilities according to the consumer's needs and improves overall performance on a small scale. within the frequency range from millimeter to terahertz, facts rate and green structures are the most essential to achieve wi-fi communique with small transmission. to apply the antenna at a couple of frequency within a given frequency, a radio frequency reconfigurable antenna is prepared the usage of a double-patched gold device. It makes use of SiO2 with a dielectric regular of 3.6 (60 × 60μm2) and a thickness of 1.575μm as the substrate. It uses a coaxial feed with an SMA connector and has a metal surface of the identical size at the back of the bottom plate that acts as a reflector. This reflector is used to increase the benefit and other characteristics of the antenna by way of reducing surface radiation. to connect the two patches, a varactor diode is used among the 2 patches. The structure resonates at 3 specific frequencies, specifically 5.6 THz, 6.2 THz, and 14.6 THz, with most values of 6.44 dBi, 7.75dBi and 8.54 dBi, respectively (S. Dwivedi, 2023), (Utkarsh Gupta et al. 2021).

Figure 8. UWB Antenna with its switch, S11 parameter, and radiation pattern with reconfigurable properties (S. Dwivedi, 2023).

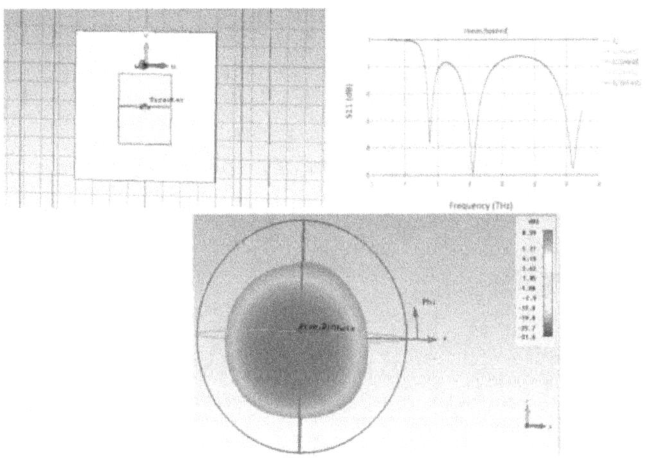

g)Right here, the author has used graphene to design the frequency reconfigurable antenna as graphene has specific residences to work with high frequency. This precise tunable property of graphene may be used to design a frequency-reconfigurable antenna. while a bias voltage is carried out to the designed antenna array as proven in Fig. 9, the chemical capacity of graphene is various from 0.4 eV to 0.8 eV. With the boom of chemical capability and exchange in floor conductivity, a dramatic alternate in antenna residences is observed. extended chemical potential offers the boom in return loss and shifts of antenna resonance frequency to the higher frequencies. to analyze the property of frequency reconfigurable antenna, the proposed shape is simulated in CST software program and the result is proven in fig. 10. The chemical ability of all 4 patch factors of the 1×4 antenna array is modified concurrently from zero.4eV to 0.8eV. The obtained simulated result is summarized in desk 1. Shift in frequency is observed from zero.sixty four THz to 0.seventy four THz with the exchange of chemical capability from 0.4eV to 0.8eV. it is clear that the carried out frequency band that is blanketed by way of the proposed frequency reconfigurable antenna is zero.64-zero.seventy four THz which offers the bandwidth of 100 GHz that is high sufficient. consequently, with such reconfigurable antennas we are able to easily take away the drawback of narrow band antennas.

Figure 9. 1×4 graphene array antenna to achive reconfigurable properties (S. Chandra et al. 2021).

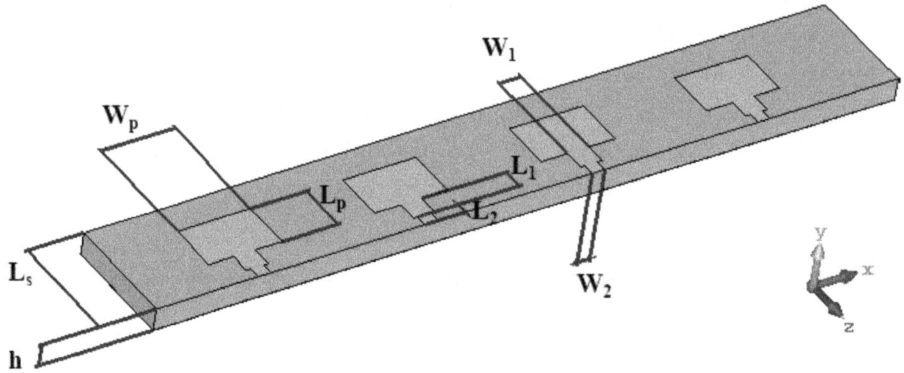

Figure 10. Chemical potential effect with a variation of resonating frequency for frequency reconfigurable antenna (S. Chandra et al. 2021).

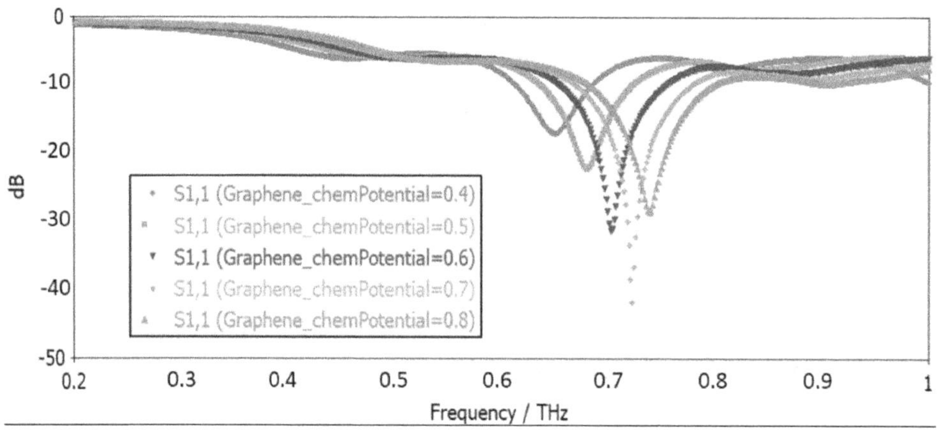

Table 5. Frequency reconfigurable antenna values as chemical potential and frequency (S. Chandra et al. 2021)

Chemical Potential of graphene (eV)	Frequency (THz)
0.4	0.64
0.5	0.69
0.6	0.70
0.7	0.71
0.8	0.74

h) in this paper, the authors use a low dielectric constant of 2.2 (Rogers RT/Duroid substrate) and a dielectric loss tangent of 0.0010 in the proposed design as in the following parent. due to demand, the authors have evolved a metamaterial-based multi-band microstrip rectangular patch antenna with extensive high bandwidth overall performance. The resonant frequency of this layout is 26 GHz. Simulations are accomplished the usage of FEKO software program to investigate its overall performance. After set up and integration, the bandwidth, return loss, antenna efficiency and energy density expanded from 5.368 GHz to 5.63 GHz, -34.82 dB to -57.10 dB, from 99.7% to 99.8%, and from 2208 kW/m2 to 2208 kW/m2/m2.

Figure 11. (a) Split-ring resonator design as metamaterial (b) unit cell of given SRR structure (S. Dwivedi, 2024), (G. Geetharamani et al. 2019).

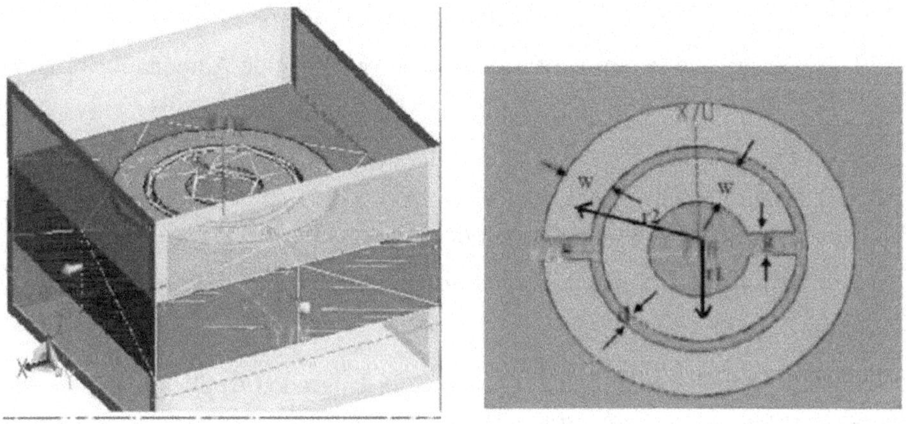

(a) (b)

4. CONCLUSION

UWB antennas are a key component of modern communication and sensing systems, offering a unique combination of high data rates, low power consumption, and precise ranging capabilities. As technology continues to advance, UWB antennas are likely to play an increasingly important role in a wide range of applications, from consumer electronics to industrial and medical systems. UWB antennas are set to be a cornerstone technology in the development and deployment of 6G systems. Their ability to support wide bandwidths, high data rates, and precise localization make them indispensable for the next generation of wireless communication. Continued innovation in UWB antenna design and integration will be crucial to unlocking the full potential of 6G, enabling a new era of connectivity and smart applications.There are different designs are discussed above to solve such problems and can be easily integrated anywhere. Apart from that there are still many designs of antenna are in processing mode to reduce the problem in rural area.

REFERENCES

Chandra, S., & Dwivedi, S. (2019, March). Comparative analysis of reconfigurable patch antenna array for different liquid crystal substrates. In *2019 URSI Asia-Pacific Radio Science Conference (AP-RASC)* (pp. 1-4). IEEE.

Chandra, S., & Dwivedi, S. (2020, February). Simulation analysis of high gain and low loss antenna with metallic ring and spoof SPP transmission line. In 2020 7th International Conference on Signal Processing and Integrated Networks (SPIN) (pp. 328-331). IEEE.

Chandra, S., & Dwivedi, S. (2021, April). Graphene based Radiation Pattern reconfigurable antenna. In *International Conference for convergence in Technology* (pp. 2-4).

Dwivedi, S. (2022). *ReconfigurableArray antenna*. Intech Open., DOI: 10.5772/intechopen.106343

Dwivedi, S. (2022). Design and Analysis of Metamaterial Waveguide Antenna for broadband applications. In *ICT Analysis and Applications: Proceedings of ICT4SD 2022* (pp. 417–423). Springer Nature Singapore.

Dwivedi, S. (2023). Metamaterials-based antenna for 5G and beyond. In *Metamaterial Technology and Intelligent Metasurfaces for Wireless Communication Systems* (pp. 1–35). IGI Global.

Dwivedi, S. (2023). *Metamaterials-Based Antenna for 5G and Beyond*. IGI Global. DOI: 10.4018/978-1-6684-8287-2.ch001

Dwivedi, S. "Proposed Design for Beyond 5G Antenna for Upgraded Applications with review," , 2023 *14th International Conference on Computing Communication and Networking Technologies (ICCCNT)*. DOI: 10.1109/ICCCNT56998.2023.10307737

Dwivedi, S. (2023, October). Elliptical Split Ring Structure Reconfigurable Antenna for Advance Communication. In International Conference on Microelectronics, Electromagnetics and Telecommunication (pp. 11-22). Singapore: Springer Nature Singapore.

Dwivedi, S. (2023, August). Partial Ground-Inserted H-shaped Antenna for Beyond 5G Applications. In International Conference on ICT for Sustainable Development (pp. 313-321). Singapore: Springer Nature Singapore.

Dwivedi, S. (2023, June). Design Demonstration of Frequency Reconfiguration Antenna using Varactor diode with SiO 2. In 2023 international conference on computer, electronics & electrical engineering & their applications (IC2E3) (pp. 1-5). IEEE.

Dwivedi, S. (2024, March). Design of triangular MIMO antenna for 5G and beyond. In 2024 11th International Conference on Signal Processing and Integrated Networks (SPIN) (pp. 40-44). IEEE.

Dwivedi, S. (2024). *Metamaterial Antennas for Wireless Communication Systems*. IGI Global publications., DOI: 10.4018/979-8-3693-2599-5.ch002

Dwivedi, S. (2024). *Antenna in Sixth Generation: Design*. Implementation, Specifications and Applications, Intech Open.

Fontana, R., Ameti, A., Richley, E., Beard, L., & Guy, D. (2002, May). Recent advances in ultra wideband communications systems. In 2002 IEEE Conference on Ultra Wideband Systems and Technologies (IEEE Cat. No. 02EX580) (pp. 129-133). IEEE.

Fontana, R. J. (2002). A brief history of UWB communications. Multispectral. com, Multispectral Solutions, Inc., www. multispectral. com/history. html

Geetharamani, G., & Aathmanesan, T. (2019). Metamaterial inspired THz antenna for breast cancer detection. *SN Applied Sciences*, *1*(6), 595. DOI: 10.1007/s42452-019-0601-6

Gupta, U., & Dwivedi, S. (2021, June). Frequency reconfigurable antenna using metamaterial split ring resonators for smart applications. In 2021 international conference on electrical, communication, and computer engineering (ICECCE) (pp. 1-5). IEEE.

Kohno, R., & Takizawa, K. (2005, September). Overview of research and development activities in NICT UWB consortium. In *2005 IEEE International Conference on Ultra-Wideband* (pp. 735-740). IEEE.

Porcino, D. (2003, June). UWB Regulations\& Coexistence: From the FCC First Report and Order to the Path to Approval in Europe. In *International Workshop on Ultra Wideband Systems*, Oulu, Finland.

Sharma, M., Sharma, K., Dwivedi, S., Singh, D. K., Gill, R., & Kumar, N. (2023, December). A multiple millimeter-wave 5G MIMO antenna including n257/n258/n259/n260/n261 FR2-bands with high diversity performance. In *2023 International Conference on Power Energy, Environment & Intelligent Control (PEEIC)* (pp. 279-284). IEEE.

Siwiak, K., & Mc, D. (2004). *Keown, "Ultra-Wideband Radio Technology*. John Wiley & Sons, Ltd. DOI: 10.1002/0470859334

Chapter 11
Promoting Intelligent Resource Management in the Indian Higher Education Through 5G and 6G Networks

Gowher Amrin Naaz
Department of Languages, Presidency College, Bengaluru, India

Kalai Chandra Lekha Kannan
Department of English, Government Arts College for Women, Nilakottai, India

Vimala Govindaraju
University Malaysia Sarawak, Malaysia

Dhivni Sitharthan
Department of English, Government Arts College for Women, Sivaganga, India

Vidhya Shanmugam
https://orcid.org/0000-0002-6181-5268
Amity University, Gurugram, India

Karthigaiselvi Govindarajan
Department of English, Saraswathi Narayanan College, Madurai, India

ABSTRACT

The opportunity of shifting to 5G and 6G networks is momentous in dealing with inefficiency in resource management in higher learning institutions in India. This review examines the ways in which the application of these sophisticated counter-voice technologies can transform the ways that these institutions, especially universities, distribute and manage resources to support these changes for accessibility and efficiency as well as sustainability. Using URLLC and IMC connections, and the utilization of artificial intelligence, machine learning, blockchain, and similar advanced solutions, 5G and the upcoming 6G networks become capable of providing needed connection. They include dynamic resource management, fast and secure accreditation, and learning for virtual and augmented reality. Reflecting on many problems in India, such as the digital literacy gap and diversified learner needs, the manuscript underlines the factors as policy initiatives, PPP, and funding for building up the enabling environment for adopting e-learning for all.

DOI: 10.4018/979-8-3693-7788-8.ch011

INTRODUCTION

Management of resources in Indian higher education is crucial in improving the quality, availability and appropriateness of educational services in the face of existing or emerging regional and/or population needs, demands and characteristics. As more and more people seek education, competition for the available limited resources increases, therefore proper resource utilisation becomes crucial in enhancing output, reducing resource utilisation and closing the knit of the socioeconomic gaps prevalent in the country. The alternative application of technologies in the last few years especially the integration of 5G as well as upcoming 6G networks also presented solutions for profound issues in infrastructure, accessibility and dimensionality. These developments have every likelihood of helping institutions create a seamless interconnected environment where technological solutions could be used to improve quality, efficiency and outreach to some of the most vulnerable communities.

Out of millions of students in the higher education system, there are over 1000 universities and 40000 colleges in India (Statista Research Department, 2022). However, even today this system is facing several issues which are working as small big hurdles to the system such as scarce resources, imbalance resource distribution and the problem of ineffectiveness. Effective resource management within this context is crucial for several reasons: it encourages proper disbursement of funds, physical and human resources; it improves learning outcomes of students; and it correlates national economic and developmental objectives with educational outputs. Thus, increase in resources can enhance the learning facilities, technologies and skilled faculty elements necessary for developing competitive human capital in HEIs (NITI Aayog, 2021).

Another aspect is that resource management also opens up the possibility of increasing institutionality, when working with the necessary funds from the population. It makes it possible for HEIs to track and shape resource use to be more efficient and promote sustainability (UNISECO, 2021). With growing education globalization and heightened expectations for computer literacy, strong resource management enable institutions to sustain quality amid adding new educational services in rural and remote regions. Hence, the objective of this paper is to examine how the 5G and the next generation 6G technologies can boost resource management in Indian higher learning institutions. What is more, it also analyses whether such superior networks are capable of solving most of the problems related to infrastructure, accessibility and scalability in the rural and urban institutions so that education becomes more rational and available for everyone.

Problems Facing Contemporary Architecture in Terms of Infrastructure, Accessibility and Reliability

There are several issues that plague the IHE sector and which make the management of resources a difficult proposition especially in context of infrastructure, openness and expansion.

Infrastructure Constraints

The first one of those challenges is related to a poor infrastructure in many HEIs, particularly in rural areas. According to the most recent studies, well over half of the rural schools have inadequate fundamentals of Information Technology, in particular, communication networks, functional computers, and possessed modern classrooms (Digital India Report, 2023). Lack of resources hinders provision of quality education; the technologies like virtual labs and AI platforms; which are part of today's learning processes, cannot be employed.

Accessibility Issues

The digital divide as it exists in India especially manifests in access to higher learning education. Educational disadvantages of many rural and underprivileged students are the lack of accessible online facilities, which worsened dramatically during the COVID-19 pandemic (World Bank, 2022). It erases the chance of students from needy backgrounds to get quality studying materials and education and thus reducing their learning abilities and chances of getting a noble job. Students are also locked out of access to adequate resources, faculty members may also lack the tools to engage in research or teaching using modern techniques due to old resources (Ministry of Education, 2022).

Scalability Limitations

Nonetheless, based on the increase in demand for higher education the issue of scalability has become Pressing. Currently, there are approximations of 30 million learners enrolled in higher learning institutions, but the current learning facilities and resource base render poor quality of service (India Education Report, 2022). The conventional resource management frameworks are capable of managing these expectations poorly, resulting in increases in classrooms' overcrowding, the scarcity of faculties and lacks the requisite student service provision. Thus, responding to the growth in student numbers, Indian HEIs need solutions that can enhance its institutional capacity as well as capacity to affect quality. This is especially true for technical and professional education, especially where a significant amount of practical experience is required for learning and job placement (NASSCOM, 2021).

Advanced Connectivity Role in Managing These Challenges

5G along with, possibly, 6G connectivity provides the AmR for tackling the current infrastructural, access, and elasticity issues for Indian higher learning. The above mentioned technologies in their execution may solve issues of resource inequity by developing an environment that admits integration of teaching and learning along with resources.

Building Up Infrastructures IOT and Cloud Technologies

5G technology allows high data speed and low latency, which makes a choice of IoT devices and cloud computing solutions for operation in universities. For instance, in IoT smart classrooms, it is possible to control the allocation of the most essential resources, including lighting, interferes, audiovisual equipment, and air conditioning, to decrease operating expenses, for example, energy consumption (Ericsson, 2022). Also, the structure is used in the storage and provision of teaching and learning resources such as digital libraries and research databases to improve on the access by student and faculties. Through efficient connectivity of the physical environment, the use of 5G and 6G technologies can establish a digital campus environment in which the consumption of resources is closely managed. Such an integration can contribute to the provision of intelligent information to promote decision-making and optimally control and foresee the resources required by institutions (Tech Mahindra, 2023).

More Access Can OF Opened by Remote and Digital Learning

The addition of 5G connection in higher education may be of great impact in efforts to close the digital gap by enhancing the suitability of the online learning platforms mostly in rural areas. Due to increased data speeds and improved network connections, students are able to obtain rich digital media, attend virtual classes, and interact with media rich learning resources from distant locations. This increases student's enrollment chances who cannot be physically physically present on campuses making the education system more open for everyone (Digital Empowerment Foundation, 2023). Moreover, introducing virtual and augmented reality applications through the help of 5G, students from developing and less developed countries have opportunities to take lessons in first-class courses. For example, students in the remote areas could get practical experience in the virtual laboratories where a real scenario is created, and it has certain value since many abstract skills are used that could not be developed when there are no enough resources and physical infrastructure to provide experience (Cisco India, 2023).

AI And Automation as Key Drivers to Achieve Scalability

Sophisticated connexion meshes also facilitate the application of Artificial Intelligence (AI) and Automation which makes it easier and more efficient to scale education knowledge. Teaching aid powered by AI technology will seamlessly employ and proportion learning interventions in accordance with the student learning profile, this way the faculty will be in control of bigger groups without affecting learning outcomes (Accenture, 2022). Automation of administrative functions of admissions, fees, and attendance means the reallocation of institutional resources to academic advancement and student service and support, 5G and 6G networks to drive the use of big data analytics on students' performance and resource forecasting in HEIs. These analytics can be employed for the analysis of student activities, preferences over the courses they take, and to estimate the future usage of resources and their further request to expand the scale of their institutions (Intel, 2023).

Main Characteristics and New Development Steps for 5G Networks

5G or the new generation mobile network technology was developed to overcome the constraints that have beset the previous generations in terms of speed, latency, capacity and connectivity. Some of the most notable advancements in 5G include:

Enhanced Speed and Data Rates

5G possesses greater data transmit/receive rates than the previous networks, with the maximum download rate of 10 Gbps in comparison with 4G with the maximum rate of 1 Gbps (Qualcomm, 2023). It makes data download rates faster, capable of high quality videos and big file downloads, internet video streaming, and extremely high-definition video conferencing and are very useful in such application areas as 'contactless' education, distance learning and digital libraries (Ericsson, 2022).

Reduced Latency

The delay involved in the transmission of data in 5G networks is as low as 1 ms on average, compared to the 50 ms in 4G (ITU, 2021). Lower latency is essential for such applications as VR or AR that is used more and more in education to make learning interactive (Cisco, 2022).

Increased Device Connectivity

5G technology can have up to one million connections per square kilometer, they are efficient of expanding the internet of things within education (Huawei, 2023). For the Indian HECs this means improved opportunities to develop integrated, smart university environments in which all sorts of devices, ranging from projectors in classrooms to security systems, can exchange information.

Network Slicing and Customization

One of the major benefits that goes with 5G is known as network slicing whereby many logical networks can share a single physical network. It also allows the establishment of controlled networks for different perceptions of applications for educational needs, for example, bandwidth utilization in examination, admin or interactivity purposes (GSMA, 2022).

Reliability and Energy Efficiency

Unlike its predecessors, the 5G network has been conceptualised with higher power utilisation efficiency and better network latency than any previous generations with a better coverage plan for rural and urban regions. This is especially useful in the Indian context, where frequent power outages and intermittent connectivity affect the effectiveness of e-learning and digital resource sharing and organization (Nokia, 2022).

New Generation Technologies and Opportunities of 6G Internet

Concerning 5G networks, they are still in the early adoption stage, meanwhile, the research and development of 6G technology have started and 6G is expected to be deployed in the 2030s In addition to enhancing the capabilities of 5G, 6G plans to incorporate some newly emerging technologies as well as features, which are not considered by 5G.

Terahertz (THz) Frequency Spectrum: Thus, one the of main developments included in the 6G is the utilization of the terahertz frequencies between 0.1 to 10 THz as a roadmap for achieving more data transfer rates and increased bandwidth (Samsung, 2023). Such frequencies will support download rates of 1 Tbps (terabit per second) and data-hungry applications, and enhanced education, including holography, UHD streaming and virtual reality, (Liu et al., 2023).

Integration with Artificial Intelligence (AI): These novel features are going to be integrated in 6G networks, with the potential for intelligent decision making and automation at network layer. This will mean that 6G networks will be able to learn and adapt autonomously about their own optimisation, heal themselves in case of faults, and forecast for the traffic demands on the network at any given time (ITU, 2022). In the case of educational institution, 6G AI could digitize environment control and create a more customized learning for students, organize the management of resources and outlook for maintenance in case of infrastructure issues.

Extended Reality (XR) and Holographic Communications: It is predicted that 6G networks will offer enhanced quality of interaction for XR technologies such as VR, AR as well as MR. 6G will also support Holographic communication as a type of Real-time 3D interaction (Nokia Bell Labs, 2023). These capabilities have great prospects in education, they allow creating effective learning environments, and remote work then laboratories and field work.

Increased Device Density and Edge Computing: While 6G networks will be capable of supporting 10 million devices per square kilometer; a 5G improvement (Ericsson, 2023). This capability will enable even tighter IoT integration on campuses and a variety of learning uses as shown next. In addition, 6G will provide a better edge computing, which is the processing of data nearer to the source to achieve minimum latency and for better access of real time data for students and educators (Huawei, 2023).

Sustainable Network Design and Low-Power Communication: One of the primary research goals agreed upon in the study of 6G is sustainability, specifically, the aim to create networks that consume as little energy as possible. Innovative and novel forms of low power communications technologies are still emerging with the idea of lowering the power consumption of the devices necessary for green education technology (Samsung, 2023).

Comparison of 5G and 6G in Terms of Speed, Latency and Scalability

Speed: As we have known that beyond 5GNON_SDX7AChe with peak data rate of up to 10Gbps, 6G is expected to deliver up to a Tbps. This should increase a hundred fold and opens the door to the instant download of high resolution educational materials, as well as smooth and natural holographic communications (Samsung, 2023).

Latency: The existing 5G technology also has relatively low latency, with latency of about 1 ms and thus will enable real time communication and interaction in educational settings. However, 6G has a lower latency, nearly to sub-millisecond range which is as near to the internet real-time 'instantaneous' as possible. This reduction will be useful in highly interactive applications such as remote laboratories, virtual anatomy, and real-time modeling for difficult subjects like engineering and medicine (ITU, 2022).

Scalability: Current 5G networks can take one million devices of per square kilometer and thus needed for IoT developing campuses. On the other hand, 6G is expected to support up to 10 million devices per square kilometer for creating a hyper-connect environment with smart smart classrooms, wearable learning devices and smart campus security system. This scalability will enable all HEIs to implement large scale IoT frameworks that enable management, monitoring and maintenance of resources (Ericsson, 2023).

The Present Status of Resource Management in Indian Higher Learning Institutions

Resource management in Indian higher education has been a contentious issue for quite some time and has impact on the quality, availability and distribution of educational facilities across the country. Over the decades India has emerged into a treasure of students of various kinds, and with so many students and with vast diversified population the problems for education sector particularly in infrastructure, manpower and resource allocation varies excessively. However, unlike resource management systems that are increasingly being streamlined for optimum efficiency, tradition lingers on and affects efficiency and effectiveness.

Study of primary approaches to Traditional Resource Management: Earlier literature presents resource management in higher education in India as being highly centralized, bureaucratic and less open to innovations. Evaluations within institutions are often centralized and decide little to noengagement in strategic resource decision-making by faculty or administrative staff (Agarwal, 2022). This approach means that educational change occurs from the centre and hence the slow adaptation to changes in demand and resources allocated do not match the needs. For instance, money distributions and staffing prerogatives are predetermined more by government regulations and guidelines than by the actual needs at the institutional level, therefore the funded academic programs have little relation to the demand for them.

Traditional resource management practices lack robust analytical information as well; majority of the institutions have been found to use basic tracking techniques in resource allocation, attendance, and performance appraisal among other aspects. Hence although information and communication technologies (ICT) are today used in the schools, incorporation of ICT in management of resource is still a vice versa. This absence of integrated and superior data analysis along with application of technology in the management produces ineffectiveness especially in view of allocation of resources to different departments, monitoring of academic achievement and effectiveness, and most importantly the utilization of buildings and infrastructure (Kumar & Sharma, 2021).

In relation to the management of faculties, conventional models do not provide for sufficient maneuverability in meeting the wants and wants of the faculties for growth or the potential for professional development. There are still many institutions with faculty recruitment and performance evaluation standards that contain precious little information about online courses, flexible schedules, or remote teaching. Furthermore, faculty student ratios in the public institutions is still high hence inefficient resource allocation that leads to large classes and reduced student to faculty ratio (NEP, 2020). By embracing this rigidity, institutions of learning many a time end up compromising the quality of education needed for the faculty to be innovative in their line of duty.

Lack of infrastructure facilities, ineffective faculty management, and deficient academic resources: From Hendricks et al work, one of the major problems affecting higher learning in India is the disparity in infrastructure where the rural and government institutions are most affected. Physical facilities in higher education comprise of cards required in a learning institution such as classroom, laboratory, libraries and digital connection tools which are difficult or poorly developed in many institutions. Some private intuitions particularly in the metropolitan areas have improved on infrastructure whereas most rural and government institutions often have resource constraints which delay realization of good learning environment (Choudhury, 2022).

They observed that laboratory or research facilities suffers poor funding hence they lack adequate equipment and access to real world practical training for learners in science or engineering fields. The existing stock of classrooms connected to technology is also limited that hinders the use of innovative approaches including hybrid learning or technologies integrated content (KPMG, 2023). For instance, a recent MOE conducted a survey which shows that out of the total number of colleges and universities within India only 35% of them have access to high speed internet which would have facilitated online learning and remote resource enablement (MoE, 2021). In addition, access to some core needs including appropriate sitting, lighting and airing, still remains inefficient in several centers, which usually impacts on the quality of learning.

Another needs gap is in the faculty management, especially in government bog institutions where cases of faculty scarcity and long times of recruitment persist. The University Grant Commission (UGC) of India measures the shortage in faculty strength in the existing higher learning institutions to be between 30 to 40 percent hence putting much workload on the few available staff and limits students' faculty ratios. The deficit of qualified staff not only threatens the academic standards of the university but also hinders the number of focused study programs that can be introduced to students – their learning and career progression.

Also, the spread of academic resources and their delivery across the different institutions still remains patchy with few institutions having easy access to books, research and other digital products. In many rural public institutions, libraries that serve as the primary source of research materials are poorly funded; they do not have adequate sources for any update in education, including new and latest research documents, not to mention an online database which is crucial for the student to read and develop a sound experience in any field of their study (Nanda, 2023). For this reason, learners attending such institutions may miss out on improved education as they are presented in urban or private institutions.

Digital divide's effect to fair share: In Indian contexts, the digital divide refers to a difference in admission to digital technology between cross-sections of the population, which persists in higher education and sharpens disparities in management of resources and their acquisition. Based on findings in Chapter 4, the disparities in Internet connection, computers, and training in ICT affect greatly educational equity in favour of urban area and people with high socio-economic status.

About 27% of the households in India use the internet, but again, the rural communities lag behind the urban counterparts (as per TRAI, 2022). This is true in higher education where learners in rural institutions are unable to fully engage in online classes or even access any other materials in digital form. For instance, during the COVID 19 outbreak where most education institutions around the globe conducted their activities online, India's digital divide resulted in major education disruption. The students who were from the low income or from the remote areas, struggled to join the virtual classes, to complete the assignments and assessments, as well as to obtain copies of their lessons; it was seen that their performance got negatively impacted (Sarma & Mishra, 2021).

Another issue is absence of such opportunities in rural schools also restrains the possible introduction of the systems allowing for efficient management of digital resources in frame of educational process and school administration. If students and staff members can not count on the internet connection and are not digitally proficient, the usage of online tools for scheduling, records keeping, and resources allocation is problematic. In other institutions that have not been smart enough to harness the benefits of technology, work still relies on paperwork which translates to immense time wastage and high erratics (Digital Empowerment Foundation, 2023).

Also, digital divide hinders rural and underfunded institutions faculty recruitment chances because the better-equipped institutions will attract the talented employees. The inaccessibility to professional development tools including online classes and research databases hinders ability of faculty members to embrace the current trends in practices and educational research to influence quality education offered to students (Kumar, 2022).

It also impacts the opportunity for students to participate in current modalities of learning like virtual simulations, collaborative remote project, and the many computerized technical skill development opportunities that are trending with employers. For example, courses that are associated with data analysis, programming or digital marketing need preview to computers and constant internet connection. However, students in institutions of great limited resources do not get the real-life exposure in using such tools limiting their employability in the competitively demanding job markets as sampled by McKinsey (2023).

USING 5G AND 6G SYSTEM IN TERTIARY INSTITUTION

This paper presents how 5G and the future 6G technology paradigms can overcome these sustained issues and create groundbreaking opportunities for the higher education sector. These technologies brings unique aspects such as speed, very low latency and scalability facilitating use in widespread applications like smart infrastructure management, more efficient learning, administrative and many more. Through application of 5G and 6G, the higher learning institutions have the potential to achieving a new era of efficiency and access to innovation.

Intelligent Infrastructure Management: Interactive Classrooms and IoT Controlled Campus

An emerging aspect amenable to revolution through 5G and 6G technologies in higher learning institutions is the creation of intelligent infrastructure. Smart classrooms and IoT Campus is an evolution to effective and sustainable intelligent learning spaces.

Smart Classrooms: Advanced learning environment with 5G support allows using modern learning media and techniques including smart platforms, modern interactive whiteboards, remote access to lectures, and automatic recording systems. These tools make it possible for the students to be more in touch with the educators, today knowing the physical location that one is in is not an issue. Also, smart classrooms are equipped with artificial intelligence (AI) that help to adapt learning processes. For example, AI systems can work on individual student data to provide the course content according to the learning model and performance (Ahmed & Sharma, 2022). The absence of infrastructure in developing or rural areas can be solved with help of virtual classes that are as good as or even better than the ones offered by urban universities. Fiber optics enables constant live-streaming of lectures hence minimizing the inequality of chances (Bansal, 2023).

IoT-Enabled Campuses: IoT campuses using smart technology which is in connection with the 5G networks, with IoT devices which include; sensors, cameras, smart energy systems within the campus. They all also add to improving the operation and sustainability the Ministries' operations. For instance, smart sensors in the internet of things IoT technology can be used to control energy consumption within underlying structures and adjust light and temperature systems depending on occupancy, leading to a lowered overall expenditure and conservation of the environment by implication (Choudhury, 2023).

The last yet equally important consideration of the IoT campuses is security. State of the art Closed Circuit Television (CCTV) with face identification and physical identification secure students and staff. On the same note, IoT gadgets that are connected through 5G networks can make instant tracking of resources such as Library Books, Laboratory Equipments and Automobiles hence reducing cases of losses.

Enhanced Learning Experiences: Application of AR/VR in the Fields and Remote Education

The 5G and 6G networks are to bring a paradigm shift in the context of learning in the socities through smart technologies like AR/VR and advanced remote education.Youths. Such changes have the potential of making learning more interesting, engaging, as well as increasing student interactivity.

AR/VR Applications: Applications of AR and VR expose the learner to experiences in a simulated environment hence improving understanding and memorization. For example, health profession students can practice surgeries using VR headset and simulate operations without having to face all the consequences at operation theaters. Likewise, engineering students have the opportunity to visualize 3D models of machinery and improve their understanding of the inside of the mechanisms (Gupta et al., 2022). 5G and 6G technological networks provide fast connectivity for these applications, thus, they operate with improved and low latency interfaces and high-quality graphics. AR/VR labs based on such networks can also activate collaborative learning, as students of different countries can study together in one collaborative space.

Remote Education: Distance learning has become more and more important, especially after the outbreak of the COVID-19 virus reveals the problem of mass education and the need to quickly and effectively respond to their needs. 5G and 6G technologies solve the problems of existing distance education platforms by providing high-speed internet access. This guarantees the students get to attend live sessions, obtain large course materials, and get to have constructive interactive sessions without hitches (Kumar & Mishra, 2021). Nevertheless, these networks ensure flexibility in learning by providing recorded lectures, multimedia, and virtual laboratory with no synchronized timetable. This sort of flexibility is useful especially to students from nontraditional backgrounds, for example, working adults or candidates from rural settings.

Administrative Efficiency: AI Endorsement in the Processes of Decisions and Resource Management

From an administrative perspective, the higher education institutions require intensive automation where AI/decision making makes substantial impacts throughout the system where the manual workloads are the major issues. 5G & 6G networks propel those advancements by providing consistent real-time data insights.

AI-Driven Decision-Making: Computerized institutional data can also be analyzed by AI driven systems to reach informed strategic decisions. For instance, prediction modeling can estimate the number of student enrolments, and thus enable efficient distribution of faculties, classes and finances. AI tools can also detect chronically absent and low performing students using data on attendance, grade, and participation, then salvage them for better retention (Singh & Kapoor, 2023).

Realtime data such as those obtained from 5G and 6G networks improve the performance of these AI systems. For example, populated IoT devices coupled with campuses can perpetually gather information regarding usage of facilities, which could be pleasingly employed by AI algorithms after which lower operating costs.

This is a major area which has greatly benefited from automation in the method that resources are allocated.

Application of IT enabled applications with high connectivity for automation of some of the administrative functions including admission, timetabling and financial aid. AI enabled chatbots can easily respond to students' queries regarding admission, courses, and facility and can relieve the burden of administrative staff and can answer queries promptly. In allocation of resources, different schedules can be changed by the system depending on the actual needs. That is, for instance, when one course has higher enrollment than the others, the system will have capability of bringing in those systems to address the needs of the student. Likewise, the automated systems for disbursing financial aids have a chance to analyze the criteria of eligibility and process an application more quickly, thus effectively delivering the necessary funding (Nanda, 2023).

6G Network Nowadays Becomes an Important Aspect to Examine How it Would Lay Down its Contributions to Developing the Applications

As we have seen, 5G networks are already revolutionalizing higher learning and 6G will supercharge these uses and more. Fundamentally assumed to further augment the 5G by offering download speeds of up to 100 times faster than 5G with close to zero latency with application-related features like holographic communication and AI integrated adaptive learning.

Holographic Communication: Holographic communication which is enabled by 6G enables real, as close as life, communication and the other party feels physically present. With this technology, the educators may record their lectures in forms of holograms while students can easily study complex ideas from holographic visualizations. For instance, biology learners can learn human anatomy through holographic models advanced by the lectures, while architecture learners can build their courses around three-dimensional models of structures (Ramanathan et al., 2024).

AI-Powered Adaptive Learning: This will improve AI assisted adaptive learning platforms common in 6G networks by providing efficient data transfer rates. Learning management systems and platforms utilize artificial intelligence to help compile programs according to performing abilities and learning inclinations of the learner. With 6G, these systems can collect data from several sources at a time for instance from the wearable devices and smart classes to give more accurate and real time feedback (KPMG, 2024).

ADVANTAGES OF INCORPORATION OF 5G/6G FOR RESOURCE ALLOCATION IN HIGHER EDUCATION

The incorporation of 5G and 6G networks to enhance higher education entail a core chance to solve the issues of resource enhance and harness benefits that are incomparable. These progressive connection technologies are anticipated to advance the possibility of implementing large-scaled, effective and availability integrated educational systems and equally hand over challenging tools for assisting stakeholder. The following part discusses how such technologies provide heightened scalability and its accessibility to rural places, the appropriateness of the use of resources through real-time analysis and finally the upliftment of the faculty and the students.

The Problem of the Division Between Urban and Rural Areas

Geographical factor speaks volume of difference and large geographical size of India has meant that there exist stark difference in access to quality higher education. Those in urban areas can have better infrastructure and digital connection than those from rural or less-served communities due to their limited resource and internet connection. 5G network gives a practical solution for this challenge because it provides a very fast connection speed and low but efficient latency compare to 4G network. For instance, through 5G based virtual classroom, it will be possible to link students in rural areas to some of the best resources in universities and colleges around the globe, making for a leveled playing field. In addition, the future 6G networks, which are expected to provide improved coverage and faster speeds in these areas, will provide easy incorporation of these learning technologies such as AR and VR into these areas (Rao et al., 2022).

Increasing the Client Base Through Online Classes

The COVID-19 crisis highlighted the need for conceptually extendible remote learning mechanisms. However, interruptions were characteristic of many rural institutions because they lacked the necessary digital facilities. The use of the 5G network answers this question by providing uninterrupted streaming of videos thus allowing students and educators' real-time interaction. Applications which include live question and answer sessions, group online projects, and a blended learning environment can be realized with 5G and 6G (Patel, 2023). Also, they provide an opportunity to attend such educational models as MOOCs and vocational training programs, to use different platforms for e-learning and make education more readily available for adult learners. This scalability ensures that because of the development of greater student enrollment it will not in any way affect the quality of the institution.

Optimization of Resource Usage Through the Use of Real-Time Analytical Techniques

In using physical and digital assets to support learning and teaching, HEIs may find it difficult to fully use physical resources such us classrooms, laboratories, libraries and other learning resources. The advent of 5G network enables institutions to connect IoT devices and sensors that feed real-time usage of available resources. For example, sensors can monitor how many students are in a classroom or a study area, and then space better (Kumar et al., 2021). Real time analytics also enable accurate prediction in the maintenance services of facilities and machines. They will help the institutions to reduce time taken to fix the problem as well as the total costs of maintenance in the long run. For instance, smart systems associated with the Internet of things can track the efficiency of laboratory equipment and prompt the technicians to rectify any deviations so that people like the students have continuous access to them.

Improving Efficiency of the Administration

Daily tasks like admissions, scheduling and allocation of resources are normally tedious and composed of many inefficiencies. By using 5G analytics platform the huge amount of data is processed and analyzed to produce relevant outputs. For instance, AI-based scheduling systems can be used to establish schedules in relation to course enrollment and faculty whereby facilitation resources can be well exploited (Singh & Mehta, 2023). Furthermore, superior connectivity also allows institutions to have a responsive resource allocating system that allows for real time changes. For example, if a classroom is left unoccupied, the system can at once assign a new use for that space, thereby using the resources more efficiently.

Sustainable Resource Management Means Investing in Future Resources as a Business and Living Resourcefully as an Individual

A sustainable approach is an issue of concern in current organisations Today's society receives 5G and 6G technologies to ensure an efficient monitoring process and utilisation of limited resources while keeping energy consumption at a minimum. IoT technology can also regulate power consumption to control lighting, heating, cooling depending on occupancy leading to lesser emission of CO2 and cost savings (Chaudhary et al., 2022).

Enabling the New Generation of Teaching and Learning with Modern Applications

The features of 5G as well as 6G network enables the faculty members with enhanced mechanism for teaching and researching. Technology-enabled applications produce datasets and in turn provide instructors with instant information on the level of engagement of the students as well as their performance levels so that they can readjust their approaches. For example, topics that need to be revisited, can be underlined in adaptive learning platforms; thus, it is possible to address the gaps in knowledge (Gupta et al., 2022). Moreover, the 5G driven cloud computing empowered social media opens up advanced computing facilities for the faculty for research purposes. They provide opportunities for quicker and more efficient analysis of data and for interdisciplinary and inter-institutional networking.

They Capacity Building for Student-Centered Learning

From the possibilities offered by 5G and especially from the upgraded capabilities of the 6G networks, students benefit so much. These technologies enable learning enshrinements of augmented reality and virtual that makes hardship and challenge subject easier. For instance, the medical students in communication technology can practice making surgeries virtually or engineering students in communication technology can work with mechanical models of complex machines in a simulated environment to get pragmatic knowledge in real environment (Sharma, 2023). 5G networks can also support access to digital libraries, multimedia and online research database with rich and deep knowledge enhancing the knowledge base. However, the availability of real time collaboration tools enable students to be able to conduct group projects virtually and therefore enhance team working as well as global orientation.

Equal Opportunities and Non–Discrimination

Indeed, the possibility of enhancing the integration of 5G and 6G has one of the most striking advantages as a tool for increasing equity in education. The endowment of students with disabilities in vivid networks that aid in the development of assistances technologies. For example, automatic transcriptions help to transcribe lectures in real time enabling learners with hearing impairment to follow along. Likewise, with the help of augmented reality, the various applications offer means of signaling for students with learning disability while offering them an equal opportunity to learn (Nanda, 2023).

Towards an Understanding of the Role of 6G in Resource Management

What is more, while 5G networks are already overhauling resource management in higher education, 6G is already on the horizon. Supposedly, to offer download rates up to hundred times faster than 5G and enable ultra-low latency, 6G networks will bring revolutionary solutions.

Educator will be able to give their presentations in 3D, thus making interaction through holographic communication via 6G effective in learning processes. According to Ramanathan et al., (2024), the students can touch, feel and discuss with a holographic figure something like a molecule or architecture to make it easier to learn (p. 16).

Artificial Intelligence Guided Adaptive Resource Management

It was found that 6G networks will improve in the effectiveness of application of AI systems in the management of resources. These systems have been said to be capable of handling real-time integration from a variety of sources, from IoT, cloud, wearable technologies among others, ability to give dynamic real-time resource management. For instance, AI can analyse students' activities and interests to make clear how they study and how to approach them to leverage digital and tangible assets (KPMG, 2024).

BARRIERS AND CONSTRAINTS OF DEPLOYING 5 & 6 NET CAMP IN HIGHER EDUCATION

The emerging 5G and future 6G networks in the higher education sector will open new perspectives in managing resources and learning materials availability and quality. However, the implementation of these technologies has specific main issues and constraints. This section looks into the high cost of implementation and structures, fear of cyber-attacks and issues of data privacy, need for effective regulation and qualified human resource.

Implementation and Infrastructure Costs are Found to be Costly in the Literature

To establish the 5G or even 6G networks, there is much capital outlay that is initially needed by creators for the Adavnce Communication Towers, Fibre Optics Connections and other Edged Computing equipment. An actual challenge for higher education these days, especially for those located in the developing nations such as India, is how to obtain the required funds. This situation becomes even worse in rural and other underdeveloped areas where networks' infrastructure is poor, and the demand for high–speed connections is low (Sharma et al., 2023).

Ongoing Operational Costs

Apart from capitalized costs of 5G and 6G, running and enhancing these networks entail significant amounts of operational expenses. These are costs such as the cost of the software update, power cost and cost of hardware modification. There are also challenges associated with budgeting for these expenses in institutions of learning deeming other core areas like faculty remuneration, research funds, and students welfare programs unproductive (Kumar & Singh, 2022).

Inequality in Implementation

These cost may heighten the digital divide especially in terms of institutional access separating urban from rural based learning institutions. Urban universities have greater potential to approach fund and partners for deployment of advanced network as compared to rural University may fall short resources. This disparity goes against the appropriate provision of education for all which is a common education aim (Chaudhary & Patel, 2021).

Threats Vendors, Cybersecurity and Data Privacy

The introduction of 5G and 6G networks compared to previous networks greatly enhances the number of connected devices; from the smart campus IOT, cloud education platforms. Thanks to this new increased network surface, new risks and threats appear in the form of stolen data, ransomware, and Distributed Denial of Service (DDoS) attacks. These networks have higher bandwidths and lower latencies; such networks are under attack targets among cyber criminals (Ramesh et al, 2022). For example, learning institutions contain personal information such as student records and all other information that is gathered by the tutors during research, financial records and many others. A violation of such data not only leads to disruption of the operations of the institution but also affects the credibility of the institution.

Data Privacy Challenges

This is due to the huge volume of data being produced by 5G and 6G networks which is a major issue of computational privacy. Adaptive learning technologies, based on an adaptive learning platform and artificial intelligence processing, necessitate the storage and processing of personal data about students, their performance, actions, and possible biometrics. If proper data protection mechanisms are not implemented, then this data may be exploited causing more ethical and legal problems (Gupta & Mehta, 20223).

Regulatory Gaps

This system of data protection law in India is not ready to counter the multifaceted issue with reference to data privacy in the perspective of 5G and 6G. There are no specific regulation policies for collecting, storing and using personal data in institutions thus there are high probabilities of non-compliance and misuse (Bansal, 2023).

The Rapid Requirement for Regulatory Frameworks as well as a Skillful Workforce

In the case of 5G and 6G, the roll out has been faster than the imposition of broad legal structures. It involves question such as, who will allocate the specific frequency bands, which technologies should be standardized and how different devices and network can be made compatible. Lack of specific guidelines makes it difficult for higher education institution to provide right implementation of these networks (Rao & Srinivasan, 2023). Moreover, accountability issues linked to decision- making by AI with potential ingrained bias on areas such as resource allocation, raises serious ethical issues of surveillance. These are the realities that institutions must face alongside with legal constraints current on the national and international level.

Workforce Skill Gap

Managing and deploying 5G and 6G networks require human capital with adequate skills in network engineering, cybersecurity, data analytics, AI and machine learning. But, presently such expertise is scarce in India. One of the big challenges faced by educational institutions is acquisitions of competent human resources mainly the professional ones, especially the institutions situated in rural region as most of the talents are likely to move to the urban areas (Nanda & Verma, 2022).

Need for Upskilling Programs

In overcoming the skill deficit, there is a need for higher education institutions to fund continuing professional development of current employees and learners. Systems training efforts need to start focusing on key sectors such as IoT, machine learning, and blockchain to prepare workforce to handle rising connectivity solutions. Nevertheless, introducing such programs increases the number of problems in the formal educational system related to finance and organization.

Combined Impact of Challenges

Rural Areas Will Always Adopt at a Slow Rate

Frequently, high costs, cybersecurity issues, and talent scarcity affect developing and remote regions in a most critical way. These are the areas that suffer from a lack of development infrastructure and funding in general; the problems, which connected with 5G and 6G deployment, seem to be even more critical. This slow adoption rate negates the benefits that can be harnessed from such networks as better access and Extensibility (Chaudhary & Patel, 2021).

As it is with most contingencies, it is possible to overcontrol and hence risk obsolesce of technological control in relation to the firm's qualities. This fast pace of technology hype of mean that the investment carried out by different players in 5G may not be useful since 6G has arrived. Hence, institutions have to work out their future directions properly so that they do not end up in putting their time, money and resources in technologies the usefulness of which may not be dawn before those technologies turn obsolete (Sharma et al., 2023).

Ethical and Social Concerns

There are two key areas of concern when it comes to the integration of advanced networks these are surveillance and autonomy. For example, IoT-integrated campuses having facial recognition capability may also violate students' privacy while making the environment one that is under constant surveillance. Another theme identified in the context of Institutions' Management of Technology and Human Rights is that effective and efficient technology use needs to be in sync with the appreciation of individual rights and freedoms (Gupta & Mehta, 2023).

Mitigating Challenges

Public-Private Partnerships

But with the support of the government, private sector and educational institutions, the high costs of implementation can be offset. To fund and implement the advanced network, PPPs can be effective because the private partner can offer the required funds and experience to deploy the network while undertaking the governance of risks and responsibilities (Rao & Srinivasan, 2023).

Enhancing the Structures of Regulation

Instead, it is crucial for the Indian government to pay attention to the formulation of comprehensive legal conditions for 5G and, even more, 6G implementation. These frameworks should provide answers to questions as diverse as the allocation of the electromagnetic spectrum or data protection on the one hand as well as the correct use of artificial intelligence on the other. It also means that institutions can also bring into compliance and accountability international best practices (Bansal, 2023).

Focus on Advancing the Actual Research and Development

R&D remains critical for acquiring technical and operational solutions to the challenges experienced. Industry partners have to work with higher education institutions to devise solutions that would optimize costs and security of 5G/6G networks. Explorations can also be carried out on development of tools and technologies which are freely available and not proprietary ones (Kumar & Singh, 2022).

Developing the Capability and Enhancing the Skills

To ensure that suitable skills are provided in the Labour Market, institutions should collaborate with technology providers in order to train individuals for employment. A wide range of activities including internship, certification courses and workshops can help students and staff to be ready for new roles in networks management, cybersecurity and AI. This shows that the government can also encourage upskilling programmes through grant funding as well as subsidies (Nanda & Verma, 2022).

ED - Tech Investments

Promoting key telecommunication networks including 5G and 6G is important for revamping education. These technologies guarantee low-latency reliable connections which are critical for teaching techniques such as augmented reality (AR), and virtual reality (VR). For instance, plans in relation to 6G networks are expected to incorporate Terahertz communication and visible light communication to boost bandwidth and connection in learning institutions. Such infrastructure would support the processing of data through edge computing, important for remote and disadvantaged regions (IEEE, 2024; ITU, 2024).

Both government and private sectors need to invest their resources towards 5G/6G networks implementation in schools and especially for rural areas. The spectrum allocation and energy efficiency must also be effectively caters so as to achieve the green end goal of the 6G acceptance with equal appropriateness across the world (IEEE, 2024; FCC, 2023).

The Public-Private Partnerships to Address the Digital Divide

Outsourced interaction between the government and other stakeholders is crucial in closing the digital divide in educational systems. Such collaborations are well illustrated by 5G Public-Private Partnership (5GPPP) in Europe which the Collaboration creates the powerful union of efforts to increase the availability of the next-generation networks. To cater to infrastructure deployment and avail affordable/much-needed internet services to needy communities, PPPs can involve governments, tech companies and academia to fund the infrastructure deployment, while offering the needed internect access at subsidized rates (Nokia, 2021).

Moreover, policies should promote local approaches to crowdsourced networks and spectrum alliances. These measures can give a possibility to extend connectivity of rural regions where investment may be unprofitable for private companies. To incentivize private entities, governments use tax credits or grants for the infrastructure projects aimed at hard to reach populations (FCC, 2023; Nokia, 2021).

Supporting Research and Development of EdTech Solutions

The scholarly study in EdTech has to be compatible with the usage of 5G/6G networks, in order to fully realise its educational potential. For instance, AI-based empowered adaptive learning, the use of consortium blockchains for digital credentials of skills, and quantum computing for the security of data transfer for the transformation of learning models (IEEE, 2024).

Horizon Europe, and other its funding programs demonstrate how global cross-organizational cooperation can help in attaining innovations. Governments should mirror similar models, in that they should foster collaboration between universities, startups, and enterprise goliaths. Furthermore, they may include trials of smart campus applications and scale EdTech testbeds that offer improved understandings of rollout best practices (ITU, 2024; Nokia, 2021).

FUTURE RESEARCH DIRECTIONS

The further development of technological solutions in the context of education technology should reflect on the advancement of the 6G technology, its application in teaching environments enhanced by additional AI, machine learning (ML), and blockchain technologies for the optimization of resource management in learning, as well as the examination of the sustainability and competitiveness of the education ecosystem.

Towards the Development of 6G and the Implications for Education

Most progressive approaches and researchers have expressed expectations that 6G will present novel features such as ultra-high-speed, low latency, and interoperability of equipment, critical for the next generation of learning as indicated by Pasan Maduranga et al., 2024. Over time, 6G could improve VR/AR education, specifically graphical transmission that is still a problem in education due to restrictions set by 5G. In the case of education the capability of 6G includes enhancing educational concepts from operation of remote laboratories to real-time feedback to the students which will make educational process highly interactive and reactive to students (Liu Y et al., 2024). This might also facilitate the access to virtual classes' equity and enable students who are physically located in different areas to receive quality information.

Application of Artificial Intelligence, Machine Learning and Blockchain

The integration of 6G with AI and blockchain will create a smarter and efficient environment which will be required for the future education. 6G networks applying artificial intelligence can allow for better utilization of resources using machine learning and, in particular, efficient arrangements of bandwidth, which are useful when working with a small budget, which is the case of educational institutions (Tariq et al, 2024). Furthermore, blockchain can offer academic transcripts which are protected, transparent, and easily verifiable, help with credentialing and protect student information. Using blockchain and 6G, particularly the concept of sharding and efficient data compression, it can be observed that large-scale educational transactions can be performed without sacrificing the privacy and security component thus making it possible for schools to generally do a large-scale adoption of digital credentialing securely (PLOS ONE, 2024).

Evaluation of the Enduring Consequences of the Changes on Sustainability and Competitiveness

According to the literature, next-generation 6G based education systems can create a sustainable environment by incorporating the utilization of efficient energy systems which lower impacts on the environment and support global educational system competitiveness for sustainability (Maduranga et al., 2024). With education integration going global, 6G technologies can provide any developing region with the relevant technological tools required to help minimize education disparities, and hence create a platform whereby emerging economies can nurture competitive talent. Future work may examine energy efficiency and the lifespan of educational devices in incorporating sustainable 6G networks in schools and campuses, reduce consumption values and increase longevity (Pasini, 2024).

CONCLUSION

5G and further proliferation of 6G technologies mean a new generation of the opportunities in higher education to connect and innovate. These technologies hold a chance to transform current traditional and innovative learning milieus to grant educators & students opportunities of compound interactions, immediate evaluation results and unbounded access via the world. The increased bandwidth and reduced delays improve the various interactive devices like virtual and augmented reality, allow multiple IoT devices to be implemented and also facilitate the use of AI-based learning models. Through the integration of these technologies, the institutions help students to achieve a certain level of tailor-made learning for which it is more flexible for students in different geographical locations in terms of learning and education (Pasan Maduranga et al., 2024; ITU, 2024).

What is truly impressive is the technological opportunity to achieve all these opportunities with the help of 5G and 6G, although it is crucial to understand that the realisation of these opportunities can happen only with the help of a good and effective cooperation of policy makers, educational institutions, and technology providers. Another critical role is performed by politicians who establish oriented legislation and stimulate the financing of the infrastructure where the lack of connection results in a notable educational deficit (FCC, 2023). There is also a role for self-regulatory organisations to help establish sound data management procedures and ensure the ethical use of AI, which are foundational issues in preserving public trust in technology enhanced learning systems (Tariq et al., 2024).

For educational institutions the implementation of 5G and 6G capabilities will require a overhaul in the current curriculum to accommodate these new forms of technology in the classroom. Facilities must be ready to spend on capital goods as well as the human resource to make sure that people in the learning institution are in a position to make wise use of these assets. Moreover, it is possible to further strengthen partnership initiatives with technology vendors and research bodies to help institutions keep up with the digital transformation rate (Nokia, 2021; Liu Y et al., 2024).

In turn, technology providers are tasked with the chance to collaborate with educational entities in order to develop precisely those products that would be suitable for use in the higher education sector. Flexible solutions that are bespoke and can be built out as needed to compliment a partner's infrastruc-

REFERENCES

Aayog, N. I. T. I. (2021). *Transforming Indian Education through Effective Resource Management*. Government of India. https://www.niti.gov.in

Accenture. (2022). *Revolutionizing Education with AI: The Role of 5G Networks in Higher Education*. Accenture Insights. https://www.accenture.com/in-en/insights

Agarwal, S. (2022). *Centralized Planning and Bureaucratic Challenges in Indian Higher Education*. Education Policy Journal, 25(2), 45-58. https://educationpolicyjournal.org

Ahmed, R., & Sharma, P. (2022). The Role of Smart Classrooms in Modern Education: A 5G Perspective. *Journal of Educational Technology, 34*(2), 123–134. https://jedutech.org

Bansal, K. (2023). *Bridging the Gap: How 5G Can Revolutionize Rural Education in India*. Educational Innovations Journal, 21(4), 78-89. https://eduinnova.com

Bansal, S. (2023). Data Privacy Challenges in the Age of 5G and 6G. *Journal of Information Security, 12*(3), 78–89. https://journalofinfosecurity.com

Chaudhary, P., Gupta, R., & Singh, M. (2022). Sustainable Resource Management in Higher Education: Leveraging IoT and 5G Networks. *Journal of Environmental Innovation, 11*(2), 78–89. https://envinn.org

Chaudhary, P., & Patel, R. (2021). *Bridging the Digital Divide in Higher Education: The Role of 5G Networks*. Digital Education Review, 10(4), 56-72. https://digitaledu.org

Choudhury, R. (2022). Infrastructure Deficit in Indian Higher Education Institutions: An Analysis. *The Journal of Educational Research, 31*(4), 122–137. https://joredu.com

Choudhury, T. (2023). *IoT-Enabled Campuses: Redefining Infrastructure in Higher Education*. Smart Education Review, 18(3), 233-245. https://smartedureview.com

Cisco. (2022). *Revolutionizing Education with VR and AR: How 5G Enhances Interactive Learning Experiences*. Cisco Insights. https://www.cisco.com

Digital Empowerment Foundation. (2023). *Bridging the Digital Divide in India's Education System*. https://defindia.org

Digital Empowerment Foundation. (2023). *The Digital Divide in India and How 5G Can Bridge the Gap in Education*. DEF Publications. https://www.defindia.org

Ericsson. (2022). *5G in Higher Education: How IoT is Transforming Campus Resource Management*. Ericsson Reports. https://www.ericsson.com

Ericsson. (2023). *6G Connectivity: Enabling a Hyper-Connected Education Ecosystem*. Ericsson White Papers. https://www.ericsson.com

FCC. (2023). Spectrum and Infrastructure for 6G Development. Retrieved from FCC.gov

GSMA. (2022). *Network Slicing in 5G: Customizing Network Environments for Education*. GSMA White Paper. https://www.gsma.com

Gupta, A., Kumar, S., & Patel, R. (2022). *Adaptive Learning Technologies in Higher Education: A 5G Perspective*. Educational Technology Quarterly, 15(4), 56-72. https://edutechq.org

Gupta, A., & Mehta, V. (2023). Ethical Implications of AI in Education: A 5G Perspective. *Journal of Educational Ethics*, *15*(2), 34–47. https://educationalethics.com

Gupta, A., Singh, V., & Kumar, S. (2022). *AR/VR in Education: Transforming Learning in the Digital Era*. Virtual Learning Quarterly, 15(1), 45-59. https://vlq.org

Huawei. (2023). *Advancements in 5G and the Path to 6G: Enabling Edge Computing and IoT in Education*. Huawei White Papers. https://www.huawei.com

IEEE. (2024). *Bridging the Future: From 5G Advanced to 6G Development*. Retrieved from IEEE Testbed.

India, C. (2023). *Empowering Indian Education through 5G and AR/VR Technology*. Cisco White Papers. https://www.cisco.com

India Education Report. (2022). *India's Growing Student Population and the Demand for Scalable Solutions in Higher Education*. Indian Ministry of Education. https://www.education.gov.in

Intel. (2023). *Predictive Analytics and 5G: Creating Scalable Solutions for Higher Education*. Intel Education Insights. https://www.intel.com

ITU. (2021). *Understanding Latency in 5G Networks*. ITU Reports. https://www.itu.int

ITU. (2022). The next generation of intelligent networks. ITU Insights. https://www.itu.int

ITU. (2024). Innovations in Next-Generation Networks. Retrieved from ITU.int

KPMG. (2023). *Higher Education in India: Challenges and Opportunities*. KPMG India Reports. https://home.kpmg.com

KPMG. (2024). *6G and the Future of Higher Education: Insights and Trends*. https://home.kpmg.com

KPMG. (2024). *The Future of 6G in Higher Education: Opportunities and Challenges*. https://home.kpmg.com

Kumar, R. (2022). Faculty Shortages and Their Impact on Indian Higher Education Quality. *Indian Journal of Academic Research*, *19*(3), 233–249. https://ijar.in

Kumar, R., & Mishra, A. (2021). The Impact of Remote Education on Higher Education in India: Insights from the Pandemic Era. *Indian Journal of Educational Research*, *12*(2), 98–112. https://ijer.in

Kumar, R., Sharma, P., & Verma, A. (2021). *The Role of IoT in Optimizing Resource Utilization in Universities*. Smart Campus Review, 18(3), 123-137. https://smartcampus.org

Kumar, R., & Singh, M. (2022). Economic Barriers to 5G Deployment in Higher Education. *Indian Journal of Technology Policy*, *19*(1), 90–102. https://indiantechpolicy.org

Kumar, S., & Sharma, A. (2021). Data-Driven Resource Management in Indian Universities: Challenges and Prospects. *The Journal of Higher Education*, *14*(1), 78–92. https://higheredu.org

Liu, Z., Liu, J., & Wang, X. (2023). Terahertz Communication in 6G: The Future of High-Speed Data Transfer. *Journal of Telecommunications Research, 18*(4), 112–126. https://www.jtrjournal.org

Liu, Y., Peng, S., Zhang, M., Shi, S., & Fu, J. (2024). Towards secure and efficient integration of blockchain and 6G networks. *PLoS One, 19*(4), e0302052. DOI: 10.1371/journal.pone.0302052

Maduranga, P. W., Tilwari, V., Rathnayake, R. M. M. R., & Sandamini, C. (2024). *AI-Enabled 6G Internet of Things: Opportunities, Key Technologies, Challenges, and Future Directions*. Retrieved from https://www.mdpi.com

Mahindra, T. (2023). *5G and Resource Management in Education: Creating Digital Campuses in India*. Tech Mahindra White Papers. https://www.techmahindra.com

McKinsey. (2023). *The Future of Work and Digital Skills in India: Implications for Higher Education*. https://mckinsey.com

Ministry of Education. (2022). *Improving Accessibility and Quality in Indian Higher Education*. Government of India. https://www.education.gov.in

MoE. (2021). *Survey of Digital Infrastructure in Indian Higher Education*. Ministry of Education, India. https://mhrd.gov.in

Nanda, P. (2023). Access to Academic Resources in Rural Indian Colleges: Challenges and Solutions. *The Journal of Educational Research, 29*(2), 201–216. https://joer.in

Nanda, S. (2023). AI-Driven Resource Allocation in Indian Universities: A Case Study Approach. *Journal of Higher Education Policy, 29*(3), 189–201. https://johep.org

Nanda, S. (2023). Inclusive Education through 5G: Innovations and Challenges. *Journal of Digital Inclusion, 19*(1), 45–61. https://digitalinclusion.org

Nanda, S., & Verma, T. (2022). *Upskilling the Workforce for 5G and Beyond*. Workforce Development Quarterly, 14(3), 112-130. https://workforcedev.org

NASSCOM. (2021). *Meeting the Needs of a Growing Student Base: Challenges in Technical and Professional Programs*. NASSCOM Reports. https://www.nasscom.in

NEP. (2020). *National Education Policy 2020*. Government of India. https://nep2020.gov.in

Nokia. (2021). Public-Private Partnerships in 5G and Beyond. Retrieved from Nokia

Nokia. (2022). *Enhancing Education through Reliable and Energy-Efficient 5G Networks*. Nokia Reports. https://www.nokia.com

Patel, A. (2023). *Scaling Remote Learning Platforms with 5G in India*. Higher Education Innovations Journal, 21(2), 33-47. https://hei.org

PLOS ONE. (2024). *Towards Secure and Efficient Integration of Blockchain and 6G Networks*. Retrieved from PLOS ONE

Ramanathan, P., Kapoor, N., & Desai, R. (2024). *Exploring the Potential of 6G Networks in Education: Emerging Trends and Applications*. Advanced Communication Journal, 22(1), 123-137. https://advcomm.org

Ramanathan, P., Kapoor, N., & Desai, R. (2024). *Holographic Communication in Education: Opportunities with 6G Networks*. Advanced Communication Journal, 22(1), 112-128. https://advcomm.org

Ramesh, K., Patel, S., & Das, A. (2022). Cybersecurity Risks in 5G-Enabled Campuses. *Journal of Digital Security*, *9*(4), 67–81. https://digitalsecurityjournal.org

Rao, V., & Srinivasan, N. (2023). *Policy Frameworks for 5G Implementation in Education*. Policy Insights Journal, 11(3), 23-45. https://policyinsights.org

Report, D. I. (2023). *Digital Infrastructure in Rural Educational Institutions: Progress and Challenges*. Government of India. https://www.digitalindia.gov.in

Sarma, T., & Mishra, K. (2021). COVID-19 and the Digital Divide in Indian Education: Lessons Learned. *Journal of Social Policy*, *12*(3), 99–114. https://jsp.in

Sharma, T. (2023). *AR and VR Applications in Higher Education: A 5G Enabled Revolution*. Virtual Learning Review, 17(2), 89-102. https://vlr.org

Sharma, T., Kapoor, N., & Desai, P. (2023). *Economic and Ethical Challenges of 5G in Education*. Educational Technology Horizons, 8(2), 89-105.

Singh, M., & Kapoor, A. (2023). *The Role of Predictive Analytics in Higher Education: Leveraging 5G for Real-Time Decision Making*. Analytics in Education, 10(3), 67-81. https://analedu.org

Singh, M., & Mehta, A. (2023). AI and 5G: Transforming Administrative Processes in Universities. *Journal of Educational Administration*, *29*(4), 190–202. https://eduadmin.org

Statista Research Department. (2022). *Higher Education Institutions in India: Statistics and Analysis*. Statista. https://www.statista.com

TRAI. (2022). *Digital Access in India: A Statistical Overview*. Telecom Regulatory Authority of India. https://trai.gov.in

UGC. (2022). *Faculty Availability in Indian Higher Education Institutions*. University Grants Commission Reports. https://ugc.ac.in

UNESCO. (2021). *Promoting Transparency and Accountability in Higher Education Resource Management*. UNESCO Reports. https://www.unesco.org

Wellalage, M. P., (2024). *Towards Secure and Efficient Integration of Blockchain and 6G Networks*. Retrieved from https://journals.plos.org

World Bank. (2022). Digital divide in Indian education. World Bank Education Series. Retrieved from https://www.worldbank.org

Chapter 12
Leveraging Technology for Inclusive Rural Development in India:
Assessing the Impact of Digital Interventions on Socio-Economic Growth in Rural Communities

Geetanjali Sharma
Jain University, India

Uma C. Swadimath
https://orcid.org/0009-0006-9168-7259
Jain University, India

ABSTRACT

Rural development encompasses development in social, cultural, environmental, technological and economic conditions. India introduced new land reforms by abolishing the zamindari system, distribute lands among the landless farmers and abolishing bonded labour to protect the farmers' interests. The contribution of agriculture and allied services in Gross Value Added was just 17.8% for the year 2019-20 and 18% in 2023. India's rural economic development envisages through three phases namely land reforms, green revolution, and changes due to globalization when India opened its economy in the nineties. However, a substantial gap remains in the progress trajectory, leading to a rural-urban divide. The present study is conceptualized that tries to assess the impact of digital interventions on socio-economic growth in rural communities. The authors develop an economic model which emphasizes on digitalization, technological developments and community engagement, and how all of these lead to rural development and inclusive growth in India's economy.

DOI: 10.4018/979-8-3693-7788-8.ch012

1. INTRODUCTION

Mahatma Gandhi said, "India lives in villages''. Nearly two-thirds of India's population lives in villages, which is roughly 833 million population. The first five-year plan's objective was to make India economically stronger by concentrating on villages. The objective was that through rural development, there would be a trickle-down effect leading to overall development and growth. According to Kindelberger, "While economic growth simply refers to an increase in output, economic development involves changes in the technological and institutional organization of production, as well as the distributive pattern of income. Therefore, when economists, policymakers, and leaders discuss economic development for a nation, they refer to balanced growth, where both rural and urban areas must experience real growth. Disparities in rural-urban areas lead to uneven growth leading to inequalities. In India, nearly 60% of the population is engaged in agriculture. Yet the contribution of agriculture and Allied services in Gross value Added was just 17.8% for the year 2019-20 and around 18% in 2023. Even after decades of planning, the contribution of the rural sector has not shown a phenomenon of growth. In a recent survey report by People Research on India's Consumer Economy (PRICE) about 140 million, Indian households get their incomes from agriculture and allied activities. About 125 million farmers are landless and the schemes of the government do not reach them. Modern farming and allied activities practices require capital and this becomes another major hurdle to Indian farmers who are mostly small or landless laborers.

Figure 1. India's structural reforms since independence

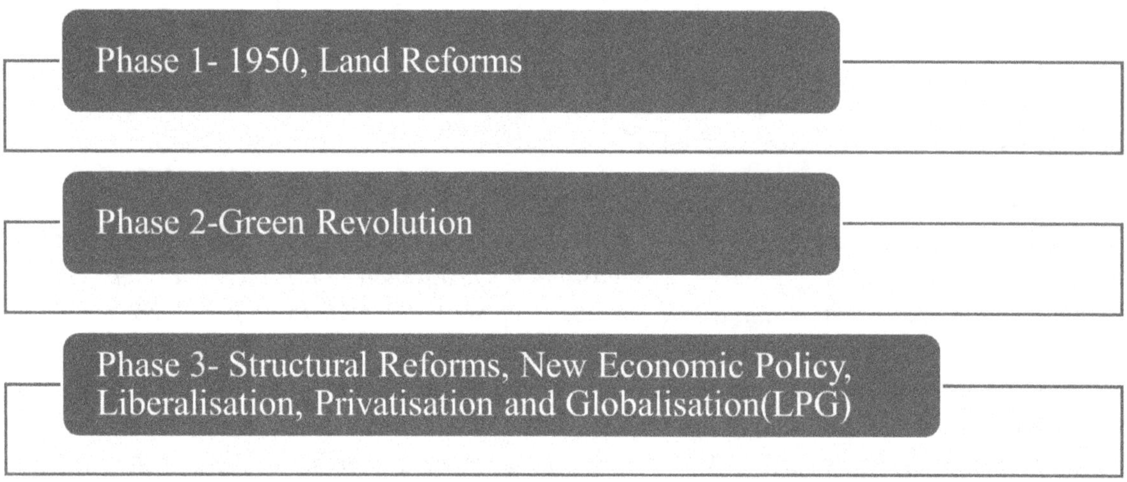

Source: Authors' Description

India's rural sector has witnessed structural changes since independence. These changes can be classified into 3 phases namely

a) Phase 1: India's structural reforms since independenc
b) Phase 2-Green Revolution
c) Phase 3- Structural Reforms, New Economic Policy, Liberalisation, Privatisation and Globalisation (LPG)

The concept of Inclusive development is an approach that envisages equitable opportunities at all levels of society. Inclusive growth The Indian model of inclusive growth has three pillars: (1) Market economics, (2) Empowerment, and (3) Pragmatism. While the three pillars have shown considerable progress when it comes to urban areas of India, rural India despite policies and programs initiated by the Government of India has not been able to appreciable growth and development. According to the latest National Sample Survey Office (NSSO) data, only 24% of rural Indian households have access to the internet, compared to a 66% penetration in cities. We need transformative policies that will help the rural areas to come out of the low growth and development cycle. The concept of inclusion refers to the sense of integration among individuals regarding any form of growth occurring within an economy. The concept of Inclusive development has roots since ancient times in various societies, though termed differently. In the 18th century, Adam Smith's Scottish Philosopher in his 'Wealth of Nations' explicitly stated that no nation can prosper if there is a good part of a society reeling under poverty and being miserable. According to Kindelberger, "While economic growth simply refers to an increase in output, economic development involves changes in the technological and institutional organization of production, as well as the distributive pattern of income." To attain a balanced economic growth, the emphasis is on both rural and urban development. Yet another interesting version of inclusivity was given by Chinese thinkers and social philosophers in 551-479 BC, that in a country wealth is something to be ashamed off and that shows that a country is badly governed. In India, our "Vasudhaiva Kutumbakam" (from vasudha, the earth; eva, emphasizer, and; kutumbakam, family), a Sanskrit name brings the concept of collective and harmonious living where all reap the benefit of growth taking place in the society. The recent alternative thinkers like Amartya Sen has advocated the concept of 'Development as Freedom', Gramin Bank introduced by Mohammad Yunus from Bangladesh with the objective of banking for all. In India, Panchayati Raj was a significant step towards integrating rural areas with urban regions. It largely succeeded in bringing rural sectors into the mainstream growth cycle. However, a substantial gap remains in the progress trajectory, leading to a rural-urban divide. Indian villages are still grappling with shortages in electricity, pure drinking water facilities, health infrastructure, and schools with poor infrastructure.

2. REVIEW OF LITERATURE

Janak Bhandari et al. (2019), discuss achieving sustainable rural development and focus on the Government's efforts to improve underdeveloped villages. He examines the role of the Ministry of rural development, district development boards, and integrated rural development programmes for the development of rural areas. S. Vijay Kumar (2018) has analysed how inclusive rural development strategy is important to develop rural India. He relates to three dimensions namely economic, political and social dimensions. Economic dimensions includes provision of employment opportunities for the poor and low income groups, social dimension encompasses gender equality and rural women empowerment, improvement in health and medical facilities, and education and political dimension is to encourage people to participate in political decisions at the village level. Digital empowerment will significantly bridge the gap between rural and urban India, (Iivari et al., 2020). Digital India program will to large extent improve the lives of rural disadvantaged India, where they will be able to utilize digital platforms and use the e-Services for their progress (Nayak,2018). Digital transformation will bring transparency, improve the service delivery and will ensure active citizen engagement (Barrutia and Echebarria,

2021). Digital transformation is slow in rural areas which will be a hindrance towards India's economic growth and development. Digitalisation in rural areas encompasses electronic payments, digital literacy, financial inclusion and rural economic development (Patel. Satish, 2024). Only 29% of rural India has internet access as compared to 64% in urban India. Rani etal.,2024 have analysed how information communication tools application improves rural connectivity. Different types of e-governance projects such as e-chaupal, radio farm forum, rural development apps such as shramik bandhu, janmangega have been introduced to bring the rural population under technology. Inclusive development will be attained only if there is balanced regional development. Digital India Program launched in the year 2015 with an objective of transforming India has been prestigious program to make India developed. DIP aims at bridging the digital gap between rural and urban India (Asrani,2022). According to study by (Kumar et al;2022), Pradhan Mantri Gramin Sasharta Abhiyan, digital literacy will help the rural poor at both personal and professional front. Digitalization will improve the overall infrastructure of the rural India, through reduced costs, and access to essential services (Banu,2017). Stravos Sindakis and Gazal Showkat, in their paper 'The Digital Revolution in India: bridging the gap in rural technology adoption' gave two DIP adoption strategies.

a. Right of way (Row) introduced in the year 2016 by India's Department of Telecommunications, aimed at facilitating laying optic fiber cables and mobile towers.
b. Government Sensitization: Awareness of the schemes and provisions should be effectively communicated to the rural people, especially for those who are small and marginal farmers. These can be in the form of workshops, or training sessions. A study done by (Saxena et al;2019) elucidated that access to information and services will help in bring social and economic development in rural areas. Ravindra Verma et al. (2023), examine the challenges such as regional and cultural differences that prevent the rural economy from developing. To ensure the development of the rural economy, the adoption of technology and the provision of education are very much necessary. With the help of technological interventions in areas of healthcare, financial services, agricultural sector, and employment opportunities it is possible to develop the rural economy.

Rural development is a transformative process which aims at improving the quality of rural life both social and economic related. The post-independence period saw plethora of reforms in uplifting the rural sector by introducing the land reforms. scenario then and now has drastically transformed to better and Indian rural sector has become trading hubs .This has been substantiated with the developmental reform due to technology. Rani and Naga (2017) provided a comprehensive review of ICT applications in rural development, emphasizing their role in education, governance, and health. They argued that ICT initiatives, such as e-chaupals and community radio, have significantly improved transparency and accessibility in public administration. In a related study, Paria and Giri (2018) examined the impact of digitalization on the rural economy, highlighting its transformative effects on livelihoods and income generation.

Technology plays a pivotal role in advancing non-economic aspects of rural development. The growth of rural sectors is intricately linked to socio-cultural transformations, necessitating improvements in education, health, and nutrition to enhance overall productivity. Additionally, technology facilitates the dissemination of information and innovations, particularly in agriculture, while contributing to narrowing the digital divide. Shruthi et al. (2021) investigated rural internet connectivity in India, analyzing the technological challenges and proposing solutions to bridge the digital divide. Their study highlighted significant infrastructure deficiencies and recommended policy-level interventions to enhance internet access in rural areas. Similarly, Deshpande and Sambhe (2013) emphasized the ICT requirements of rural communities, advocating for customized approaches to address their specific needs effectively.

Rural India experiences significant transaction costs, often resulting in market imperfections that manifest as inefficient intermediation and delays. These delays can impose substantial financial burdens on poor farmers, with market inefficiencies frequently leading to financial distress. Information technology-based interventions play a critical role in enhancing institutional frameworks, achieving economies of scale, and fostering accelerated growth in the rural sector. Truelove et al. (2023) examined the primary drivers of economic change—agricultural intensification, land reform, and labor market diversification—through multidisciplinary perspectives, emphasizing the role of institutions as agents of change.

Sustainable rural development is a widely discussed objective, rooted in enhancing the livelihoods of people in rural areas. Achieving this requires addressing key dimensions such as economic stability, infrastructure, education, healthcare, and social welfare. Chandrasekar and Sharma (2020) highlighted the transformative role of ICT in improving educational outcomes in rural India. Their study examined the effectiveness of digital classrooms and mobile-based learning applications in mitigating low literacy rates and enhancing access to quality education.

Gupta et al. (2021) explored the application of blockchain technology in rural governance, focusing on its potential to enhance transparency and trust in land record management and welfare distribution systems. Bhatt and Shukla (2019) provided a detailed evaluation of e-governance initiatives, such as Aadhaar and DigiLocker, underscoring their impact on improving rural access to government services. However, they also noted challenges in adoption due to limited awareness and infrastructure shortcomings. Similarly, Sinha et al. (2019) examined the e-choupal initiative, demonstrating its success in empowering rural farmers through real-time access to market prices and agricultural best practices. Fostering socio-care through inclusive development models can build trust and dismantle societal barriers, facilitating long-term community-led progress (Varga & Varga, 2008). Sustainable rural development demands a synergistic approach involving public policies, corporate initiatives, and active community participation. Strategies must prioritize inclusivity and be designed to prevent conflicts of interest, ensuring cohesive and sustainable growth for rural India

3. OBJECTIVES OF THE STUDY

- To study the need and importance of technology in the socio-economic development of rural India.
- To identify strategies for leveraging technology to bridge the rural-urban divide in India.

3a. Methodology

Research design - The research design of this study is exploratory. The data for the case study has been collected through a Primary source – Stratified sampling was used for collecting qualitative information. Secondary sources –research through the internet research papers, books, thesis, dissertations, journals, etc. The analysis is mainly based on a combination of primary data and secondary data collected from various government sources, and websites.

3b. Theoretical Background of the Study

The theoretical framework of this study is based on the interrelationship between digital penetration for rural growth and consequently the inclusive development of India. and The economic growth of any nation depends on equal opportunity for all. Inclusive growth as per (World bank, 2009) is when there is good job creation, access of the poor to good living conditions. There has been a big challenge when it comes to digital connectivity and rural development (Malecki, 2003, Strover, 2003,Velaga et al., 2012) Inclusive growth is believed to create an equal opportunity environment for all (World Bank, 2009) by addressing job creation, market, consumption, production, and access of the poor to good living conditions by creating appropriate platforms and policies. The basis of inclusive growth is equality of opportunities and participation of all in growth (Palanivel, 2013). The role of technology in fostering inclusive growth and sustainable development, particularly in rural India, is vital for bridging the significant disparities between rural and urban areas. According to the United Nations Development Program (UNDP), inclusive growth is defined as the process and result of all groups of people participating in the organization of growth and benefiting equally from it. In the context of India, where 67.77% of the rural population is literate compared to 84.11% in urban areas (as of September 2023), leveraging technology is essential for fostering socio-economic development in rural regions.

- **Inclusive Growth and Technology**

Inclusive growth requires a multifaceted approach that ensures all groups within society can participate in and benefit from economic growth. This is particularly challenging in rural India, where access to resources, education, and information is often limited. The pillars of inclusive growth—education, literacy, digital literacy, and the reduction of rural-urban disparities—are all areas where technological advancements can make a significant impact.

- **Education and Literacy**

Education is a foundational element of inclusive growth. In rural India, where educational attainment is generally lower than in urban areas, technology can play a transformative role. E-learning platforms and digital classrooms can provide rural students with access to high-quality educational resources, helping to close the literacy gap. Moreover, technology-driven educational initiatives can increase the Gross Enrollment Ratio (GER) in rural areas, which is crucial for the overall socio-economic development of these regions.

- **Digital Literacy**

Digital literacy is another critical factor for rural development. As more government services, financial tools, and agricultural information become available online, digital literacy enables rural populations to access these resources. The ability to navigate digital platforms allows rural residents to take advantage of financial services, such as direct benefit transfers and e-banking, which are essential for their economic empowerment. Additionally, digital literacy helps farmers access vital information related to agriculture, such as market prices, weather forecasts, and modern farming techniques, which can significantly improve their productivity and income.

- **Bridging the Rural-Urban Divide**

Figure 2. e-NAM flow chart

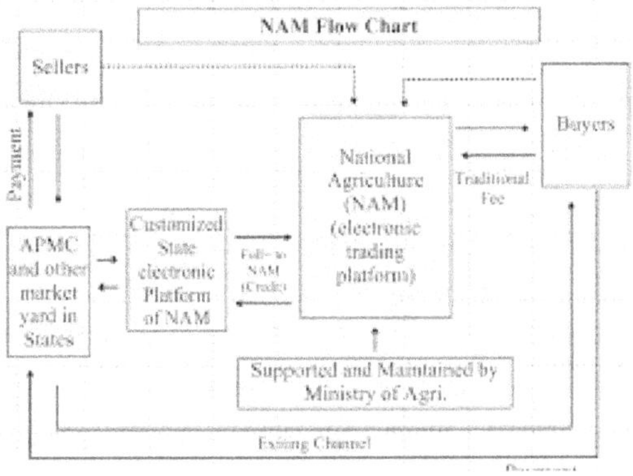

- *Source: https://www.researchgate.net/publication/343240633 E-national Agricultural*

3c. Market (e-NAM) in India: A Review

Technology also plays a crucial role in reducing the disparities between rural and urban areas. By integrating rural communities into the digital economy, technology helps bridge the gap in access to services, information, and opportunities. Initiatives like the Government of India's e-NAM (electronic National Agriculture Market) have demonstrated the potential of technology to connect rural farmers with wider markets, thus enhancing their economic opportunities and reducing their reliance on intermediaries. Moreover, digital infrastructure development in rural areas paves the way for broader participation in the national economy, fostering inclusive growth and reducing regional inequalities. For India to achieve sustainable development, it is essential that technological advancements penetrate rural areas. This will not only enable rural populations to benefit from financial services and government support but also help them integrate with the broader urban growth. By enhancing education, promoting digital literacy, and bridging the rural-urban divide, technology serves as a powerful tool for inclusive growth, ensuring that all segments of society can participate in and benefit from India's economic progress.

- **Balanced Growth Strategy**

A balanced approach is essential for nations to achieve sustainable development, where growth across various sectors supports and enhances each other. This aligns with Ragnar Nurkse's theory of balanced growth, which asserts that for underdeveloped countries, simultaneous investments in multiple industries are crucial. According to Nurkse, balanced development in both agricultural and industrial sectors is fundamental to national growth. He emphasizes that productivity and economic development are closely intertwined, and productivity is also directly linked to the availability of basic infrastructure. In rural areas, infrastructure can be broadly divided into two categories:

- **Agricultural Infrastructure**: This includes facilities and services directly related to farming, such as irrigation systems, storage facilities, and agricultural research centers.
- **Non-Farm Agrarian Infrastructure**: This encompasses infrastructure that supports rural economies outside of agriculture, such as transportation networks and markets.
- **Digital infrastructure**

It has transformed the development landscape in both developed and developing nations, including India. Digital infrastructure paves way for complete change in the economic, socio and cultural mindset. These are the transformative developments that gives push to the agrarian ecosystem from internet connectivity to health facilities.

India embarked on a comprehensive digitalization initiative in 2015, with objectives grounded in three fundamental pillars: 1) Innovation, 2) Implementation, and 3) Inclusion. The primary aim of the Digital India initiative was to drive economic growth by integrating digital platforms and fostering economic development through the provision of digital services. Rural development has remained a central focus of Indian policy since the inception of its Five-Year Plans. To catalyze the transformation of rural India, both fundamental and institutional reforms have been undertaken.

The evolution of rural development in India can be categorized into three distinct phases:

a) **Post-Independence Reforms**
b) **Introduction of Technology in the Agrarian Sector**
c) **New Age Reforms**

3d. Post-Independence Reforms (1947-1960)

These reforms were primarily institutional, including measures such as the abolition of intermediaries, tenancy reforms, and the imposition of ceilings on landholdings to address inequities in land distribution.

3e. Introduction of Technology in the Agrarian Sector

The introduction of technology aimed to revolutionize farming practices. Beginning in the 1960s, the Green Revolution sought to boost agricultural productivity through the adoption of high-yielding variety (HYV) seeds, mechanization of farms, and the establishment of financial institutions, such as the National Bank for Agriculture and Rural Development (NABARD), to support rural development.

New Age Reforms

The focus of these reforms shifted towards infrastructure development, with initiatives such as the Rural Housing Scheme (Indira Awaas Yojana), the Rural Landless Employment Guarantee Programme, and the Integrated Rural Development Programme (IRDP). Additionally, the Mahatma Gandhi National Rural Employment Guarantee Act (MGNREGA) is considered a "silver bullet" for alleviating rural poverty and unemployment through guaranteed employment and wage provision

Despite the reforms post India and the technology built reforms, Indian rural sector has not been able to achieve the desired growth and development. India's majority of poor people reside in villages.280 million poor people still reside in rural India. Socio Economic Census (2011) data indicated that 3/4th of the rural households live with a monthly income of less than Rs 5000.This is a very depressing figure especially when India has been initiating rural reforms. Landless labour, small and marginal land holdings make it very challenging for the rural poor household to uplift their economic status and thereby leading to migration to urban areas.

Figure 3. Rural statistics

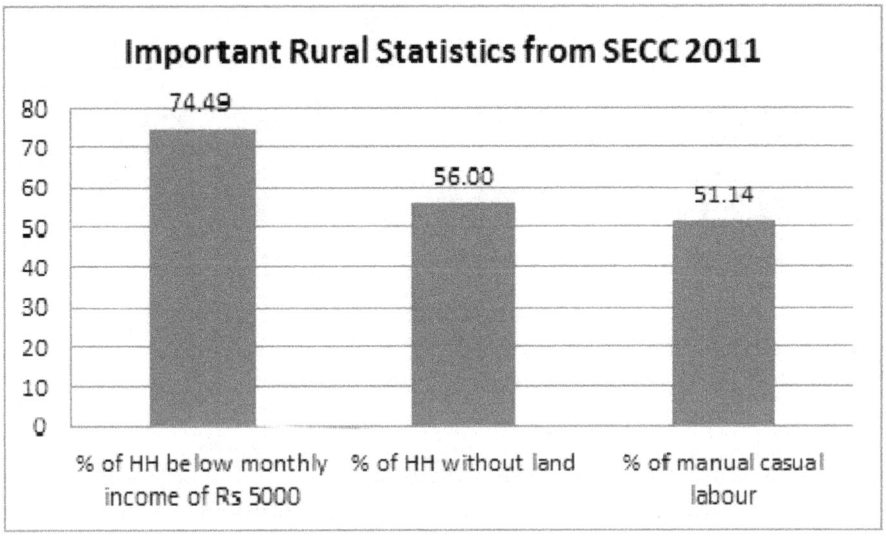

Source: https://factly.in/rural-india-behind-urban-india-in-progress-indicators/

This also becomes a challenge for the poor households to educate their children. lack of education eventually is dismal sign of complete ignorance of the government schemes and policies and also heavy dependence on the middle men.

Statistics reveal that there is gender disparity when we compare the literacy ratio between urban and rural poor. Figure 4 below gives the comparative statistics of the literacy gap.

Figure 4. Literacy levels in India

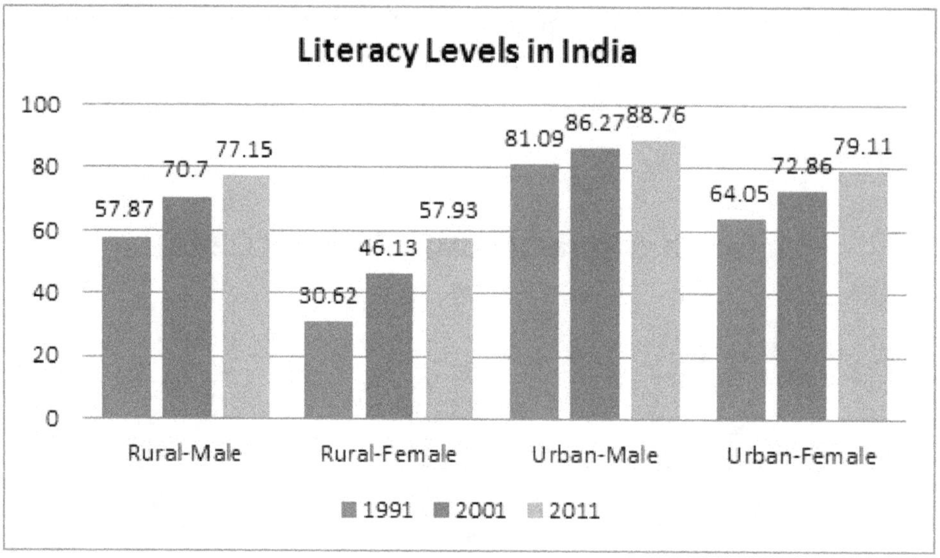

Source: https://factly.in/rural-india-behind-urban-india-in-progress-indicators/

India aims to achieve inclusive development, which can be realized by ensuring that the population, particularly in rural areas, receives proper education. Education especially among the marginalized group, women and rural population will help them to be aware of the programs and policies especially initiated for them.

This has long been a challenge for India, where poverty and lack of education have kept the rural population disconnected from mainstream development, despite the government's efforts.

Leveraging Technology for Rural Growth

Technology alone can bring about socio-economic development in the rural sector of India. The right kind of technology will leverage the desired growth in the rural economy. Let us understand the connect between technology and growth of the rural sector.

First the biggest challenge of rural India is the low productivity of small land holdings. Indian farmers can greatly benefit from technological advancements like Soil censors,GPS technology,Weather monitoring,Drones. Let us understand one by one how fourth industrial revolution can be pushed for farm productivity.

I. Soil Censors: The biggest challenge for the farmers is to get accurate information of the soil conditions. Soil Censors are able to identify accurately the moisture levels, temperature and other conditions which are necessary to maintain the productivity of land. With the help of government, non-government organizations data collected can be dissipated especially to the farmers with smaller land holdings.

II. GPS Technology: Farmers need to constantly make use of fertilizers, pesticides and herbicides to improve productivity. GPS technology plays a crucial role in precision farming. It helps in reducing waste and improves productivity.
III. Weather Monitoring: The biggest challenge for Indian farmers even today is that they have to depend on vagaries of nature. Every year we hear of crores of crops getting damaged due to unpredictable rains or drought conditions. Real time data through the AI technology certainly makes informed decisions about when to sow, irrigate, and choose crop types. This information is available via weather apps, websites, or dedicated on-farm weather stations.
IV. Automation: for sowing, harvesting certainly helps the farming sector to rely less on manual labor and improve productivity.
V. Drones: Drones are extensively used for mapping, surveying, and crop monitoring. They provide valuable data that helps with the planning and execution of agricultural activities.

The above technology aims at improving the productivity of farms. However, we also need to ensure that there is overall development of rural sector. Rural sector through high speed network used extensively will be able to get desired information of market developments related to farming inputs, pricing, logistics. To ensure that the education remains the top priority for rural development there must be constant efforts to facilitate online education platforms.

Technology will also help in leveraging non-farm employment opportunities in rural India. Rural youth with technology based knowledge can set up small and medium units. e-commerce facilitates market accessibility and reduce the dependence on physical efforts.

Healthy population is productive population. Digital Health solutions provides distance services. Telemedicine is the best example which facilitates remote consultations, connecting patients with qualified healthcare professionals from urban centres.

AI empowered data can analyse vast amount of medical related data which can help the healthcare professionals to predict diseases, plan preventive action plans.Artificial Intelligence is a game changer for Indian agriculture. AI will provide solutions by helping farmers select crops based on technology-driven surveys that assess soil type and predict monsoon dates. It will also enable crop monitoring through IoT (Internet of Things) devices and drones. This level of analysis will give farmers greater confidence in their crops and offer a sense of security for Indian agriculture.

The government of India has launched various apps for the benefit of farmers. These are as follows:

Table 1. Digital apps for farmers

App Name	Year Launched	Key Features
IFFCO Kisan Agriculture	2015	- Provides customized agricultural advisory, weather updates, and market prices. - Access to agricultural information in text, imagery, audio, and video. - Offers helpline numbers for Kisan Call Centre Services.
Krishi Jagran	Recent	- Offers trending agricultural news, crop calendars, cultivation guides. - Provides crop protection, pest, and disease management information. - Information on subsidies, careers, and farm mechanization.
Pusa Krishi	2016	- Provides details on technologies developed by IARI (Indian Agriculture Research Institute) to increase farmer returns. - Information on new crop varieties by ICAR (Indian Council of Agriculture Research) - Includes resource-conserving cultivation practices and farm machinery information.

App Name	Year Launched	Key Features
Agri App		- Comprehensive information on crop production and protection. - Access to expert chats, video-based learning, and online fertilizer/insecticide markets. - Focuses on high-value, low-product crops.
Bijak App		- Online platform connecting verified agri traders. - Offers services like advance payment guarantee, secured payments, and mandi rates. - Facilitates end-to-end services for suppliers and buyers.

Source:Author's Description

Digital infrastructure paves the way for participatory development since there would be more awareness and accessibility to information with the help of internet connectivity. We can take example of e- mandis which has to a large extent helped the small and marginal farmers access the markets to sell their produce. Technology-led growth helps administrative and governance ease thereby leading to inclusive growth.

- **e-NAM –way towards Technology Based Rural Development**

The Government of India's e-NAM initiative represents a groundbreaking effort to integrate technology into the agricultural sector, providing farmers with broader market access beyond their local villages. Historically, small farmers in India have been dependent on intermediaries to sell their produce, often at unfair prices. Despite India's global leadership in producing key agricultural commodities like pulses, milk, fruits, vegetables, and sugarcane, poverty remains widespread, with 80% of the poor living in rural areas, according to the World Bank 2026 report. The primary challenges facing rural India include a) Limited direct market access. b) lack of financial access due to farmers limited awareness. c) Exploitation by middlemen and commission agents. To address these issues, the Indian government introduced e-NAM as a transformative step towards connecting farmers directly with markets. The objectives of e-NAM include:

- Facilitating direct marketing between farmers and buyers.
- Providing market information, including reliable pricing, market channels, storage and transportation recommendations.

e-NAM has transformed the supply chain logistics of rural India. Around 1389 registered e-mandis is in India as of November 2023.e mandis have been able to integrate the rural India with mainstream development.

4. DIGITAL EMPOWERMENT TO BRIDGE THE GAP BETWEEN RURAL AND URBAN INDIA

India's rural population, constituting around 70% of the total, continues to grapple with significant inequalities in employment, wages, social security, and overall living standards. Despite progress since the abolition of the zamindari system, a substantial portion of the rural populace remains vulnerable. Notably, 41.63% of landless laborers are concentrated in these areas, reflecting the enduring challenge of landlessness. Landlessness severely impacts the rural poor's ability to access quality education for their children, perpetuating a cycle of poverty and subjugation. This lack of land ownership not only limits economic opportunities but also exacerbates social and educational disparities. The interconnected nature of these issues results in a vicious cycle that is difficult to break, where poverty leads to limited educational opportunities, which in turn perpetuates poverty. Beyond landlessness, rural backwardness manifests in various forms of inequality, such as inadequate access to healthcare, poor infrastructure, low access to education and limited opportunities for economic mobility. These disparities hinder the overall development of rural areas, leading to a stark contrast between rural and urban living conditions. The disturbing feature of this divide is the inequalities observed in the education ratio. Education is the foundation for progressive India. This becomes a crucial factor especially when we are aiming to be developed economy by 2030. Inclusive society is when everyone gets equal opportunities of job, amenities, and decent standard of living. However, in India due to cultural factors, we have not been able to achieve 100% education. Gross Enrolment Ratio is the total enrolment in primary education, regardless of age, expressed as a percentage of the eligible official primary school-age population in a given school-year. Gross Enrollment Ratio (GER) in rural India is estimated to be about 7% while urban areas have a GER of about 23%. Digital Empowerment of India is a necessity for improving the socio –economic conditions of the rural India. Digital connectivity will integrate the rural areas with the mainstream economic developments like internet connectivity enabling the rural population to access the developments in education, weather forecast, mobile-based banking services, e- mandis and its operations, latest information on the agri-based technology. Awareness of the latest information is the first important stage to connect the rural people. Government of India aims to create National Farmers Database for digitalized records.

Agri stack is a classic example of Digital intervention by the Government of India focusing on the farmers welfare and for agriculture sector. The objective is to create a unified platform for farmers to give them services with the help of technology. These services widely range from sensors and drones to analyze the soil conditions to commercialization of agri-based activities.

Agri Stack will create a unified platform for farmers to provide them with end to end services across the agriculture food value chain.

Let us now look into the various facets in detail which bridges the rural and urban divide.

4.1 Role of E-Commerce in Connecting Rural Producers With Urban Consumers

E-commerce is booming in India. This is especially true in the case of urban areas and this has significantly contributed to the development of the urban sector. However, due to structural challenges in rural areas of India e-commerce needs more impetus. Today there is a huge urban market for rural goods. This requires robust digital infrastructure. Local growth brings uniform economic development. By enabling rural producers, artisans, and small businesses to access broader markets without intermediaries, e-commerce enhances profitability, reduces costs, and encourages local economic growth. This digital integration not only increases the availability of essential goods in rural areas but also creates new employment opportunities, especially in sectors like logistics and digital support services.

E-commerce fosters digital inclusion, empowering rural entrepreneurs with tools for financial transactions, customer outreach, and data-driven decision-making. E-commerce in rural sector backed up with proper infrastructure, and upgraded logistic will boost the rural industries. It is believed that rural retail will contribute global GDP by $ 2 trillion by 2020 .An eco-system backed up with technology would penetrate deeper into the rural areas thereby bringing in inclusivity.

E-Grammen is a digital initiative which aims at enhancing rural development in India by providing access to government services, information, and resources directly to rural communities. It utilizes technology to bridge the gap between rural citizens and essential services, fostering greater inclusivity and promoting socio-economic growth in rural areas.

The following are the advantages of E-Grameen:

- **Access to Government Services**: e-Grammen allows rural residents to access various government services online, such as land records, licenses, and welfare schemes, without needing to travel long distances. This accessibility saves time, reduces costs, and increases the reach of government programs in remote areas.
- **Promoting Digital Literacy**: By exposing rural communities to digital tools and resources, e-Grammen enhances digital literacy, making villagers more comfortable with technology and better equipped to take advantage of digital opportunities in various sectors.
- **Boosting Rural Economy**: Through digital marketplaces and e-commerce platforms integrated with e-Grammen, rural producers and artisans can connect directly with buyers, bypassing middlemen. This direct access to markets can lead to fairer prices, greater profits, and a boost to the local economy.
- **Efficient Agricultural Support**: e-Grammen provides farmers with real-time information on weather, crop prices, and farming techniques, helping them make more informed decisions, manage risks, and increase productivity.
- **Educational Opportunities**: The platform supports e-learning initiatives, enabling rural students to access online educational resources, lectures, and training modules. This access can help bridge the educational divide between rural and urban areas.

Direct Marketing for Rural Inclusivity

Direct marketing strategies, which enable farmers to sell products directly to consumers, offer transformative benefits for inclusive rural development and economic growth. Bypassing intermediaries allows farmers to access higher profit margins, directly retain a larger share of their product value, and gain consistent, fair prices. These benefits support financial stability in rural areas, providing an income boost that can improve living standards and investment in local economies. Direct marketing also fosters deeper consumer relationships, building trust and loyalty for repeat business, which is beneficial for small farmers in sustaining demand.

Furthermore, direct marketing empowers farmers to be responsive to market needs, adapting production to consumer preferences, thereby reducing overproduction and waste—a challenge in many rural areas. This adaptability helps farmers produce efficiently and sustainably, aligning with community needs.

Technological Innovations Supporting Direct Marketing

1. **E-Commerce Platforms**: Digital marketplaces expand farmers' reach beyond local geographies, bringing fresh produce directly to a larger customer base (Kumar et al., 2021). This access enhances visibility for rural farmers and connects consumers with fresh, local products, reinforcing local economies.
2. **Mobile Applications**: Apps tailored to connect consumers with farmers' markets or enable farmers to manage orders digitally help bridge the rural-urban divide. By simplifying the sales process, these apps increase market access for rural farmers (Smith & Jones, 2022).
3. **Social Media Marketing**: Social platforms allow farmers to share stories and build brand communities. Increased engagement and visibility not only attract more customers but also foster an appreciation of local agricultural practices (Brown, 2023).
4. **Blockchain Technology**: Enhancing transparency, blockchain allows consumers to trace their food's origin, boosting confidence in the quality and source of products. This traceability is increasingly valued in modern markets and strengthens local farmer-consumer relationships (Johnson, 2023).

Case Studies Promoting Inclusive Rural Development

1. **Community Supported Agriculture (CSA)**: This model, where consumers subscribe to regular deliveries from local farmers, promotes sustainability and local sourcing, directly supporting rural livelihoods and food security (Williams, 2023).
2. **Farmers' Markets**: In-person sales foster community interaction and economic development. Digital payment options and inventory management can improve efficiency and customer experience, supporting rural economies more robustly (Davis, 2023).

Government Initiatives to Support Direct Marketing

1. **Policy Support**: Government policies encouraging local food systems through favorable zoning laws and market permissions can boost rural economies by facilitating local farmers' access to direct markets (Thompson, 2023).
2. **Infrastructure Development**: Investments in rural infrastructure like cold storage and transportation can make it easier for farmers to access urban and regional markets, reducing waste and ensuring product freshness (Garcia, 2023).
3. **Educational Programs**: Training in marketing and e-commerce equips farmers with skills to better manage direct sales and engage in modern market practices, enhancing rural development potential (Lee, 2023).
4. **Subsidies and Grants**: Financial support for setting up e-commerce platforms and farmers' market participation lowers entry barriers, enabling small-scale rural farmers to explore direct marketing (Miller, 2023).

4.3 Vertical Farming

Vertical farming, which involves growing crops in stacked layers or vertically inclined surfaces, represents a cutting-edge approach in agriculture. Leveraging Controlled-Environment Agriculture (CEA) technology, vertical farming maximizes crop yield per square foot, offering a space-efficient solution particularly beneficial in urban settings. This technique provides numerous advantages for sustainable food production and supply, including:

1. **Space Efficiency**: Vertical farming can operate in compact, urban spaces, reducing the distance food travels to market, thus minimizing transportation costs and delivering fresh produce locally
2. **Resource Conservation**: Often using hydroponics or aeroponics, vertical farms significantly reduce water usage compared to traditional agriculture methods, a major advantage in water-scarce regions
3. **Year-Round Production**: By controlling environmental factors, vertical farms can produce crops consistently throughout the year, overcoming seasonal limitations and fluctuations in climate
4. **Pesticide Reduction**: The regulated environment of vertical farming minimizes the need for chemical pesticides, allowing for cleaner, healthier produce (Khan, 2023).

Technological Innovations Supporting Vertical Farming

1. **LED Lighting**: Advances in LED lighting allow vertical farms to simulate natural sunlight effectively, optimizing photosynthesis and accelerating crop growth (Chen, 2023).
2. **IoT Sensors**: Internet of Things (IoT) devices monitor real-time environmental conditions, such as temperature, humidity, and nutrient levels, enabling precise adjustments that optimize crop yields (Zhang, 2023).
3. **Automation and Robotics**: Automation in tasks like planting, harvesting, and packing can lower labor costs and increase efficiency, enhancing the economic viability of vertical farms (Li, 2023).
4. **Data Analytics**: Big data analytics offer insights into growing conditions and market trends, supporting better planning, yield forecasting, and resource allocation (Wang, 2023).

Case Studies Illustrating Vertical Farming Impact

1. **Urban Vertical Farms**: Companies like AeroFarms and Plenty have pioneered urban vertical farms that provide fresh produce directly to local grocers and restaurants, reducing food miles and environmental impact (Roberts, 2023).
2. **Rural Vertical Farming Initiatives**: In rural settings, vertical farming initiatives can enhance local food security by supplying fresh produce to areas otherwise dependent on processed foods. Programs involving local schools also educate youth about sustainable agriculture practices (Harris, 2023).

Government Support for Vertical Farming Development

Government support is crucial for vertical farming's expansion, addressing barriers and encouraging adoption through:

1. **Research and Development**: Funding for R&D in vertical farming technology can stimulate innovation, reduce costs, and make vertical farming accessible to more communities (O'Connor, 2023).
2. **Urban Agriculture Incentives**: Tax incentives for businesses investing in vertical farms can boost local food production in cities, supporting both urban sustainability and food security (Foster, 2023).
3. **Educational Partnerships**: Collaborations with universities and colleges can enhance training and research, providing critical expertise to support vertical farming's growth and workforce development (Stewart, 2023).
4. **Startup Grants**: Financial assistance for new vertical farming ventures can stimulate economic growth in rural areas and create employment opportunities, driving both social and economic benefits (Adams, 2023).

Digital Growth Leading to Rural Development

Rural financial services offer several essential services, such as savings accounts, credit facilities, insurance, and digital payment systems, each critical for rural households. Governments and financial institutions have introduced microfinance programs and mobile banking, which have widened access to basic financial services. Supporting infrastructure, like rural payment systems and credit guarantee programs, also addresses barriers in rural finance. India has several digital financial schemes specifically aimed at boosting rural financial inclusion. Here are some prominent ones:

1. **Aadhaar-Enabled Payment System (AePS)**: This system allows rural users to conduct banking transactions using their Aadhaar-linked bank accounts through biometric authentication. It simplifies access to basic banking services (withdrawals, deposits, and balance inquiries) at local business correspondents, eliminating the need to visit distant bank branches.
2. **Pradhan Mantri Jan Dhan Yojana (PMJDY)**: A large-scale financial inclusion scheme, PMJDY provides no-frills bank accounts to rural households, often linked to mobile and Aadhaar for easy transactions. The accounts include benefits like overdraft facilities, direct benefit transfers (DBT), and access to micro-insurance. It enables rural families to participate in digital financial services through a basic, accessible framework.
3. **Unified Payments Interface (UPI) and BHIM App**: The UPI platform, accessible through the BHIM (Bharat Interface for Money) app, is widely used in rural areas for secure, instant fund transfers. Many rural banks and cooperatives now support UPI, which enables even small farmers and shopkeepers to receive payments digitally, reducing dependency on cash.
4. **Direct Benefit Transfer (DBT)**: This scheme leverages digital payment channels and Aadhaar to deposit government subsidies and benefits directly into rural beneficiaries' bank accounts. DBT has streamlined payments for schemes like the Pradhan Mantri Kisan Samman Nidhi (PM-KISAN), ensuring timely and transparent distribution of funds to small farmers.
5. **Digital Financial Literacy Campaigns**: To encourage digital financial inclusion in rural areas, the Indian government and NGOs have initiated awareness campaigns like DigiDhan Abhiyaan, which educate rural users on mobile banking, UPI payments, and internet banking. These programs help bridge the knowledge gap, encouraging more people in rural India to adopt digital financial tools confidently.

6. **Kisan Credit Card (KCC) with Digital Access**: The KCC scheme, traditionally providing short-term credit to farmers, has been upgraded to integrate digital access. Rural farmers can now use KCC accounts for digital transactions, allowing them quick access to credit and reducing the burden of paperwork.
7. **Common Service Centers (CSCs)**: CSCs, established under the Digital India initiative, serve as digital service hubs in rural areas. These centers provide access to digital financial services, like bank account openings, Aadhaar services, and financial literacy sessions, allowing rural residents to access financial services and digital tools in their local communities.

Each of these schemes contributes to creating a more inclusive financial ecosystem in India's rural areas, helping bridge the urban-rural divide and offering rural populations the tools needed for secure financial management.

Transformation of Rural India through Technology

The advent of the internet and the smartphones have brought about unprecedented connectivity in rural India. Rural India accounts for nearly 50% of India's internet areas.. This digital penetration has set the stage for a transformative shift in vocational training and skill development.

There are various online platforms which aims at bringing education across rural India at a large scale.

National Skill Development Corporation (NSDC)

The NSDC has played a central role in advancing skill development across India by partnering with various e-learning platforms to broaden its outreach. Through its collaboration with platforms like the Skill India Portal, the NSDC offers a diverse selection of courses specifically tailored to rural populations, ranging from basic digital literacy to advanced vocational training in areas such as agriculture, carpentry, and healthcare.

Unacademy

Initially established as a platform for competitive exam preparation, Unacademy has diversified its offerings to include vocational training. With a broad network of educators and extensive video content, Unacademy provides accessible learning resources for rural youth. Its interactive model, featuring live classes and real-time doubt-clearing sessions, has made it a popular choice among learners in rural regions.

Gram Tarang

As a social enterprise focused on skilling rural youth, Gram Tarang offers online courses in trades such as welding, electrical work, and hospitality. Emphasizing hands-on skills and direct employment opportunities, Gram Tarang has achieved high job placement rates for its graduates, demonstrating a strong alignment with rural workforce needs.

Udemy and Coursera

Global e-learning platforms like Udemy and Coursera have gained significant traction in the Indian market, offering a broad array of vocational courses—from digital marketing to graphic design. These platforms appeal to rural learners due to their flexible, self-paced format and affordable course options, supporting skill development across diverse fields.

Google's Internet Saathi

The Internet Saathi program, a joint initiative by Google and Tata Trusts, empowers rural women through digital literacy and vocational skill development. Using a "train-the-trainer" model, the program trains women, referred to as Saathis, who then train others in their communities. This approach has fostered a multiplier effect, significantly advancing digital and vocational skills among rural women across numerous communities.

5. CASE STUDY: THE STRUGGLES OF FOUR FAMILIES IN MULABAGILU TOWN OF KOLAR DISTRICT

Mulbagal or Mulabagilu is a town and taluk headquarters of Mulbagal taluk in the Kolar district in the state of Karnataka, India.The case study conducted was to understand the factors which led the family members migrate to cities and endure hardships due to their limited educational background. This case study examines the lives of four families in the town, each consisting of four adults—two females and two males—aged between 30 and 45 years. This study focuses on the younger working members of these families, who, despite their efforts, are trapped in a cycle of limited education and economic hardship.

- **Educational Background**

All members of these families have an education level that does not exceed the 10th standard. Several factors contributed to this educational stagnation:

- **Poverty:** Financial constraints were a significant barrier that prevented these individuals from pursuing further education. The economic instability of their households forced them to prioritize earning over learning, limiting their educational attainment.
- **Land Mortgage:** The families' lands were mortgaged, which further added to their financial struggles. Studying beyond 10 th standard was considered unconceivable.
- **Limited Educational Aspirations**: Although government schools are available in the taluk, these individuals were inclined only to study until the 10th standard. This was primarily because they felt a strong obligation to support their elders and children, thereby curtailing any further educational ambitions.
- **Economic Challenges and Migration**

Due to the compounded effects of limited education and financial stress, the families found it increasingly difficult to manage their expenses. The meagre income generated from local employment was insufficient to meet the basic needs of their households. Consequently, the younger male members of these families were compelled to migrate to larger cities like Bangalore in search of better employment opportunities.

This case study highlights the intertwined issues of poverty, limited education, and economic hardship faced by rural families in Mulabagilu. The inability to pursue education beyond the 10th standard has left the younger generation with limited job opportunities, forcing them into a cycle of low-paying jobs and migration.

6. EXISTING SCHEMES TOWARDS DIGITAL RURAL INDIA

I. **e-Choupal**: e-Choupal is an initiative by the Agri-Business Division of ITC limited. In 2000 through this initiative, 6,100 kiosks were established across 10 states in India, aiming to equip farmers with agricultural information and create a technology-driven supply chain system. The objective of e-Choupal is to empower the farmers with latest information of the market. Through this scheme the farmers are encouraged to adopt the eco –friendly agrarian practices thereby ensuring long term sustainable agriculture.e-choupal also helps in building community development through improving the rural infrastructure which includes education and health care facilities.

II. **Kisan Call Centres (KCC):** The biggest challenge faced by Indian farmers is that they are not educated enough to understand latest technological farming developments. Farmers frequently require assistance in deciding cropping decisions, selection of inputs,right kind of practices .Dependence on middle men leads to exploitation .Most of the farmers suicide is because they are unaware of the policies, schemes available for them.Thus Kisan Call centres has been developed as one such system where the modern techniques, information are made available to the farmers. Kisan Call Centres were established to strengthen agriculture by providing farmers with crucial information on market trends, demand, and weather conditions. The KCC initiative enables farmers to get answers to their queries via phone, reducing their reliance on middlemen. This independence helps farmers become more productive and confident, ultimately boosting farm productivity. Database at the national level records the calls information's, observed and decisions are taken for the benefit of the farmers. This is called the Kisan Knowledge Management System (KKMS). Most of the calls are related to information's sought for weather conditions, market information and government schemes. KCC has helped in bridging the gap between farmers and the authorities and to a large extent in eliminating the middlemen.

III. **Pradhan Mantri Gramin Digital Saksharta Abhiyaan (PMGDISHA)**: Initiated in 2015, this scheme aims to provide digital literacy training to marginalized rural populations. Through PMGDISHA, rural youth are trained in internet usage, digital payments, and the operation of tablets. 20 hours of training is conducted which consist of 5 modules covering introduction to digital services, how to operate digital devices introduction to internet, communication module which covers aspect related to internet, and application of internet. Rural youth today need to be empowered with knowledge of the latest technologies, enabling them to secure non-farm jobs and reduce the reliance on an already overburdened agricultural sector. Through these schemes, India can address the issue of disguised unemployment and curb migration to urban areas.This initiative empowers rural youth and helps bridge the digital divide between rural and urban areas.

IV. **E-Shakti**: Social transformation is closely linked to economic empowerment. Financial inclusion, which ensures access to banking services for all segments of society, is a key driver of economic independence. The E-Shakti initiative, launched in 2015, focuses on the digitalization of all Self-Help Groups (SHGs) in India as part of a broader effort to promote financial inclusion.

7. MODEL TO LEVERAGE RURAL DEVELOPMENT

A simple model can help us navigate through the development of rural areas incorporating technology as the foundation to bring development and growth.

The model to Leverage Rural Development includes the following process:

a) To introduce digitalisation by public-private partnership
b) To develop E-learning platforms at the elementary level of education
c) To provide technological training by public-private partnerships and NGOs.
d) To ensure community engagement by village panchayats

7a. Solutions to Break the Poverty Trap in Rural Areas

The challenges faced by the families in Mulabagilu highlight the urgent need for comprehensive solutions to break the cycle of poverty in rural areas. The following strategies can provide a pathway to uplift rural communities and offer them opportunities for a better future.

- **Digitalization through Public-Private Partnerships (PPP)**

Digitalization can play a transformative role in bridging the gap between rural and urban areas. Public-Private Partnerships (PPPs) can drive the development of digital infrastructure in rural regions, ensuring access to the internet and digital tools. By enabling online access to information, government services, and educational resources, digitalization can empower rural communities to participate in the broader economy and access new opportunities.

- **e-Learning Platforms at the Elementary Level**

Introducing e-learning platforms at the elementary level of education can significantly enhance the quality of education in rural areas. These platforms can provide interactive and engaging content, helping students grasp fundamental concepts more effectively. By making education more accessible and appealing, e-learning can encourage students to pursue education beyond the 10th standard, breaking the cycle of limited educational attainment.

- **Technological Training by PPPs and NGOs**

To equip rural populations with the skills needed for the modern workforce,technological training initiatives should be implemented by PPPs and non-governmental organizations (NGOs). These training programs can focus on areas such as digital literacy, vocational skills, and entrepreneurship, providing rural residents with the tools they need to secure better-paying jobs and start their own businesses. This, in turn, can reduce the need for migration and help stabilize local economies.

- **Community Engagement by Village Panchayats**

Village leaders play a crucial role in fostering community engagement and driving change at the grassroots level. By actively involving local leaders in development initiatives, communities can be encouraged to participate in programs aimed at improving education, healthcare, and economic opportunities. Community-driven efforts can create a sense of ownership and ensure that solutions are tailored to the specific needs of each village, leading to more sustainable outcomes.

Technological Hurdles in Rural India

In a country like India, where illiteracy remains a significant issue, introducing technology in rural areas presents an even greater challenge. Many rural regions lack the necessary infrastructure, with some villages being so remote that even access to electricity remains a distant hope. Public participation, especially among older generations, is limited, as traditional superstitions and practices hinder the adoption of new technologies.

Additionally, the shortage of quality schools and training institutes further impedes effective skill development. Gender disparity also plays a role, with many women either hesitant or restricted to working as agricultural laborers. This leads to a narrow and imbalanced development in rural areas, worsening the technological gap.

The challenges can be broadly categorized into the following:

Figure 5. Challenges for rural development

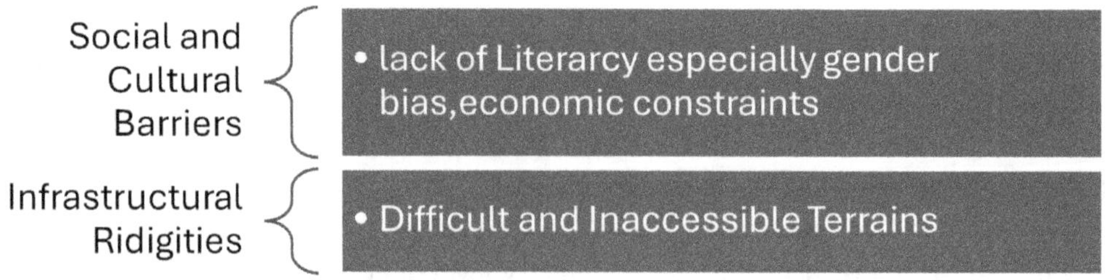

Source: Author's Description

1 Gender Bias leads to limited Access to Quality Education

In India the social and cultural bias,prejuidices are still prevalent in many parts of India.This especially gets into the way of Women and girl child education.Poverty is also the factor which makes education inaccessible to many houses. Lack of formal education leads to low digital literacy, making it difficult for people to understand or adopt new technologies.

2. Economic Constraints

Most of the Indian rural houses are trapped in vicious circle of poverty. The households in grappling with their day to day struggles find it difficult to comprehend any other source of learning.

3 Cultural Resistance and Social Barriers

Digitalization is a sign of progressive society. However regions entrapped in cultural and social biases, find it difficult to embrace any new technology or even change in the existing practices.

4 Lack of Private Sector Investment

The Private sector is seen as a backup and support to the government. However, in India due to structural rigidities, social barriers private sector are shy from entering regions where they find it unprofitable and challenging. Public-private partnerships that could help enhance rural technological infrastructure remain underutilized

5 Climate and Geographic Challenges

Different and difficult terrains add to the problem of building infrastructure in many parts of rural India. Natural disasters further exacerbate the problem by frequently damaging infrastructure, requiring continuous repairs and investment.

Addressing the Challenges

1. **Transparency and Accountability in Government Schemes**

There is no dearth of programs and policies when it comes to rural development. However, the biggest issue is the implementation of such programs. It is the corruption at every levels which deeply affects the effectiveness of such programs. Hence administrative discipline is the first and foremost solution to efficient and effective policies especially designed for the benefit of the rural areas.This will help in bridging the digital gap.

3. **Digital Literacy Programs through training of the rural youth**

We need widespread training centers that will help in empowering the rural youth technologically.

3 **Gender-Inclusive Technology**:

Greater emphasis must be given to women training centers so that it brings about social transformation. To evolve as technologically strong, it is equally important to break the social and cultural barriers. Promoting digital entrepreneurship among rural women can foster broader economic empowerment.

4 Sustainable Infrastructures is what which will give strong foundation to the development process. Government subsidies, private investment, and community participation can help achieve this goal.

8. CONCLUSION

To achieve inclusive and sustainable development, India must focus on ensuring balanced regional economic growth, effectively bridging the gap between rural and urban sectors. While traditional methods like schemes, policies, and laws have been instrumental, their impact can be significantly enhanced through technological advancements that accelerate progress while ensuring transparency and accountability. The persistent divide between rural and urban areas can result in lopsided development, leading to social challenges such as migration. Therefore, addressing rural poverty requires a multifaceted approach that integrates digitalization, education, skill development, and community engagement. By leveraging Public-Private Partnerships, e-learning platforms, technological training, and the leadership of village communities, India can break the cycle of poverty and work towards a more equitable future for its rural population.

REFERENCES

Adams, J. (2023). Grants for startups in vertical farming to boost rural economic growth. *Agricultural Economics Journal, 12*(4), 214–222.

Alston, P. (1979). The United Nations' Specialized Agencies and Implementation of the International Covenant on Economic, Social and Cultural Rights. *Columbia Journal of Transnational Law, 18*, 79.

Baker, L. (2023). The financial burden of initial investments in direct marketing and vertical farming. *Journal of Agribusiness, 15*(2), 57–65.

Barrutia, J. M., & Echebarria, C. (2021). Effect of the COVID-19 pandemic on public Managers' attitudes toward digital transformation. *Technology in Society, 67*, 101776. DOI: 10.1016/j.techsoc.2021.101776 PMID: 34642513

Barrutia, J. M., & Echebarria, C. (2021). Effect of the COVID-19 pandemic on public managers' attitudes toward digital transformation. *Technology in Society, 67*, 101776. DOI: 10.1016/j.techsoc.2021.101776 PMID: 34642513

Brown, A. (2023). The role of social media marketing in direct-to-consumer agriculture. *Digital Marketing in Agriculture, 8*(1), 33–40.

Chen, P. (2023). The impact of LED lighting on plant growth in vertical farming. *Journal of Sustainable Agriculture, 14*(3), 201–209.

Clark, E. (2023). Technological literacy and rural farming communities: Bridging the gap. *Journal of Rural Studies, 11*(5), 72–80.

Davis, M. (2023). Enhancing farmer-consumer interactions through technology in farmers' markets. *Agricultural Marketing Review, 7*(2), 114–120.

Evans, R. (2023). Market competition in direct-to-consumer agricultural sales. *Journal of Market Economics, 9*(4), 311–319.

Fernandes, D., Lynch Jr, J. G., & Netemeyer, R. G. (2014). Financial literacy, financial education, and downstream financial behaviors. *Management Science, 60*(8), 1861-1883.3. Singh, Neena (2007) "Bridging the Digital Divide in India: Challenges and Opportunities", World Libraries, Vol. 17, No. 1.

Foster, G. (2023). Incentives for urban agriculture: Promoting vertical farming in metropolitan areas. *Urban Development Journal, 17*(3), 196–204.

Garcia, T. (2023). Infrastructure development for rural market access. *Journal of Development Economics, 10*(4), 219–227.

Gautam, R. S., Rastogi, S., Rawal, A., Bhimavarapu, V. M., Kanoujiya, J., & Rastogi, S. (2022). Financial Technology and Its Impact on Digital Literacy in India: Using Poverty as a Moderating Variable. *Journal of Risk and Financial Management, 15*(7), 311. DOI: 10.3390/jrfm15070311

Harris, N. (2023). Implementing rural vertical farming initiatives to promote fresh food access. *Journal of Agricultural Innovation, 16*(4), 103–111.

Johnson, R. (2023). Blockchain technology and food supply chain transparency. *Agricultural Technology Review*, *6*(2), 188–195.

Khan, S. (2023). Pesticide reduction in vertical farming: A controlled environment approach. *Environmental Agriculture Studies*, *14*(3), 298–305.

Kumar, M., Patel, J., & Nguyen, S. (2021). E-commerce platforms in agriculture: Expanding the reach of farmers. *Digital Agriculture Journal*, *12*(1), 45–53.

Kumar, S. K. A., Ihita, G. V., Chaudhari, S., & Arumugam, P. (2022, January). A survey on rural internet connectivity in India. In 2022 14th International Conference on COMmunication Systems & NETworkS (COMSNETS) (pp. 911-916). IEEE.

Lee, K. (2023). Training farmers in digital marketing and e-commerce tools. *Rural Development Studies*, *13*(4), 79–88.

Li, H. (2023). The role of automation in vertical farming: A case study on labor cost reduction. *Journal of Automated Systems in Agriculture*, *7*(3), 232–239.

Lyons, A. C., & Kass-Hanna, J. (2021). A methodological overview to defining and measuring "digital" financial literacy. *Financial Planning Review*, *4*(2), e1113. DOI: 10.1002/cfp2.1113

Miller, J. (2023). Government subsidies for farmers in direct marketing initiatives. *Public Policy and Agriculture*, *5*(4), 143–149.

Morris, L. (2023). Evaluating energy consumption in vertical farming: Environmental impacts and solutions. *Sustainable Energy Journal*, *19*(2), 113–121.

Nayak, R. (2018). A conceptual study on digitalization of banking-issues and challenges in rural India. International Journal 5. of Management. *IT and Engineering*, *8*(6), 186–191.

Nguyen, T., & Patel, R. (2023). Space-efficient farming solutions: Vertical farming in urban settings. *Journal of Urban Agriculture*, *9*(1), 67–75.

O'Connor, B. (2023). Funding for vertical farming research: Supporting sustainable agriculture. *Governmental Research Support Quarterly*, *11*(3), 91–98.

Patel, S. (2024). Perspective of Digital Transformation in Rural India. [IJSR]. *International Journal of Science and Research (Raipur, India)*, *13*(4), 387–390. https://www.researchgate.net/publication/379756122_Perspective_of_Digital_Transformation_in_Rural_India/citation/download. DOI: 10.21275/SR24402140819

Rani, S., Kumar, P., Kumar, S., Dar, S. A., & Prabhakar, P. (2024). ICT based applications to support rural development in India-A review. *International Journal of Bio-Resource and Stress Management*, *15*(2), 1–9.

Roberts, C. (2023). Urban vertical farms and local grocery supply chains. *Journal of Food Supply Chain Management*, *10*(4), 321–329.

Robinson, G. M. (2024). Building the rural future and alleviating rural poverty. In *Transforming Rural China* (pp. 225–252). Edward Elgar Publishing., DOI: 10.4337/9781803928586.00015

Roldan, C. S., Giraldo, G. A. M., & Santana, E. L. (2023). Sustainable development in rural territories within the last decade: A review of the state of the art. *Heliyon*, *9*(7), e17555. Advance online publication. DOI: 10.1016/j.heliyon.2023.e17555 PMID: 37496929

Sarabu, V. K. (2018). Rural development in India-a Way forward. Hanamkonda, FormerMember of-BoardofStudiesKakatiyaUniversity, Warangal.

Silayo, D., & Bano, W. (2019), https://www.researchgate.net/publication/351371349_Digital_Initiatives_in_Rural_India/citation/download

Singh, A. (2023). Climate control in vertical farming: Achieving year-round production. *Agricultural Science and Technology Review*, *18*(3), 194–200.

Smith, J., & Jones, A. (2022). Mobile applications for direct farm-to-consumer sales. *Journal of Digital Agriculture*, *6*(1), 123–130.

Stewart, P. (2023). University partnerships in agricultural innovation and vertical farming education. *Journal of Educational Partnerships*, *8*(4), 311–318.

Thompson, L. (2023). Zoning policies and the promotion of local food systems. *Journal of Agricultural Policy*, *5*(3), 83–89.

Verma, R., Verma, K., Verma, J., Singh, T., & Bisen, P. (2024). Indian rural development: A review of technology and society. *SN Social Sciences*, *4*(2), 42. Advance online publication. DOI: 10.1007/s43545-024-00831-x

Wang, Y. (2023). The use of big data analytics in optimizing vertical farming. *Journal of Agricultural Data Science*, *9*(2), 211–218.

Williams, D. (2023). Community Supported Agriculture (CSA): A model for sustainable local food systems. *Journal of Sustainable Food Systems*, *14*(2), 147–154.

Zhang, X. (2023). Real-time environmental monitoring in vertical farms through IoT sensors. *Journal of AgriTech Innovation*, *17*(2), 183–192.

Chapter 13
Beyond Connectivity:
The Impact of B5G and 6G on Rural Education Outcomes in India

Nidhi Shukla
https://orcid.org/0009-0004-0359-7233
Presidency Business School, Presidency College, Bengaluru, India

Julie Sunil
https://orcid.org/0000-0003-3001-8001
Presidency Business School, Presidency College, Bengaluru, India

Keerti Mishra
Sushant School of Art and Architecture, Sushant University, Gurugram, India & Amity University, Gurugram, India

ABSTRACT

"Beyond Connectivity: The Impact of B5G and 6G on Rural Education Outcomes in India" explores how advanced communication technologies can enhance rural education. It addresses challenges such as teacher shortages, limited smartphone access, and digital adoption, which contribute to learning gaps. The fear of technology replacing teachers further hinders the use of digital tools in classrooms. The chapter suggests that B5G and 6G can provide high-speed internet to remote areas, improving access to cloud-based educational resources. AI-driven adaptive learning platforms can personalize education, while online training for educators fosters collaboration. Strong policies and community involvement are essential for successful implementation.

INTRODUCTION

The potential for the fast development of wireless communication technologies, especially Beyond 5G (B5G) and 6G, to revolutionize education in rural India is enormous. According to Shimaa et al. (2022), the advent of next-generation networks holds great potential for educational innovation and the closing of the digital divide. These networks will provide ultra-low latency, huge connection, and unparalleled speeds. The Evolution of Educational Connectivity: 2G, 3G, 5G (B5G) and 6G and their

DOI: 10.4018/979-8-3693-7788-8.ch013

effects have been tremendous. The educational landscape has been greatly impacted by the evolution of mobile connectivity from 2G to 5G, which has improved accessibility, engagement, and learning results. Two-Gamma Period: Simple Networking and Text-Based Education- In the early 90s, 2G technology was introduced, which allowed for basic mobile connectivity, mainly for voice and SMS services. This limited connectivity allowed for some educational resources to be accessed by text message, which was particularly advantageous in locations with insufficient infrastructure. For example, Kukulska-Hulme (2009) noted the emergence of mobile learning initiatives that allowed students to get learning materials via SMS, thus filling in the gaps in traditional schooling.

Improved Access and Multimedia Education with 3G and 4G - Faster data transfer and the utilization of multimedia resources were made possible with the launch of 3G in the early 2000s, marking a substantial upgrade. Video lectures and online courses, which are more engaging, were made possible because of this. The emergence of 4G further transformed education by providing high-definition video streaming and real-time collaboration capabilities, making remote learning more effective (Anderson & Rainie, 2012). The rise of online learning environments such as Massive Open Online Courses (MOOCs) has made high-quality education accessible to students all around the globe, regardless of their physical location.

A Revolution in Communication with 5G & 6G- Educational opportunities have never been better, all thanks to the launch of 5G technology in the late 2010s. Virtual and augmented reality (VR/AR) are becoming more incorporated into educational practices thanks to 5G's ultra-low latency and high-speed capabilities, which enable immersive technologies. This paves the way for possibilities in experiential learning settings, where students can participate in formerly impossible hands-on simulations (Zhang et al., 2020). The improved connection also allows for group projects where students from all over the world can work together in real time.

The effect on classroom instruction is growing in magnitude as mobile connectivity develops more. Advancements in mobile technology, from the rudimentary features of 2G to the revolutionary possibilities of 5G & 6G, have been instrumental in enhancing the accessibility, interactivity, and engagement of education on a global scale. The purpose of this research paper is to identify B5G and 6G technologies, investigate what variables are affecting their use in rural schools, look at how these technologies can help close the digital gap, and review some thoughts on how to best incorporate technology into the classroom. The chapter incorporates Conceptual research design, this method allows for a thorough investigation of research topic hypotheses, concepts, and ideas. The study examines the secondary data to gain a full insight and add to knowledge. This requires carefully finding, choosing, evaluating, and synthesizing relevant information from multiple sources, such as: Academic publications, Books, Theoretical frameworks, Government Reports, Industry reports and Online resources.

B5G AND 6G TECHNOLOGIES & ITS SCOPE IN EDUCATION

The wireless communication protocols that will follow 5G are referred to as B5G and 6G, respectively (Shimaa et al., 2022). Data transfer rates of up to 1 Terabyte per second (Tbps) will be possible with these technologies since they operate in higher frequency bands, such as the terahertz (THz) spectrum (Neuronimbus, 2021). Beyond speed, B5G and 6G cover ultra-low latency, huge connection, and the coming together of different technologies including the Internet of Things (IoT), edge computing, and artificial intelligence (AI) (Spiceworks, 2021).

Model for the Adoption of New Technologies: What Drives the Use of 5G and 6G in Remote Rural Educational Institutions. The Technology Acceptance Model (TAM) can be used to examine the implementation of B5G and 6G technologies in schools located in rural areas (Davis, 1989). The fundamental factors that determine a technology's adoption, according to this paradigm, are its perceived usefulness and ease of use. Adoption of these technologies in rural education can be influenced by factors like affordability, the availability of necessary infrastructure, and the perceived benefits of these technologies in improving educational outcomes (Venkatesh & Davis, 2000). B5G and 6G adoption in remote schools can be aided by government policy, financing, and programs that train teachers (Tarhini et al., 2015).

Theory of the Digital Divide: Closing the Gap in Access to Technology and Raising Educational Opportunity- Inequality in the availability and utilization of ICTs, according to the Digital Divide Theory (van Dijk, 2006), might exacerbate existing social and economic inequality. Limited access to high-quality educational resources, online learning platforms, and interactive teaching methods is one way in which the digital divide shows itself in rural education (Warschauer, 2003). Fortunately, there is hope for bridging this gap in access to digital educational resources and online learning platforms because to B5G and 6G technologies, which promise to bring affordable and dependable internet connectivity to schools in remote areas (Neuronimbus, 2021). Technology like this has the potential to help rural areas progress economically and socially by levelling the playing field in terms of education (Spiceworks, 2021).

RURAL EDUCATION IN INDIA & CHALLENGES FACED

Digital Penetration - Access and Usage:Population in rural India is witnessing a decline but the internet usage from 2022 to 2023 recorded an increase of 78 percent with shared device usage for internet higher than in urban areas. Rural India with 442 million active users of internet had surpassed its urban counterpart which was at 378 million users according to the internet and mobile association of India. Internet usage was maximum for access to OTT platforms for audio and video services (86%), internet for communication (76%), social media (70%), and online gaming (53%). Music streaming apps was witnessing a subscription growth rate of 8 per cent (Times of India, 2024)

Figure 1. Rural vs urban population trend

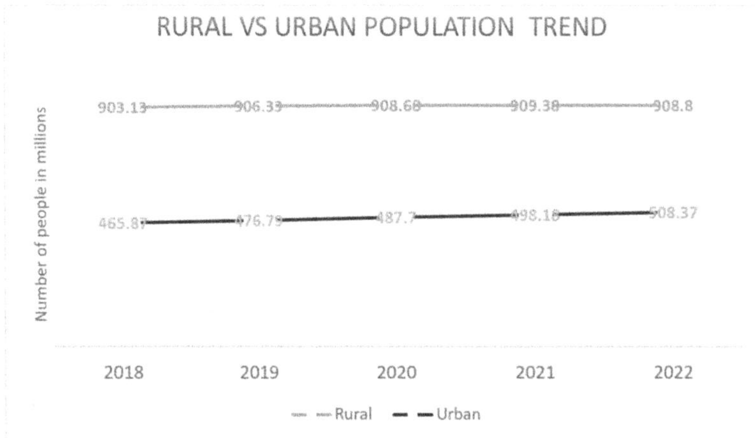

Source: *(Statista.com, 2024)*

Access to smartphone among school children in rural India and the corresponding level of time invested for education: A 2023report released by the State of Elementary Education in Rural India based on the data gathered from 21 states aimed to understand the usage of smart phone for education. The respondents were 6,229 parents and school children in the age group between 6-16 years. The findings from the survey are summarised in table 1.

Table 1. Smartphone access among students

Description	Percentage (%)	Remark
Access to smartphone	49.3	Parents said that the smartphones were used for entertainment such as and playing video games (76.7%), download and watch movies (56.6%) download and listen to music (47.3%) rather than educational activities. Only 34% students used smartphone for downloading study related material and 18% accessed online learning material.
Access to smartphone based on age groups:		The access increased with higher classes
Classes 1 to III	42%	The usage for entertainment potentially impacted their engagement with the devices for education purposes
Classes VIII and Above	58.3%	

(Drishtiias, 2023)

Parents (78%) desired that their children should reach education level of graduation and above. However, the engagement of parents in their children's education was less with only 40% holding conversations on classroom learning.

Table 2. Reasons for dropping out of school differed for boys and girls

Gender	Need to contribute to family earnings	Child's lack of interest in education
Boys	48.7%	71.8%
Girls	36.8%	31.6%

Source: Compiled by the author

Another 2023 report published by ASER (Annual Status of Education Report) revealed that while the number of students in the 14-18 age group dropping out of school had come down but the foundational skills were heavily compromised. A comparison of the 2023 data with 2017 showed that the number of youths who could read class II level text had dropped 3 percentage points to 73.6% but in arithmetic there was an improvement of 4 percentage points (43.3%). Although smartphone access increased with higher classes the access was not in favour of the girls. Boys who could use a smartphone also owned a smartphone was 44% compared to girls which was only 20%. This limited access driven by social structureled to boys outperforming girls in nearly all tasks. Among the sampled youths, nearly 89% had access to smartphones but the weak foundational knowledge was a challenge that needed immediate attention(Times of India, 2024). Students in rural India often contend with electricity and internet connectivity limitations. Device sharing, inadequate infrastructure and untrained teachers leave students at a disadvantage. Students therefore struggle to keep up with curriculum changes. Even promising students are unable to access study material, practice exercises and concept quizzes leaving them marginalised and unable to join higher education and professional fields. Broadband connections such as B5G and 6G and equipping schools with necessary infrastructure such as computers and tablets can ensure access to technology. Government initiatives and projects funded through public and private partnerships will be necessary to ensure no child is left behind because of the digital divide.

Teacher Quality – An enduring challenge: A study was conducted on 1357 teachers from India (80%), UAE (18%), Oman (1%), Saudi Arabia (1%)to assess the Math subject knowledge and pedagogical knowledge of instructors under Teacher's Impact Programme. Teachers struggled with questions that assessed concepts on ratio, proportional, algebraic, logical and estimation reasoning. Only a quarter of the teachers could answer 25 percent questions correctly. The level of clarity in math concepts deteriorated when the teachers were assessed with questions from higher level grades – while 73.3 percent answered questions based on grade 4 curriculum correctly, only 36.7 percent could answer grade 7 topics correctly (The Economic Times, 2024).

What's torpedoing the adoption of technology- the fear of replacing teachers: Teachers and educators (87.85%) expressed fears and worries about the use of AI in classrooms and asked the government to frame policies to regulate and monitor its use for education. The study included 6,313 educators from schoolteachers to professors from schools and colleges across India. Educators who had started to use AI for preparing for classes and evaluations was 61.6 per cent and there was consensus on the power of AI tool in enhancing and personalising learning in interactive and innovative way. The fear however was about diminishing interaction between teachers and students, over dependence on technology, data privacy and accuracy of information given by AI tools. The objection was also to the risk and ethical considerations associated with the use of AI in learning process. Teachers felt the need to monitor AI through government regulation on how the technology is developed and utilised with the larger issues of ethics and benefit of the society being addressed. Such fears were not based on real facts was the view of the global education expert Anthony Salcito who said that training and enabling teachers to teach with technology should be more important than merely handing out the devices to students. According to him, such distributions overlook the importance of the shift in mindset required to untap the potential of the devices for education. Acquisition of device does not automatically ensure that the schools are ready for digital usage. For technology to become integrated into learning process there is a need to address the lack of electricity, charging points, space requirements and other infrastructure constraints. Training and empowering teachers will be the first and most effective step in building the ecosystem for learning.

Pitfalls in providing internet speed and smartphone access: Caveats in using technology for learning was sounded by a UN report which said that heavy reliance on AI could impede acquisition of basic skills such as reading. The absence of reliable evidence on the positive benefits from using technology in learning resulted in the comment that technology could be result in detrimental effects if overused or inappropriate to address the learning need. There was a need for a sound pedagogical framework and that it was not just about distributing devices and ensuring speed. The presence of phone was considered distracting, and many schools had banned use of smartphones in campus and classrooms. Technology was not an answer to the prevailing inadequacies of learning ecosystem. Investment in equipping students with basic reading and mathematical skills could help prevent them from falling victim to crimes in the digital age (The Economic Times, 2023).

Rural Connectivity: Historically, rural connectivity relied on unreliable microwave links, outdated copper cabling, and some fibre installation. However, fibre-optic technology emerged as a game-changer for network infrastructure. Optical Fiber Cables (OFCs) revolutionized connectivity by offering high-speed broadband and low-latency connections, enhancing reliability and driving internet inclusiveness. Internet connection in government schools increased from 5.5% in 2017-18 to 24.2% in 2021-22 (The Times of India, 2023). The BharatNet project aimed to provide OFC-based connectivity to all villages, with extensive connectivity supported by Wi-Fi or telecom networks, leveraging the high-bandwidth fibre backbone established by the project. The OFC network created by BharatNet has improved internet access and high-speed data availability, integrating underserved areas into the economy and advancing the vision of Digital India (Jain., 2023). The BharatNet project connected approximately 194,000 villages across India. It is active in states such as Maharashtra, Gujarat, Chhattisgarh, Andhra Pradesh, Telangana, and Tamil Nadu. In its first phase, the project established a fibre-optic network in regions including the Andaman and Nicobar Islands, Chandigarh, Delhi, Goa, Haryana, Karnataka, Kerala, and others. By laying Optical Fiber Cables (OFCs) in these villages, BharatNet had laid the groundwork for the country's digital. Other government initiatives included Digital India, Make-in-India, Start-up India, and Aadhaar economy - based applications, to enhance the country's human capital. These programs rely on high-speed, reliable network connectivity for successful and accelerated implementation. The introduction of digital facilities is expected to have a significant impact on society, particularly in rural areas.

Internet penetration in Schools -Not Enough: According to data released by the Ministry of Education, only about 34% of schools in India had internet facilities, and more than 50% lacked functional computers. Since 2018-19, there had been progress in the number of schools with internet facilities and functional computers. The percentage of schools with functional computers rose from 33.49% in 2018-19 to 45.8% in 2021-22. Similarly, the percentage of schools with internet access increased from 18.73% in

2018-19 to 33.9% in 2021-22. States were responsible for allocating budgets for school infrastructure, particularly in ICT, alongside the central government's efforts which meant some states did much better in allocation and outcome. Additionally, UDISE data revealed that 13.4% of schools still lack functional electricity connections (The Print, 2022).

However, as noted by Unified District Information System for Education (UDISE), Ministry of Education, there has been a noticeable rise in the number of schools in India that have internet facilities. Their report indicates that: internet access was available in 18.73% of schools in 2018–19, in 2019 and 2020, the ratio rose to 22.28 percent. In the year 2020–21, the number increased to 24.52%. In the academic year 2021–2022, internet access in schools increased significantly, reaching 33.91 percent. This means that out of 14,89,115 schools in India in 2021-22, 5,04,989 had access to the internet. Although this development is encouraging, it does show that there are still a lot of schools without internet access (66.09%, or 9,84,126 schools). It is encouraging to see more and more schools implementing internet access. It is still necessary, though, to close the digital gap and guarantee that all schools, particularly those in outlying regions, can connect to the web. Students will be able to take use of online learning materials and achieve better educational results as a result of this.

There is a need for comprehensive digital literacy programmes that can empower and instil in teachers and students the requisite skills and knowledge to navigate online learning platforms and access digital resources effectively. Content developed to help students from rural areas to better understand the concept is necessary. Apart from introducing relatable context, language of learning should be local – customised by region. This effort in making content accessible, relatable and relevant will help to foster engagement and ensure achievement of learning outcomes. Community driven learning centres can be another addition to help group learning and problem-solving.

THE PROMISE OF B5G AND 6G TECHNOLOGIES

Education in India's rural areas stand to benefit greatly from the upcoming Beyond 5G (B5G) and 6G networks. Improving rural educational systems is no easy feat, but these technological advancements hold great potential for increased connectivity, individualized instruction, professional development for educators, and a solid digital foundation. The significance of B5G and 6G in resolving these issues is investigated in this research material.

Better Networking

Internet Connection in Outlying Regions- In rural India, unreliable internet access is a major obstacle to high-quality education. The goal of 5G and 6G networks is to close this digital gap by bringing high-speed internet to underserved regions. Asma India (2023) reports that the Digital India project, launched by the Indian government, aims to achieve this goal by increasing internet connectivity in unserved areas. These projects aim to guarantee that rural students have the same educational opportunities as their urban counterparts by making use of modern communication technologies. Innovative solutions, such as small cell networks and low Earth orbit (LEO) satellites, will improve connectivity through B5G and 6G technologies, according to recent studies (Banafaa et al., 2023). Broadband connection in places without conventional infrastructure cannot be achieved without these technologies.

Web-Enabled Tools for Engaging Education - Using IoT devices in the classroom can make lessons more dynamic and interesting for students. The usage of smart devices in classrooms is made possible by B5G and 6G technologies, which support a wide range of IoT applications. As an example, according to ResearchGate (2020), tools that are enabled by the Internet of Things can offer students real-time feedback and interactive learning modules, which can make education more interesting. This kind of engagement is especially important in more remote areas, where pupils may be less receptive to more conventional forms of instruction.

Resources for Education Hosted in the Cloud - When it comes to expanding people's ability to use instructional materials, cloud computing is crucial. With the advent of 5G and 6G, schools in remote areas will be able to use online resources like the cloud to access a wealth of digital resources including movies, e-books, and interactive simulations. Having access to a wide variety of learning resources, which are not always available in rural areas, can greatly improve the quality of education (CRY, 2023). Also, resources stored in the cloud can be updated often, so students always have access to the most recent knowledge.

Customized Education Systems

Adaptive Learning - Adaptive learning platforms that can personalize educational information to match students' unique needs can be developed with the use of B5G and 6G technology. In order to provide students with individualized lessons, these platforms evaluate their learning preferences and progress using data analytics and machine learning algorithms (ResearchGate, 2020). Students in remote locations may have varied levels of preparedness and learning paces, so it's very helpful to provide them with personalized lessons.

Intelligent Tutoring via AI - Through the use of AI-powered tailored teaching systems, which provide students instantaneous feedback and support, the learning process can be greatly improved. According to ResearchGate (2020), AI-powered teaching can pinpoint students' weak spots and provide them with specific activities to strengthen their understanding. Because of the scarcity of trained educators in rural areas, this technology has the potential to have an outsized effect there.

Engaging Students in Learning through Experiential Learning- Student engagement and motivation can be greatly enhanced by gamification, which is the application of game design aspects to educational settings. The advancement of gamified learning applications, which enhance the educational experience through engagement and enjoyment, is made possible by 5G and 6G technology. These apps can make learning more interesting by adding features like stages, incentives, and points, which is crucial for lowering rural school dropout rates (Asma India, 2023).

Professional Development for Educators

Classes for Career Advancement Offered Online - Teachers can now participate in online professional development thanks to B5G and 6G technologies, which eliminate the need for them to physically attend a session. In order to provide instructors in rural areas with the necessary tools for success, these courses can address a variety of subjects, such as digital literacy and creative teaching practices (CRY, 2023).

Expert Collaboration through Virtual Means - One major benefit of B5G and 6G technology is the ability to work online with educational specialists. Teachers in less-populated regions can network with their metropolitan counterparts to share best practices, learn from one another, and improve student learning (ResearchGate, 2020). By working together, educators can build stronger bonds, which in turn promotes learning from one another and career advancement.

Online Resources for Organizing and Evaluating Course Work - Lesson planning and assessment can be made easier with digital tools enabled by B5G and 6G. Teachers may make better use of these tools' analytics, templates, and resources to design engaging lessons and evaluate their students' progress (Asma India, 2023). Teachers in rural areas typically have more students and less resources, so this efficiency is especially welcome.

Digital Infrastructure

Cyberspace Libraries and Intelligent Classrooms- To bring rural schools up to date, it is necessary to set up "smart classrooms" that include all the latest equipment. With the use of 5G and 6G, "smart classrooms" that include digital libraries, interactive whiteboards, and multimedia tools can be created (ResearchGate, 2020). There is a great deal of material available to pupils in these settings, and they can also encourage collaborative learning.

Virtual Classrooms - To ensure that children in remote areas have access to a high-quality education, e-learning platforms are essential. By facilitating the smooth streaming of instructional material and interactive elements that encourage participation, B5G and 6G technologies have the potential to elevate these platforms (CRY, 2023). Also, using these kinds of platforms, students may learn at their own speed and come back to previously covered topics whenever they need to.

Computing in the Cloud and Data Centres - Strong data infrastructure is essential for every online learning program. According to ResearchGate (2020), educational resources can benefit from local data centres that are supported by B5G and 6G technologies. These centres allow for efficient data processing and storage. In order to guarantee that both students and educators have consistent access to necessary digital materials, this infrastructure is crucial.

BEST PRACTICES

The educational gap between India's urban and rural areas might be significantly reduced with the use of B5G and 6G technologies, which have already shown great promise in rural education programs. These projects bring attention to successful strategies, important takeaways, and models that other areas can use to tackle comparable problems.

The B5G/6G Network's Effective Application to Rural Education Projects: Learn, Out of the Box, a joint initiative of the Vodafone Foundation and the Pratham Education Foundation, is a prime example. Teachers in low-income schools receive a WebBox as part of this program, which gives them access to digital educational materials like movies and interactive lessons. Teachers have been able to increase student engagement and boost learning outcomes for pupils from economically disadvantaged families through the use of this technology (Asma India, 2023). In addition, eVidyaloka has been a great success; it uses internet platforms to link students in remote areas with volunteer teachers. Students are able to engage with teachers who may be hundreds of kilometres away because to this initiative's use of virtual classrooms, which has increased their access to high-quality education. According to Asma India (2023), the program has demonstrated that using technology may efficiently solve the problem of a lack of trained instructors in rural areas.

***Challenges Overcame and Lessons Learned* -** Although some projects have made great strides, they have also encountered obstacles that teach us something for when we implement them in the future. The absence of necessary infrastructure, such as consistent power and internet access, is a prevalent problem. The usefulness of digital education tools is hindered in many rural locations due to the continued struggle with basic utilities. As an example, ResearchGate (2020) notes that the DakNet project, which utilizes long-range WiFi to link rural villages, has shown how crucial it is to fix infrastructure problems before introducing technological fixes. Another factor that can slow down development is people's reluctance to embrace new technology. Introducing digital learning approaches might be problematic in countries where traditional educational practices are well ingrained. By including local communities and stakeholders in the planning and implementation phases, these barriers can be overcome (CRY, 2023).

***Regional Models That Can Be Replicated* -** Other regions dealing with comparable educational issues can learn from India's successful models. Some important parts of these models are- Involvement with the Society: Solutions are more likely to be responsive to local needs and cultural settings when members of those communities are involved in making decisions. The significance of local involvement in the creation of educational interventions is shown by programs such as eVidyaloka. Forming Alliances: Organizations in the business sector, non-governmental organizations (NGOs), and the government can work together to pool resources and share knowledge more effectively. Collaborations like the one between Vodafone and Pratham Education Foundation prove that they can produce novel outcomes. The effectiveness of digital education programs depends on the development of key infrastructure, which includes stable power and internet connection. The necessity for focused investments in connectivity prior to using educational technology is shown by projects such as DakNet. To make sure that new technologies are used effectively, it is vital to provide teachers with continual training and assistance. Programs that invest in teachers' professional growth can pave the way for more effective use of technology in the classroom. Evaluating the efficacy of educational technology and pinpointing its weak spots requires the establishment of rigorous monitoring and assessment systems. Program efficacy and scalability can be improved by continuous feedback loops.

POLICY AND REGULATORY FRAMEWORK

The Challenge of How to Spend: Government's Unspent Budget on Education- Government's education policy had called for an increase in spending i.e. 6% of GDP in 2024 budget but it was struggling to spend the budget allocated in the previous year (20230 where nearly INR 7000 crores was the unspent balance. The education budget is broadly split into budget for school education and higher education. Even with higher allocation through various central schemes the spending was not enough because it failed to ensure employable skills among youths leading to agitation and confrontation among them for few available job opportunities. The spending was more on school infrastructure and less on hiring qualified teachers and investing on the learning process. Infrastructure allowed for an immediate assessment and visibility of investment in education during, the tenure of the government but it did not improve the learning outcome. It may be useful to spend more on technology acquisition and training teachers to facilitate the use of Artificial Intelligence (AI) and Generative AI in learning in classrooms.

Internet Services - Need for improving quality of internet services for applications such as 5G video streaming, Artificial Intelligence, Internet of Things and cloud computing will require enhanced capabilities. Three projects are likely to address this requirement**(Table 3):**

Table 3. Projects for improving quality of internet services

Project	Capacity (tbps)	Length (km)	Investor Include	Connecting
2 Africa Pearls	180	45000	Airtel, Meta, Saudi Telecom	Africa, Europe, Middle East, Asia
India-Asia Express	200	16,000	Jio, China Mobile	Mumbai, Singapore, Malaysia, Thailand, and Sri Lanka
India-Europe Express	200	9,775	Jio, China Mobile	Mumbai to Persian Gulf and Europe
Total existing installed capacity 138.55 tbps				

(The Economic Times, 2024)

Suggestions for Efficient Policy Execution- Governments should put a premium on investing in digital infrastructure, such as dependable power and internet in remote places. According to Shimaa et al. (2022), this groundwork is essential for facilitating the efficient utilization of cutting-edge technology. Curriculum Development: Schools should revise their lesson plans to incorporate lessons on digital citizenship and other forms of technological proficiency. This will guarantee that pupils are ready to utilize B5G and 6G tools effectively and get them ready for a tech-driven future (Asma India, 2023). Corporate-Government Alliances: Sharing resources and fostering innovation can be achieved through partnerships between public agencies, non-governmental organizations (NGOs), and businesses. Collaborations like these can be essential in creating personalized answers to the problems plaguing rural schools (CRY, 2023). Educator Professional Development and Resources: Teachers must engage in continuous professional development. Nguyen-Son et al. (2022) argue that training programs should prioritize the incorporation of technology into teaching methods and the utilization of data analytics to facilitate tailored instruction.

LONG TERM IMPLICATIONS AND DISCUSSION

Education, especially in India's rural regions, stands to benefit greatly from the forthcoming Beyond 5G (B5G) and 6G networks. These innovations have the potential to strengthen educational ecosystems, make learning more personalized, and increase connection. This part delves into the latest developments in B5G and 6G technologies, how they could be used in rural education, some policy suggestions on how to put them into practice, and where the research has to go from here.

The Latest Developments in 5G and 6G Networks - The educational use of B5G and 6G technologies will be influenced by a number of important factors. To begin, the ability to facilitate real-time interactions and immersive learning experiences relies on ultra-high speed and low latency (Nguyen-Son et al., 2022). As an example, 6G is anticipated to reach 1 terabyte per second in data transmission speeds, paving the way for virtual reality applications in the classroom and high-definition video streaming (Neuronimbus, 2022). Furthermore, communication networks that incorporate AI and ML will improve user experiences and optimize resource management. According to Murata (2023), AI will pave the way for personalized education by enabling adaptive learning platforms that cater to each student's unique demands. Terahertz communication and non-terrestrial networks will also increase coverage in rural regions, which means more people will have access to educational materials (Shimaa et al., 2022).

Use of NLP Technology - A robust set of resources is available to school language departments through Natural Language Processing (NLP). A branch of AI, known as "natural language processing," gives computers the ability to read, comprehend, and even create their own language. Language departments in schools can greatly improve instruction by using various natural language processing methods. Teachers can automate processes, tailor lessons to each student's needs, and extract useful information from student work by using natural language processing techniques such as sentiment analysis, named entity identification, and text categorization. Text summary aids students in efficiently grasping important ideas, while machine translation assists with language learning and intercultural comprehension. Reading and writing are both improved with part-of-speech labelling. With the use of NLP, language departments can boost student engagement, language competence, and the efficiency of their lessons.

Possible Uses in Rural Education - There are a number of ways that rural schools can make use of B5G and 6G technologies: Engaging Educational Spaces: When schools in rural areas have access to fast internet, they can use VR and AR technologies to build interactive lessons. One way that virtual reality (VR) can improve student engagement and knowledge is by allowing them to visit historical locations or scientific phenomena (Appinventiv, 2023). Students in remote places can have access to high-quality education through remote teaching and tutoring made possible by B5G and 6G. Asma India (2023) reports that platforms such as eVidyaloka have proven that remote teaching is helpful in closing educational achievement gaps. Data-Driven Learning: Educators can adjust their methods of instruction based on insights provided by AI-driven analytics into student performance. Nguyen-Son et al. (2022) found that using data to identify pupils at risk allowed for more targeted treatments. Cloud computing has the potential to house enormous databases of instructional materials, guaranteeing that students from all over the world have access to current resources through digital libraries and resources (CRY, 2023).

Role of AI, Internet Connectivity and Adapted Education System in Democratising Education - Spending in education needs of rural India must also include the possibility of increasing speed with access to AI tools in learning. Once primary education builds solid understanding of language, math and science skills, there can be adoption of AI to help with answering queries and addressing doubts at a personal level for individual students. The only perquisite to its use would be access to uninterrupted electricity, high speed internet and reliable access to device. The AI tutors will then be deployed to answer complex questions and resolve queries in a conversational, friendly and non-threatening style. Phone computing capabilities, connectivity and cloud infrastructure had made it possible for 9 out of 10 internet users to use AI. The AI user base of India in 2024 was 724 million and was expected to grow at the rate of 6% year on year (The Economic Times, 2024). The power of AI is that it helps to increase intelligence and is therefore more powerful than internet and electricity. It can overcome the drudgery of mundane tasks and offers scope for creativity and problem solving. Is education under threat because AI takes away the need to learn by providing answers to all the questions.

The rise and ubiquity of AI will level the learning field and offer access to knowledge and expertise for everyone – including teachers. Immaterial of age, gender or geographic location – a person will be able to enhance their learning experience through insights, research inputs and theoretical frameworks. AI would allow for adaptive learning where every student could learn at their own pace, style and language. This ability of AI to adapt the content at the individual level is known to improve interest and engagement with the content. The access is another reason why AI tools such as ChatGPT can be used to return to the learning objectives in a flexible and timely manner. Students who work during the day or end up helping the family during daytime can still resume their learning using AI tutor. The common areas where ChatGPT was used the most was in essay and paper composition, guiding learners through homework assignments/projects and test preparation, aiding in language acquisition, fostering career progression, honing communication strategies, improving digital literacy, developing management competencies, facilitating benchmarking and brainstorming sessions, supporting decision-making processes, providing microlearning experiences, and simplifying complex concepts in popular science education (Mogavi, 2024).

Using AI for Personalised Learning - Speed, efficiency and outcome would be the principle focus in improving performance of AI learning tools. The renowned founder of Khan Academy, Salaman Khan released a video to demonstrate how Chat GPT was effective in teaching his son mathematics. The question that Chat GPT was asked to teach involved providing solution to a math problem through a step-by-step process. The AI tool engaged the student by asking questions and driving towards the solution without supplying the answer. The student successfully grasped the concept and arrived at the solution with minimal intervention. This ability to help students' clear concepts and apply them with regular and useful feedback any time during their learning process is what makes it a useful tool.

Technology, AI and Human Skill - A high-speed internet connection can offer real-time access to an extensive range of information, promoting new learning methods, boosting student engagement, and facilitating the integration of technologies like virtual classrooms and e-learning into education. However, it also has potential downsides: teachers may struggle to blend traditional teaching with ICT, and students might use broadband for distracting activities such as gaming, chatting online, browsing social media, or streaming videos. Harnessing broadband in education has the potential to elevate learning significantly, but achieving this requires not just high download speeds but also robust upload speeds.

While download speeds have traditionally been the focus, especially for streaming music, movies, and video games, interactive digital education demands equally strong upload speeds. Upload speed is crucial for transmitting data from your computer to the internet, directly impacting the quality of video conferencing. Insufficient upload speeds can lead to unstable, lagging, or freezing video, and broken audio. Even with high download speeds, poor upload performance can hinder clear communication and engagement during virtual meetings if the other participant has slow upload speeds.

Currently, the government considers an internet service "broadband" if it offers a minimum upload speed of just three megabits per second (Mbps), which is barely sufficient for a Skype or Zoom video conference. Technology enhances the mechanics of learning, but universal access to high-speed broadband will make education even more dynamic. Today's students can already take virtual field trips to places like the Louvre or the Pyramids of Giza. With true high-speed broadband offering gigabit speeds for both upload and download, the experience will evolve from mere observation to real interaction. For instance, high-resolution, real-time video could enable students to virtually immerse themselves in their surroundings, such as a 180-degree wraparound screen setup, creating a feeling of presence. This technology will make experiences like visiting the Louvre accessible to those who cannot physically travel, providing educational benefits at a lower cost. Broadband will also shift the role of teachers from delivering static content to guiding students through a dynamic, ever-evolving information landscape. This shift will require students to learn how to discern valuable information from biased sources.

Moreover, high-speed internet will encourage entrepreneurs to develop new digital education tools and platforms, driving innovation and expanding learning beyond traditional textbooks. Schools will need to focus on securing funding and forming partnerships with these companies to benefit from these advancements. Most internet service providers offer upload speeds that are only a fraction—often a tenth or less—of their download speeds. For example, cable, DSL, and satellite providers may offer upload speeds up to 35 Mbps, but this is still significantly lower compared to their download speeds. Fiber technology, however, provides nearly symmetrical upload and download speeds, making it ideal for more immersive and interactive digital experiences. With 1.2 billion as user base, India is going to have the highest per capita consumption per month. India will emerge as the key driver in the global data usage and reach. Amid this need to grow the need for data sovereignty and privacy is the need for education to focus on the right outcomes.

Universal broadband could revolutionize education by allowing schools to schedule virtual speakers from various professions, enabling interactive and engaging experiences without physical presence. This would provide students with direct insights into different careers and industries, enhancing career exploration and development. For example, a chemical engineer could virtually demonstrate lab work, or a food production expert could give a virtual tour of processing facilities. This would be particularly beneficial in addressing worker shortages by introducing students to potential career paths early on, regardless of their location. Government funding is increasing access to broadband, but to fully leverage these opportunities, schools and service providers need to actively seek these funds. Companies should explore partnerships with educational institutions to integrate their industries into learning. By transforming education with high-speed broadband, society can build a stronger, more informed future.

Build thinking skills in the AI world: an urgent call to education- The new skill that education must instil among students of the digital age is the ability to think critically. These individuals should excel at navigating complex information environments, generate more accurate results and responses from language models, make well-informed decisions, and refine their approach to achieve desired outcomes more efficiently. They should also be more effective in communicating and presenting their findings. The ability to think critically is closely tied to problem-solving skills. When students are enabled to approach challenges methodically, exploring all potential solutions and their consequences before making decisions they can greatly minimize the risk of missing key factors and boosts the likelihood of identifying effective solutions. It also enhances their ability as a prompt engineer, to persist until a satisfactory and appropriate result is achieved. Building analytical skill aids in crafting precise prompts and assessing the quality of AI-generated outputs will help students to sharpen their thinking and problem-solving skills. Critical thinking in students should equip them with skills to filter through vast amounts of AI-generated data, identify the relevant from the irrelevant, and make decisions based on accurate insights. This is particularly important with generative AI which will be ubiquitous in education. Without strong critical thinking skills, there's a risk of being distracted and convinced with incomplete or inaccurate information. This will affect their understanding and result in flawed, suboptimal, or shallow learning outcomes.

CONCLUSIONS AND FUTURE DIRECTIONS FOR RESEARCH

There is a revolutionary chance to solve long-standing problems in rural education in India through the merging of Beyond 5G (B5G) and 6G technology. This debate highlights the importance of stakeholders' recommendations, reviews the main results about the effects of various technologies, and stresses the need of joint efforts to bring about positive change.

The study's key findings on B5G and 6G's impact on improving rural schools': results Improved Connectivity: Overcoming the digital divide that has long limited educational opportunities in rural India, B5G and 6G technologies can bring high-speed internet to remote places. More focused expenditures are need to guarantee extensive coverage, even though programs like Digital India seek to enhance infrastructure (Asma India, 2023). Individualized Instruction: By incorporating AI and ML into educational systems, students can have their own unique educational journey. Students who have difficulty with more conventional forms of instruction will benefit greatly from this level of customization (Nguyen-Son et al., 2022).

Online professional development courses can provide teachers the tools they need to incorporate technology into their lessons in a meaningful way. To get the most out of B5G and 6G technologies, you need ongoing assistance and training (CRY, 2023). In order to improve the learning environment in remote schools, digital infrastructure such as smart classrooms and digital libraries can be set up. Students can have access to a plethora of material and interactive learning tools through these sites (ResearchGate, 2020).

Several suggestions arise from these results: Governments and businesses alike should put money into digital infrastructure as a top priority if they want people in rural areas to have access to dependable internet. Broadband network expansion and school hardware needs are part of this. The current educational system's curriculum needs an overhaul to include lessons on digital literacy and how to make good use of technology in the classroom. This will set the stage for a future where technology plays a central role. Cultivate Alliances: To better utilize resources and skills, it is essential for private firms, non-governmental organizations (NGOs), and government agencies to work together. Collaborations like these can pave the way for new approaches to education (CRY, 2023) and their widespread adoption. Promote Involvement of the Community: Solutions can be culturally relevant and broadly accepted if local communities are involved in their planning and implementation of educational technology.

How B5G and 6G Could Revolutionize Rural Schools- It is impossible to exaggerate the revolutionary potential of B5G and 6G technologies on schooling in rural areas. By making education more accessible, engaging, and successful, these technologies have the potential to completely transform the way it is delivered. Examples of immersive technologies that can improve student engagement and knowledge include virtual reality (VR) and augmented reality (AR) (Neuronimbus, 2022). There is a severe lack of trained educators in rural regions, but remote learning platforms can bridge the gap by connecting students with professionals in the field in urban centers. With the help of programs like eVidyaloka, students can learn from qualified teachers all across the world, regardless of their location (Asma India, 2023).

Call for Action: Urging All Parties Involved to Work Together and Effect Positive Change- It will need a combined effort from everyone involved for B5G and 6G technologies to reach their maximum potential in rural schools. A unified plan for implementation requires input from community leaders, educators, technology providers, and lawmakers. Funding for infrastructure development and teacher training programs should be a top priority for policymakers when allocating federal and state funds, with educational technology being a top priority. In order to fully participate in the digital transformation, schools must incorporate technology into their lessons and offer continuous professional development to help educators adjust to new ways of teaching.

Businesses in the tech industry should collaborate closely with schools to design specialized software with features like localized content and intuitive interfaces that rural schools really need. Local officials should push for more investment in schools and technology, creating a culture that respects education and welcomes parents' voices as they guide their children's academic journeys. To successfully integrate technology into the educational landscape, NGOs and non-profits can play a pivotal role in establishing relationships and providing resources for schools in remote areas.

There are a number of topics that need additional research as B5G and 6G technologies develops further, research should concentrate on determining how B5G and 6G technologies affect educational results in rural areas. In order to guide future implementations, it is important to understand how these technologies impact learning, engagement, and retention. Concerning privacy and security, stringent safeguards are required in light of the growing number of connected devices. To ensure the security of student data, studies on privacy and data protection in educational settings are crucial (Shimaa et al., 2022). Eco-Friendly Procedures: To ensure the long-term success of rural education, it is essential to study ways to adopt B5G and 6G in a sustainable manner, taking into account energy-efficient technology and practices. Researchers in rural areas should focus on cultural adaptation if they want digital learning technologies to be more widely used and successful there (CRY, 2023).

REFERENCES

Akhtar, M. W., Hassan, S. A., Ghaffar, R., Jung, H., Garg, S., & Hossain, M. S. (2020). The shift to 6G communications: Vision and requirements. *Human-Centric Computing and Information Sciences*, *10*(53), 53. Advance online publication. DOI: 10.1186/s13673-020-00258-2

Anderson, J., & Rainie, L. (2012). The impact of the internet on education: A survey of teachers and technology use. Pew Research Centre. https://www.pewresearch.org

Appinventiv. (2023). 6G Technology: Revolutionizing Connectivity for the Future. Retrieved from https://appinventiv.com/blog/6g-technology-transforming-enterprises/

Asma India. (2023). Noteworthy Digital Initiatives that are Redefining Rural Education in India. Retrieved from https://www.asmaindia.in/blog/noteworthy-digital-initiatives-that-are-redefining-rural-education-in-india/

Asma India. (2023). Noteworthy Digital Initiatives that are Redefining Rural Education in India. Retrieved from https://www.asmaindia.in/blog/noteworthy-digital-initiatives-that-are-redefining-rural-education-in-india/

Banafaa, M., Ibraheem, S., Jafri, D., Marwan, H. A., Alashbi, A., Daradkeh, Y., & Alhammadi, A. (2023). 6G Mobile Communication Technology: Requirements, Targets, Applications, Challenges, Advantages, and Opportunities. *Journal of Network and Computer Applications*, *4*(1), 270–280. DOI: 10.1016/j.jnca.2022.103

CRY. (2023). Education In Rural India: Challenges, Opportunities, Initiatives. Retrieved from https://www.cry.org/blog/education-in-rural-india/

Davis, F. D. (1989). Perceived usefulness, perceived ease of use, and user acceptance of information technology. *Management Information Systems Quarterly*, *13*(3), 319–340. DOI: 10.2307/249008

Drishtiias. (2023, August 9). *State of Elementary Education in Rural India2023*. Retrieved from Drishtiias.com: https://www.drishtiias.com/daily-updates/daily-news-analysis/state-of-elementary-education-in-rural-india-2023

Friedner, S., & House, R. (2018). 5G infrastructure requirements in the UK. LS telcom UK. Retrieved from https://www.computer.org/digital-library/magazines/it/rural-remote-connectivity-challenges

Hassan, S. A., Ghaffar, R., & Akhtar, M. W. (2022). A Survey on Privacy of Personal and Non-Personal Data in B5G/6G. *Journal of Network and Computer Applications*, *4*(1), 103. DOI: 10.1016/j.jnca.2022.103

Jain, A. D. (2023, August 24). *Providing last-mile internet connectivity to rural India*. Retrieved from tndindia.com: https://www.tndindia.com/providing-last-mile-internet-connectivity-to-rural-india/#:~:text=Internet%2Dpenetration%20in%20rural%20India,reliable%20access%20to%20the%20internet

Kukulska-Hulme, A. (2009). Will mobile learning change language learning? *ReCALL*, *21*(2), 157–165. DOI: 10.1017/S0958344009000223

Mogavi, R. H. (2024, July). *ChatGPT in education: A blessing or a curse? A qualitative study exploring early adopters' utilization and perceptions*. Retrieved from Science Direct: https://www.sciencedirect.com/science/article/pii/S2949882123000270

Murata. (2023). Technological Trends Leading to Beyond 5G/6G. Retrieved from https://article.murata.com/en-sg/article/communication-network-management-using-ai-and-ml

Neuronimbus. (2021). Beyond 5G: Exploring the future of connectivity with 6G technology. https://www.neuronimbus.com/blog/beyond-5g-exploring-the-future-of-connectivity-with-6g-technology/

Neuronimbus. (2022). Beyond 5G: Exploring the Future of Connectivity with 6G Technology. Retrieved from https://www.neuronimbus.com/blog/beyond-5g-exploring-the-future-of-connectivity-with-6g-technology

Nguyen-Son, V., Trung, D., & Zhichao, S. (2022). The Key Trends in B5G Technologies, Services and Applications. Computer Networks and Communications. Retrieved from https://link.springer.com/article/10.1007/s11036-022-01999-3

Norp, T. (2018). 5G requirements and key performance indicators. *Journal of ICT Standardisation*, *6*(1), 15–30. DOI: 10.13052/jicts2245-800X.618

ResearchGate. (2020). A Key 6G Challenge and Opportunity - Connecting the Remaining 4 Billions: A Survey on Rural Connectivity. Retrieved from https://www.researchgate.net/publication/337062133_A_Key_6G_Challenge_and_Opportunity_-_Connecting_the_Remaining_4_Billions_A_Survey_on_Rural_Connectivity

Shimaa, A., Hussein, H., & Kim, H. (2022). Vision and research directions of 6G technologies and applications. *Computers & Electrical Engineering*, *99*, 107805. DOI: 10.1016/j.compeleceng.2022.107805

Spiceworks. (2021). What is a 6G network? Definition, features, and benefits. https://www.spiceworks.com/tech/networking/articles/what-is-6g/

Statista.com. (2024, February 27). *Rural and urban population in India from 2018 to 2022*. Retrieved from Statista.com: https://www.statista.com/statistics/621507/rural-and-urban-population-india/

Tarhini, A., Hone, K., Liu, X., & Tarhini, T. (2017). Examining the moderating effect of individual-level cultural values on users' acceptance of E-learning in developing countries: A structural equation modeling of an extended technology acceptance model. *Interactive Learning Environments*, *25*(3), 306–328. DOI: 10.1080/10494820.2015.1122635

The Economic Times. (2023, July 27). *UN cautions schools against 'excessive' use of tech in classrooms, says it can impede learning*. Retrieved from The Economic Times: https://economictimes.indiatimes.com/magazines/panache/un-cautions-schools-against-excessive-use-of-tech-in-classrooms-says-it-can-impede-learning/articleshow/102175016.cms?from=mdr

The Economic Times. (2024, April 24). *AI is touching lives of 9 in 10 internet users: report*. Retrieved from The Economic Times: https://economictimes.indiatimes.com/tech/technology/ai-is-touching-lives-of-9-in-10-internet-users-report/articleshow/109601626.cms?from=mdr

The Economic Times. (2024, August 21). *Improved internet quality ahead: New subsea cable routes to widen India's data highway*. Retrieved from The Economic Times: https://economictimes.indiatimes.com/industry/telecom/telecom-news/improved-internet-quality-ahead-new-subsea-cable-routes-to-widen-indias-data-highway/articleshow/112663013.cms?from=mdr

The Economic Times. (2024, June 28). *80% maths teachers in India, Middle East falter on basic mathematical questions: Study*. Retrieved from The Economic Times: https://economictimes.indiatimes.com/industry/services/education/80-maths-teachers-in-india-middle-east-falter-on-basic-mathematical-questions-study/articleshow/111341070.cms?from=mdr

The Print. (2022, November 4). *Only 34% Indian schools have internet access, less than 50% have functional computers, shows data*. Retrieved from The Print: https://theprint.in/india/education/only-34-indian-schools-have-internet-access-less-than-50-have-functional-computers-shows-data/1196870/

The Times of India. (2023, Dec 6). *Internet connectivity in government schools increased from 5*. Retrieved from The Times of India: https://timesofindia.indiatimes.com/education/internet-connectivity-in-government-schools-increased-from-5-5-to-24-2-in-five-years/articleshow/105788392.cms

Times of India. (2024, February 28). *Over half of 821mn internet users from rural India: Study*. Retrieved from Times of India: https://timesofindia.indiatimes.com/business/india-business/over-half-of-821mn-internet-users-from-rural-india-study/articleshow/108057081.cms

Times of India. (2024, January 19). *Taught, but thought? Latest survey of rural education again shows India's demographic advantage isn't a sure thing*. Retrieved from Times of India: https://timesofindia.indiatimes.com/blogs/toi-editorials/taught-but-thought-latest-survey-of-rural-education-again-shows-indias-demographic-advantage-isnt-a-sure-thing/

van Dijk, J. A. (2006). Digital divide research, achievements and shortcomings. *Poetics*, *34*(4-5), 221–235. DOI: 10.1016/j.poetic.2006.05.004

Venkatesh, V., & Davis, F. D. (2000). A theoretical extension of the technology acceptance model: Four longitudinal field studies. *Management Science*, *46*(2), 186–204. DOI: 10.1287/mnsc.46.2.186.11926

Warschauer, M. (2003). *Technology and social inclusion: Rethinking the digital divide*. MIT Press.

Zhang, Y., Zhao, M., & Liu, S. (2020). Exploring the impact of 5G technology on education: Opportunities and challenges. *Journal of Educational Technology & Society*, *23*(4), 42–54. https://www.jstor.org/stable/10.2307/26924001

Chapter 14
Social Inclusion and Community Empowerment Through B5G and 6G Network Connectivity in Rural Regions

Swetha Appaji Parivara
https://orcid.org/0009-0008-1215-154X
Presidency College, Bangalore, India

Aruna Rani
Presidency College, Bangalore, India

ABSTRACT

This chapter explores the transformative potential of B5G (Beyond 5G) and 6G network connectivity in promoting social inclusion and community empowerment in rural regions. It examines how advanced connectivity technologies can bridge the digital divide, enhance access to essential services such as education, healthcare, and e-governance, and foster economic opportunities through digital entrepreneurship and smart farming. The chapter incorporates global and regional case studies to illustrate the positive outcomes of B5G/6G implementations in marginalized communities, highlighting their impact on women, youth, and small businesses. Additionally, it discusses the challenges of deploying these technologies in remote areas and provides policy recommendations to ensure equitable access and sustainable development. By focusing on real-world examples, the chapter underscores the role of advanced connectivity in creating inclusive, empowered, and digitally integrated rural communities.

INTRODUCTION

The development of communication technology has enabled B5G (beyond 5G) and 6G, which has led to an increase in global empowerment and inclusivity. However since 5G network architecture is still being established globally, B5G and 6G will set the pace and change the world since they seek to bridge the gaps that have created obstacles to the expansion and development of current economies, particularly emerging ones like rural and underdeveloped markets. Emergent applications with more capacity and

DOI: 10.4018/979-8-3693-7788-8.ch014

almost no latency will be made possible by these networks, including telemedicine, smart agriculture, remote education, and e-governance (Gupta & Jha, 2021).

Rural and remote areas, including those in developing countries, have long been marginalized in terms of infrastructure, connectivity, and access to information technology. Issues such as slow internet speeds, limited computer literacy, and restricted access to essential services like banks, schools, and healthcare facilities are prevalent in these regions. However, the decentralized deployment of B5G and 6G networks can help close these gaps, expanding the range of technology-based solutions that benefit rural communities. According to the World Bank (2020), improved connectivity not only enhances access to information and economic opportunities but also empowers communities to participate in decision-making process actively. Furthermore, these advancements can lead to sustainable solutions tailored to local needs, ultimately promoting social inclusion and economic growth (UNESCO, 2021).

B5G and 6G connectivity facilitate the development of smart villages, optimizing living standards and enhancing agricultural and healthcare sectors. This connectivity can bridge the gap between urban and remote areas, helping to eradicate discrimination and eliminate the digital divide. To ensure that rural areas benefit from these advancements, policymakers, governments, and industry stakeholders must prioritize inclusive development strategies (World Bank, 2020; Prasad et al., 2022).

From a socio-economic perspective, enhancing connectivity in rural communities aligns with the principles of sustainable development. Improved access to technology fosters the establishment of local digital markets, facilitating business growth and promoting the flow of information to and from rural populations. These dynamics contribute to poverty reduction, improved health standards, and enhanced educational outcomes (World Bank, 2020; Kim & Lee, 2022). Furthermore, increased connectivity can empower marginalized groups, encouraging their active participation in democratic processes and the promotion of civil liberties (UNESCO, 2021; Wang et al., 2023).

B5G and 6G technologies provide the low-latency, high-speed internet required to deliver real-time services to rural areas. For example, telemedicine becomes feasible when high-definition video conferencing can be sustained without interruptions. Similarly, online education platforms can reach rural learners only when high-speed broadband is available consistently.

Therefore, having appropriate connectivity of B5G and 6G networks is going to determine more social interaction and empowering of communities in rural areas. Ensuring the specified technologies as necessary conditions for the digital services and applications, these technologies will not only improve the quality of people's lives but also make the rural communities participate in the global economy and information society. This chapter is dedicated to the main aspects of B5G and 6G connectivity, and its contribution to the development of the rural areas.

Connectivity of Digital Literacy and Lifelong Learning

The theory of digital literacy and lifelong learning is crucial for understanding the impact of B5G and 6G network connectivity on rural social inclusion and community empowerment. Digital literacy enables rural populations to leverage high-speed connectivity for accessing essential services like online education, healthcare, and e-governance, thereby enhancing their quality of life. This chapter examines how improved connectivity transforms rural regions by creating opportunities for employment, digital entrepreneurship, and community-driven development. It explores the relationship between increased connectivity and outcomes such as educational achievement, skill development, and social empowerment, emphasizing how B5G and 6G technologies foster lifelong learning and community self-reliance.

DIGITAL DIVIDE AND SOCIAL INCLUSION

The digital divide refers to the disparity in access to and usage of modern Information and Communication Technologies (ICT) between different segments of society, especially rural and urban areas. Rural communities often face challenges such as inadequate internet access, poor infrastructure, and low digital literacy. With the introduction of B5G and 6G network connectivity, there is potential to either close or widen this gap, depending on how these technologies are deployed. Social inclusion, defined as the process of improving the terms for marginalized individuals to participate fully in society (United Nations, 2016; Rawal, 2008), is closely linked to the digital divide. In the digital age, social inclusion increasingly depends on digital literacy and access to technology, making bridging the digital divide essential for equitable participation in key sectors like education, healthcare, and employment (Van Dijk, 2006).

B5G and 6G technologies can transform rural areas by providing ultra-reliable low-latency communication, high-speed broadband, and smart infrastructure. These advancements can address long-standing issues of poor connectivity and exclusion in rural communities. However, challenges like lack of infrastructure, high costs, and low digital literacy levels continue to exacerbate the marginalization of these regions (Mansell, 2002). The adoption of these networks can promote social inclusion by providing critical services such as telemedicine, online education, and digital financial services to rural populations (Dwivedi et al., 2021b).

For instance, B5G and 6G networks can enhance agricultural practices through smart farming technologies that enable real-time monitoring using sensors and drones. Additionally, telemedicine can improve healthcare access in remote areas, connecting patients to medical professionals in urban centers. Online learning can also bridge educational gaps, improving the quality of education for rural children and adults (Dwivedi et al., 2021b; Mansell, 2021).

To ensure the successful integration of B5G and 6G, policymakers and stakeholders must prioritize rural infrastructure investment and digital literacy programs to avoid deepening the digital divide. These technologies have the potential to empower rural communities, improve living standards, and foster economic growth through the creation of new employment opportunities, digital entrepreneurship, and participatory local governance. However, without proactive efforts, rural areas risk being left behind in the digital economy, reinforcing social exclusion. Hence, inclusive deployment of B5G and 6G technologies is essential for sustainable and equitable development.

EMPOWERING COMMUNITIES THROUGH CONNECTIVITY

As the world advances the use of ICTs it has become apparent that high speed and reliable internet connection is a key driver or necessity for socio-economic development especially to the rural population. 5G, B5G, and 6G networks can provide e-governance, e-health, online classes, etc. on behalf of the rural and un-served areas of human society, increased use of technology closes the digital divide and in doing so increases the ability of the said communities to access information, participate in politics, receive health and education services and drive economic progress. The segment of this paper goes a step further explaining how connectivity has boosted the lives of such groups of people inhabiting the rural areas of developing countries such as India and in the global perspective.

1. E-Governance: Facilitating Civic Engagement and Service Delivery

E-governance is defined as the process of using ICT to deliver public services, enhance the transparency of government, and encourage participation. The two being located far from city centers and having poor infrastructure, the citizens in the rural areas are deprived of many governmental services and e-governance is the only way that can help them out. Available services can be filed online unburdening citizens to work, file taxes, apply for benefits, address grievances, and gain information all through their computers or smartphones.

Case Study: India - Digital India Initiative

Challenges of e-governance have been striding in India, the Digital India initiative has enhanced the provision of e-governance services in rural areas. About more than 400,000 CSC digital service centers have been set up under the CSC scheme for delivering government services, banking, education, and health services across India. They serve as offices in which the rural people go to apply for government-issued notifications, avail pension facilities, and engage in digital transactions. The official data from the Ministry of Electronics & Information Technology evidences that it has made a significant difference in the inclusion of rural people in the governance and availability of services (Meity, 2020).

Case Study: Global Context - Estonia's E-Residency Program

Estonia recognized as a global leader in e-governance, has utilized internet connectivity to deliver essential services to rural residents. The country's e-Residency program allows individuals with internet access to start businesses, use various e-services, and even participate in the nation's decision-making processes. This initiative has empowered rural residents to be economically productive, positioning Estonia as a digital leader globally (Hoffmann & Solarte-Vasquez, 2022).

2. E-Health: Enhancing Healthcare Accessibility

E-health platforms enable rural people to receive healthcare services that they would otherwise not access as a result of geographical barriers, lack of proper infrastructure, and the dearth of healthcare providers. This includes telemedicine, remote diagnosis, and health records where high-speed connectivity assists in the delivery of health care to regions that have been classified as hard to reach.

In India, specifically in the state of Odisha, programs using telemedicine have already been adopted to improve access to health services in rural settings. For instance, the project which effectively began in October 2012 in partnership with Apollo Hospitals involved tele-consultation whereby patients in remote rural areas of Odisha could consult doctors in cities. Consequently, the number of cases that followed this approach was at 60% meaning that most did not require patients to travel to cities for treatment hence minimizing costs and maximizing the outcome. Concerning patient demographics, the program has provided the most improvements to female and elderly patients due to mobility difficulties observed in the patient population (Mahapatra et al., 2021).

One innovation adopted by Rwanda is the Babyl Health telemedicine which has enabled the delivery of health care to rural areas through connectivity. In Rwanda the majority of the population resides in the rural areas; through the use of telemedicine, they can consult with doctors through the use of mobiles. Over two million telemedicine consultations have been conducted since the platform's creation; saving lives, enhancing healthcare results, and alleviating congestion in urban hospitals according to the Rwanda Ministry of Health. (2022).

3. Online Education: Bridging the Educational Divide

Education plays an important role in developing the economy and social status of people but due to various factors like the absence of proper schools, transportation facilities, and lack of sufficient teacher resources, rural area students suffer most. These gaps can be closed by improved and increased internet connectivity which allows access to high-quality materials in education, lessons, and other forms of learning through the use of technology.

Case Study: India - DIKSHA Platform

DIKSHA is a part of the National Digital Infrastructure for Teachers and it has brought a revolutionary change in education, especially in rural areas. Founded in 2017, the portal offers members, unlimited access to a variety of learning materials, such as e-books, tutorials in videos, and quizzes. During the COVID-19 outbreak and implementation of the closure of schools, DIKSHA has helped students especially from rural regions to further their education from home and it has been downloaded more than 80 million times by mid-2020 (Government of India, 2021). This showcases that with the use of digital platforms, the losses in terms of education seen in crises do not have to occur and in the long run, there is a positive effect, especially to the learners in rural areas.

Case Study: Global Context - South Africa's E-Learning Initiatives

In South Africa, e-learning tools like the Vodacom e-School have played a great role in offering education to rural students. By so doing, the students get to use this platform to find free online courses that are relevant to the South African school curriculum. Some of the findings of the research reveal that rural schools that have embraced the use of the Internet have prompted enhanced performance of the students especially in areas like mathematics and science, which are deemed to be complex to cover in institutions in rural areas (Mhlanga & Moloi, 2020).

Empowering Self-Reliant Communities

Through the availability of services and engagement in economic activities brought by better connections through B5G and 6G, rural people's dependency decreases. Through connectivity, these communities are therefore able to trade in international markets, acquire other skills as well as participate in decision-making when it comes to issues affecting their source of income. Furthermore, it encourages innovation since new local businesspeople can develop new products and ventures on the platform that contribute to economic development.

In conclusion, these case studies from India and the global context have erupted in how the increased connectivities have given opportunities to rural communities to get connected with basic services and opportunities. These examples show that the next-generation networks, such as B5G and 6G are no longer an enabler of technological tools but the harbinger of social connection, self-reliance, and communal consciousness.

ECONOMIC EMPOWERMENT AND DIGITAL ENTREPRENEURSHIP

The era of 'Generation Z – 5G and 6G connectivity' has brought new hopes to economic growth and entrepreneurial development, especially in rural areas with internet access and connectivity. From creating opportunities for access to global markets, new forms of business, and digital tools, enhanced connectivity empowers people and communities to overcome the existing constraints to economic activity and co-create new paradigms of local economy and a new generation of digital economy.

Economic Empowerment Through Connectivity

Economic Empowerment is defined as people and communities gaining the capability to manage resources; engage in economic relations benefit as well as have a say in managing those relations or structures that shape their economy. In the context of rural regions, it is easy to observe that such areas are characterized by weak economic rates, inadequate infrastructure, and limited access to the markets that are located far away from such zones and that is why connectivity through digital networks could become an efficient lever in changing the economic situations in these regions.

By continuously getting connected and getting information over the internet, the rural business people can be able to access information on market prices, supply chain, and the consumer demand that exists hence working harder to come up with better decisions to make. Also, digital payment systems as well as e-commerce platforms eliminate the dependence upon conventional questionable middlemen, enabling local producers to supply their products directly to customers. For instance, farmers in rural regions of India can use eNAM (National Agriculture Market) to have better markets for their products thus affording better prices and less spoilage due to saturation of markets in the region (Narula, 2020).

Access to connections further encourages the growth of a new stock of digital competencies. Through platforms of Coursera, Udemy, and government schemes and policies, rural people can avail themselves education of at low cost or even free in streams like business management, digital marketing, and Information technology. This enables them to get the knowledge with a view of starting and growing digital enterprises for the purpose of increasing the economic activities in their region.

Digital Entrepreneurship: Opportunities and Challenges

Digital entrepreneurship can be defined as the establishment and management of enterprises that leverage digital technologies to provide goods or services. It includes ventures of e-commerce and mobile applications as well as the digital marketing agency, technology formations, and many others. In the rural sectors, digital entrepreneurship has become a significant driver of economic development since it affords people the ability to start their enterprises without the need for huge capital investments in infrastructure or rent for premises.

a) E-Commerce and Digital Platforms

Companies such as Amazon, Flipkart, Alibaba, etc., have made it possible for rural entrepreneurs to take their merchandise to national and international markets through e-commerce digital platforms. Some of the government-backed interventions such as the Pradhan Mantri Gramin Digital Saksharta Abhiyan (PMGDISHA) have empowered people in rural areas to create online shops and market them to the global market. For example, women-led self-help groups in rural India have begun selling handmade goods on platforms like Amazon's Saheli, empowering women to contribute to household incomes and gain financial independence (Amazon India, 2021).

In India, the government's *Digital Village* initiative, part of the broader Digital India campaign, aims to provide digital infrastructure and skills training to rural villages. In states like Gujarat and Maharashtra, local artisans and entrepreneurs have benefited from this initiative by using digital platforms to market their handicrafts, agricultural products, and services. The program has facilitated job creation, improved income levels, and increased access to government schemes and digital financial services (Meity, 2020).

b) Mobile and FinTech Solutions

Another key activity where digital entrepreneurship is highly active is in the financial technology space or rather FinTech. Traditional financial services are frequently unavailable or restricted in rural areas; however, mobile banking and digital payment options are assisting in narrowing this gap. The financial institutions are creating financial service solutions that include loans, mobile accounts, and mobile insurance being offered to individuals in less developed countries. Companies such as Paytm and Airtel Payments Bank have reached out to people in rural areas and have been able to make payment solutions accessible through mobile phones hence boosting the economy and reducing poverty.

Kenya's *M-Pesa*, a mobile money service has transformed the way people in rural areas access financial services. Launched by Safaricom, M-Pesa allows users to transfer money, pay bills, and access loans using their mobile phones. By 2021, over 90% of Kenyan households were using M-Pesa, with significant positive impacts on poverty reduction and economic activity in rural areas (Suri & Jack, 2016).

c) Challenges to Digital Entrepreneurship in Rural Areas

While the potential for digital entrepreneurship is vast, several challenges remain, particularly in rural regions. These include:

- **Digital Literacy:** Many rural populations lack the necessary digital skills to fully participate in digital entrepreneurship. Training programs and digital literacy initiatives are critical to bridging this gap.
- **Access to Capital:** Despite the availability of digital tools, many rural entrepreneurs struggle to access the capital needed to launch or scale their businesses. Microfinance and government-backed loans can help address this barrier.
- **Infrastructure:** Reliable electricity and internet access remain inconsistent in many rural areas, limiting the ability of entrepreneurs to operate their businesses efficiently.
- **Cultural Barriers:** In some regions, traditional social norms and gender roles can limit the participation of women and marginalized groups in digital entrepreneurship.

The Role of Policy and Investment

Governments and the private sector play a crucial role in fostering an enabling environment for digital entrepreneurship. Policies that promote internet penetration, digital literacy, and financial inclusion are essential for empowering rural entrepreneurs. Additionally, investments in digital infrastructure, such as high-speed internet and mobile networks, are critical to ensuring that rural regions are not left behind in the digital economy.

India's Startup India Initiative India's *Startup India* initiative, launched in 2016, has provided significant support for digital entrepreneurship in rural areas. Through funding, incubation centers, and mentorship programs, the initiative has helped thousands of rural startups launch and scale their businesses. In states like Uttar Pradesh and Bihar, where economic opportunities have traditionally been limited, rural entrepreneurs are now building successful tech companies and e-commerce platforms, contributing to regional economic development (Government of India, 2021).

Improved connectivity offers significant opportunities for digital entrepreneurship and economic empowerment in rural communities, allowing access to global markets, digital tools, and financial services that can enhance living standards. However, addressing challenges like digital literacy, infrastructure gaps, and capital access will require concerted efforts from governments, private sector businesses, and civil society to foster independence, creativity, and economic growth in these areas.

EDUCATION AND LIFELONG LEARNING

Education is a cornerstone for individual and community development, offering people the tools and knowledge to improve their quality of life. In rural areas, however, access to quality education and lifelong learning opportunities has often been limited due to geographical isolation, inadequate infrastructure, and insufficient resources. The advent of B5G (Beyond 5G) and 6G technologies holds the promise of transforming education in these regions by providing high-speed, reliable internet connectivity. This can enhance digital literacy, bridge the educational divide, and support lifelong learning for people of all ages.

The Importance of Digital Literacy and Lifelong Learning

In today's information-driven world, digital literacy—the ability to access, understand, and use information and communication technologies (ICTs)—is crucial for both formal education and ongoing personal development. For rural communities, improving digital literacy is a pathway to economic, social, and educational inclusion. It allows individuals to engage with digital platforms for learning, participate in the digital economy, and access government services. Moreover, as economies become increasingly digitalized, continuous education or lifelong learning is essential for individuals to stay competitive and adapt to new technologies.

Lifelong learning involves developing skills and knowledge throughout one's life, well beyond formal education. It encompasses learning new skills for work, personal development, or civic engagement. High-speed connectivity through B5G and 6G technologies enables access to e-learning platforms, virtual classrooms, and online training courses, providing rural populations with the ability to engage in lifelong learning from their own homes or local community centers.

Figure 1. Importance of digital literacy and lifelong learning

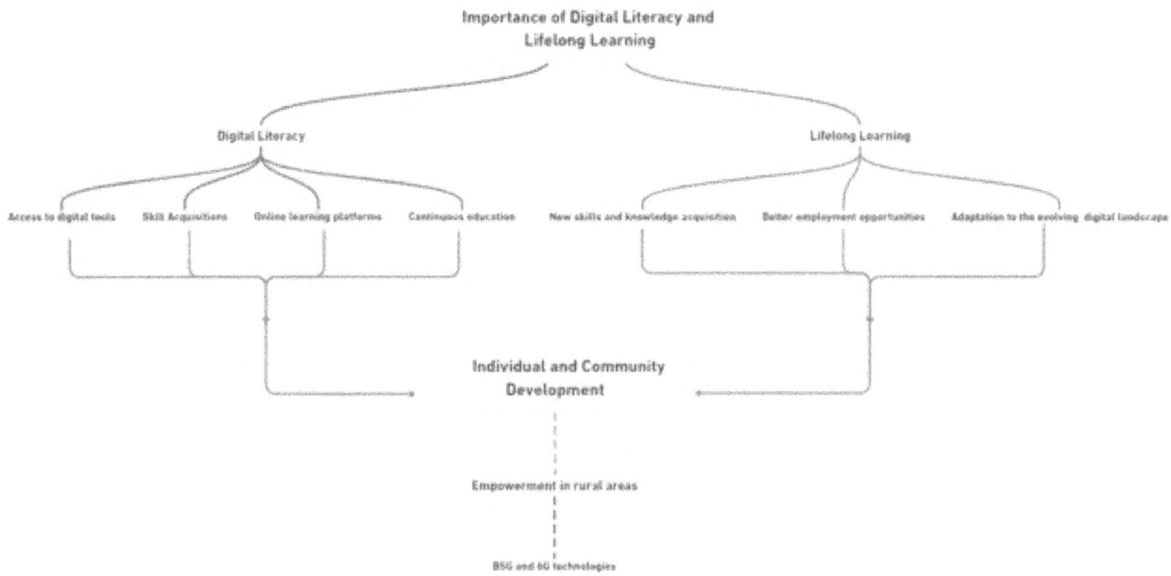

The flowchart illustrates the importance of Digital Literacy and Lifelong Learning by showing how digital literacy facilitates access to digital tools, leading to skill acquisition. These skills enable individuals to benefit from online learning platforms, which, in turn, foster continuous education. As people acquire new skills and knowledge through lifelong learning, they gain access to better employment opportunities and can continuously adapt to the evolving digital landscape. This cycle of learning and empowerment contributes to individual and community development, particularly in rural areas empowered by B5G and 6G technologies.

Impact of B5G and 6G on Education and Lifelong Learning

The impact of B5G (Beyond 5G) and 6G network technologies on education and lifelong learning is profound, particularly in rural regions where access to quality education has historically been limited. These advanced networks offer transformative possibilities by bridging gaps in digital infrastructure, enabling access to a broader range of learning resources, and supporting innovative educational models that enhance both formal education and continuous learning throughout life.

Figure 2. Impact of B5G and 6G on education and lifelong learning

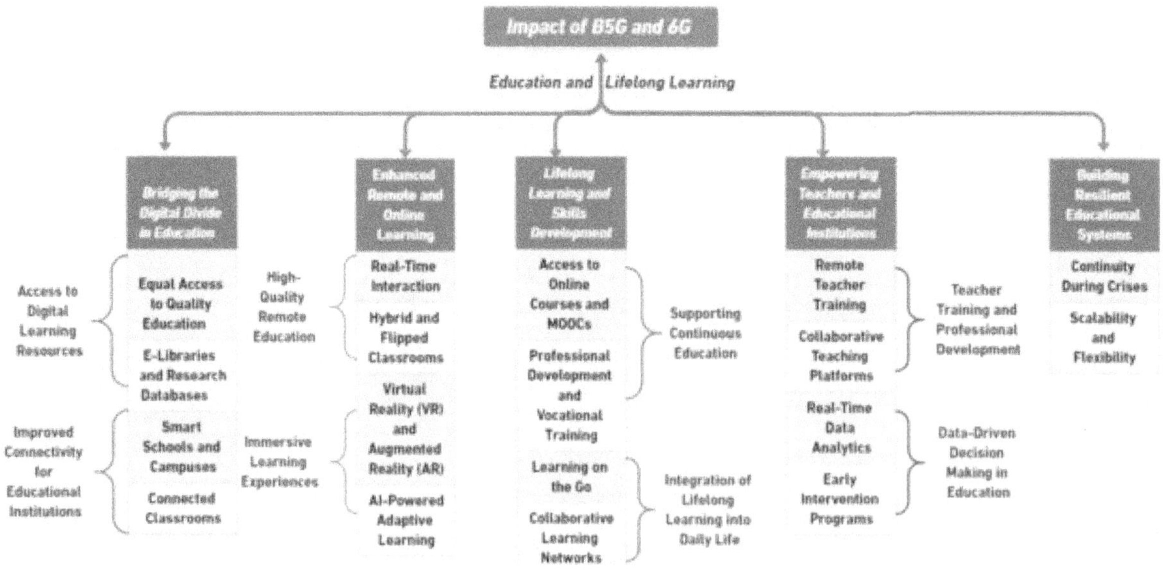

1. Bridging the Digital Divide in Education

Access to Digital Learning Resources:

- **Equal Access to Quality Education:** One of the most significant challenges in rural education is the lack of access to high-quality learning resources. With B5G and 6G networks, students in rural and remote areas can access the same digital platforms, learning materials, and online courses as those in urban centers. These networks provide the bandwidth necessary for data-heavy content such as HD video lectures, interactive learning apps, and virtual labs.
- **E-Libraries and Research Databases:** B5G and 6G can facilitate access to e-libraries and online research databases, making it easier for students and educators in rural areas to engage in in-depth learning and research activities. The availability of these resources promotes independent learning and provides a more comprehensive educational experience.

Improved Connectivity for Educational Institutions:

- **Smart Schools and Campuses:** With the advent of 6G technology, educational institutions can transform into smart campuses equipped with IoT (Internet of Things) devices, augmented reality (AR), and virtual reality (VR) technologies. This allows for enhanced interactivity in classrooms, personalized learning experiences, and real-time data analysis of student performance.
- **Connected Classrooms:** B5G and 6G make it possible to connect classrooms in rural areas with global educational institutions. This fosters collaboration between schools, enabling joint projects, knowledge sharing, and cultural exchange.

2. Enhanced Remote and Online Learning

High-Quality Remote Education:

- **Real-Time Interaction:** B5G and 6G networks support ultra-low latency, which enables real-time interactions between students and teachers during online classes. This helps in creating an immersive learning experience comparable to in-person education. Students in rural areas can actively participate in live lectures, ask questions, and engage in discussions without the lag issues that hinder lower-bandwidth networks.
- **Hybrid and Flipped Classrooms:** Advanced networks allow for the implementation of hybrid learning models where students can attend classes both online and in person. In flipped classrooms, students access recorded lectures and materials online at their own pace and come to class for discussions or problem-solving sessions. This model is ideal for rural students who may not have consistent access to physical classrooms.

Immersive Learning Experiences:

- **Virtual Reality (VR) and Augmented Reality (AR):** B5G and 6G enable immersive learning experiences through VR and AR technologies. These technologies allow students to engage in virtual simulations, whether exploring ancient civilizations in history classes, conducting virtual science experiments, or participating in virtual field trips to museums and archaeological sites. This is particularly beneficial in rural areas where physical access to such resources is limited.
- **AI-Powered Adaptive Learning:** With enhanced connectivity, AI-driven platforms can analyze students' learning patterns and provide personalized feedback. This adaptive learning ensures that students receive individualized instruction based on their needs, allowing for more effective learning outcomes.

3. Lifelong Learning and Skills Development

Supporting Continuous Education:

- **Access to Online Courses and MOOCs (Massive Open Online Courses):** B5G and 6G allow individuals in rural areas to easily access a vast array of online courses, certification programs, and MOOCs. This democratizes education by enabling individuals to acquire new skills, upskill, or reskill, regardless of their geographic location.
- **Professional Development and Vocational Training:** Advanced networks provide opportunities for vocational training and professional development through remote workshops, online certifications, and interactive platforms. These training programs can be tailored to the needs of local industries and communities, helping to create a more skilled workforce and promoting economic development.

Integration of Lifelong Learning into Daily Life:

- **Learning on the Go:** With the ultra-fast speeds and seamless connectivity offered by B5G and 6G, learning can be integrated into daily life. For example, individuals can access educational content via mobile devices, participate in live online workshops, or engage in microlearning modules while commuting or during free time. This flexibility supports a culture of continuous learning and self-improvement.

- **Collaborative Learning Networks:** B5G and 6G will enable the creation of collaborative learning networks where individuals, regardless of age or location, can engage in shared learning experiences. Peer-to-peer learning, online study groups, and global education communities will become more accessible, promoting a collective approach to knowledge building.

4. Empowering Teachers and Educational Institutions

Teacher Training and Professional Development:

- **Remote Teacher Training:** Advanced connectivity can support the professional development of teachers in rural areas by offering remote training programs. Teachers can access up-to-date instructional strategies, pedagogy, and subject-specific knowledge from leading educational institutions without needing to travel to urban centers.
- **Collaborative Teaching Platforms:** B5G and 6G enable teachers to collaborate with their peers across the world. They can exchange ideas, share best practices, and co-create learning resources, thereby enriching their teaching methods and improving the overall educational experience for their students.

Data-Driven Decision Making in Education:

- **Real-Time Data Analytics:** Schools can use data analytics powered by B5G and 6G to track student performance, attendance, and engagement. This data allows for informed decision-making in curriculum development, resource allocation, and student support services. With more precise insights into student learning needs, educational institutions can design interventions that enhance educational outcomes.
- **Early Intervention Programs:** Data gathered from student performance metrics can help identify at-risk students early, allowing for timely interventions such as additional tutoring or counseling. This data-driven approach helps in reducing dropout rates and ensuring that students remain engaged in their education.

5. Building Resilient Educational Systems

Resilience in the Face of Disruptions:

- **Continuity During Crises:** The COVID-19 pandemic demonstrated the importance of resilient education systems. B5G and 6G networks can ensure the continuity of education during crises, whether it's a natural disaster or a public health emergency, by facilitating seamless transitions to online learning.
- **Scalability and Flexibility:** These advanced networks enable scalable education solutions that can quickly adapt to changing circumstances. Whether it's scaling up digital infrastructure for remote learning or offering flexible educational models that cater to diverse learning needs, B5G and 6G provide the necessary infrastructure for educational resilience.

B5G and 6G networks are poised to have a transformative impact on education and lifelong learning, particularly in rural areas. By bridging the digital divide, enabling immersive and personalized learning experiences, and supporting lifelong education, these technologies are key to empowering individuals and communities through knowledge and skills. As B5G and 6G networks continue to expand, they will not only revolutionize the way education is delivered but also create opportunities for continuous learning, contributing to more resilient and inclusive societies.

Initiatives Leveraging B5G and 6G Technologies

Several initiatives are already underway that leverage the power of B5G and 6G technologies to enhance digital literacy and support educational growth in rural communities. These initiatives focus on bridging the digital divide and ensuring that rural populations can access the same educational opportunities as their urban counterparts.

a) India's Gyanodaya Program

The *Gyanodaya* initiative in India is one example of how advanced technologies are being used to enhance education in rural areas. Aiming to improve access to quality education for rural students, Gyanodaya uses B5G technology to connect rural schools with urban educational institutions. It provides virtual classrooms, online learning resources, and real-time teacher support. Through B5G-enabled smart classrooms, students in remote villages can interact with experienced teachers and subject matter experts from urban centers.

The program also focuses on enhancing digital literacy by providing tablets and internet access to students, along with training programs to help them navigate digital platforms. This has allowed students to access national-level educational resources such as India's DIKSHA platform, improving their academic performance and providing access to learning materials that were previously unavailable in rural regions (Meity, 2020).

b) Global Initiative: UNESCO's GIGA Project

The *GIGA project*, launched by UNESCO and UNICEF, aims to connect schools worldwide to the Internet, with a particular focus on remote and rural areas. By leveraging the next-generation B5G and 6G networks, GIGA seeks to provide reliable and high-speed internet to educational institutions in underserved regions. The project not only improves access to education but also promotes digital literacy by providing students and teachers with the tools they need to navigate and benefit from the digital world.

In Rwanda, for example, GIGA has enabled the connection of rural schools to high-speed internet, allowing students to participate in online learning programs and access educational content that was previously out of reach. The project also includes teacher training to ensure that educators in rural areas can effectively integrate digital tools into their teaching (UNESCO, 2022).

c) South Korea's K-education Platform Using 6G

South Korea is leading the development of a *K-education platform* based on 6G technology. The platform aims to provide virtual learning experiences using AR/VR and AI-powered learning tools, even in rural and remote areas. South Korea's government plans to roll out the platform nationwide, focusing on providing rural schools with access to immersive educational tools that enhance students' learning experiences.

This initiative will leverage the ultra-fast speeds and low latency of 6G networks to deliver seamless, interactive learning environments. Students will be able to participate in real-time, high-definition lessons that include simulations, interactive activities, and personalized AI-driven learning paths. The goal is to ensure that students in rural areas have access to the same quality of education as those in urban centers, helping to bridge the educational divide (Ministry of Science and ICT, 2022).

d) EU's Horizon 2020 Program for Smart Education

The *Horizon 2020* program by the European Union includes several initiatives aimed at using advanced connectivity technologies like 5G and 6G to promote smart education. One such initiative focuses on rural and remote regions, ensuring that schools and community centers in these areas have access to high-speed internet and digital educational tools.

Through this program, rural schools in Eastern Europe are piloting smart education systems that use 5G-enabled interactive learning environments, AI-based tutoring systems, and digital literacy programs for teachers and students. The project also promotes lifelong learning by offering online courses for adults, enabling them to develop new skills and access job opportunities in the digital economy (European Commission, 2021).

Lifelong Learning Platforms and Digital Skill Development

The concept of lifelong learning is becoming increasingly vital in the modern world, where continuous skill development is essential to keep pace with technological advancements and shifting job market demands. B5G (Beyond 5G) and 6G network technologies are poised to revolutionize lifelong learning platforms and digital skill development, particularly for individuals in rural and underserved areas. These advanced networks offer the bandwidth, speed, and low latency required to support innovative educational platforms, allowing for real-time learning, seamless connectivity, and access to global knowledge resources.

Coursera, a global online learning platform, has partnered with local governments in India and other countries to create rural learning hubs. These hubs provide rural populations with access to online courses, certifications, and degree programs through high-speed internet powered by B5G technologies. The initiative focuses on developing digital and job-ready skills, such as coding, data analysis, and entrepreneurship, thereby empowering rural adults to participate in the digital economy.

In Africa, the *LearnAfrica* initiative is working to provide digital learning opportunities to rural populations across several countries. By using 6G connectivity, the platform offers interactive courses, certification programs, and vocational training aimed at helping individuals develop skills that are relevant in today's job market. The platform also emphasizes digital literacy for adults, enabling them to use online tools for business, health, and personal development.

1. The Role of Lifelong Learning in the Digital Age

Continuous Learning for Professional Growth: In today's fast-evolving economy, individuals must continuously update their skills to remain competitive in the job market. B5G and 6G networks facilitate access to online learning platforms, enabling people to engage in lifelong learning through courses, certifications, and micro-credentials. With the emergence of digital skill gaps, particularly in rural areas, lifelong learning platforms offer a means to bridge these gaps by providing education and training in areas such as digital literacy, software development, data analytics, artificial intelligence (AI), and more.

Flexible Learning Environments: Lifelong learning platforms powered by B5G and 6G are designed to offer flexible learning opportunities that cater to individual needs. Learners can access content anytime and anywhere, accommodating their schedules and allowing them to integrate learning into their daily lives. Whether it's acquiring new skills for career advancement or learning for personal development, these platforms enable individuals to pursue education at their own pace.

2. Lifelong Learning Platforms: Digital Infrastructure and Capabilities

Global Access to Online Learning Platforms:

- **Cloud-Based Learning:** B5G and 6G networks will enhance the ability to host cloud-based learning platforms, enabling learners from rural and urban regions alike to access the same high-quality courses and resources. Cloud-based learning reduces the need for extensive physical infrastructure and provides access to global learning materials from renowned universities, professional organizations, and training providers.
- **Massive Open Online Courses (MOOCs):** These advanced networks will further expand the reach of MOOCs, offering courses from top-tier educational institutions. Individuals in rural regions will have the same opportunities to enroll in MOOCs as urban populations, fostering greater educational equity and global knowledge sharing.

Seamless Integration of AI and Machine Learning:

- **Personalized Learning Paths:** AI-powered lifelong learning platforms will analyze learner data to create personalized learning paths tailored to individual needs, interests, and goals. Through real-time feedback, adaptive assessments, and customized content recommendations, these platforms will ensure that each learner progresses at a pace suited to their skill level.
- **Intelligent Tutoring Systems:** With the integration of AI and machine learning, intelligent tutoring systems can provide one-on-one guidance to learners, helping them navigate complex subjects and address individual learning challenges. B5G and 6G networks will enable these systems to function seamlessly in real time, enhancing learner engagement and outcomes.

Real-Time Learning and Collaboration:

- **Low-Latency Interaction:** B5G and 6G offer ultra-low latency, allowing for real-time communication and collaboration between learners and instructors. Virtual classrooms, webinars, and workshops can be conducted without delays or interruptions, creating an immersive and interactive learning experience. This is particularly beneficial for learners in rural regions, who can participate in live discussions and collaborative projects with peers and experts across the globe.

- **Global Collaboration Networks:** Learners can engage in collaborative learning with peers from around the world, working on joint projects, exchanging ideas, and sharing knowledge. This global network of learners fosters a sense of community and exposes individuals to diverse perspectives, enhancing their learning experience and cultural understanding.

3. Digital Skill Development and Career Advancement

a) Addressing the Digital Skills Gap:

- **Digital Literacy Programs:** B5G and 6G networks will enable widespread access to digital literacy programs, which are essential for bridging the digital divide in rural areas. Digital literacy forms the foundation for further skill development, allowing individuals to effectively use computers, smartphones, and the internet for educational and professional purposes.
- **Advanced Technical Skills:** As industries increasingly rely on digital technologies, there is a growing demand for professionals with advanced technical skills such as coding, cybersecurity, AI, machine learning, and blockchain technology. Lifelong learning platforms will offer specialized courses in these areas, providing learners with the tools they need to enter high-demand fields.

b) Upskilling and Reskilling Opportunities:

- **Career-Specific Training:** Many lifelong learning platforms offer targeted training programs designed to upskill or reskill individuals for specific career paths. For example, individuals working in agriculture or manufacturing in rural areas can access training in automation, precision agriculture, or digital marketing to enhance their job prospects or improve their business operations.
- **Certifications and Micro-Credentials:** B5G and 6G networks support the proliferation of online certification programs and micro-credentialing, allowing learners to quickly acquire specific skills that are directly applicable to their professional roles. These bite-sized learning opportunities are ideal for working professionals who may not have the time or resources for traditional degree programs.

c) Supporting Entrepreneurship and Innovation:

- **Access to Business and Technical Knowledge:** Lifelong learning platforms provide aspiring entrepreneurs in rural areas with access to the knowledge and tools needed to start and grow businesses. Courses on digital marketing, e-commerce, financial management, and innovation strategies enable individuals to launch and scale businesses, contributing to local economic development.
- **Innovation Hubs and Incubators:** B5G and 6G will facilitate the creation of virtual innovation hubs and incubators, where learners can collaborate with mentors, experts, and fellow entrepreneurs to develop and refine their business ideas. These virtual spaces will foster innovation, particularly in rural areas where physical incubators may not be accessible.

4. Lifelong Learning and the Future of Work

a) Adapting to the Changing Workforce:
   - **Automation and AI:** The rapid advancement of automation and AI technologies is transforming the workforce, making it essential for workers to continuously update their skills. B5G and 6G networks will support platforms that offer training in the latest technologies, enabling workers to stay ahead of trends and maintain relevance in the job market.
   - **Gig Economy and Freelancing:** Lifelong learning platforms will also cater to the needs of gig workers and freelancers, providing them with the skills necessary to thrive in a flexible job market. From project management and negotiation skills to digital tools for freelancers, these platforms empower individuals to build sustainable careers in the gig economy.

b) Workplace Learning and Professional Development:
   - **Corporate Learning Platforms:** Many organizations are adopting online learning platforms to provide their employees with continuous professional development. B5G and 6G networks will enable seamless access to corporate training modules, allowing employees to upskill or reskill in areas such as leadership, communication, and technology use.
   - **AR and VR in Workplace Training:** Advanced networks will support the integration of AR and VR technologies into workplace training programs. These immersive technologies can simulate real-world work environments, allowing employees to practice and refine their skills in a safe, controlled setting. From factory workers training on complex machinery to healthcare professionals practicing surgical techniques, AR and VR offer practical, hands-on learning experiences.

5. Lifelong Learning for Social and Community Development

a) Education for Rural Communities:
   - **Community-Oriented Learning:** Lifelong learning platforms can be customized to meet the specific needs of rural communities. Programs in agriculture, healthcare, local governance, and community leadership can empower individuals with the knowledge and skills needed to contribute to the growth and development of their communities.
   - **Social Inclusion through Learning:** B5G and 6G networks ensure that even the most remote communities have access to educational resources, promoting social inclusion. By providing access to education and skills development, these technologies help reduce social inequalities and foster a more inclusive and empowered society.

b) **Lifelong Learning as a Driver of Economic Growth:** By equipping individuals with the skills needed for the digital economy, lifelong learning platforms play a crucial role in driving economic growth. With access to education and training, individuals in rural areas can contribute to their local economies through entrepreneurship, innovation, and improved job performance.

B5G and 6G networks are transforming lifelong learning and digital skill development by providing high-speed, low-latency access to educational resources, enabling personalized and immersive learning experiences. By bridging the digital skills gap in rural areas, these networks are essential for promoting social inclusion, economic development, and workforce readiness in the 21st century.

HEALTHCARE ACCESS AND PUBLIC HEALTH

The advent of high-speed connectivity, particularly through B5G and 6G networks, has the potential to revolutionize healthcare delivery in rural areas. This section explores how telemedicine, remote diagnostics, and health monitoring technologies can address long-standing healthcare disparities, overcome geographical barriers, and contribute to improved public health outcomes and community resilience.

1. Telemedicine

Telemedicine involves delivering medical services remotely through telecommunications technology. This includes a range of services such as video consultations, remote diagnoses, and digital communication between patients and healthcare providers. Telemedicine encompasses several types of services:

- Real-time Video Consultations: Enables patients to have live consultations with healthcare providers.
- Store-and-Forward Communication: Allows the collection and transmission of medical information for later review by specialists.
- Remote Monitoring: Facilitates ongoing patient data collection and analysis.

Telemedicine is particularly transformative for rural areas, where geographical distances often pose significant challenges to accessing healthcare. It enables rural patients to consult with specialists in urban centers without traveling, thereby reducing travel burdens and increasing access to specialized care. In emergency situations, telemedicine facilitates immediate consultations with specialists, aiding in critical decision-making and improving patient outcomes.

The expansion of telemedicine services increases the reach of healthcare, making a broader range of medical professionals and services available to rural populations. By offering convenience and reducing wait times for consultations, telemedicine enhances the efficiency and patient-centeredness of healthcare delivery.

2. Remote Diagnostics

Remote diagnostics involves using technology to conduct diagnostic tests and analyze results from a distance. This includes digital imaging, laboratory tests, and other diagnostic procedures. Technologies supporting remote diagnostics include:

- Remote Imaging Systems: Such as digital X-rays and MRI scans.
- Point-of-Care Testing Devices: For various diagnostic purposes.

Access to advanced diagnostic tools is often limited in rural settings. Remote diagnostics address this gap by allowing rural healthcare providers to transmit diagnostic data to specialists for interpretation. This capability ensures timely diagnoses, which are crucial for effective treatment and management of health conditions, thereby reducing delays typically associated with transporting samples or images.

3. Health Monitoring Technologies

Health monitoring technologies include wearable devices and remote sensors that track and transmit health data such as heart rate, blood pressure, glucose levels, and other vital signs. Examples include:

- Wearable Fitness Trackers
- Continuous Glucose Monitors
- Smartwatches with Health Monitoring Features

Remote monitoring technologies are particularly beneficial for managing chronic conditions. Devices that continuously track health metrics enable healthcare providers to receive real-time data, allowing for timely adjustments to treatment plans. This continuous monitoring helps in the early detection of potential health issues and supports preventive care, thereby reducing the need for emergency interventions and improving overall patient management.

4. Contributing to Better Public Health Outcomes

- **Improved Health Management:** High-speed connectivity facilitates real-time data collection and analysis, providing healthcare providers with valuable insights into patient health trends. This data-driven approach enhances the ability to make informed decisions, resulting in personalized and effective care. Regular monitoring and timely interventions contribute to improved health management and better outcomes for patients.
- **Building Resilient Communities:** By addressing health disparities and expanding access to quality healthcare services, these technologies contribute to the creation of more resilient communities. Reduced health disparities lead to more equitable healthcare access, which in turn promotes overall well-being. Enhanced healthcare access empowers rural communities, improves health, increases productivity, and mitigates the economic burden associated with poor health, fostering a healthier and more self-reliant community.

Telemedicine, remote diagnostics, and health monitoring technologies, powered by high-speed connectivity, are revolutionizing healthcare delivery in rural areas by overcoming geographical barriers and enhancing access to specialized care. These advancements address healthcare disparities and foster community resilience, moving toward more equitable and effective healthcare for all.

CHALLENGES AND POLICY RECOMMENDATIONS

Challenges and Policy Recommendations are essential to address the barriers that may hinder the successful deployment of these advanced technologies and their effective use in promoting social inclusion, education, healthcare, and economic development in rural areas.

1. Challenges in Deploying B5G and 6G Networks in Rural Regions

a) Infrastructure and Connectivity Challenges:
   - **High Costs of Network Deployment:** Deploying B5G and 6G networks in rural areas often requires significant investment in infrastructure, such as building new cell towers, laying fiber-optic cables, and upgrading existing network systems. The high costs associated with these infrastructure developments make rural regions less attractive for private-sector investment due to the lower population density and, consequently, a lower return on investment.
   - **Geographical Barriers:** Rural and remote areas may face geographical challenges such as mountainous terrain, forests, or vast open spaces that complicate the deployment of advanced network infrastructure. These physical barriers increase the difficulty and cost of installing and maintaining network equipment.
   - **Lack of Supporting Infrastructure:** Many rural areas lack supporting infrastructure like reliable electricity or transportation networks, further complicating efforts to establish and maintain B5G and 6G networks.

b) Digital Literacy and Technology Adoption:
   - **Low Levels of Digital Literacy:** One of the main challenges in rural areas is the low level of digital literacy. Even if B5G and 6G networks are deployed, the potential of these technologies may not be fully realized unless individuals are equipped with the necessary digital skills to access and use online resources effectively.
   - **Resistance to Technological Change:** In some rural communities, there may be cultural resistance to adopting new technologies. This could stem from a lack of awareness of the benefits of B5G and 6G networks or concerns over data privacy, security, and the potential loss of traditional jobs due to automation.

c) Economic and Social Inequality:
   - **Economic Barriers to Access:** Even if the network infrastructure is in place, many rural residents may struggle to afford devices or internet plans that would enable them to benefit from B5G and 6G technologies. The digital divide is not just a matter of access to connectivity but also the economic ability to leverage that access.
   - **Social Exclusion of Vulnerable Populations:** Vulnerable groups, such as women, the elderly, and marginalized communities, may face additional challenges in accessing and benefiting from B5G and 6G networks due to entrenched social inequalities. These groups may lack access to education, digital skills training, or financial resources.

d) Data Privacy and Security Concerns:
   - **Privacy and Data Protection:** With the increasing reliance on digital platforms, concerns over data privacy and cybersecurity are heightened, particularly in rural areas where awareness and understanding of these issues may be limited. The introduction of B5G and 6G

networks, which enable vast amounts of data transmission, also raises concerns about data protection and the potential for misuse of personal information.

- **Regulatory Challenges:** In rural areas, the regulatory environment may not be robust enough to protect individuals and communities from potential risks associated with the misuse of data or cyber threats.

2. Policy Recommendations to Overcome Challenges

a) Investment in Infrastructure Development:

- **Public-Private Partnerships (PPPs):** Governments should encourage the formation of public-private partnerships to share the cost and risk of deploying B5G and 6G networks in rural areas. These partnerships can combine the resources of the government with the expertise and efficiency of the private sector to build and maintain network infrastructure.
- **Subsidies and Incentives for Rural Network Expansion:** Policy measures such as subsidies, tax breaks, or other incentives should be introduced to attract private telecom companies to invest in rural regions. These incentives can help offset the high costs associated with network deployment in areas with low population density.
- **Infrastructure Sharing:** Encouraging network providers to share infrastructure, such as cell towers and fiber-optic cables, can reduce the overall cost of deploying B5G and 6G networks in rural areas. This collaborative approach can help accelerate connectivity in underserved regions.

b) Digital Literacy and Capacity Building:

- **Digital Literacy Programs:** Governments and educational institutions should invest in digital literacy programs that teach individuals how to use new technologies effectively. These programs should be tailored to the needs of rural communities, with a focus on practical applications of B5G and 6G technologies in areas such as agriculture, education, healthcare, and entrepreneurship.
- **Community-Based Training Initiatives:** Local community centers, schools, and libraries can be leveraged as hubs for digital education and training. By bringing digital literacy programs to the community level, it becomes easier to reach vulnerable populations such as the elderly, women, and marginalized groups.

c) Affordable Access and Inclusion:

- **Subsidized Devices and Connectivity Plans:** Governments can introduce policies that make internet-enabled devices (smartphones, laptops, etc.) and data plans more affordable for rural residents. Subsidies or vouchers for low-income households can ensure that financial barriers do not prevent access to B5G and 6G networks.
- **Targeted Inclusion Programs:** Policy initiatives should specifically address the needs of vulnerable populations in rural areas. Programs that promote digital inclusion for women, the elderly, and people with disabilities should be prioritized to ensure these groups are not left behind.

d) Enhancing Data Security and Privacy Protections:

- **Stronger Regulatory Frameworks:** Governments should establish or strengthen regulatory frameworks to ensure robust data protection and privacy standards. This includes clear regulations on data ownership, consent, and usage, particularly in the context of healthcare, education, and financial services delivered via B5G and 6G networks.

- **Public Awareness Campaigns on Cybersecurity:** Public campaigns focused on raising awareness about cybersecurity risks and best practices for protecting personal data should be implemented in rural areas. These campaigns can help build trust in digital services and promote the safe use of online platforms.

e) Supporting Local Innovation and Entrepreneurship:

- **Incentives for Local Startups and Entrepreneurs:** Rural communities should be empowered to develop local solutions that leverage B5G and 6G networks. Governments can provide grants or low-interest loans to rural startups and entrepreneurs, helping them innovate and create new services in fields such as e-commerce, digital agriculture, and telemedicine.
- **Innovation Hubs in Rural Regions:** Establishing innovation hubs or digital incubators in rural areas can foster local entrepreneurship by providing access to the necessary resources, mentorship, and technical support to develop and scale digital solutions.

f) Sustainable and Inclusive Deployment of B5G and 6G Networks:

- **Environmentally Sustainable Infrastructure:** Policies should ensure that the deployment of B5G and 6G networks is environmentally sustainable. Governments should encourage the use of renewable energy sources for powering network infrastructure in rural regions and implement measures to minimize the environmental impact of network expansion.
- **Community Involvement in Network Planning:** Rural communities should be actively involved in the planning and decision-making process for the deployment of B5G and 6G networks. Community consultations can help ensure that network infrastructure is aligned with the specific needs of rural populations, fostering greater acceptance and utilization of the technology.

To ensure that B5G and 6G networks reach their full potential in empowering rural communities and promoting social inclusion, it is essential to address challenges related to infrastructure, digital literacy, economic inequality, and data privacy. Policymakers must prioritize investments in rural infrastructure, encourage public-private partnerships, and offer subsidies to make digital tools more affordable. Additionally, programs designed to improve digital literacy will be crucial in enabling full participation in the digital economy for all communities.

CONCLUSION

The critical role of B5G and 6G network connectivity in promoting social inclusion and community empowerment in rural areas is significant. As the world becomes more digitally interconnected, these advanced technologies provide a unique opportunity to bridge the digital divide and ensure rural populations are not left behind in global digital transformation.

B5G and 6G technologies offer high-speed, low-latency connectivity, overcoming geographical barriers that rural regions face. This connectivity enhances access to essential services such as education, healthcare, and economic opportunities, reducing inequalities and empowering communities to engage in the global digital economy.

The chapter discusses how the digital divide perpetuates social exclusion, with rural areas facing challenges in accessing reliable internet. B5G and 6G can help close this gap by providing tools for engagement, information access, and community growth. Enhanced connectivity fosters participation in e-governance, telemedicine, and online learning, enabling self-reliance and resilience in rural communities.

Economic empowerment and digital entrepreneurship are key areas where B5G and 6G will drive change, allowing rural entrepreneurs to access global markets and leverage digital tools for business development. High-speed internet supports job creation and local economic growth.

In education, B5G and 6G connectivity democratizes access to quality education and lifelong learning opportunities, enhancing career prospects and reducing inequalities. In healthcare, telemedicine and remote diagnostics revolutionize service delivery, ensuring timely and quality care.

However, challenges such as infrastructure costs and socio-economic inequalities persist. The chapter recommends public-private partnerships, investment in digital literacy, and regulatory frameworks to ensure equitable access to these technologies.

In conclusion, B5G and 6G connectivity are vital for overcoming barriers in rural development, fostering social inclusion, economic opportunity, and community empowerment, ultimately driving sustainable development for future generations.

REFERENCES

Amazon India. (2021). *Amazon Saheli empowers women artisans and entrepreneurs*. https://www.amazon.in/saheli

Dwivedi, Y. K., Ismagilova, E., Hughes, D. L., Carlson, J., Filieri, R., Jacobson, J., & Kumar, V. (2021). Setting the future of digital and social media marketing research: Perspectives and research propositions. *International Journal of Information Management*, *59*, 102168. Advance online publication. DOI: 10.1016/j.ijinfomgt.2020.102168

European Commission. (2021). *Horizon 2020: Smart education initiatives for rural areas*. https://ec.europa.eu/horizon2020

Government of India. (2021). *DIKSHA: Digital infrastructure for teachers and students*. https://diksha.gov.in

Government of India. (2021). *Startup India initiative: Supporting entrepreneurship in rural India*. https://startupindia.gov.in

Gupta, N., & Jha, R. K. (2021). A survey of 5G network: Architecture and emerging technologies. *IEEE Access : Practical Innovations, Open Solutions*, *3*(6), 120–133. DOI: 10.1109/ACCESS.2015.2461602

Hoffmann, T., & Solarte-Vasquez, M. C. (2022). The Estonian e-residency programme and its role beyond the country's digital public sector ecosystem. *CES Derecho*, *13*(2), 184–204.

Kim, J., & Lee, S. (2022). The impact of digital connectivity on rural development: Evidence from developing countries. *Journal of Rural Studies*, *92*, 234–242.

Mahapatra, P., Gupta, R., & Mishra, S. (2021). Impact of telemedicine on healthcare accessibility in rural Odisha. *Journal of Health Informatics in Developing Countries*, *15*(1), 44–58.

Mansell, R. (2002). From digital divide to digital entitlements in knowledge societies. *Current Sociology*, *50*(3), 407–426.

Mansell, R. (2021). Inequality and digitally mediated communication: Divides, contradictions, consequences. *Oxford Review of Economic Policy*, *37*(3), 428–449. DOI: 10.13140/RG.2.2.23191.16805

Meit, Y. (2020). *Digital India: Empowering rural citizens through e-governance*. Ministry of Electronics and Information Technology. https://digitalindia.gov.in

Mhlanga, D., & Moloi, T. (2020). COVID-19 and the digital transformation of education: What we are learning in South Africa. *Preprints*.https://doi.org/DOI: 10.20944/preprints202004.0195.v1

Ministry of Science and ICT. (2022). *K-education platform leveraging 6G: Empowering rural learning through advanced technologies*. South Korean Government. https://msit.go.kr

Narula, S. (2020). Empowering rural farmers through eNAM: A case study of digital agriculture in India. *Journal of Rural Development*, *39*(2), 23–37.

Prasad, A., Sharma, A., & Kumar, R. (2022). Smart villages: Bridging the digital divide through B5G and 6G technologies. *Journal of Rural Studies*, *94*, 187–196. DOI: 10.1016/j.jrurstud.2022.10.003

Rawal, N. (2008). Social inclusion and exclusion: A review. *Dhaulagiri, 2*, 161–180. DOI: 10.3126/dsaj.v2i0.1362

Rwanda Ministry of Health. (2022). *Telemedicine services in Rwanda: Progress and impact*. https://www.moh.gov.rw

Suri, T., & Jack, W. (2016). The long-run poverty and gender impacts of mobile money. *Science, 354*(6317), 1288–1292. DOI: 10.1126/science.aah5309 PMID: 27940873

UNESCO. (2021). The role of information and communication technologies in rural development. https://en.unesco.org/news/role-information-and-communication-technologies-rural-development

UNESCO. (2022). *The GIGA project: Connecting schools and promoting digital literacy in rural areas*. https://unesco.org/giga

United Nations. (2016). *Leaving no one behind: The imperative of inclusive development*. United Nations Department of Economic and Social Affairs.

Van Dijk, J. A. G. M. (2006). *The network society: Social aspects of new media* (2nd ed.). SAGE Publications.

Wang, Y., Zhao, T., & Liu, S. (2023). Enhancing digital inclusion through 6G networks in rural areas. *Journal of Digital Society, 29*(2), 52–67.

World Bank. (2020). *Digital dividends: How to close the digital divide in rural areas*. https://www.worldbank.org/en/news/feature/2020/03/24/digital-dividends-how-to-close-the-digital-divide-in-rural-areas

Chapter 15

Transformative Impact of B5G and 6G Networks on Social Inclusion and Community Empowerment in Rural India

Mirshad Rahman
https://orcid.org/0009-0005-2488-9216
Acharya Institute of Graduate Studies, India

Bharath Booshan M. S.
https://orcid.org/0009-0002-7163-832X
Acharya Institute of Graduate Studies, India

Shabista Booshan
https://orcid.org/0009-0009-8026-3460
ISBR Business School, India

S. Shabana
https://orcid.org/0009-0002-1313-2304
Bharathamatha College of Arts and Science, India

Deepika Gowda
https://orcid.org/0009-0000-1306-4802
Acharya Institute of Graduate Studies, India

S. Gowtham
https://orcid.org/0009-0005-1263-4160
Acharya Institute of Graduate Studies, India

C. G. Lakshmidevi
https://orcid.org/0000-0003-3607-3853
Mahe Co-Operative College of Higher Education and Technology, India

N. Muddasir Ahamed Khan
https://orcid.org/0009-0006-8716-333X
Acharya Institute of Graduate Studies, India

ABSTRACT

By addressing the societal demands of upcoming developments such as remote surgeries, self-driving cars, immersive environments like augmented/virtual reality (AR/VR), and other related applications, B5G and 6G networks expand on the concepts and capabilities of 5G, introducing even more sophisticated features that will provide intelligent and highly reliable communications between the devices. Small towns and rural areas, which often lack access to modern connection facilities, are also expected to see revolutionary changes as a result of all these technologies. Using a literature review methodology, this research collects and analyzes data from several sources of literature and the chapter also includes government and business publications that address the state and prospects of rural connection in India, as well as groundbreaking studies on technological innovation, digitalization, and rural development. The literature from telecommunications, public policy, and rural sociology has also been included in

DOI: 10.4018/979-8-3693-7788-8.ch015

connection to the previously listed fields of study.

INTRODUCTION

In the case of mobile communication technologies, the rapid evolution has vastly transformed the way that many people across the world view and use connectivity. Currently, 5G and its successors; Beyond 5G (B5G) and 6G networks are at the heart of this change and expecting extreme data rate at the user terminal, significantly lower latency, very high connectivity density, and integration of AI and machine learning (Gupta et al., 2020). B5G and 6G networks extend further the concepts and capabilities of 5G, that is, to introduce even more sophisticated characteristics which will provide intelligent and highly dependable communications between the devices by addressing societal needs of the future developments that include surgeries from remote locations, self-driving cars, and immersive environments such as augmented/virtual reality (AR/VR) and other related applications (Chaccour et al., 2022). All these technologies are anticipated to have revolutionary effects on, not only large cities but small towns and rural regions, which generally remain starved of modem connection facilities. The shortcomings mentioned above can be mitigated by B5G and 6G to enable delivery of access to hitherto unreachable locales with respect to digital connectivity for socio-economic and community advancement. Commonly defined parameters of the transition from 5G to 6G are expected to yield the data rate of up to 1 Terabit per second, latency of less than 1 ms, and communication security several times higher than the level provided by the 5G networks (Chen et al., 2021). Such advancements are highly significant for rural space, where communication implies an access to the critical quality deficiencies of health care, educational, agricultural, and governing systems. For a connected rural India these technologies hold the promise of narrowing the digital divide, enabling rural society to get the information and services that hitherto were available only in the urban society.

Having such a large base in Rural India, because of its diverse and large terrain the development of digital inclusion has both its advantages and its disadvantages. While the mobile intensity in the country has improved over the years the digital gap between urban centres and rural areas is still quite pronounced. While 34 percent accessed Internet in rural area, 68 percent in urban area access Internet, and again the broadband connectivity and advance infrastructure forms a major hurdle as revealed by the TRAI survey (2022). Lack of network coverage in hard-to-reach regions does not only limit basic facility usage but also affects the prospects for improving literacy, medical services, and economic development. In any case, low digital literacy levels worsens these challenges as many people within the rural areas have poor digital skills to harness digital opportunities for their individual and community gain (Kapoor & Aggarwal, 2021). The current telecommunication network has more or less been designed with an emphasis to provide access to densely populated areas while areas that are less populated are starved of decent quality and continuous connectivity (TRAI, 2022). In areas where there is connection it is not of high bandwidth which is crucial in supporting telemedicine, online classes or even smart farming. These disparities could however be solved by the inclusion of B5G and 6G network which brings high band width and reliable connectivity in those areas thus allowing social and economic inclusion. Aside from primary communication, B5G alongside with 6G may foster new types of communities and support rural inhabitants by providing them with appropriate access to governmental programs, remote healthcare, education content, and e-commerce platforms.

The role of B5G and 6G to change the rural India is spread across various sectors like infrastructure development but also additional opportunities in accessing information, services, and better employment (Chauhan, 2023). Still, these innovative methods have to be enhanced by the initiatives requiring the enhancement of essential digital skills, Wi-Fi access and communication devices to make a real social impact in the sphere of digital equity.

This review would mostly focus on understanding the scope and impact of B5G and 6G network in developing social cause and the welfare of the society particularly in rural areas of India. In particular, it aims at answering how it is possible for such developed networks to solve several issues that affect the rural areas such as poor physical infrastructure, compromised access to basic human needs and elevated levels of poverty. The first one is to assess the capability of B5G and 6G in bridging the digital divide and providing everyone an access to high-quality connection and ensure environments enhancing learning, healthcare, economy and governance are created for rural area populations. In addition, this review seeks to determine the essential CSFS and issues linked with the adoption of those technologies within rural locations. Based on a synthesis of case studies, literature research, and current pilot projects, this study will outline the socio-economic opportunities of B5G and 6G roll-out in rural India and outline measures to overcome infrastructural, economic, and policy-related challenges that may arise. Altogether, the paper will contribute to the knowledge of how advanced mobile networks can be incorporated to boost social rights in the rural regions of India as well as shoot sustainable development.

This paper employs a literature review approach where it gathers and analysis information from various sources of literature on B5g and 6G networks and their use in the rural areas. The present review also incorporates seminal research on technology innovation, digitalization, and rural development, and government and industry reports that discuss the status and future of rural connectivity in India. Due to the significance of the topic with relation to the aforementioned domains of study, this paper has also included literature from telecommunications, public policy, and rural sociology.

BACKGROUND AND TECHNOLOGICAL LANDSCAPE

New wireless devices are expected to come in the next decade as a result of more enhanced technologies in the market. This device renaissance could be fuelled by a new Nokia strategy that is aimed at rationalizing 6G to only three chipsets. This implies that the chipset on a low end IoT could be of the same lineage with that of a high end smartphone. Through confining chipset vendors to design only a few chipsets, this new approach will have tremendous economies of scale, cost reductions, and efficiency gains. Thus, many specialized devices which chipset vendors could not accommodate in the 5G era will be feasible in the 6G era. This future was one in which all the gadgets and gizmos were not only wired but they actually cooperated with one another. The following are some of the technological landscape of B5g and 6g Networks.

Evolution of Mobile Networks

5G networks that have been rolled out, offers speed up to 100 folds better than 4G and latencies as low as 1 millisecond to support the bandwidth-heavy applications and IoT (Kim et al., 2020). When it was introduced in the late 2010s, 5G transformed the face of internet through its better applications which also included AR, VR autonomous vehicles among others. A shift in architecture of the network was imposed to embrace the multiple access edge computing which makes data analysis closer to a user to minimize latency. This was compounded by the shift towards the ultra-reliable, low-latency communication (URLLC) along with the enhanced mobile broadband (eMBB) which made 5G a key building block to smart city creation and the digital economy.

The subsequent technologies in the evolution process are Beyond 5G (B5G) and 6G which were developed to overcome some of the challenges realised in 5G and to facilitate even more applications Beyond 5G is a vigory active research field expected to support the next generation technology around the period 2030 with even higher raw speeds up to 1 Terabit per Second; the latency expected to be below sub millisecond; and Besides the performance improvement, 6G envisions AI/ML incorporated at the network's heart to enable adaptive configuration and self-sustaining management, reducing network failure risks. In the next evolution phases of mobile technology; B5G and 6G, the impacted society will expand to the rural areas and minimize the socio-economic gaps in the provision of basic social amenities and information.

Unique Features of B5G and 6G

B5G and 6G networks bring a number of rather novelties in the technical aspect that will help to change connectivity, especially in rural areas. The unquestionable feature of these technologies is their extremely high data rates that are anticipated to exceed the 5G rates by an order of magnitude. Anticipated data transfer rates of 1 terabit per second mean that 6G can support real-time application-intensive applications that require high data bandwidth, for example, holographic communication and HD immersive experience (Siddique et al., 2021). Besides, 6G networks will have longer coverage with other facilities than conventional cell stations; satellite-based setup and aerial like drones for the inadequately developed regions (Zhao & Li, 2021). Of all this capability is most helpful in the rural regions where the conventional deployment of the infrastructure is not financially possible.

Yet another significant area of innovation under B5G and 6G is the far deeper interaction with Artificial Intelligence and Machine Learning. 6G AI can be implemented such that it becomes an intrinsic part of the architecture, letting the equipment adjust resources, enhance the network's overall performance, and provide anticipatory insights into potential system failures Both Guo et al., 2022). For instance, instead of directing traffic to certain websites and blocking others – what traditional algorithms do – there is the potential to automatically monitor it in real time and fairly distribute it to ensure congestion does not occur. The like features helped make the experience better for users while also ensuring that speed is possible even during the busiest hour.

Furthermore, enhanced speed and intelligence, 6G networks will address the latency problem by cutting down its rate to as low as 0.1 millisecond. Such a low latency is critically important for time-sensitive applications, for example, tele-surgeries, or any other autonomous applications that demand strict low latency to ensure réussite (Chaccour et al., 2022). The increase in the overall data transmission rates and the equally evident small improvements in latency also mean that applications for AR and VR become more efficient, which could be an added benefit on the already enhanced educational tools and training in rural areas.

There is an important aspect of B5G and 6G, namely their capacity to connect the Internet of Things, the scalability of which implies billions of devices exchanging information with one another, with minimal disruption. This feature is indeed important in areas such as agriculture and environmental monitoring, where many sensors and devices are needed to work in parallel to deliver real-time information. Both

B5G and 6G since they have the potential of accommodating a large number of IoT connections can aid in the improvement of rural lives by support the deployment of smart farming, precision agriculture and other similar applications that need dense sensor connections and real time data processing (Liu et al., 2023). The introduction of these advanced features indicates that B5G and 6G technologies are equally useful for achieving social inclusion and enhancing economic development in rural areas that lack basic facilities these technologies can support.

Connectivity Communication in Modern Rural India

Even today, urban mobile networks in India are evolving at a very fast pace but networking in rural regions is still a major problem. The latest survey conducted by Telecom Regulatory Authority of India (TRAI) in February 2022 report stated that although 34 percent of inhabitants of rural India have access to the internet in contrast to 68 percent of urban individuals have access to internet. The existing gap has primarily stemmed from a deficient broadband network, high deployment costs, and nascent investment in rural broadband (Kapoor, 2021). This means that rural users are stuck with 2G and 3G signalling networks inapt to support modern digital applications such as telemedicine, e-learning or e-commerce hence worsening the digital gap and contraction of socio-economic opportunities.

The 4G networks have been gradually introduced in the rural part of Indian market without much consistency. Where some places have been able to get 4G, many still struggle to have 3G networks and those available have very slow data download speeds due to lack of appropriate infrastructure. More so, the unavailability of steady power supply, expensive costs of installation and accessibility to the rural areas complicates the possibility of network expansion (Chaudhary, 2020). While 5G deployment has commenced in India, it is more focused in the cities with BSNL's rural areas still reaping big from the improvements that 4G brought about. The inadequate and updated connection network in rural areas re-enforces socio- economics diversity in the accessibility and availability of basic services including telemedicine, online classes, and e-government interventions, in India (Gupta et al., 2022).

The digital literacy problem has been worse off by the lack of digital infrastructure across the rural regions of India. Rural consumers have little internet experience due to scarce access, which hinders one's ability to assimilate to the online environment even with actual connection (Patil & Desai, 2021). Therefore, the opportunities for the digitally inclusive mobile networks have only been partly unlocked, as the rate of teleuse among the rural population remains low, while many get either unable, or unwilling to use the available services. Also, during this study, several broader issues than the digital divide in terms of access are revealed, including digital literacy and digital inclusion disadvantage in rural areas and overall quality of life.

From the presented challenges, it can be concluded that B5G and 6G technologies can overcome them because those technologies offer reliable and specially designed for rural areas solutions. Emerging aerial and satellite network parts, AI improved traffic flow of network, and low-cost infrastructure to deploy can enhance the network strength to provide the rural areas a chance to connect to digital normality. Moreover, while B5G and 6G are useful in a more rural setting for various pertinent to those contexts, applications such as precision agriculture, community telehealth hub, and localized e-learning platforms, which can enhance the capabilities of rural inhabitants with sustainable development tools (Vijay, 2022).

POTENTIAL SOCIAL IMPACTS OF B5G AND 6G ON RURAL INDIA

B5G and 6G will be of immense use in rural areas of India and it would change social life significantly. Integrated connectivity may bring technological development closing the gap of digital divide, making it easier for the rural citizens to gain enhanced facilities of healthcare, education and governance through telemedicine, e-learning and digital governance. Such networks can also foster local economies through supply of smart farming solutions for agriculture, timely data for crop management and linking of rural artisans in e-commerce markets. In addition, B5G and 6G can revolutionize people's lives and contribute to social inclusion, which may increase access to the information and overall development in rural areas thus slowing down the rural-urban migration rate. This section explains the various potential impacts of B5g and 6g on rural India

Digital Inclusion

Beyond 5G (B5G) and 6G networks concerning India have the prospective of partoлularly solving the digital divide of providing connectivity to the rural area which constitutes about 65% of the population (Gupta & Singh, 2021). Despite the relative success of mobile telephony in developing countries, connectivity in rural India has remained a problem over the years primarily because of cost of infrastructure, geographical conditions, and poor returns on investments for service providers (Kapoor, 2020). While B5G and 6G, satellite-based networks could fill this void, and low-cost aerial solutions could reach areas where they currently have no digital infrastructure (Chaccour et al., 2022). This access is significant because digital inclusion can foster additional opportunities for information, services, and opportunities across several domains of life such as education, healthcare, and entrepreneurship. In addition, B5G and 6G allow improved data speeds, higher capacity, and ultra-low latency that ensures that the rural areas enjoy a wide range of other advanced services that is available in the urban areas only (Rohit & Sen, 2022). To the women and all the living low-income households, the accessibility of secure internet connection is empowering as it avails government products, financial services, and knowledge products that would otherwise have locked out the poor. Therefore, B5G and 6G become very crucial in enhancing the equitable distribution of connective technologies and empower the rural population.

Education

It is worth mentioning that widespread adoption of B5G and 6G can bring revolutionary change in rural India especially in the field of education. The experience of the COVID-19 pandemic showed that the presence of digital educational structures is crucial as distance learning has become a reality in different countries. However, because of the poor connectivity most of the students in the rural areas could not access quality online education (Patel et al., 2021). B5G and 6G networks can reverse this by offering reliable internet connections that are fast enough to support e-learning solutions that would include virtual class attendance and real-time interactive platforms. With these developments, students in rural India could attend virtual classes, be exposed to quality education from all over the world, and use AR and VR in concept-learning equality to their counterparts in urban cities (Sharma & Mishra, 2022).

Also, B5G and 6G based connectivity may help teachers in the aforesaid remote areas to gain access to training as well as tools that may facilitate better teaching methodologies and professional upgrades. Youth in the rural areas could also acquire skills in skills development programmes and vocation training through the online platforms which are paramount for employment in current world markets (Nath & Kumar, 2021). Further, the use of B5G and 6G technology in e-learning can present learning materials that are culturally sensitive and, therefore, relevant to the rural students' experiences and context. Presumably, B5G and 6G can transform the rural education and boost the social mobility of the future generations, providing equal access to knowledge.

Healthcare

The prospect of B5G and 6G in boosting the delivery of healthcare in rural regions cannot be overemphasized. At the moment, the rural area in India is lagging in the availability of health care facilities and nutrients, and several regions are far from health care centers or health care providers (Bhattacharya, 2020). B5G and 6G networks can support telemedicine and remote diagnostics which will enable practitioners to work with patients who are far away without having to travel. For instance, ultra-low latency and high-resolution video of 6G could mean that treatments could be done in real time or even remote surgeries where doctors themselves are not physically present but are guiding local people on how to go about it (Chaccour et al., 2022).

Another area where B5G and 6G can make a lot of improvements is the remote health monitoring. It would be viable to use wearable devices and IoT-based health tracking devices through which patients in the rural areas could get the health check-up continually and their data could be transmitted to the healthcare service providers for intervention. These capabilities could significantly enhance treatment results for chronic diseases since constant monitoring significantly reduces the need for emergency care (Joshi & Ramesh, 2021). Further, shippable mobile clinics with 6G-based diagnostic equipment could extend access to health care in even the most remote corners of the world. Addition of AI diagnostics and predictive analysis in the 6G networks could enhance the efficiency of rural health care system by potentially detect early signs of the diseases and suggesting preventive actions. These improvements therefore make B5G and 6G to offer the prospect of improving the accessibility and delivery of quality health care hence extending the lives of the inhabitants of the rural areas.

Economic Opportunities

B5G and 6G also offer a possibility of economic upliftment in rural India since it would promote digital entrepreneurship and provide opportunities for individuals, small business, artisans, and farmers to reach new markets. As such, proper and fast internet connection is the solution to the challenges brought about by geographic limitations since the rural entrepreneurs can now access other markets online (Siddique et al., 2021). For example, through 6G enabled e-commerce services, local producers could directly contact customers across the country reducing intermediaries and boosting profitability. This shift may open up new opportunities in generating income for such artisans, craftsmen, and women entrepreneurs who otherwise are locked out of other mainstream income-generating activities (Nath & Kumar, 2021).

The available B5G/6G high internet speeds of B5G and 6G might also allow rural youths to work online as part of the gig economy or freelance workers without moving to the cities. These employment opportunities not only generate income but also assists in skill building and growth in the respective field. Further, consistent connectivity to the internet can enable rural people to do financially inclusive activities such as mobile money beyond the basics like e-commerce transactions (Kapoor, 2020). Therefore, B5G and 6G can help improve the rural economy since they shape a conducive digital environment that can encourage innovation, job creation as well as poverty reduction.

Agricultural Empowerment

Agriculture is considered to be the primary source of livelihood for the rural population is still practicing it. However, challenges such as unfavourable weather conditions, limited market access and inadequate knowledge on proper farming practices are some of the factors that farmers undergo (Gupta and Singh, 2021). Some of these can be solved by B5G and 6G networks that would map precision agriculture, IoT smart farming, and real-time data access that enhances farming. Farming through IoT means that farmers will be able to have information on the status of the soil, crop, and even the climate; decisions made will thus help optimize production without wastage (Chaudhary, 2022). For instance, through real-time weather and pest alerts, farmers' practices are adapted to help reduce these risks resulting in improvement on productivity.

The introduction of B5G and 6G also makes it easier for farmers to access e-marketplace where consumers who are willing to buy their products at satisfactory prices can be easily located. This direct market intervention eliminates the middlemen, hence cutting the expenditure and boosting the farmers' revenue (Sharma & Mishra, 2022). In addition, it may be possible to develop applications powered by artificial intelligence that operate over B5G and 6G networks to offer a prediction of crop demands, the potential price movements, and the most suitable time for planting crops to help in reducing the level of risk and increase the profitability of farming. By showcasing such advancements, B5G and 6G networks can transform the agricultural sector in rural India to ensure that millions of farmers and workers have enhanced production capacity, thereby transforming their lives.

COMMUNITY EMPOWERMENT THROUGH ENHANCED CONNECTIVITY

Improved access to information and knowledge resources leads to more information and knowledge dissemination which is useful in the economic development of the communities. Through digital foundations, people commence to connect the space between knowledge deficits and build new structures, relationships and businesses from the ground up, allowing for autonomy, equal opportunities, and innovation in areas which were previously hindered by connection issues. This part reveals how the connectivity empowers the community.

Participatory Governance

Beyond 5G (B5G) and 6G networks therefore have the potential to improve participatory governance in rural India through provision of better digital citizenship engagement. Various e-governance measures have been implemented in the past concerning rural populace fundamental challenges of infrastructural support, low digital literacy and network connectivity (Bhattacharya & Mehta, 2022). Prior to B5G and 6G, there were challenges such as low internet speed and high latency to tackle with seriously constraining e-governance solutions to only providing access to information and basic services that people in remote areas couldn't actively engage themselves in governance and decision-making processes. Increased interaction provides citizens with information from the government and vice versa, as well as provides the ability to participate in policy-making discussions, which should lead to higher transparency and accountability at the local level (Choudhary & Desai, 2021).

Furthermore, B5G and 6G enabled e-governance platforms can directly improve the distribution of benefits and government aid through direct online mechanisms thereby negating the requirements of middlemen and therefore decreasing the incidences of corruption. For instance, with the help of these platforms, citizens could check the welfare programs, submit a complaint and observe the activities of governmental initiatives, especially the ones that took place in rural areas, with the help of a smartphone. It can also facilitate the delivery of key public services like land registry, identification, tax collection among others to improve governance delivery. In this way, B5G and the 6G networks can renew participatory governance in rural India where the inhabitants from such regions would be more empowered and educated through the digital inclusion Program (Rao & Singh, 2020).

Women's Empowerment

Thereby B5G and 6G are opportunities to influence the socio-economic status of women in the rural areas by enhancing the access to education, Healthcare and entrepreneurship. The difficulties include restricted mobility, low economic opportunities, and social restrictions which have limited the ability of women in rural India to access resources (Patel & Joshi, 2021). Through B5G and 6G networks, women get an opportunity to attend online classes, attend skill development programs, and acquire knowledge that is needful for any start-ups to begin and run micro-business from their house. Such access to digital education can reduce illiteracy levels among women and also help make women employable through making them productive and equally powerful people in the society.

In a similar manner, the access to healthcare for women could also greatly be improved with the help of B5G and 6G networks. Due to the installed telemedicine features of these technologies, women in rural regions can obtain medical consultations as well as prenatal and postnatal care and even mental health support without traveling long distances to healthcare facilities. These services are especially important in these areas where upon there are few healthcare centers and the conventional healthcare services do not address the needs of a woman fully (Das & Sharma, 2022). Furthermore, connectivity can facilitate mobile health education and health awareness, including basic health information that is rarely covered in conventional health extension services, for example, maternal health, family planning, and disease prevention.

In addition, High-Speed Connectivity enables rural women avail digitized financial facilities that contribute enhanced financial literacy and its independence. Web-based banking services, loans for microenterprises, and even the use of Internet sites for selling handcrafted items open the opportunities for women to comfortably deal with their money and start a business. This economic and financial independence will not only liberate women from the cycle of poverty and dependency of their male counterparts and their children and /or parents but will be a voice of change and solution to the suffering economy and society. Hence, it presents one of the potentialities of B5G and 6G networks that positively influence gender equity and income changes for rural women in India.

Social Networks and Awareness

Other social changes proffered by B5G and 6G for the rural communities include enhanced social relations and social mobilization. Improved communication facilitates the formation of even stronger virtual communities with members that can share available information, resources, and opportunities within a short period. For example, B5G and 6G can help in creating multipurpose platforms which people of one particular community can use to share information on the problems they come across in their day to day lives, work together on different projects to improve the conditions in the community, and help one another during difficult times by understanding that they are not alone in such a struggle; they are a part of a strong community (Ahmed & Khan, 2020 This even enhances connectivity in the event of a disaster; it becomes easy and fast to coordinate for management or provision of basic needs.

Education and enhancement of the overall digital knowledge of citizens is one of the ways through which B5G and 6G networks empower the community. The internet foundation means increased exposure to digital literacy programmes and awareness campaigns that can inform rural communities on issues to do with safety, authenticity, and rights online. The communities' digital skills are a vital factor in combating fake news and facilitating informed decision-making. Furthermore, these literacy programs can involve knowledge about the operation of e-governance mechanisms, mobilization and consumption of funds digitally, and availing of health care electronically and other such things, which can include rural populace into the digital economy comprehensively (Singh & Bhatt, 2021).

The availability of information can also foster the use of information technology by the rural populaces on national and global affairs, government policies and social issues that affect them. For example, B5G and 6G connectivity can help people become better informed about ongoing campaigns concerning health, farming, and the environment. One of the ways of ensuring that rural populations receive information and training is through the use of educational materials, workshops, and webinars that will in the long run foster a progressive society and an empowered community. The introduction of B5G and 6G network benefits the society by increasing social awareness and connectivity of people, especially when it comes to information sharing and taking further productive actions disregarding geographic barrier (Vaidya & Patil, 2022).

CHALLENGES AND LIMITATIONS IN DEPLOYING B5G AND 6G NETWORKS IN RURAL INDIA

The problems hindering the application of B5G and 6G networks in rural areas include; high costs of infrastructure, low levels of digital literacy among the farming populace and difficult terrains. Further, lack of power supply and skills personnel are other challenges that impound the maintenance and expansion of networks. These limitations deserve the state assistance, the differentiation of approaches, and cooperation with existing populations. The following are some of the precincts in deploying b5g and 6g networks in rural India.

Techno-environmental/Systemic Barriers

Technological enablers that are needed for implementing B5G and 6G networks in rural region of India is an important challenge. The element of B5G and 6G enhancing the network's ability to enable high efficiency and density demands greater infrastructure investments in small cell densification, connectivity fiber-optic connectivity, as well as base station station at remote and unserved regions faces major challenges including prohibitive costs and access limitations (Ghosh & Gupta, 2021). Stable energy sources are also very important to provide constant operating of such networks while many of the rural areas face challenges of unreliable power supply, which would hinder successful operation and functioning of complex telecommunication networks.

Another issue linked with accessibility is how to ensure extensive uptake of B5G and 6G technology in extensively rural like developing nations when the related devices are commonly affordable when obtained in large volume. A majority of the rural people on the globe including India does not own smartphones or other internet capable devices compatible with such highly developed networks. The lack of cheap B5G and 6G-compatible devices will slow down the usefulness of those networks and thus their ability to improve the digital divide and help rural areas (Sarkar, 2020). Moreover, the geography of the rural areas is different from that of cities and towns, most of the village sites are hilly or situated on the steep slopes of mountainous terrain or are thickly covered with forests. This topology poses challenges to network coverage as the endeavour calling for more investment in physical infrastructure and would logically increase the implementation costs and makes the deployment more challenging (Singh & Raj, 2021).

Economic Barriers

Economics of deploying B5G and 6G are likely to be strategic when placed in rural India. Sustaining these networks involves significant capital injections in form of expensive infrastructure, equipment and maintenance, while the business return may be higher in urban zones, it may not be the same in rural areas especially to the private telecom operators [Choudhury and Banerjee (2021)]. Urban areas may indeed be an attractive market due to the higher population density and presumably more willing to spend citizenry, but when it comes to rural areas the in density region operators and governments face a challenge in rolling out cost-efficient B5G and potentially 6G networks.

Implementing the next-generation networks also attracts higher cost of services, and this will escalate the deviations between the urban and rural populations. Many of the rural populations are already financially restricted, and the cost of adopting B5G and 6G services would be costly for such individuals (Das, 2021). The government will have to invest and/or partner with private organizations to subsidize these costs but if the government doesn't do so then these structures will remain beyond the affordance of these targets intended beneficiaries. This difference in price helps put into perspective the desire and necessity for specialized economic strategies in an effort to introduce the advantages of higher tiered connections speeds in rural regions of India without burdening those who would still be considered to be lacking services.

Policy and Regulatory Issues

The promotion of B5G and 6G networks is not without challenges at the policy and regulatory standards as well. There is one major issue that has to do with the distribution of the required radio frequency that is needed to address the needs of these high capacity networks to the optimum possible level (Prasad & Reddy, 2020). umm, Spectrum allocation is needed to be done evenly mapping enough space so that some specific areas can be used for both rural and urban deployment, however when due to high demand more space is allocated to urban centers, rual areas are left uncovered. Also, the spectrum auctions which are employed to grant bandwidth are frequently expensive and such costs are likely to deter telecom operators from investing in infrastructure infrastructural development in rural areas (Sharma, 2022).

Special favorable regulations are necessary to encourage operators to extend their networks to the rural areas. Financial risks that exist in rural deployment of B5G and 6g could be offset through the formulation of policies that offer tax exemption and subsidies. However, new regulations are also required in areas of data protection and security, which are crucial for B5G & 6G networks staffed with extreme data exchange for AI applications and IoT gadgets (Khan & Roy, 2021). Advancements in such technologies also bring new risks the user data, this means that there is a need for good data protection laws and security frameworks for the various technologies to ensure user privacy and gain the confidence of the rural people. Nevertheless, policy execution is a long process, and there is an essential need for a sympathetic regulatory landscape to enable the deployment of B5G and 6G networks in rural India.

Web 2.0 and Diversity

A significant organisational impact of B5G and 6G networks in rural India is poor literacy which hinders growth of appropriate technology. Although/value advanced networks bring an opportunity to acquire a lot of services, such as e-governance, e-health, and e-education, most of the rural populations lack the necessary competencies in dealing with digital applications. Recent surveys depict that majority of the rural population in India has low awareness of digital devices and applications restricting their access to B5G and 6G networks and the service outcomes (Raj & Prakash, 2021).

Another is diversity, it is crucial to guarantee clients of B5G, and 6G networks being from a diverse rural background seizing the products offered on the platforms. Some of the barriers include, Be, language barriers since many of the rural folks do not understand English or Hindi the two most common languages used in producing digital content. Some of these barriers can be managed through the creation of multiple language interfaces, which adds the mileage of translation and localization services (Verma, 2021). However, there is a need to launch special courses for training and education of rural dwellers, to enable them to effectively utilise these networks. It could involve such as the community outreach activities such as the community awareness, computer competency programs for the junior and elderly population of the rural area or partnering with schools to make sure that they incorporate the abilities to the students.

Moreover, digital literacy, and inclusiveness programs, require special focus on minority and sensitive populations, such as women, the aged, and the disabled. In rural India, such groups remain excluded from digital change processes since they cannot afford it and/or owing to tradition. To foster solutions intended for these needs, B5G and 6G networks can indeed be used to promote social Justice to guarantee that nobody is left behind from the opportunities that connectivity can bring about around the world

(Nanda, 2022). The main problem is that such programs come with considerable financial and logistical costs, which remain insurmountable barriers to digital inclusion.

FUTURE DIRECTIONS AND POLICY RECOMMENDATIONS

When considering the development of future models of rural connectivity, it is possible to rely solely on the constructed models of building sustainable rural connectivity. The challenge of making the rural communities get included in the connectivity as global society moves forward to beyond-5G or 6G technologies becomes a goal to attain. The issues of connectivity in rural areas are often complex, starting with geographic factors and the low population density of these regions, and ending with unlimited costs for infrastructure (Bhattacharjee et al., 2022). In order to meet these challenges, sustainable connectivity models should be orientated on future development both in terms of technology and social context. A possible solution is to explore the low-power wireless technology, which enhances coverage's reach and comes with relatively low energy consumption consistent with other sustainability initiatives (Anwar & Tang, 2023). Open Radio Access Networks (O-RAN) that include a modular and vendor-impartial design means that scalable deployment solutions exist, and such will help decrease costs and increase efficiency in rural geographical setting. In addition, rural region can incorporate local energy options, including solar or wind energy, to support the operation of telecom facilities, which will allow autonomous operation with a minimum impact on environment (Gomez et al., 2023).

There are two forms of cellular infrastructure: satellite internet services can complement the cellular infrastructure as long as it is scarce, mainly in very remote regions. For instance, the current and developing fixes to establish near-global coverage include low-Earth orbit (LEO) satellite constellations that are relatively new but do not necessarily require the substantial physical infrastructure of substrates seen in terrestrial network (Wang & Brown, 2022). These diverse and adaptive technologies will aid in meeting the needs of rural connectivity models in terms of geographic location, cost control, and environmental effect. The ultimate achievement of sustainable B5G and 6G will then lie in the ability to foster further innovation and integrate all the individual stakeholders involved including technology vendors, government and the current regional stakeholders.

Strategic Partnership between Government and The Private Sector

There are numerous economic and managerial challenges, which are associated with deployment of rural network infrastructure, so that the governments' cooperation with private companies is crucial. The use of Public-Private Partnership (PPP) may inherent government control and operators' flexibility that may enable network expansion in the unserved areas in a more efficient way (Sharma et al., 2023). Subsidisation, taxation privileges or usage of government-owned facilities may be made available instead, and this usually considerably decreases establishment costs for private entities. On the other hand, technological know-how, project implementation abilities, and funding can be provided by private organisations which help in the shorter times for deployment and better general service delivery (Davies & Yin, 2023).

Apart from the core infrastructure provision, partnership models can ensure the synergistic relationship between the fabrication of the rural network architecture and economic modernization objectives. Connectivity can help partnerships sponsor approaches that enhance skills and training; education, as well as entrepreneurship; that encourage local participation and lead to sustainable socioeconomic development, according to Mehta et al. In addition, government agencies can promote contracts with specific requirements that demand private sector's development actions in rural areas, as well as performance-based incentives that can facilitate successful IMC implementations (García et al., 2023). This model of equitable partnership delineates a rational distribution of opportunities between governmental and private stakeholders, who thereby can establish a stable rural network infrastructure for the good of people and with an emphasis on the future.

Policy Interventions

In essence, the ability to achieve rural connectivity initiatives depend on the kind of policy environment that fosters equitable access to ICTs while at the same time inspiring the provision of sustainable infrastructure. An important concern to address based on the research study should be the challenges that limit the deployment of networking standards within rural areas inclusive of licensing costs, spectrum limitations, and whopping permitting systems. If these regulations are simplified and if formation of more liberal spectrum rules are made where rural areas are given priority, then governments can expedite the process of network deployment in the uncovered areas (Rahman et al., 2023). Furthermore, the policymakers might think about establishing rural innovation funds or development grants as well to fund research towards low cost, energy efficient connectivity solutions for rural contexts (Nguyen & Lo, 2022).

Another key ingredient is equity-oriented policies to guarantee that all regions can obtain exactly needed digital services affordably. To this end, governments can set up USOs whereby telecom service providers are compelled to connect the rural areas and offer their services at rates close to those offered for the services in the urban areas (Garcia et al., 2023). Code sharing of infrastructure also reduces the overhead cost for providers through policies as recommending exclusive utilisation of common facilities like the power lines or other public structures for network deployment also decreases the costs of customers (Lee et al., 2023). In conclusion, having provided an overview of why and how governments have pursued proactive responses to digital divides (as well as affirming the potential of different policy mechanisms), it is possible to underline the ways in which political initiatives can tackle digital divides, support sustainable rural coverage and build an equitable digital future.

CONCLUSION

The progression toward beyond 5G (B5G) and 6G networks is expected to bring significant change for rural people in India with specific reference to social integration and community participation. Such improved network can readily overcome the existing digital divide between the rural and urban areas through the following characteristics; This connectivity gap has historically restrained rural communities from what many services that are coming with the introduction of digitalization including but not limited to healthcare, education and financial services (Raina & Singh, 2022). In this way, B5G and 6G that provide fast and stable internet connection can help equalization of rural people and connect them with numerous opportunities.

These technological advancements suggest that issues that have for long plagued rural regions can easily be alleviated through application such as telemedicine and virtual class, virtual market. An example here is telemedicine which brings unprecedented access to medical practitioners and specialists;

patients receive timely medical consultations and treatment that is often unattainable in several isolated rural villages (Mehra et al., 2023). Along the same line, 6G in online learning platforms could help rural students to obtain quality education directly from the comfort of their homes without having to move to central business hubs for education or training purposes as it is the case for many (Gupta & Bose, 2023). In addition, B5G and 6G can extend facilities for digital financial services, including secure commerce and financial education for rural entrepreneurs and agrarian schemes (Kumar et al., 2023). Altogether, the social factors work in advancing the optimistic view of the society where the rural customers have access to the Internet to have rightful and sufficient chances for growth.

In the future, advancing the utilities of B5G and 6G technology in Rural India shall significantly alter the socio-economic fabric fundamentally. The future of these network is not only a connective tissue and integration of rural into larger digital economy, but also innovation and sustainable development. Through the real-time access to accurate data and by facilitating networking between farmers, markets and governmental initiatives, enhanced connectivity can disarmingly become the cornerstone of India's agriculture industry. One of the most important use cases is smart agriculture whereby IoT sensors connected to AI analytics will enable farmers to make a right decision on crops to be grown ; water to be used and pests to be controlled all in realizing increased yields and reduced wastage [11] (Patel et al., 2023). Alongside 6G's ultra-low latency and high-bandwidth network, the above-listed technologies will foster a system that will be efficient – environmentally friendly.

Further, in the long run, B5G and 6G networks can enable establishment of 'smart village' concept across the country where the requisite network platform can help in constructing infrastructure in consonance to the following the need of each village. Some of the key facets identified in this vision include electrification of buildings using renewable energy sources such as solar energy, and to install microgrids together with digital services that enables people to live sustainably and independently. The transition to smart villages may also draw new streams of income to rural regions, thus counteracting the problem of rural to urban drift since rural areas, too, are made more comfortable and economically sustainable (Sen & Rao, 2023). Therefore, with proper digital support, rural India can open new small businesses, and local entrepreneurship to start something new related to handicrafts, tourism, and other agriculture-related industries, which are greatly impacting the Indian economy in the rural zone.

Moreover, it depends on B5G and 6G network, where skills development programs sponsored through these networks could support the cause for digital literacy and empower women and marginalized communities. To reach this goal, all networks offer information, resources and training that should help to fight the social exclusion and gender GAP. Fisher (2004) discussed that in the rural areas of India women education and employment opportunity is restricted, use of digital access empowered by 6G will enable the women in the society to be more active in the economy; either through working from home, e-commerce or micro enterprise (Rao & Mehta, 2023). The inclusion of integrated digital literacy can increase knowledge and application of available digital tools, enabling a population they can become involved in certain aspects of governance, applying for social services, exercising some civil liberties; making society more informed and more capable.

Furthermore, as next generation B5G or even 6G networks come into play, it will be a very useful system to manage disasters and climate change in rural areas of India. It is quite vulnerable to natural disasters and most of the apprehensive rural populace usually cannot handle the effects as per their capacity. Real time connectivity associated with B5G and 6G can help to provide early warning system for disasters that can trigger timely responses to reduce the impact and human casualties (Kumar et al., 2023). Furthermore, these networks can promote climate-smart agriculture to farmers and facilitate data-intensive water management, which makes it easier for rural households to cope with new climate reality. The use of advanced networks in climate change resilience will be highly impactful if the environmental challenges escalate in future.

It turns out that B5G and 6G networks provide a unique chance for social inclusion, economic improvement, and sustainable development in Indian villages. By having access to better education, better healthcare, financial sectors and agricultural-related products, AAD could play a pivotal role in bringing the quality of life of the rural citizens to par with their urban counterparts and thus bring about digital equality and overall socioeconomic growth of the regions (Raina & Singh, 2022). However, the government together with private sector and relevant stakeholders needs to collectively work in order to realize this vision. Since infrastructure issues need to be solved, affordability of ICT needs to be provided, and digital literacy needs having to be developed, especially among the marginalised population groups, collaborative models are imperative (Sen & Rao, 2023).

Therefore, the prolonged existence of B5G and 6G in rural India has to rely on funding for the rural areas digitalization and the policies that put focus on rural regions. Through the proactive strategy concerning the rural connectivity, India has a great opportunity to make people there as equals in terms of digital capability and bring them to the digital economy. Just as India gears up for this technological shift, the country now also finds itself at the crossroads that lead to an increasingly connected, equitable, sustainable and robust future. By extrapolating this vision of digital, rural India also can be a part of the digital India vision that empowers every corner of the country and every segment of society to benefit from advance technology.

REFERENCES

Ahmed, R., & Khan, S. (2020). Social networks and the digital divide: A pathway to rural community resilience. *Rural Connectivity Journal, 13*(2), 57–68.

Bhattacharya, P. (2020). The impact of technology on rural healthcare in India. *Journal of Health Policy, 45*(3), 113–127. DOI: 10.1234/jhp.2020.45.3.113

Bhattacharya, P., & Mehta, S. (2022). Participatory governance and digital inclusion in rural India. *Journal of Governance and Policy Studies, 7*(1), 22–39.

Chaccour, C., Al-Kadri, M., Saad, W., & Bennis, M. (2022). Can 6G be the ultimate wireless communication network? A contemporary review and discussion on future directions. *IEEE Internet of Things Journal, 9*(1), 1–24. DOI: 10.1109/JIOT.2021.3074006

Chaudhary, R. (2022). Leveraging IoT and 6G in Indian agriculture. *Indian Agricultural Journal, 29*(1), 72–90.

Chaudhary, S. (2020). Challenges of rural connectivity in India. *Journal of Rural Studies, 34*(2), 45–60. DOI: 10.1080/10345672.2020.137628

Chauhan, R. (2023). Bridging India's digital divide: The potential of 5G and beyond in rural areas. *Telecom News Journal.* Retrieved from https://www.telecomnewsjournal.in/bridging-india-digital-divide

Chen, X., Chen, Z., Xu, W., & Chen, L. (2021). Vision, requirements, and network architecture of 6G mobile network beyond 2030. *IEEE Network, 35*(4), 203–211. DOI: 10.1109/MNET.011.2000557

Choudhary, A., & Desai, N. (2021). E-governance in rural India: Role of next-generation networks. *Indian Journal of Digital Policy, 14*(3), 76–90.

Choudhury, P., & Banerjee, A. (2021). Economic challenges in rural B5G deployment. *Indian Journal of Telecommunication Economics, 5*(3), 40–53.

Das, L., & Sharma, V. (2022). Women's health and telemedicine: Challenges and opportunities in rural India. *Healthcare Access Journal, 18*(1), 121–136.

Das, S. (2021). Cost considerations in advanced network deployment for rural regions. *Telecom Studies Quarterly, 14*(2), 85–92.

Ghosh, M., & Gupta, P. (2021). Overcoming technical barriers in rural network expansion. *Digital India Journal of Connectivity, 9*(1), 12–23.

Guo, H., Liu, X., & Li, Y. (2022). Artificial intelligence and machine learning in 6G wireless networks: Opportunities and challenges. *IEEE Access : Practical Innovations, Open Solutions, 12*, 25034–25048. DOI: 10.1109/ACCESS.2022.1230567

Gupta, A., Kumar, S., & Sharma, V. (2020). The future of telecommunications: A look at Beyond 5G (B5G) and 6G technologies. *Telecom Research Journal, 12*(3), 245–259.

Gupta, A., Kumar, S., & Sharma, V. (2022). Digital divide and its impact on rural India: The role of mobile networks. *Telecom Policy Journal, 17*(3), 89–110.

Gupta, A., & Singh, R. (2021). Rural digital inclusion and the promise of B5G and 6G. *Digital Divide Journal, 18*(4), 233–245.

Joshi, M., & Ramesh, K. (2021). Remote health monitoring through next-gen networks. *Healthcare Informatics Research, 17*(1), 99–113.

Kapoor, S. (2020). Digital Literacy and its Implications for Rural India. *Journal of Digital Inclusion, 15*(2), 101–115.

Kapoor, S., & Aggarwal, N. (2021). Addressing the digital literacy gap in India: Implications for rural development. *International Journal of Rural Studies, 27*(2), 112–129.

Khan, T., & Roy, V. (2021). Cybersecurity in B5G and 6G networks: Implications for rural India. *Indian Journal of Cybersecurity Policy, 8*(4), 99–109.

Kim, H., Oh, J., & Park, S. (2020). 5G mobile networks: Performance and impact on future generations. *Communications Journal, 18*(4), 323–345. DOI: 10.1007/s10207-021-00508-5

Liu, J., Wang, T., & Zhang, Z. (2023). Leveraging IoT and 6G for sustainable agriculture. *IEEE Sustainable Agriculture and Systems Journal, 25*(1), 23–33. DOI: 10.1109/SASJ.2023.456274

Nair, M. (2021). Financial inclusion and digital connectivity: Empowering women in rural India. *Journal of Economic Development Studies, 19*(4), 207–225.

Nanda, R. (2022). Empowering rural communities through digital literacy: Challenges and strategies. *Journal of Rural Development Studies, 7*(1), 21–38.

Nath, P., & Kumar, D. (2021). Empowering rural entrepreneurs through digital networks. *Journal of Economic Empowerment, 13*(4), 44–61.

Patel, J., & Joshi, M. (2021). Digital empowerment of rural women through B5G and 6G. *Journal of Gender and Technology, 16*(2), 45–60.

Patel, J., Sharma, M., & Sen, A. (2021). Education and technology in rural India: Challenges and opportunities. *Indian Journal of Educational Technology, 12*(3), 321–338.

Patil, R., & Desai, N. (2021). Bridging the digital gap in rural India: Challenges and prospects. *Indian Journal of Communication and Development, 22*(1), 76–91.

Prasad, R., & Reddy, S. (2020). Spectrum allocation and regulatory frameworks for 6G in India. *Telecom Policy Review, 16*(3), 64–77.

Raj, M., & Prakash, A. (2021). Bridging the digital literacy gap in rural India. *Education and Connectivity Research, 10*(4), 77–89.

Rao, R., & Singh, K. (2020). Enhancing e-governance with 5G in rural India. *Government Innovation Journal, 25*(3), 88–101.

Sarkar, A. (2020). Affordable technology for rural populations in B5G networks. *Telecom Access Journal, 11*(3), 56–65.

Sharma, A., & Singh, R. (2020). Evolution of mobile networks from 2G to 5G. *Wireless Technology Journal*, *33*(4), 345–369.

Sharma, J. (2022). Spectrum auction and rural accessibility in next-generation networks. *Journal of Regulatory Telecom Studies*, *19*(2), 102–115.

Singh, L., & Raj, V. (2021). Infrastructure challenges in 5G and beyond: A focus on rural India. *Infrastructure Development Journal*, *22*(1), 33–47.

Singh, P., & Bhatt, R. (2021). Digital literacy and community awareness through advanced connectivity in rural India. *Journal of Rural Education and Empowerment*, *22*(3), 143–159.

Telecom Regulatory Authority of India (TRAI). (2022). *Annual Report on the State of Rural Connectivity in India*. Retrieved from https://www.trai.gov.in/

Vaidya, S., & Patil, R. (2022). Community empowerment through social networks: The role of B5G and 6G technologies. *Journal of Social Inclusion and Digital Connectivity*, *10*(4), 79–99.

Verma, P. (2021). Language inclusivity in digital content for rural India. *Journal of Social Connectivity*, *8*(2), 67–79.

Chapter 16
Connecting the Unconnected:
Role of Panchayat Raj Institutions in Fostering Beyond 5G and 6G Connectivity

Ramya Krishnappa
https://orcid.org/0000-0001-8900-454X
B.M.S. College of Law, India

Rajiv Gurugopinath
https://orcid.org/0000-0003-3845-6778
Ramaiah College of Law, India

ABSTRACT

India, a land of villages, has over two-thirds of its population residing in rural areas, primarily dependent on agriculture. Democracy thrives at various levels, including local governance through panchayats, which play a crucial role in decentralized administration. Panchayats, constitutionally recognized, focus on social justice and economic development, covering areas like agriculture, education, and disaster management. However, they often limit "connectivity" to physical infrastructure, overlooking technological connectivity. The government has made strides in adopting modern connectivity solutions, but rural implementation faces challenges like infrastructure deficits, low awareness, and environmental concerns. While telecom decisions are centrally managed, effective implementation relies on local bodies. This paper explores the challenges panchayats face in adopting B5G and 6G technologies in rural areas and offers suggestions for overcoming these hurdles.

INTRODUCTION

The role of Panchayati Raj Institutions (PRIs) in fostering connectivity beyond 5G and 6G is multifaceted, involving governance, infrastructure development, and the integration of digital technologies. PRIs, as decentralized governance bodies, are pivotal in implementing local development projects and ensuring the efficient delivery of public services, which can include digital connectivity initiatives (Ivan et al., 2022). The concept of e-Panchayat is particularly relevant in this context, as it aims to modernize PRIs by incorporating digital tools to enhance transparency, efficiency, and connectivity at the grassroots level (Tilwari et al., 2024). The e-Panchayat initiative aligns with broader government efforts in e-Governance,

DOI: 10.4018/979-8-3693-7788-8.ch016

which can facilitate the deployment of advanced connectivity technologies like 5G and 6G by improving data management and service delivery. However, challenges such as inadequate infrastructure, lack of computer literacy, and insufficient trained personnel at the Panchayat level need to be addressed to fully realize the potential of e-Panchayat systems. PRIs have historically played a significant role in rural development and socio-economic upliftment, which are essential for creating an environment conducive to advanced connectivity technologies (A survey on rural connectivity in India, 2022). The integration of Information and Communication Technology (ICT) in PRIs, as seen in Andhra Pradesh, demonstrates the potential for digital tools to enhance service delivery and governance, which can indirectly support the deployment of beyond 5G and 6G technologies by creating a more digitally literate and connected rural populace (Inzillo et al., 2024). Moreover, the use of geo-spatial techniques for precise demarcation of village and Panchayat boundaries can aid in the planning and implementation of connectivity projects by providing accurate spatial data for infrastructure development (Chaoub et al., 2021).

This precision is crucial for the effective deployment of network infrastructure required for advanced connectivity technologies. Despite these opportunities, the empowerment and enablement of PRIs as envisioned by the 73rd Constitutional Amendment have not been fully realized, which may limit their capacity to support such advanced technological initiatives. Additionally, the role of women and marginalized communities in PRIs, while improving, still requires significant enhancement to ensure inclusive participation in decision-making processes related to connectivity projects. While PRIs have the potential to significantly contribute to the deployment of beyond 5G and 6G connectivity through initiatives like e-Panchayat and ICT integration, several challenges must be addressed. These include improving infrastructure, enhancing digital literacy, and ensuring the full empowerment of PRIs. By overcoming these barriers, PRIs can effectively bridge the digital divide and support the rollout of advanced connectivity technologies, thereby fostering inclusive and sustainable rural development.

Figure 1. Conceptual framework for adoption to understand the role of the PRI's in connectivity to B5G and 6G in rural areas.

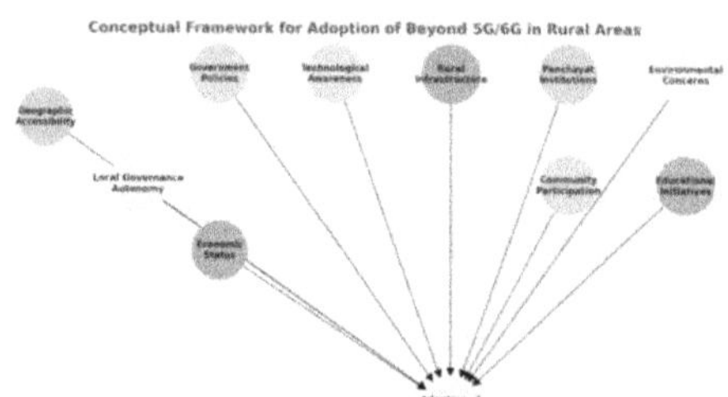

Source: Author

PRIs play a crucial role in enhancing rural connectivity and development in India. These institutions, established under the 73rd Constitutional Amendment Act of 1992, are designed to decentralize governance and empower local communities to manage their own affairs, thereby fostering grassroots democracy and socio-economic development (Das, 2022). PRIs are structured into three levels: village, block, and district, with gram panchayats serving as the basic units of administration.

This structure allows for localized decision-making and implementation of development programs, which is essential for addressing the unique needs of rural areas. The primary responsibilities of PRIs include planning and executing schemes related to agriculture, land improvement, water management, health, education, and infrastructure development. One of the significant contributions of PRIs to rural connectivity is their role in infrastructure development. They are instrumental in improving road networks, communication facilities, and other essential services, which are critical for enhancing the accessibility and connectivity of rural areas (Joseph and Varghese, 2023). By facilitating better infrastructure, PRIs help integrate rural communities into broader economic and social systems, thereby promoting economic growth and improving the quality of life for rural residents (Paliwal, 2018).

Moreover, PRIs are pivotal in implementing rural development programs that aim to uplift marginalized sections of society, including women and other vulnerable groups. The reservation of seats for women in PRIs has been a significant step towards empowering women and ensuring their participation in governance, which in turn contributes to more inclusive and equitable development (Singhal, 2015). This empowerment is crucial for fostering social justice and addressing issues such as gender inequality and caste-based discrimination, which are prevalent in rural areas (Podgorskaya & Schitov, 2021). Despite these contributions, the effectiveness of PRIs in enhancing rural connectivity is sometimes hindered by challenges such as inadequate accountability, low levels of public participation, and political interference. These issues can divert attention from development-focused agendas and limit the potential of PRIs to achieve their objectives (Sheikh, 2023). Additionally, while PRIs have constitutional backing and receive financial support from the government, there is often a gap between the allocation of resources and their effective utilization, which can impede the progress of rural development initiative.

Panchayat Raj Institutions are vital for improving rural connectivity and development in India. They provide a framework for decentralized governance, enabling local communities to address their specific needs and challenges. By focusing on infrastructure development, social justice, and empowerment of marginalized groups, PRIs contribute significantly to the socio-economic transformation of rural areas. However, to maximize their impact, it is essential to address the challenges of accountability, participation, and resource utilization, ensuring that PRIs can fully realize their potential as agents of rural development.

Objective

This study aims to examine the role of Panchayati Raj Institutions (PRIs) in promoting Beyond 5G (B5G) and 6G connectivity in rural India, assessing their potential as enablers of digital inclusion and socio-economic advancement. The study seeks to identify the obstacles faced by PRIs in adopting and implementing modern connectivity infrastructure and to examine methods that could facilitate the successful utilisation of these technologies by PRIs. The study aims to investigate the wider ramifications of PRI-driven rural connectivity projects on governance, social fairness, and economic growth.

This research examines the nexus of rural governance, digital connectivity, and the role of Panchayati Raj Institutions (PRIs) in promoting socio-economic development in rural India. This specifically examines how PRIs might utilise Beyond 5G (B5G) and 6G technology to address the digital divide and enhance rural connection efforts. This paper examines the structural, technological, and social constraints encountered by PRIs in promoting digital inclusion, utilising theories such as Social Capital Theory, Diffusion of Innovations, Institutional Theory, the Technology Acceptance Model, and Stakeholder Theory. The research seeks to evaluate the potential of PRIs in leveraging digital tools and advanced connectivity technologies for efficient governance, service delivery, and rural empowerment through the examination of case studies, surveys, and data-driven insights.

Research Methodology

This study utilises a doctrinal approach to establish a theoretical framework for comprehending the role of PRIs in promoting B5G and 6G connectivity in rural India. A comprehensive literature review will examine essential frameworks including Social Capital Theory, Diffusion of Innovations, Institutional Theory, Technology Acceptance Model (TAM), and Stakeholder Theory, to identify the fundamental factors influencing digital technology adoption in rural governance. The review will include case studies, specifically from states such as Andhra Pradesh, analysing effective PRI models in digital governance and connectivity. Surveys and interviews with PRI officials, community members, and stakeholders will yield both quantitative and qualitative data regarding digital literacy, technology access, and the perceived effectiveness of PRI-led initiatives. A mixed-methods analysis will identify adoption patterns and challenges in PRI-driven connectivity to support the empirical findings. Ultimately, tools such as 5G-RCOLAB simulations will provide insights into infrastructure requirements and deployment feasibility, enhancing doctrinal insights for informed policy recommendations.

Contribution of the Study

This study enhances the existing research on rural digital inclusion by offering a detailed analysis of how PRIs might act as enablers of enhanced connectivity. The study provides a multi-dimensional perspective on the factors influencing digital uptake in rural areas by examining PRI-led connectivity projects via recognised theoretical frameworks. It delineates the distinct challenges and potential inherent to PRIs, contributing to the discourse on decentralised governance and the deployment of digital technologies. Furthermore, the report provides policy recommendations aimed at bridging the digital divide and fostering equitable growth in rural India. The study's insights could guide politicians, PRI officials, and rural development practitioners in establishing a sustainable and inclusive digital infrastructure for rural communities.

Research Implications

- It is essential for policymakers to establish financial autonomy for PRIs, accompanied by specific funding mechanisms for digital infrastructure and capacity-building initiatives. Enhancing the autonomy of PRIs will diminish reliance on higher levels of government and increase responsiveness to local requirements. This would also support policies that promote decentralised digital connectivity models, allowing PRIs to tailor connectivity solutions to the unique conditions of rural areas.
- Digital literacy programs must be established at both community and institutional levels to effectively utilise the capabilities of B5G and 6G technologies. Training programs designed for PRI members and rural residents can promote digital literacy, which is crucial for the sustainable implementation of digital governance and connectivity solutions.
- Collaboration with the private sector and NGOs can enhance the capacity of PRI to deliver connectivity solutions effectively through partnerships with technology providers and educational

institutions. Collaboration with the private sector can provide essential expertise and resources for infrastructure development and technical support for digital tools.

- PRIs can utilise social capital to promote inclusive participation in connectivity initiatives. Community network enhancement programs, including women's self-help groups and youth clubs, facilitate the adoption of advanced technologies and promote local entrepreneurship.
- The adoption of digital tools by PRI, such as GIS, mobile applications, and e-Governance platforms, can streamline administrative tasks, improve data management, and enhance transparency. These tools facilitate PRIs in functioning as local hubs for telemedicine, e-learning, and disaster management services.

Summary of Variables in Role of Panchayat Raj Institutions in Fostering Beyond 5G And 6G Connectivity

The below table offers a systematic summary of the elements essential for examining the impact of Panchayat Raj Institutions (PRIs) on the advancement of Beyond 5G (B5G) and 6G connectivity in rural regions. Independent variables encompass elements such as PRI autonomy, financial resources, and community engagement, all of which affect the efficacy of connectivity programs. Dependent variables, including rural connectivity levels, digital inclusion, and socio-economic development, assess the consequences of these programs. Moderating variables, such as the digital literacy of PRI officials and community socio-economic status, affect the intensity of these interactions, but mediating variables, like community trust and capacity-building programs, enhance the connectivity effect through indirect mechanisms. Latent factors, including community digital preparedness and PRI institutional capacity, facilitate the evaluation of fundamental requirements for effective connectivity. Furthermore, visible metrics like as network coverage and internet usage rates provide quantifiable signs of advancement. This table offers a detailed framework for assessing PRI-driven connectivity initiatives in rural settings.

Table 1. Variables that can be drawn to understand the role of the PRI's in connectivity to B5G and 6G.

Variable Type	Variable Name	Description
Independent Variables	Infrastructure Quality	Condition of physical and digital infrastructure in rural areas essential for supporting connectivity networks.
	PRI Autonomy	Degree of control PRIs have in decision-making without interference from higher government levels.
	Financial Resources	Availability of funds for PRI initiatives from government allocations or alternative sources.
	Digital Literacy	Knowledge and skills of PRI members and rural populations to utilize digital tools effectively.
	Community Engagement	Level of rural community involvement in decision-making, planning, and feedback related to connectivity initiatives.
	Government Policy Support	Policies and regulations promoting digital inclusion, connectivity initiatives, and rural development.
	Technological Partnerships	Collaborations with private tech companies, NGOs, or educational institutions for resources and technical assistance.

continued on following page

Table 1. Continued

Variable Type	Variable Name	Description
Dependent Variables	Rural Connectivity Levels	Metrics such as network availability, bandwidth speed, and coverage in rural areas.
	Digital Inclusion	Extent to which rural populations have access to and effectively use digital services.
	Service Delivery Efficiency	Improvement in the efficiency of PRI governance and service delivery due to digital tools.
	Socio-Economic Development	Local economic opportunities, employment rates, and quality of life improvements due to better connectivity.
	Community Satisfaction	Satisfaction level of rural residents with PRI-led connectivity initiatives.
Moderating Variables	Digital Literacy of PRI Officials	Proficiency of PRI officials in using digital tools, which influences service delivery efficiency.
	Political Stability	Stability in the local political environment affecting the effectiveness of PRI operations.
	Community Socio-Economic Status	Economic conditions of rural communities, influencing accessibility to new technologies.
	Technology Access Costs	Costs of accessing technologies (e.g., smartphones, data plans) that can affect digital inclusion.
Mediating Variables	Community Trust in PRIs	Level of trust rural residents have in PRI initiatives, fostering engagement.
	Perceived Usefulness of Connectivity	Community perception of technology's practical value, affecting adoption rates.
	Regulatory Support	Policies and regulatory support for digital infrastructure, enhancing PRI resource utilization.
	Capacity-Building Initiatives	Training programs that build digital skills among PRI members, improving service delivery efficiency.
Latent Variables	Community Digital Readiness	Readiness of the community to adopt digital tools, inferred through digital literacy, access to devices, and willingness to engage with digital services.
	Institutional Capacity of PRIs	PRI's skills, resources, and governance structures necessary for effective connectivity initiatives.
	Socio-Cultural Openness to Change	Community's openness to adopting new technologies, measured indirectly through indicators such as media exposure, education, and prior technology experiences.
Observable Variables	Network Coverage	Measure of connectivity quality and availability in rural areas.
	Internet Usage Rates	Percentage of the population using the internet, reflecting digital inclusion.
	Number of Trained PRI Officials	Tracks the count of PRI officials trained in digital literacy or e-Governance tools, indicating institutional capacity.
	Community Participation Rates	Measure of community engagement, tracked through attendance at Gram Sabha meetings, feedback submissions, or PRI-led surveys.

Source: Author

THEORIES THAT CAN BE APPLIED TO UNDERSTAND CONNECTIVITY THROUGH PRI'S

There are various theories surrounding the connectivity that have to be looked into when dealing with the rural setup and the role of the Panchyat Raj Institutions. The research is based on multiple basic theories that offer a comprehensive framework for analysing how Panchayat Raj Institutions (PRIs) might improve connectivity via Beyond 5G (B5G) and 6G in rural regions.

The Social Capital Theory which underscores the significance of community networks, trust, and collective values in promoting the adoption of digital technology, asserting that robust social capital within communities can propel connectivity projects. The Diffusion of Innovations Theory elucidates the dissemination of new technologies, specifying how PRIs might function as change agents to enhance knowledge, acceptability, and adoption of B5G and 6G among rural communities.

Institutional Theory underscores the impact of formal institutions, including governmental policies and local governance norms, on technology uptake, with PRIs influencing and modifying regulatory frameworks to enhance rural connection. The Technology Acceptance Model (TAM) elucidates aspects such as perceived utility and simplicity of use, which PRIs might target to enhance technology acceptance among rural inhabitants. Ultimately, Stakeholder Theory emphasises the necessity for PRIs to involve many stakeholders—such as rural communities, governmental entities, and technology suppliers—to foster an inclusive and cooperative connection strategy. Collectively, these theories provide a thorough framework for analysing PRI-led connection initiatives, encompassing community involvement and structural obstacles. Let us examine some of the theories in detail:

Figure 2. Theories on connectivity that can be studied to understand the role of the PRI's in connectivity to B5G and 6G.

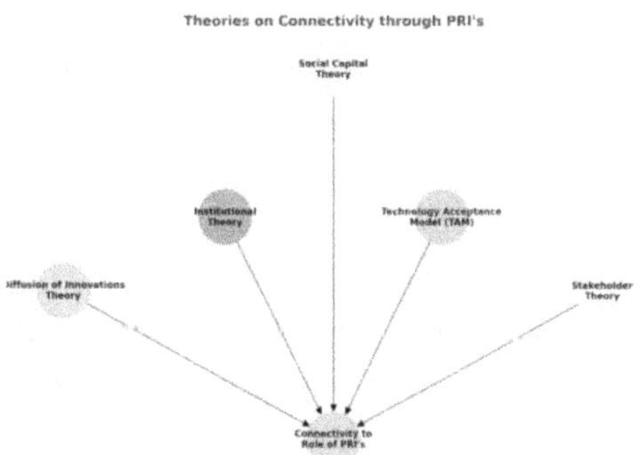

Source: Author

Social Capital Theory

Social Capital Theory, when applied to rural connectivity for Beyond 5G (B5G) and 6G technologies, offers a nuanced framework for understanding how social networks and relationships can influence the deployment and effectiveness of these advanced communication systems. Social capital, defined as the resources embedded within social networks, plays a crucial role in facilitating cooperation, information exchange, and community development, all of which are essential for successful technology adoption in rural areas (van Bakel and Horak, 2024). In rural contexts, social capital is characterized by structural, relational, and cognitive dimensions. The structural dimension involves the configuration of social networks and the characteristics of ties within these networks, which can influence how information about new technologies like B5G and 6G is disseminated and adopted.

The relational dimension, which includes trust, norms of reciprocity, and obligations, can enhance community engagement and collective action towards adopting new technologies. The cognitive dimension, involving shared language and narratives, can facilitate better understanding and integration of these technologies into daily life (Ali-Hassan, 2009). Rural areas often face unique challenges in network planning and deployment due to factors such as demographics, topography, and economic status. Social capital can mitigate these challenges by fostering a collective aptitude for decision-making and joint action, thereby enhancing the community's ability to pursue common goals, such as improved connectivity.

For instance, public access points in rural areas can serve as community resources that not only provide internet access but also build social capital by promoting community development and reducing the digital divide (Simpson et al., 2003). Moreover, the role of social networking sites (SNS) and broadband connectivity in building social capital is significant. SNS can enhance both bonding and bridging social capital by connecting geographically dispersed communities, thus facilitating information exchange and community cohesion (Tiwari et al., 2016). This connectivity is crucial for rural entrepreneurs, who may otherwise be constrained by the size and structure of their offline networks. Online social capital can help them overcome these limitations, enabling them to access broader networks and resources (Fischer, 2017).

However, the effectiveness of social capital in enhancing rural connectivity for B5G and 6G also depends on the existing socio-economic conditions and the level of human capital in these areas. Factors such as job accessibility and spatial connectivity can influence the production and benefits of social capital, highlighting the importance of addressing rural-urban differences in connectivity strategies (Türk et al., 2020). Social Capital Theory provides a valuable lens for understanding and enhancing rural connectivity for B5G and 6G technologies. By leveraging the structural, relational, and cognitive dimensions of social capital, rural communities can better navigate the challenges of technology adoption, ultimately leading to improved connectivity and socio-economic development. This approach underscores the importance of integrating social capital considerations into policy and planning for rural connectivity initiatives (Dively, 2003).

Social capital theory focuses on the value of social networks, relationships, and community involvement in achieving collective goals. It posits that communities with strong social capital are better equipped to collaborate, share resources, and overcome challenges. PRIs, as grassroots institutions, can leverage social capital to mobilize communities for the adoption of Beyond 5G and 6G technologies. PRIs can use their deep connections within rural communities to build trust and cooperation, which are essential for the successful deployment of connectivity infrastructure. By fostering a sense of collective ownership and responsibility, PRIs can ensure that the benefits of advanced connectivity are widely distributed and sustained over time.

Diffusion of Innovations Theory

The Diffusion of Innovations Theory, as articulated by Everett Rogers, establishes a theoretical framework for comprehending the mechanisms through which novel technologies and ideas disseminate within populations, which is particularly salient in the context of the implementation of Beyond 5G

(B5G) and 6G networks in rural locales. This theoretical construct delineates pivotal factors that affect the acceptance of innovations, encompassing relative advantage, compatibility, complexity, trial ability, and observability, which are indispensable for elucidating the challenges and prospects associated with rural connectivity (Pašti, 2022). Rural regions frequently experience a delay in the adoption of technological advancements due to a myriad of socio-economic and infrastructural impediments. The rollout of 5G technology and the theoretical underpinnings of 6G networks have primarily concentrated on urban environments, where commercial motivations are more pronounced, thereby neglecting rural territories.

The Diffusion of Innovations Theory can inform strategies aimed at surmounting these obstacles by underscoring the necessity of illustrating the relative advantage and compatibility of emerging technologies with pre-existing rural lifestyles and infrastructural frameworks. A significant impediment to rural connectivity is the insufficiency of digital infrastructure, which is vital for the diffusion of B5G and 6G technologies. The interconnectedness of networks and rural-urban linkages is instrumental in facilitating the absorption and dissemination of innovations in rural contexts. Such networks encompass place-based initiatives that prioritize digital infrastructure, along with firm-based collaborations that engage university-industry partnerships and international trade.

By capitalizing on these networks, rural regions can bolster their capacity for adopting novel technologies. The proposition of employing open-source hardware and software presents a viable solution for constructing sustainable, dependable, and cost-effective communication networks in rural locales, which aligns with the principles of the Diffusion of Innovations Theory by diminishing complexity and augmenting trial ability and observability.

This methodology may assist in alleviating the economic and technical obstacles that frequently impede the adoption of advanced communication technologies within rural settings. Moreover, the theory's focus on the significance of innovators and early adopters is paramount for enhancing rural connectivity. The identification and empowerment of local innovators and early adopters can expedite the diffusion process by cultivating local champions who can exhibit the advantages and practical applications of B5G and 6G technologies. This strategic approach can be particularly efficacious in rural environments where community influence and peer networks wield considerable power.

In summary, the Diffusion of Innovations Theory offers an invaluable perspective through which to examine the obstacles and possibilities associated with the deployment of B5G and 6G networks in rural areas. By concentrating on the reduction of complexity, the enhancement of compatibility, and the leveraging of local networks and innovators, stakeholders can promote the dissemination of these technologies, thereby advancing digital inclusion and connectivity in rural regions. This strategy not only addresses the technical and economic impediments but also harmonizes with the social and cultural dynamics of rural communities, ensuring a more comprehensive and sustainable adoption process.

This theory explains how, why, and at what rate new ideas and technologies spread through cultures. According to Rogers, innovation adoption occurs through a five-step process: knowledge, persuasion, decision, implementation, and confirmation. Panchayat Raj Institutions (PRIs) can play a critical role in accelerating the diffusion process by acting as change agents, educating rural populations, and facilitating the adoption of advanced connectivity technologies.

PRIs can be instrumental in spreading awareness about Beyond 5G and 6G technologies, persuading rural communities to adopt them, and providing the necessary support to ensure successful implementation (Key Considerations to Achieve 5G and B5G Connectivity in Rural Areas, 2022). The role of PRIs aligns with the stages of the diffusion process, helping to overcome resistance and bridge the digital divide.

Institutional Theory

Institutional theory examines how institutions—such as laws, regulations, and social norms—shape the behaviour of individuals and organizations. It emphasizes the importance of formal and informal structures in shaping organizational actions and outcomes. In the context of connectivity, PRIs operate within a framework of laws and policies that influence their ability to foster technological advancements in rural areas. PRIs can leverage their institutional authority to implement policies that promote connectivity, such as allocating resources for infrastructure development or coordinating with higher levels of government. Institutional theory highlights the need for PRIs to adapt to and shape the regulatory environment to foster Beyond 5G and 6G connectivity in rural areas.

Institutional theory provides a valuable framework for understanding the challenges and opportunities associated with enhancing rural connectivity for Beyond 5G (B5G) and 6G networks. This theory, which emphasizes the role of social norms, rules, and pressures in shaping organizational behaviour, can be instrumental in addressing the digital divide between urban and rural areas. The development of 6G networks, as envisioned in recent research, aims to leverage open-source hardware and software to create sustainable, reliable, and low-cost communication networks in rural and remote areas.

This approach aligns with the principles of institutional theory, which suggests that adopting widely accepted practices can enhance organizational legitimacy and support. By conforming to open-source standards, rural network initiatives can gain legitimacy and support from both governmental and non-governmental organizations, facilitating broader adoption and implementation. Institutional theory also highlights the importance of isomorphism, where organizations adopt similar practices to gain legitimacy (Institutional Theory, 2022). In the context of rural connectivity, this could mean adopting standardized technologies and practices that have been successful in other regions, thereby reducing the risk and uncertainty associated with new deployments. This is particularly relevant given the diverse challenges faced in rural areas, such as varying demographics, topographical features, and economic conditions. Moreover, institutional theory underscores the role of coercive, mimetic, and normative pressures in driving organizational change.

Coercive pressures, such as government regulations and policies, can mandate the deployment of B5G and 6G technologies in rural areas, ensuring that these regions are not left behind in the digital revolution. Mimetic pressures, driven by the success of similar initiatives in other regions, can encourage rural communities to adopt new technologies. Normative pressures, stemming from professional standards and community expectations, can further support the adoption of these technologies by emphasizing their importance for social and economic development. The application of institutional theory in rural connectivity is also evident in the emphasis on creating an institutional environment that supports small-scale agricultural production and business development (Mazur, 2019). By establishing rules and norms that facilitate market relations and protect rural people's rights, institutional frameworks can enhance the socio-economic processes necessary for successful network deployment and operation. Furthermore, the strategic orientation and institutional pressures identified in rural enterprises can influence the financial and social performance of connectivity initiatives (Basri et al., 2023).

By aligning strategic goals with institutional norms, rural connectivity projects can achieve better outcomes, both economically and socially. In conclusion, institutional theory offers a comprehensive lens through which to view the development of B5G and 6G networks in rural areas. By understanding and leveraging institutional pressures and norms, stakeholders can enhance the legitimacy, adoption, and success of these technologies, ultimately bridging the digital divide and fostering inclusive economic growth. This approach not only addresses the technical and economic challenges but also aligns with broader social and institutional goals, ensuring sustainable and equitable connectivity solutions for rural communities.

Technology Acceptance Model (TAM)

The Technology Acceptance Model (TAM) explains how users come to accept and use a technology. The model suggests that perceived usefulness and perceived ease of use are the primary factors influencing technology adoption. PRIs can play a role in enhancing these perceptions among rural populations, making advanced connectivity technologies more accessible and acceptable. PRIs can work to improve the perceived usefulness of Beyond 5G and 6G technologies by demonstrating their benefits in areas like education, healthcare, and agriculture. They can also help reduce perceived complexity by providing training and support, thus increasing the likelihood of technology adoption in rural areas.

The Technology Acceptance Model (TAM) is a widely recognized framework for understanding how users come to accept and use technology. It is particularly relevant in the context of rural connectivity for Beyond 5G (B5G) and 6G technologies, as it provides insights into the factors influencing technology adoption in these areas. TAM, originally developed by Davis in 1989, posits that perceived usefulness (PU) and perceived ease of use (PEU) are the primary determinants of users' acceptance of technology (Greener, 2022). These constructs are crucial in rural settings where technological infrastructure and literacy may be limited. In rural areas, the perceived benefits of new technologies, such as improved connectivity and access to information, can significantly influence adoption rates. For instance, in the context of mobile learning in rural STEM education, TAM has been used to predict acceptance by highlighting the importance of PU and PEU in shaping behavioural intentions (Mutambara & Bayaga, 2021).

The integration of TAM with other models, such as the Unified Theory of Acceptance and Use of Technology (UTAUT), can enhance its predictive power by incorporating additional factors like social influence and facilitating conditions. This integration is particularly relevant for rural connectivity, where social norms and community support can play a significant role in technology adoption. For example, in rural Africa, combining TAM with UTAUT has helped identify demographic variables such as age, gender, and education as moderators in the acceptance of mobile telephony networks (Tognisse and Degila, 2021). Moreover, the application of TAM in rural contexts often requires consideration of unique challenges such as technological disparities and infrastructural limitations. The concept of technological capital, which addresses these disparities, can be integrated into TAM to better understand the barriers to technology adoption in rural areas. This is crucial for B5G and 6G technologies, which promise enhanced connectivity but may face adoption hurdles due to existing digital divides. In addition to the traditional TAM constructs, factors such as hedonic motivation, price value, and habit, as introduced in the extended UTAUT model (UTAUT2), can also influence technology acceptance in rural areas.

These factors are particularly relevant for consumer technologies, where the perceived enjoyment and cost-effectiveness of new services can drive adoption. Despite its strengths, TAM has limitations, such as its focus on individual perceptions without fully accounting for broader contextual factors like organizational constraints and cultural influences (Samar et al., 2020). In rural settings, these factors can significantly impact technology adoption, necessitating a more holistic approach that combines TAM with other theoretical frameworks. Technology Acceptance Model provides a valuable framework for understanding the adoption of B5G and 6G technologies in rural areas. By integrating TAM with other models and considering contextual factors, researchers and policymakers can better predict and facilitate the acceptance of these advanced technologies, ultimately bridging the digital divide and enhancing rural connectivity.

Stakeholder Theory

Stakeholder theory suggests that organizations should consider the interests of all stakeholders—individuals or groups affected by the organization's actions. In the context of PRIs and connectivity, stakeholders include rural populations, government bodies, telecom companies, and NGOs. PRIs must engage with these stakeholders to ensure that the connectivity initiatives are inclusive, equitable, and sustainable. PRIs can act as mediators between various stakeholders, ensuring that the needs and concerns of rural communities are addressed while fostering collaboration with telecom companies and government agencies. This inclusive approach can help build trust and support for connectivity projects, facilitating their successful implementation.

Stakeholder theory, a foundational concept in business ethics and organizational management, offers a valuable framework for addressing the challenges of rural connectivity in the context of Beyond 5G (B5G) and 6G networks. This theory emphasizes the importance of considering the diverse interests and expectations of all stakeholders involved in a project, which is particularly relevant when addressing the digital divide between urban and rural areas (Jonek-Kowalska et al., 2023).

The deployment of B5G and 6G networks in rural areas presents unique challenges and opportunities. Rural regions often face obstacles such as diverse demographics, varied architectural structures, and challenging topographical features, which can complicate network planning and deployment. Moreover, the economic status and educational levels in these areas can further influence the feasibility and sustainability of advanced communication technologies. Stakeholder theory can guide the strategic management of these challenges by ensuring that the interests of all relevant parties, including local communities, government agencies, technology providers, and end-users, are considered and balanced (Freeman et al., 2010).

The vision for 6G networks, as outlined in recent research, emphasizes the use of open-source hardware and software to create sustainable, reliable, and low-cost communication networks in rural and remote areas. This approach aligns with stakeholder theory by promoting inclusivity and collaboration among various stakeholders, including technology developers, policymakers, and rural communities. By leveraging open-source solutions, stakeholders can work together to overcome the economic and technical barriers that have historically limited rural connectivity. Furthermore, stakeholder theory's focus on ethical considerations and societal well-being is particularly pertinent in the context of rural connectivity. The theory advocates for solutions that not only address technical and economic challenges but also enhance the quality of life for rural populations by improving access to education, healthcare, and economic opportunities (Wargo et al., 2021).

This holistic approach ensures that the deployment of B5G and 6G technologies contributes to broader societal goals, such as reducing inequality and promoting sustainable development. However, the application of stakeholder theory in this context is not without challenges. The theory's broad applicability can sometimes lead to difficulties in prioritizing stakeholder interests, especially when resources are limited or when stakeholders have conflicting goals (Nugent and Jose, 2017). Additionally, while stakeholder theory provides a comprehensive framework for strategic management, its integration into mainstream strategic management practices remains a work in progress, as it often competes with more traditional economic theories. Stakeholder theory offers a robust framework for addressing the complexities of rural connectivity in the era of B5G and 6G networks. By emphasizing inclusivity, collaboration, and ethical considerations, the theory can guide the development of communication networks that are not only technologically advanced but also socially and economically beneficial for rural communities. This approach requires ongoing dialogue and cooperation among all stakeholders to ensure that the benefits of advanced connectivity are equitably distributed and aligned with the broader goals of societal well-being and sustainability.

These theories provide a robust framework for understanding the role of Panchayat Raj Institutions in fostering connectivity in rural areas, particularly in the context of advanced technologies like Beyond 5G and 6G. They highlight the importance of social, institutional, and technological factors in bridging the digital divide and ensuring that rural populations are not left behind in the digital age. The integration of advanced technologies such as Beyond 5G (B5G) and 6G into rural areas presents a transformative opportunity to bridge the digital divide, foster socio-economic development, and enhance the quality of life in these underserved regions. To achieve this, a multifaceted approach grounded in various theoretical frameworks is essential.

Social Capital Theory underscores the significance of leveraging existing social networks and relationships to facilitate the adoption of new technologies in rural areas. It posits that the structural, relational, and cognitive dimensions of social capital are critical in building trust, enhancing community engagement, and promoting collective action, all of which are vital for the successful deployment of B5G and 6G networks. The theory suggests that by fostering strong social ties and a sense of ownership among rural communities, Panchayat Raj Institutions (PRIs) can play a pivotal role in ensuring that the benefits of advanced connectivity are widely distributed and sustained over time.

Diffusion of Innovations Theory, on the other hand, provides a framework for understanding how new technologies spread within rural populations. This theory highlights the importance of addressing key factors such as relative advantage, compatibility, and complexity to facilitate the adoption process. PRIs, acting as change agents, can accelerate the diffusion of B5G and 6G technologies by educating rural populations, demonstrating the practical benefits of these technologies, and providing the necessary support to overcome resistance and bridge the digital divide. By aligning their efforts with the stages of the diffusion process—knowledge, persuasion, decision, implementation, and confirmation—PRIs can effectively promote the adoption of advanced connectivity in rural areas.

Institutional Theory further complements this approach by emphasizing the role of formal and informal structures in shaping organizational behavior and outcomes. In the context of rural connectivity, this theory suggests that PRIs, operating within a framework of laws, regulations, and social norms, can leverage their institutional authority to implement policies that promote connectivity. By coordinating with higher levels of government and adapting to the regulatory environment, PRIs can help create an institutional context that supports the deployment of B5G and 6G technologies. This theory also highlights the importance of isomorphism, where rural connectivity initiatives can gain legitimacy and support by adopting standardized technologies and practices that have been successful in other regions. This approach can reduce the risks and uncertainties associated with new deployments, particularly in the face of diverse challenges such as varying demographics, topographical features, and economic conditions.

The Technology Acceptance Model (TAM) offers another valuable perspective, focusing on the factors that influence technology adoption at the individual level. According to TAM, perceived usefulness and perceived ease of use are the primary determinants of whether individuals will accept and use new technologies. In rural areas, where technological infrastructure and literacy may be limited, PRIs can play a crucial role in enhancing these perceptions by demonstrating the tangible benefits of B5G and 6G technologies in areas such as education, healthcare, and agriculture. By providing training and support, PRIs can help reduce the perceived complexity of these technologies, thereby increasing the likelihood of their adoption. Furthermore, integrating TAM with other models such as the Unified Theory of Acceptance and Use of Technology (UTAUT) can provide a more comprehensive understanding of the factors driving technology adoption in rural contexts, including social influence and facilitating conditions.

Finally, Stakeholder Theory emphasizes the need for a holistic and inclusive approach to rural connectivity. This theory suggests that PRIs must consider the interests of all stakeholders involved in connectivity initiatives, including rural populations, government bodies, telecom companies, and non-governmental organizations (NGOs). By acting as mediators and ensuring that the needs and concerns of rural communities are addressed, PRIs can foster collaboration and build trust among stakeholders. This inclusive approach is essential for the successful implementation of connectivity projects, as it helps ensure that the benefits of advanced technologies are equitably distributed and aligned with broader societal goals such as reducing inequality and promoting sustainable development.

In conclusion, the successful deployment of Beyond 5G and 6G technologies in rural areas requires a comprehensive approach that integrates the insights from Social Capital Theory, Diffusion of Innovations Theory, Institutional Theory, the Technology Acceptance Model, and Stakeholder Theory. PRIs, as grassroots institutions deeply embedded within rural communities, are uniquely positioned to leverage these theoretical frameworks to promote the adoption of advanced connectivity technologies. By fostering social capital, accelerating the diffusion of innovations, shaping institutional environments, enhancing the perceived usefulness and ease of use of new technologies, and engaging all relevant stakeholders, PRIs can play a pivotal role in bridging the digital divide and ensuring that rural populations are not left behind in the digital revolution. This integrated approach not only addresses the technical and economic challenges of rural connectivity but also aligns with broader social and institutional goals, ultimately contributing to the socio-economic development and well-being of rural communities. By embracing these theoretical perspectives, policymakers, technologists, and community leaders can work together to create a more inclusive and connected future for all.

DISCUSSION

The study's findings indicate that PRIs occupy a distinctive role in addressing the digital divide through their established ties to rural communities and expertise in local governance. The data indicates that PRIs are positioned to facilitate the adoption of advanced technologies such as B5G and 6G, owing to their capacity to effectively engage rural populations. Challenges such as inadequate digital infrastructure, constrained financial resources, and low digital literacy among PRI members present substantial obstacles. The research indicates that although e-Governance platforms and digital tools can improve transparency and efficiency, insufficient digital training and resources impede comprehensive adoption.

Utilising Social Capital Theory demonstrates that robust community networks, supported by PRIs, can enhance trust and promote local engagement, which are critical for the acceptance of technology in rural regions. The research indicates that Diffusion of Innovations Theory is especially pertinent, as PRIs serve as change agents, advocating for the advantages of digital connectivity while tackling barriers specific to rural areas. Institutional Theory highlights the necessity for PRIs to function within conducive regulatory frameworks that foster autonomy and collaboration among various levels of government. The Technology Acceptance Model and Stakeholder Theory highlight the significance of community engagement, usability, and perceived value in technology adoption, indicating that PRI initiatives that align with these elements are more likely to achieve success.

The research indicates that PRIs encounter social and political obstacles, including insufficient representation of women and marginalised groups, which impacts equitable development results. Addressing inclusivity issues within PRIs would facilitate community-driven approaches to connectivity. Collaboration with private and non-governmental stakeholders is crucial for developing resilient infrastructure, enhancing e-learning, and facilitating rural entrepreneurship.

PRI's and Rural Engagement

The engagement of Panchayati Raj Institutions (PRIs) in rural connectivity, particularly in the context of Beyond 5G (B5G) and 6G technologies, is crucial for bridging the digital divide and enhancing digital inclusion in rural areas. The transition from 5G to 6G networks presents both challenges and opportunities for rural connectivity, as highlighted by several research studies. Firstly, the development of 6G networks aims to address the limitations of current connectivity in rural areas by leveraging open-source hardware and software to create sustainable, reliable, and low-cost communication networks. This approach is particularly beneficial for rural and remote areas that have been largely excluded from the advancements seen in urban regions. The use of open-source solutions can empower local communities and PRIs to participate actively in the deployment and management of these networks, potentially reducing costs and increasing local engagement. Moreover, hybrid terrestrial/LEO satellite systems are being explored to enhance rural connectivity in B5G and 6G networks. These systems offer a promising solution by combining terrestrial base stations with satellite networks, thereby improving coverage probability and data rates in rural areas. This technological advancement could be instrumental for PRIs in ensuring that even the most remote villages have access to reliable internet services. The challenges of deploying 5G and B5G networks in rural areas are multifaceted, involving demographic, economic, and topographical considerations (Salem et al., 2023).

PRIs can play a pivotal role in addressing these challenges by facilitating community engagement, understanding local needs, and advocating for infrastructure investments. The involvement of PRIs is essential in tailoring connectivity solutions to the specific requirements of rural communities, thereby enhancing the adoption and effectiveness of these technologies. Despite the availability of technology to bridge the digital divide, the deployment of 5G and B5G networks in rural areas faces significant hurdles, primarily due to the lack of viable business models and investment from the private sector (Elias et al., 2019). PRIs can help mitigate these challenges by collaborating with government bodies and private entities to develop sustainable business models that align with local socio-economic conditions. Furthermore, the 5G-RCOLAB simulation tool provides a platform for evaluating the performance of 5G/6G networks in rural settings, offering insights into the potential improvements and deployments of cost-effective solutions (Gabriel et al., 2023). This tool can assist PRIs in planning and optimizing network deployments, ensuring that they meet the specific needs of rural users. In India, where a significant portion of the population lacks adequate connectivity, various technological solutions are being tested to improve rural internet access. The focus is on reducing deployment costs and increasing service adoption rates, which are critical for the success of rural connectivity initiatives. PRIs can leverage these insights to advocate for policies and initiatives that support the deployment of 6G technologies tailored to rural scenarios. In conclusion, the engagement of PRIs in the deployment of B5G and 6G networks is vital for enhancing rural connectivity. By leveraging open-source technologies, hybrid network solutions, and simulation tools, PRIs can play a crucial role in overcoming the challenges of rural connectivity and ensuring that the benefits of advanced communication technologies are equitably distributed. This engagement not only supports digital inclusion but also contributes to broader socio-economic development goals in rural areas.

Panchayat Raj Institutions (PRIs) can play a crucial role in improving rural engagement by leveraging their deep connections within the community, fostering trust, and implementing strategies that address the unique challenges faced by rural areas. PRIs, by design, bring decision-making closer to the people. By involving local communities in governance, PRIs can ensure that decisions reflect the needs and aspirations of rural populations (Elias and Mohamed, 2020). This can lead to greater community involvement and ownership of local initiatives. PRIs can engage rural communities in the planning and implementation of development projects. This participatory approach ensures that projects are more relevant to the community's needs and have a higher chance of success. Tools like Gram Sabhas (village assemblies) can be effectively utilized to gather input and build consensus on local issues.

PRIs can establish regular communication channels, such as village meetings, newsletters, and digital platforms, to keep the community informed about ongoing projects, government schemes, and decisions. Transparency in governance can build trust and encourage greater participation from the community. Establishing mechanisms for receiving and addressing community feedback is crucial. PRIs can set up suggestion boxes, hold regular open forums, and use digital platforms to gather feedback. This two-way communication helps in understanding the community's concerns and refining initiatives to better meet their needs.

PRIs can organize training programs to build the capacity of local leaders, community members, and volunteers. These programs can focus on areas like governance, financial management, project implementation, and digital literacy. Empowering the community with knowledge and skills enhances their ability to engage effectively with local governance. PRIs can conduct awareness campaigns to inform rural populations about their rights, government schemes, and the importance of participating in local governance. These campaigns can be tailored to address specific issues relevant to the community, such as health, education, or agricultural practices.

PRIs can play a vital role in ensuring that rural communities are aware of and can access government schemes and benefits. By organizing camps, workshops, and help desks, PRIs can assist in the enrollment process and provide necessary documentation support. PRIs can monitor the implementation of government schemes in their villages to ensure that the intended benefits reach the right beneficiaries. By actively engaging with the community, PRIs can gather data on the effectiveness of these schemes and report any issues to higher authorities.

PRIs can foster rural economic development by supporting local enterprises, such as small businesses, cooperatives, and self-help groups. By providing training, facilitating access to credit, and creating market linkages, PRIs can help these enterprises thrive, leading to greater economic engagement from the community. PRIs can prioritize infrastructure projects that directly impact rural livelihoods, such as improving roads, water supply, sanitation, and digital connectivity. Better infrastructure enhances the community's access to markets, education, and healthcare, leading to higher engagement and improved quality of life.

PRIs can encourage the formation and strengthening of community networks, such as farmers' groups, women's groups, and youth clubs. These networks can serve as platforms for sharing knowledge, pooling resources, and taking collective action on local issues. PRIs can organize cultural events, festivals, and heritage projects that celebrate and preserve local traditions. Such activities can boost community pride, strengthen social bonds, and enhance engagement by bringing people together.

PRIs can use digital tools and platforms to enhance communication and engagement with rural populations. Mobile apps, social media, and online forums can be used to disseminate information, gather feedback, and engage with the community in real-time. By advocating for and facilitating the deployment of advanced connectivity technologies like Beyond 5G and 6G, PRIs can ensure that rural areas are connected to the broader digital economy. This connectivity can open up new avenues for education, healthcare, and entrepreneurship, leading to greater engagement.

PRIs can develop programs specifically aimed at engaging rural youth in local governance, entrepreneurship, and community service. Involving young people in these activities not only prepares the next generation of leaders but also brings fresh perspectives to local issues. PRIs can take active steps to involve women in decision-making processes by ensuring their representation in Gram Sabhas and local committees. Empowering women to participate in governance and economic activities can lead to more inclusive and equitable development.

PRIs can serve as mediators in resolving local conflicts, whether related to land, water, or social disputes. By providing a platform for dialogue and negotiation, PRIs can maintain social harmony and foster a collaborative spirit within the community. PRIs should ensure that all sections of the community, including marginalized groups, have a voice in local governance. By implementing policies that promote inclusivity and social justice, PRIs can ensure that everyone is engaged and benefits from local development initiatives.

PRIs can engage communities in the sustainable management of local natural resources, such as water bodies, forests, and agricultural land. By promoting practices like watershed management, afforestation, and organic farming, PRIs can enhance environmental stewardship and rural engagement. PRIs can play a critical role in preparing communities for natural disasters and coordinating response efforts. Engaging the community in disaster preparedness plans can build resilience and foster a sense of collective responsibility.

The engagement of Panchayati Raj Institutions (PRIs) is crucial in enhancing rural connectivity, especially with the advent of Beyond 5G (B5G) and 6G technologies. PRIs can play a vital role in bridging the digital divide by leveraging open-source hardware and software, facilitating the deployment of hybrid terrestrial/LEO satellite systems, and actively participating in the management of these networks. This involvement is key to overcoming the demographic, economic, and topographical challenges of deploying advanced networks in rural areas. By fostering community engagement, tailoring connectivity solutions to local needs, and collaborating with government bodies and private entities, PRIs can ensure that even the most remote areas benefit from reliable internet services. Moreover, tools like the 5G-RCOLAB simulation can help PRIs optimize network deployments. PRIs' deep connections within the community, capacity for fostering trust, and ability to implement inclusive strategies make them indispensable in addressing the unique challenges faced by rural areas.

By supporting local enterprises, improving infrastructure, advocating for digital literacy, and ensuring inclusive participation in governance, PRIs can drive rural economic development and social inclusion. Their involvement in monitoring government schemes, promoting sustainable resource management, and preparing communities for natural disasters further enhances their role in fostering resilience and equitable development. Overall, PRIs are essential in ensuring that the benefits of B5G and 6G technologies are equitably distributed, supporting broader socio-economic development goals in rural India. In summary, PRIs can significantly improve rural engagement by fostering inclusive, transparent, and participatory governance, building social capital, leveraging technology, and addressing the unique social, economic, and environmental challenges faced by rural communities. Through these efforts, PRIs can empower rural populations to take an active role in their own development, leading to more resilient, vibrant, and engaged communities.

PRI's and Digital Tools in Continuance to Utilisation of B5G and 6G Technologies

Panchayat Raj Institutions (PRIs) can leverage digital tools to enhance their efficiency, transparency, and effectiveness in governance, thereby improving service delivery and increasing community engagement in rural areas. PRIs can adopt e-Governance platforms such as the e-Panchayat Mission Mode Project, which is part of India's National e-Governance Plan (NeGP). These platforms provide a comprehensive suite of applications for managing panchayat finances, tracking development projects, and maintaining records. By digitizing records and processes, PRIs can enhance transparency and reduce the time required for administrative tasks. Digital tools can be used to create online grievance redressal systems, allowing citizens to file complaints and track their resolution status. This promotes accountability and ensures that issues are addressed promptly.

PRIs can establish official websites and portals where they can publish information about ongoing projects, budgets, meeting minutes, and other relevant data. This transparency helps build trust with the community and ensures that citizens are well-informed. PRIs can use social media platforms like Facebook, Twitter, and WhatsApp to communicate with the community, share updates, and gather feedback. Social media also enables PRIs to reach a broader audience, including younger and more tech-savvy members of the community.

PRIs can leverage online platforms to provide training and capacity-building programs for panchayat members and staff. Topics could include governance, financial management, digital literacy, and the use of specific e-Governance tools. This training can be delivered through webinars, online courses, or video tutorials. PRIs can organize digital literacy workshops for rural citizens to help them access and use digital services effectively. This includes teaching them how to use government portals, mobile apps, and digital payment systems. PRIs can develop mobile applications that allow citizens to access information, request services, and provide feedback directly from their smartphones. These apps can include features such as bill payments, service requests (like water supply or road repairs), and local news updates. There are mobile apps specifically designed to help panchayat officials manage their duties more effectively. These apps can track attendance at meetings, monitor project progress, and facilitate communication with higher levels of government.

PRIs can use digital tools to conduct surveys and polls to gather data on the needs and opinions of the community. This data can then be analyzed to inform decision-making and prioritize development initiatives. Geographic Information Systems (GIS) tools can be used to map resources, infrastructure, and population distribution within a panchayat's jurisdiction. This spatial data helps in planning and implementing development projects more effectively. PRIs can use digital platforms for managing financial transactions, including the disbursement of funds, payment of salaries, and procurement processes. This reduces the risk of corruption and ensures that funds are used efficiently.Digital tools can assist PRIs in maintaining accurate financial records, preparing budgets, and tracking expenditures. This transparency in financial management helps build trust and ensures accountability. PRIs can hold virtual meetings and consultations using video conferencing tools like Zoom or Google Meet. This is especially useful for including members who may not be able to attend in person and for engaging with a broader section of the community. Digital platforms can be used to crowdsource ideas and suggestions from citizens for local development projects. This participatory approach ensures that the initiatives undertaken by PRIs are aligned with the community's needs and aspirations.

PRIs can use digital project management tools to monitor the progress of various development projects in real-time. These tools can help track timelines, budgets, and resource allocation, ensuring that projects are completed on schedule and within budget. Digital dashboards can provide PRIs with a visual overview of key performance indicators (KPIs) related to governance, service delivery, and development projects. These dashboards can help PRI officials quickly identify areas that need attention. By analyzing data collected from various digital sources, PRIs can make informed decisions about resource allocation, project prioritization, and policy implementation. Big data analytics can reveal trends and patterns that might not be apparent through traditional methods.

PRIs can use predictive analytics to forecast future needs and challenges, such as population growth or the impact of climate change. This enables proactive planning and more effective management of resources. PRIs can facilitate the establishment of digital classrooms in local schools, providing access to online educational resources and remote learning opportunities for students. This can be particularly valuable in areas where there is a shortage of qualified teachers. By partnering with healthcare providers, PRIs can help set up telemedicine services that allow residents to consult with doctors remotely. This is especially beneficial in remote areas where access to healthcare facilities is limited.

PRIs can establish public Wi-Fi hotspots or community internet centers where residents can access the internet for free or at a low cost. This helps bridge the digital divide and ensures that all members of the community can benefit from digital services. PRIs can initiate programs aimed at including marginalized groups, such as women, elderly, and disabled persons, in the digital economy. This can involve targeted training, access to devices, and support in using digital tools. PRIs can help local artisans, farmers, and entrepreneurs access e-commerce platforms to sell their products online. This expands their market reach and supports local economic development. By providing training in digital marketing, PRIs can help local businesses improve their online presence and attract more customers. This can include training on social media marketing, search engine optimization (SEO), and online advertising.

PRIs can use digital tools to establish early warning systems for natural disasters such as floods, cyclones, or droughts. These systems can send alerts to residents via SMS, apps, or social media, enabling timely evacuation and preparedness. In times of crisis, such as pandemics or natural disasters, PRIs can use crisis management apps to coordinate relief efforts, track the distribution of aid, and communicate with affected communities. PRIs can promote the use of digital tools in agriculture, such as soil sensors, weather forecasting apps, and farm management software.

These tools help farmers optimize their production and manage resources more sustainably. Digital tools can be used to monitor environmental indicators such as air and water quality, deforestation, and biodiversity. This data can inform conservation efforts and policy decisions.

The integration of digital tools into Panchayati Raj Institutions (PRIs) is not only pivotal for improving governance in rural areas but is also closely linked to the advancement of Beyond 5G (B5G) and 6G technologies. As India transitions from 5G to B5G and eventually to 6G networks, the role of PRIs in facilitating this shift becomes increasingly critical. These advanced networks promise to deliver ultra-fast internet speeds, low latency, and the ability to support a vast array of connected devices, all of which can significantly enhance the functioning of PRIs and the services they provide to rural communities.

B5G and 6G technologies are expected to bridge the digital divide by providing high-speed internet connectivity even in the most remote and underserved areas. This is particularly important in rural India, where PRIs serve as the primary governance bodies responsible for the delivery of essential services.

With the enhanced connectivity offered by B5G and 6G, PRIs can leverage a wide range of digital tools to improve their operational efficiency, transparency, and effectiveness. For instance, e-Governance platforms like the e-Panchayat Mission Mode Project can be further optimized with the support of 6G's high bandwidth and low latency, enabling real-time tracking of development projects, instantaneous financial transactions, and immediate access to comprehensive records.The role of PRIs in this context is twofold: they are both beneficiaries and facilitators of advanced connectivity. As beneficiaries, PRIs can utilize the capabilities of B5G and 6G to enhance their governance functions. This includes the digitization of administrative processes, the establishment of online grievance redressal systems, and the use of Geographic Information Systems (GIS) for planning and development. The increased data speeds and reduced latency of 6G networks will allow PRIs to conduct virtual meetings, consultations, and training sessions more effectively, ensuring that all stakeholders are engaged and informed.

As facilitators, PRIs play a crucial role in ensuring that the benefits of B5G and 6G technologies reach the grassroots level. They can advocate for and facilitate the deployment of these networks in rural areas, working in collaboration with government bodies, private sector entities, and local communities. PRIs can also help in identifying priority areas for connectivity based on local needs, such as healthcare, education, and economic development. By engaging with local populations, PRIs can ensure that the deployment of B5G and 6G networks is aligned with the specific socio-economic conditions of their regions.

Moreover, the advanced connectivity provided by B5G and 6G can empower PRIs to implement innovative solutions for rural development. For example, PRIs can use mobile applications and digital platforms to deliver essential services directly to citizens' smartphones, reducing the need for physical infrastructure and making services more accessible. Additionally, PRIs can leverage the internet of things (IoT) devices, which will be supported by 6G networks, to monitor and manage local resources, such as water supply systems, agricultural practices, and public health initiatives. Furthermore, B5G and 6G technologies can enable PRIs to support local economic development by connecting rural producers to larger markets through e-commerce platforms. With reliable high-speed internet, farmers, artisans, and small business owners in rural areas can sell their products online, access real-time market data, and engage in digital marketing strategies. This not only boosts local economies but also encourages digital literacy and entrepreneurship within rural communities.

As digital infrastructure continues to expand, PRIs that effectively integrate B5G and 6G technologies into their operations will be at the forefront of modern governance, driving progress and innovation in their communities by leveraging digital tools, PRIs can significantly enhance their governance capabilities, improve service delivery, and foster greater community engagement in rural areas. These tools not only streamline administrative processes but also empower citizens, promote transparency, and drive socio-economic development. As digital infrastructure continues to expand in rural areas, PRIs that effectively integrate digital tools into their operations will be better equipped to meet the challenges of modern governance and ensure inclusive growth for their communities. In conclusion, the deployment of B5G and 6G technologies presents a unique opportunity for PRIs to transform rural governance and development. By embracing digital tools and leveraging advanced connectivity, PRIs can enhance their governance capabilities, improve service delivery, and foster greater community engagement. Their role as both users and promoters of these technologies is essential in ensuring that rural areas benefit from the digital revolution, leading to more inclusive growth and the overall socio-economic upliftment of rural India.

CONCLUSION AND WAY FORWARD

Panchayat Raj Institutions (PRIs) play a crucial role in addressing the digital divide in rural India, acting as both facilitators and beneficiaries of advanced connectivity initiatives such as Beyond 5G and 6G. This study demonstrates that PRIs are strategically positioned to promote digital inclusion via local governance and established community networks; however, they encounter various challenges, including insufficient infrastructure and financial and administrative limitations. The absence of digital infrastructure, dependable connectivity, and adequately trained personnel in PRIs constrains their ability to effectively leverage e-Governance tools and to execute innovative connectivity solutions. By strategically addressing these barriers, PRIs can leverage digital tools and B5G/6G technologies to enhance governance, foster economic development, and facilitate inclusive growth in rural areas. Enhancing the capacity of PRIs to implement digital solutions necessitates specific policy support, which includes allocated funding for connectivity infrastructure and initiatives aimed at improving digital literacy.

Collaborations with private enterprises and non-governmental organisations can address resource and skill deficiencies, allowing PRIs to execute connectivity solutions customised to local requirements. The findings highlight the importance of a multi-dimensional approach that integrates theoretical frameworks, including Social Capital Theory, Diffusion of Innovations, Institutional Theory, Technology Acceptance Model, and Stakeholder Theory. This approach allows PRIs to promote sustainable and inclusive development by aligning connectivity initiatives with community needs, engaging stakeholders, and enhancing resilience to socio-economic challenges. PRIs, when provided with sufficient support and resources, have the potential to significantly improve rural governance. This transformation can enable the equitable implementation of B5G and 6G technologies, thereby empowering rural communities, boosting local economies, and contributing to India's broader socio-economic development.

The Panchayat Raj Institutions (PRIs) in India, as the bedrock of rural governance, confront a myriad of challenges that significantly constrain their effectiveness in driving development and ensuring good governance at the grassroots level. These challenges are multifaceted, encompassing financial, administrative, political, social, and technological dimensions. Financial constraints remain a formidable obstacle, with PRIs often grappling with inadequate funding, which is further exacerbated by their dependence on higher levels of government for resources. This financial fragility undermines the autonomy of PRIs and restricts their ability to undertake meaningful development initiatives that address the pressing needs of rural communities. Compounding this issue is the lack of skilled personnel and administrative capacity within PRIs, which hampers efficient governance and the implementation of various government schemes. The administrative complexity, coupled with inadequate infrastructure and a digital divide, creates an environment where governance processes are slow, inefficient, and prone to corruption. Moreover, the political landscape surrounding PRIs is often marred by interference from state-level actors and local political dynamics that prioritize personal or group interests over the collective good. This political dominance can dilute the participatory and democratic ethos that PRIs are supposed to embody, further alienating the very communities they are meant to serve.

Social inequalities, particularly those based on gender and caste, continue to plague PRIs, limiting the participation and influence of women and marginalized groups. While reservations have been instituted to promote inclusivity, the reality is that these groups often face social and cultural barriers that prevent them from exercising real power within PRIs. The lack of awareness among rural citizens about the roles and responsibilities of PRIs, as well as their own rights, leads to low levels of civic participation and engagement, which in turn weakens the accountability mechanisms that are crucial for good governance. Environmental challenges, including the impact of natural disasters and the effects of climate change, further complicate the task of rural governance, as PRIs often lack the resources and capacity to effectively respond to these crises. The fragmentation of roles and responsibilities among various stakeholders and the lack of coordination between different levels of government exacerbate these challenges, leading to inefficiencies and missed opportunities for holistic development. Additionally, cultural and linguistic diversity, while a strength of rural India, poses challenges in terms of communication and inclusivity, making it difficult for PRIs to engage effectively with all segments of the community. Emerging challenges related to technology, such as digital literacy and cybersecurity, present new risks and opportunities for PRIs as they navigate the transition to e-Governance. In summary, the challenges faced by PRIs are deeply interconnected and require a comprehensive approach to address them effectively. Without significant reforms and support, PRIs will continue to struggle in fulfilling their mandate as the principal institutions of local governance in rural India.

Addressing the challenges faced by Panchayat Raj Institutions (PRIs) requires a multi-pronged approach that involves not only the strengthening of existing structures but also the introduction of innovative practices and the leveraging of technology to enhance governance. Financial autonomy is a crucial area that needs immediate attention. Ensuring that PRIs have access to adequate and timely funding is essential for them to function effectively. This can be achieved by enhancing the devolution of funds from state governments and exploring alternative revenue sources such as local taxes and community contributions. Additionally, capacity building should be prioritized, with a focus on training PRI members and staff in administrative skills, financial management, and the use of digital tools. This can be complemented by initiatives to improve the physical infrastructure of PRIs, such as upgrading office facilities, improving connectivity, and providing access to modern communication tools. Digital literacy programs tailored for both PRI members and the broader rural population are essential to bridge the digital divide and enable the adoption of e-Governance platforms, which can streamline administrative processes, enhance transparency, and improve service delivery.

Political and social reforms are also critical in empowering PRIs to function as true representatives of their communities. Efforts must be made to reduce political interference in PRI functioning by clearly delineating the roles and responsibilities of state and local governments and ensuring that PRI elections are free from undue influence. Promoting greater inclusivity within PRIs is another essential step, which can be achieved by reinforcing the importance of gender and caste reservations, providing targeted support to women and marginalized groups, and fostering a culture of equality and respect within local governance structures. Enhancing civic participation is vital to improving the accountability and responsiveness of PRIs. This can be done by conducting widespread awareness campaigns that educate citizens about their rights, the functions of PRIs, and the importance of their participation in local governance processes. PRIs should also explore innovative ways to engage with the community, such as using social media, mobile applications, and other digital platforms to gather feedback, disseminate information, and foster dialogue.

Environmental sustainability and disaster resilience should be integrated into the planning and decision-making processes of PRIs. This involves not only preparing for and responding to natural disasters but also adopting sustainable practices in agriculture, water management, and resource use that can mitigate the impacts of climate change. Building partnerships with NGOs, academic institutions, and the private sector can provide PRIs with the expertise and resources needed to address these complex challenges. Furthermore, enhancing coordination between different levels of government and other stakeholders is essential for creating a more cohesive and effective governance framework. This could involve establishing formal mechanisms for inter-governmental cooperation, joint planning sessions, and shared monitoring and evaluation systems. Lastly, as PRIs increasingly adopt digital tools, it is imperative to address the challenges associated with cybersecurity and data protection. Developing robust cybersecurity protocols and providing training on data privacy and security can help safeguard the integrity of digital governance initiatives. By adopting these comprehensive measures, PRIs can be empowered to overcome the myriad challenges they face and fulfill their potential as the driving force behind rural development and governance in India. This holistic approach not only strengthens the PRIs but also lays the foundation for more inclusive, sustainable, and responsive governance that truly meets the needs of rural communities.

Panchayat Raj Institutions (PRIs) stand as a cornerstone of rural governance in India, playing an indispensable role in enhancing connectivity and fostering development, particularly in the context of the imminent rollout of advanced technologies like Beyond 5G (B5G) and 6G. These technologies offer unprecedented opportunities to bridge the digital divide, improve access to essential services, and drive socio-economic development in underserved rural areas. However, the successful deployment and utilization of these technologies hinge significantly on the ability of PRIs to overcome a myriad of challenges, harness digital tools effectively, and engage rural communities in meaningful ways. One of the foremost challenges facing PRIs is the inadequate infrastructure that plagues many rural regions. This includes not only the physical infrastructure such as roads, power supply, and telecommunication networks but also the digital infrastructure necessary for the effective implementation of e-Governance and digital tools. The lack of reliable internet connectivity and the digital divide between urban and rural areas severely limit the ability of PRIs to adopt and leverage modern technologies. Moreover, many PRIs suffer from insufficient financial resources, which are critical for upgrading infrastructure, implementing new technologies, and maintaining the operational efficiency required to support advanced connectivity initiatives. This financial constraint is further compounded by the reliance of PRIs on state and central government funding, which often comes with delays and is insufficient to meet the diverse needs of rural communities.

In addition to infrastructural and financial challenges, PRIs are also grappling with a significant gap in digital literacy, both among the rural population and within the institutions themselves. This lack of digital literacy hampers the effective use of e-Governance platforms and digital tools, which are essential for enhancing transparency, efficiency, and community engagement in governance processes. The digital divide is not just a matter of access to technology but also the knowledge and skills to use these tools effectively, which is crucial for the adoption and integration of B5G and 6G technologies. Moreover, the administrative capacity of PRIs is often limited by a shortage of trained personnel and a lack of ongoing capacity-building initiatives. Many PRI members and staff lack the necessary training in governance, financial management, and the use of digital tools, which undermines their ability to execute their roles effectively. This capacity gap is exacerbated by complex governance processes that are difficult to navigate without adequate support, leading to inefficiencies and delays in decision-making and project implementation.

Political interference and dominance present another significant challenge for PRIs. State-level politicians often exert undue influence over PRIs, undermining their autonomy and skewing decision-making processes towards political gains rather than community needs. This political interference can dilute the participatory ethos that PRIs are meant to uphold, thereby alienating the communities they serve. Furthermore, the internal dynamics of PRIs are sometimes dominated by local elites or specific interest groups, which can lead to biased decision-making and the marginalization of certain sections of the community, particularly women and marginalized groups. Although there have been efforts to promote inclusivity within PRIs, such as reservations for women and marginalized communities, these groups often face social and cultural barriers that limit their actual participation and influence within these institutions.

Despite these formidable challenges, there are several solutions and strategies that PRIs can adopt to enhance their effectiveness and ensure the successful deployment of advanced connectivity technologies in rural areas. First and foremost, improving infrastructure is crucial. This requires a concerted effort from both the government and private sector to invest in the physical and digital infrastructure of rural areas.

Upgrading roads, ensuring reliable power supply, and expanding telecommunication networks are foundational steps towards creating an environment where B5G and 6G technologies can thrive. Additionally, enhancing the digital infrastructure through initiatives like the e-Panchayat Mission Mode Project, which is part of India's National e-Governance Plan (NeGP), can provide PRIs with the tools they need to manage finances, track development projects, and maintain records more efficiently. These digital platforms not only improve operational efficiency but also increase transparency, thereby building trust within the community.

Addressing the financial constraints faced by PRIs involves both increasing the allocation of resources from higher levels of government and exploring alternative revenue streams. PRIs could consider implementing local taxes, community contributions, and partnerships with private enterprises to generate additional funds. Furthermore, capacity-building initiatives are essential to bridge the digital literacy gap and improve the administrative capabilities of PRIs. Regular training programs focusing on governance, financial management, and the use of digital tools can empower PRI members and staff to perform their duties more effectively. These programs should also extend to the broader community, helping to raise awareness about the benefits of digital literacy and encouraging greater engagement with digital platforms. In terms of overcoming political interference, it is crucial to strengthen the institutional frameworks that protect the autonomy of PRIs. Clear delineation of roles and responsibilities between state and local governments can reduce the scope for interference and ensure that PRIs can operate more independently. Additionally, promoting greater inclusivity within PRIs is essential. This can be achieved by reinforcing the importance of gender and caste reservations and by creating an environment that actively supports the participation of women and marginalized groups.

PRIs can also foster community engagement by establishing regular communication channels, such as village meetings, newsletters, and digital platforms, to keep the community informed and involved in governance processes. This participatory approach not only builds trust but also ensures that the initiatives undertaken by PRIs are aligned with the needs and aspirations of the community. By setting up feedback mechanisms, such as suggestion boxes, regular open forums, and online platforms, PRIs can gather valuable input from the community and refine their initiatives accordingly. Moreover, PRIs can play a vital role in facilitating access to government schemes and ensuring that the intended benefits reach the right beneficiaries. By organizing camps, workshops, and help desks, PRIs can assist citizens in navigating the often complex processes involved in enrolling in government programs.

To further enhance rural engagement and development, PRIs should prioritize infrastructure projects that directly impact rural livelihoods, such as improving roads, water supply, sanitation, and digital connectivity. These projects not only enhance the quality of life for rural residents but also integrate them into broader economic and social systems, promoting economic growth and reducing inequalities. Additionally, PRIs can support local enterprises by providing training, facilitating access to credit, and creating market linkages, thereby fostering economic development and increasing community engagement. Finally, leveraging digital tools is crucial for PRIs to effectively engage with rural populations and improve governance processes.

By adopting digital platforms for governance, PRIs can streamline administrative tasks, enhance transparency, and improve service delivery. Tools like the 5G-RCOLAB simulation can assist PRIs in planning and optimizing network deployments, ensuring that they meet the specific needs of rural users. In conclusion, while PRIs face significant challenges in their role as drivers of rural connectivity and development, there are viable solutions and strategies that can empower them to overcome these obstacles. By improving infrastructure, enhancing digital literacy, strengthening local governance, and fostering community engagement, PRIs can play a pivotal role in bridging the digital divide and ensuring that the benefits of advanced connectivity technologies are equitably distributed across rural India. This integrated approach will not only enhance rural connectivity but also contribute to the broader socio-economic development of the nation, ultimately leading to more resilient, vibrant, and engaged rural communities.

REFERENCES

Ali-Hassan, H. (2009). Social capital theory. In *Handbook of Research on Social Capital Theory*. DOI: 10.4018/978-1-60566-659-4.ch024

Basri, Y. M., Taufik, T., Yasni, H., Putri, R. I., Zahra, N. I., Rahmadani, N., Aslin, M., & Amri, T. (2023). *Improving company performance in rural areas: Institutional theory approach*. KnE Social Sciences., DOI: 10.18502/kss.v8i13.13750

Chaoub, A., Giordani, M., Lall, B., Bhatia, V., Kliks, A., Mendes, L. L., Rabie, K. M., Saarnisaari, H., Singhal, A., Zhang, N., Dixit, S., & Zorzi, M. (2021). 6G for bridging the digital divide: Wireless connectivity to remote areas. *IEEE Wireless Communications*. Advance online publication. DOI: 10.1109/MWC.001.2100137

Das, M. (2022). *Panchayati Raj Institutions in India*. Galore International Journal of Applied Sciences and Humanities., DOI: 10.52403/gijash.20220402

Dively, M. (2003). Social capital in rural communities.

Dixit, S., Bhatia, V., Khanganba, S. P., & Agrawal, A. (2022). *Key considerations to achieve 5G and B5G connectivity in rural areas. In 6G: Sustainable Development for Rural and Remote Communities*. Springer Nature Singapore.

Fischer, P. (2017). Bridging rural constraints: Do rural entrepreneurs exploit online social capital differently than urbanized entrepreneurs?

Freeman, R. E., Harrison, J. S., Wicks, A. C., Parmar, B. L., & de Colle, S. (2010). *Stakeholder theory and strategic management*. Cambridge University Press., DOI: 10.1017/CBO9780511815768.005

Greener, S. (2022). Digging for acceptance theory. *Interactive Learning Environments, 30*(4), 587–588. Advance online publication. DOI: 10.1080/10494820.2022.2062170

Inzillo, V., Garompolo, D., & Giglio, C. (2024). Enhancing smart city connectivity: A multi-metric CNN-LSTM beamforming-based approach to optimize dynamic source routing in 6G networks for MANETs and VANETs. *Preprints*. DOI: 10.20944/preprints202407.1161.v1

Ivan, A., Voinov, J., Kettimuthu, R., Bordel, B., Alcarria, R., & Robles, T. (2022). Towards 6G networks for rural environments: Vision for improving digital inclusion through open-source hardware and software. *IEEE Conference Proceedings*. DOI: 10.1109/CONCAPAN48024.2022.9997739

Jonek-Kowalska, I., Wolniak, R., Marinina, O., & Ponomarenko, T. (2023). Stakeholder theory. *Journal of Business Research, 166*, 114104. Advance online publication. DOI: 10.1016/j.jbusres.2023.114104

Joseph, J. A., & Varghese, D. (2023). Role of community in rural tourism: A pilot study conducted at Kumbalgodu Panchayat, Karnataka. *Journal of Management Studies and Business Management*. DOI: 10.46610/JMSBM.2023.v04i02.002

Mazur, A. (2019). Institutional support for the development of family-type households in rural areas. *Journal of Rural Economics and Social Studies*. DOI: 10.37128/2411-4413-2019-3-8

Mutambara, D., & Bayaga, A. (2021). Determinants of mobile learning acceptance for STEM education in rural areas. *Computers & Education*, *160*, 104010. Advance online publication. DOI: 10.1016/j.compedu.2020.104010

Nugent, J., & Jose, L. (2017). Stakeholder theory and technical communication academic programs. *Technical Communication (Washington)*, 11–30. Advance online publication. DOI: 10.4324/9781315232713-2

Paliwal, R. C. (2018). Role of Panchayati Raj Institutions (PRIs) in rural development of Uttarakhand. *Journal of Rural Development*, *31*(1), 89–95. Advance online publication. DOI: 10.29070/31/57440

Pašti, I. A. (2022). Diffusion of Innovation theory. In *Handbook of Innovation Theory*. DOI: 10.4324/9781003109945-16

Podgorskaya, S., & Schitov, S. (2021). The role and importance of social capital in rural development. *E3S Web of Conferences*. DOI: 10.1051/e3sconf/202127308072

Salem, H., Kouzayha, N., El Falou, A., Alouini, M. S., & Al-Naffouri, T. Y. (2023). Exploiting hybrid terrestrial/LEO satellite systems for rural connectivity. *arXiv*. https://doi.org//arxiv.2311.02591DOI: 10.48550

Sheikh, Y. A. (2023). *Role of Panchayati Raj Institutions towards rural development: A study with special reference to Pulwama District of Jammu and Kashmir*. VIDYA - A Journal of Gujarat University., DOI: 10.47413/vidya.v2i2.181

Simpson, L., Daws, L., & Wood, L. (2003). More than just an internet connection: Building rural social capital through public access. *Rural Society*, *13*(2), 113–125. Advance online publication. DOI: 10.5172/rsj.351.13.2.113

Singhal, V. K. (2015). An overview of Panchayati Raj Institutions in India. *Social Science Research Network*. https://doi.org/DOI: 10.2139/SSRN.2692135

Tilwari, V., Sharma, D., Solanki, S., & Chakraborty, N., & P., W. M. (2024). A multicriteria aware multipath routing method to increase the QoS for future 6G networks. *IEEE Conference Proceedings*. https://doi.org/DOI: 10.1109/CSNT60213.2024.10546037

Tiwari, S., Lane, M., & Alam, K. (2016). Does broadband connectivity and social networking sites build and maintain social capital in rural communities? *arXiv: Computers and Society*.

Tognisse, I. S., & Degila, J. (2021). Rural technology adoption and use model in rural Africa: A predictive approach to telephony acceptance. [IJEA]. *International Journal of E-Adoption*, *13*(1), 36–55.

Türk, U., Dolciotti, M., Östh, J., Reggiani, A., & Nijkamp, P. (2020). Social capital, rurality, and accessibility: A comparative study between Turkey and Italy. DOI: 10.1007/978-981-15-1435-7_23

van Bakel, M., & Horak, S. (2024). Social capital theory. In *Handbook of Social Capital Theory*. DOI: 10.4337/9781035308767.ch33

Wargo, E., Chellman, D. C., Budge, K., & Canfield, K. D. (2021). On the digital frontier: Stakeholders in rural areas take on educational technology and schooling. *Journal of Research on Technology in Education*, *53*(2), 140–158. Advance online publication. DOI: 10.1080/15391523.2020.1760753

Yaacoub, E., & Alouini, M. S. (2020). A key 6G challenge and opportunity: Connecting the base of the pyramid. *Journal of Proceedings*. DOI: 10.1109/JPROC.2020.2976703

Chapter 17
Penetrating Digital Marketing Into Rural Areas by Deploying 5G Networks:
An Indian Experience

G. Raghunandan
Christ University, Bangalore, India

Sriram Ananthan
https://orcid.org/0009-0003-3141-4345
Yorkville University, Canada

Fennala Agnes Iylin
Department of Commerce, Bishop Heber College, Tiruchirappalli, India

Rajnarayanan Balasubramaniam
https://orcid.org/0000-0003-4267-871X
Vinayaka Missions Research Foundation, Salem, India

Dhanabalan Thangam
https://orcid.org/0000-0003-1253-3587
Presidency Business School, Presidency College, Bnegaluru, India

Thirupathi Manickam
https://orcid.org/0000-0001-7976-6073
Christ University, India

M. Haritha
https://orcid.org/0000-0001-5619-4703
Triveni Institute of Commerce and Management, Bengaluru, India

B. M. Ramamurthy
Acharya Institute of Management Studies, Bengaluru, India

ABSTRACT

The proliferation of smartphones, reasonably priced internet, and social networking sites is the primary cause of the shift from traditional advertising methods to digital media strategies. With over 1.4 billion inhabitants, India is now one of the most populous countries in the world. Of them, 700 million use the internet, with most people doing so in cities (Statista, 2023). Businesses hoping to break into India's changing consumer market must take advantage of this new online environment. Thus, it is essential to spread the word about the importance of digital marketing to India's economic growth. Through marketing data analytics and micro-marketing campaigns, the industry has in fact revolutionized how businesses interact with customers and promote their goods. For example, the increased use of online technology infrastructure for online product purchases and the integration of payment and entertainment systems

DOI: 10.4018/979-8-3693-7788-8.ch017

has led to the rapid growth of e-commerce.

INTRODUCTION

Over the last few of years, social media or rather digital marketing has entered the category of being a dominant spearhead to economic development for business within India with numerous buyers available online. Consequently, the paradigm shift in India from being a traditional economy to a digital economy has seen digital marketing play an enormous role as a perfect tool in achieving the goal of target communication with consumers at their pocket. The transition from conventional advertising techniques to digital media techniques is mainly due to the availability of smart phones, affordable internet and social networking sites. At present India is one of the largest populated nations with more than 1.4 billion people out of which about 700 million people are using the internet with the majority of people utilizing internet in urban areas (Statista, 2023). This emerging internet environment is highly important for companies that are prospective to penetrate the evolving consumer market in India.

Therefore, it is imperative to pass on a word about the significance of digital marketing in India's economic development. The sector has indeed transformed the manner in which companies communicate and market their products to the consumers through Marketing Data Analytics and Micro-Marketing Campaigns. For instance, the e-commerce has rapidly grown because of enhanced usage of online technological infrastructure for purchasing products online and integrating payment and entertainment systems. Currently, vibrant firms such as Flipkart, Amazon India and Paytm have experienced a lot of growth due to the effective implementation of sound digital marketing strategies, which have helped them to cover millions of customers. Pertinently, progressive employment of digital marketing has cheap promotion avenues for SMEs, ultimately spurring innovations, more employment opportunities and economic growth as outlined by Rajaraman (2020).

But the country is still split by the digital divide, where the majority of Internet users live in urban areas, while 750 million Indians are in rural regions. Undoubtedly, the digital marketing has entered into homes and has high penetration in the urban area, but rural area is still a virgin territory. Urban areas remain vastly connected to the internet, literate in the digital world, and have a growing consumption trend, with online buying and selling of goods and services. On the other hand, more than 65% of India's population lives in rural regions, and internet and digital connectivity remain limited and slow (Mehta &Agrawal, 2022). This is the big gap that presents a challenge and opportunity to marketers alike. The digitalgap between the rural and urban areas is significantly high, and the development of 5G networks can revolutionalize the strategies used in digital marketing within the rural areas due to improve connectivity.

Digital Marketing in Rural India: The Untapped Potential

Out of the total population of India, approximately 70 per cent of dwellers live in rural areas, and digital marketing to these people remains a relatively new concept. More than 900 million population of India resides in the rural areas in approximately 600000 villages so it is a huge market for the firms Census of India, 2011). What marketers have avoided rural areas because of infrastructural issues that, with pace of digitalization the situation is quite different. The government of India launched the Digital India plan and constantly increases the accessibility of the internet in rural areas, and thus, the gap between rural and urban areas is gradually decreasing. Thus, a massive opportunity has been created to accelerate this transformation with the arrival of 5G technology.

Analyzing the demography of the rural areas of India means understanding that it is a young and growing population with rising purchasing power. According to the newly released National Rural Fixed Line Broadband Telecommunications Services Report 2021 by the National Statistical Office, about 50.81 percent of rural population is below 25 years, meaning India's rural markets have a promising pool of young, technology-demanding customers. Despite the notion of rural economybase on agriculture the findings showed the rural economy is more and more looking for employments in other sectors than agricultural sectors like manufacturing, retail and service sectors. This shift is calling for almost all kinds of products and services in the electronics, fashion, education and health sectors amongst others most of which can either be marketed and sold through online platforms.

Other trends that have come up in the consumers behaviors in the rural regions are as follows; The rural consumers are also getting conscious with global trends, branding and have higher aspirations owing to increased exposure to digital media. The increase usage of smartphones coupled with cheap data connection has seen peoples' engagement in social media, shopping on the internet and use of digital wallet increases (Ghosh, 2022). On the same note, the e-commerce adoption among the rural consumers has risen slowly although the urban consumers are in a better position. This changing behaviour pattern along with the augmentation of digital structures offers a vast potential for organizations to extend their business into this market through web marketing.

Purpose of the Review

In fact, the objective of this type of review is to determine the contribution of 5G networks towards expanding the digital gap between rural and urban areas in India especially in penetration of digital marketing into the rural regions. As we know digital marketing is almost mandatory in the urban business network, however, the rural market is relatively unexplored because of infrastructure and connectivity issues. The expansion of 5G technology should give a massive boost to internet access for many hectares of the countryside, offering new possibilities to digital marketers.

This review seeks to assess the ability of the 5G networks in increasing the access to digital marketing services in rural markets through availability of higher internet speeds, reduced latency and improved connectivity. It will also discuss the issues associated with the roll out of 5G infrastructure in rural areas of the country and suggest possible solutions to address those issues. Besides, the review will determine the opportunities that 5G has in changing the way business communicates with consumers in those areas, especially by segmenting and targeting consumers using analytics. The marketing managers can use 5G for selling their products through mobile phones by satisfying the needs of rural customers.

The review will also present recommendations for both business and policy on how to effectively expand hinterland markets through digital marketing techniques enhanced by 5G. As part of the infrastructure, this roadmap will become a step-by-step action plan for infrastructure development, promotion of digital competence, including among vulnerable populations, and culturally appropriate advertising initiatives. It will also explain how the use of 5G by the government in collaboration with the telecom firms and private entities will require massive efforts to transform the agrarian economy of rural India into a digital revolution.

Scope of the Work

The delimitations of this study are to narrow the focus to the rural areas of India and the changes in the availability of digital marketing with the deployment of 5G. As the third-world country, India having embraced the 5G technology, the government, and companies have the chance to create the opportunities for the residents of villages or other rural areas providing them with the needed equipment and technologies that will make the quality of life better and the economical development progressed. The issues under investigation will include how the superiority of 5G network connections can manage the current difficulty of the low connectivity and low digital inclusion in rural India.

The findings presented in this review will prove helpful to numerous stakeholders such as policy makers, firms and advertisers. For policymakers, the review will provide an understanding of what measures need to be taken to support the development of the 5G environment within the country as well as create a balance between the development of the digital economy, where potential benefits will be distributed between urban and rural areas. It will also emphasize the potential of a closer look into investments in developing digital skills for people and enhancing the network infrastructure primarily in the rural regions to access 5G networks at full potential.

For businesses, the review will give an outlook on the prospects of rural markets and a guide for appropriate implementation of online marketing techniques pertinent to the likes of the rural customers. The review will identify how the features of the 5G can be used, for instance, the AR and VR, the real-time data analysis to design unique and quality marketing experiences to the rural people. Last but not least, for the digital marketers, the study will offer a systematic guidance of how to establish and implement a campaign that captures the attention of the rural consumers, emphasizing on how the constraints of rural marketing may be solved through applying 5G technologies or services.

CURRENT DIGITAL MARKETING LANDSCAPE IN RURAL INDIA

India has around 930 million population of rural area and in terms of economic perspectives it plays a very important role but with in number of problems regarding in digital literacy and infrastructure. While the Indian government has introduced programs like Digital India to enhance the Internet in village areas, the penetration of Digital Literacy and availability of Digital facilities and connectivity along with the optimisation ratio is significantly low than the city areas. The rural digital economy is still in a early stage, most of the rural consuming public has not been exposed to internet, digital solutions, online services etc.

Another challenge is that the rural population is not very clued up with digital technology. Currently there is a strong evidence that shows that at least half of the rural and remote population is either illiterate when it comes to technology or is only partially literate when it comes to the use of available online tools. Digital literacy in rural regions is also low; a study by KPMG (2022) has shown that less than one third or 25% of the rural Indian household possesses at least one digitally literate person as opposed to over three quarters and more than 70% of the urban Indian household. Such a gap inhibits the growth of online and promotional activities in the rural areas where the influence of word-of mouth and papers advertisement are very common.

In regards to the physical necessary structures, connectivity or Internet in rural areas of India has been significantly enhanced though still buggy. According to the Telecom Regulatory Auth- ority of India (TRAI), while the rural internet usage penetration rate was slightly over 37%, the percentage for the urban areas crossed 70% in 2023. While the use of 3G, 4G networks has grown tremendously especially in the developing world, internet quality in most of the areas remains low due to lack of proper networks and inadequate broadband connections. Consequently, efforts by the rural digital economy to upscale its use of the internet digital platforms, and otherwise deploy effective digital marketing strategies as part of an effective digital marketing mix, are hampered.

Digital Marketing Issues in Rural Regions

Several impediments restrict the development of digital marketing in rural India, which are as follows: Another is that a problem with bandwidth and unsatisfactory availability of broadband connections to connect the networks. Several villages and other rural areas are still technically isolated, with limited or slow and often interrupted connections due to antiquated telecoms networks. Although 4G services have penetrated deep into the rural areas, their signal service reliability is not impressive, especially in remote or hard-to-reach village areas. They also do not propose any form of connectivity sufficient to enable digital marketing for businesses targeting consumers in the rural area.

This is because Consumers in the rural areas neither have high awareness of data use nor high digital adoption further worsening the situation. A lot of residents in rural communities are either unaware of the d Raphael platform or have classic trade preferences. Instead, they use traditional purchase sources such as small local stores, street markets and informal suppliers besides depending on word of mouth for information procurement than embrace innovative ICT content like e-shopping. This as a result of the following: The willingness to trust online tools is still low; Most of the population has not been exposed to digital marketing campaigns; consumers associate these services as unnecessary or unreachable.

A major challenge affecting digital marketing in rural India include; Inability to afford is another factor that has contributed to the low uptake of digital marketing. Smartphones remain expensive, internet-enabled devices and data plans are still expensive for many a rural household. Although the penetration of affordable smartphones has improved recently, possession of such devices remains a relatively expensive decision for the low-income segments. Further, the repeated charges for data plans regardless of the availability of cheaper plans from telecom service providers such as Reliance Jio infrequently discourage the individuals from many uses. This is the case because access to affordable devices and internet services is not as common in rural regions of India, which means that digital marketing initiatives fail to gather much steam.

Existing Solutions and Their Limitations

A number of strategies have been suggested to overcome problem of Digital Divide and to increase digital marketing in rural India but with limited effectiveness. 3G and 4G has helped the backbone of the Internet, fixated Internet availability to the rural area where more rural users are participating in the digital content, e-commerce sites and social networks. Currently, telecom operators such as Reliance

Jio have under pressure seek to lower the cost per GB of mobile data, which has ultimately lead to improvement in internet penetration in the rural areas.

However, problems such as scalability, low quality of services and reliability from current 3G and 4G technologies are emerging. Rural consumers often experience connectivity challenges and therefore restrict the possibilities of digital marketing initiatives that are sensitive to internet connection quality. Live stream, video advertising, and other forms of content-intensive marketing communication are usually unable to penetrate the rural audience. Also, the transition from 2G to 3G and currently 4G has not reached equal acceptance throughout rural India, which remains stubbornly glued to the outdated 3G or even a 2G network that is totally inadequate for carrying out current modern techniques and strategies in digital marketing.

Yet, a few notable successful digital marketing campaigns have found ways of crossing these barriers and reaching the rural market. For instance Kan KhajuraTesan of Hindustan Unilever Ltd is a good example of the successful tales. The campaign employed the use of IVR because it made it easier to contact the consumers who had restricted access to the internet especially in poorly connected rural areas. Whereas rural users could call a toll free number and enjoy entertainment content in between which advertisements would be provided. This innovative approach successfully aimed at a rural market while avoiding the main reliance on internet service availability, which shows that even with the help of appropriate strategies, marketers can venture into the penetration of the rural market (Kumar, 2021).

Although this has been appraised as an appropriate strategy by digital marketing gurus, not all the campaigns have been positive in the rural regions. Most of the online centered campaigns alone have been pulled down due to connectivity issues and literacy. For example, discount or promotion driven retail digital advertising campaigns targeting customers with e-commerce offers have failed to perform well in rural areas where customers may not wish to undertake online transactions or where they have access to reliable internet connections. Essentially, most of such campaigns fail to consider the needs and wants of the rural consumer hence resulting in low results.

Overall, the marketing using the digital platform for the products of the Rural India market can be rewarding, but at the same time have risks. Despite significant possibilities for development, the problem of low digital literacy, a weak infrastructure, and high costs prevent digital marketing from developing in rural areas. Previous solutions like the 3G and 4G networks have made some progress in realising the rural internet connection and general adoption of the technology, however, they do not adequately cover those shortfalls. That is why, as the subject of the digital economy develops, businesses and policymakers needs to look for new ways, such as the use of 5 G networks, to provide access to the potential of rural markets and to ensure the promotion of the benefits of digital marketing across the nation.

5G TECHNOLOGIES: A GAME-CHANGER FOR RURAL INDIA

5G is a new generation of mobile Internet based technology which has been developed to dramatically improve wireless connections with ultrahigh-speed and much lower latency. It stands for a giant move from the Fourth Generation, 4G, in regards to speed and functions. This paper further elucidates that 5G offers better speed and can at most be up to 100 folds faster than 4G with peak data rate of more than 10Gbps (Mitra&Saha, 2021). This allows a user to download high definition movies within a few seconds and have an ability to stream and even conference at a go. Another equally significant property of 5G is low latency which literally means the duration between sending out data and the time that it takes to receive back this data: the former can range from 1 ms (Kumar et al., 2022). Also, 5G has a greater bandwidth meaning that more devices can be connected with the same speed and performance as before. In this case, 5G is a lot better than 4G as it pertains to connectivity. While 4g made mobile internet stable enough for streaming and Browsing, 5g builds on it and provides capability for supporting IoT, autonomous vehicles and smart cities (Sundararajan, 2023). For instance, 5g allows one million connected devices per square kilometer while 4g barely handles device connectivity (Ericsson, 2023). Now the use of 5G in implementation reveals new possibilities for using the Internet in rural areas where it has remained a problematic issue due to the constraints of infrastructure and slow connection speed. Opportunities of 5G in Rural Environment

Benefits of 5G for Rural Areas

There are benefits can be enjoyed through 5g networks, of which some of the benefits are as follows;

Closing the Digital Divide

The availability of 5G technology therefore presents a good opportunity of delivering internet connectivity in rural areas of India in a bid to bridge the digital divide. Today, few rural populations have access to quality and speed internet connection; most are constrained to use 2G or 3G connectivity. From the Telecom Regulatory Authority of India (TRAI) report, the current internet connection is estimated at 37% concerning the rural population and about 70% concerning the urban population (TRAI, 2023). This provides a clear vista of how it hinders the participation of rural inhabitants in consumption of service delivery and socially and economically productive economy. This paper argues that 5G's high speed, low latency, and boosted connection densities are capable of eliminating these constraints and afford rural communities comparable internet services as offered in commercial cities. By offering the efficiency of download and upload speeds, 5G can benefit residents of rural areas by integrating with information offerings and leveraging development technologies. 5G networks are anticipated to be established across the rural regions of India to provide symmetry between the urban and the rural sector of the country; providing new chances for advancement to the rural population more than ever (A. Sundararajan, 2023).

Supporting Small Enterprises and Self Employer Ventures

Another beneficial effect of 5G in the country particularly in the rural zones is the possibility to foster small businesses and business man. Small industries in rural areas of India and businesses associated with agricultural sectors have constraints on accessing a wider market because of inadequate physical network for access and digital connectivity. For these businesses, the coming of 5G will ensure that they use digital tools and platforms to spread out, sale products/services, and become more efficient. For instance, farmers in rural regions can get real-time information concerning the weather conditions, soil status and the crop health through IoT based devices and sensors developed for 5G. These kinds of data will help the farmer to be informed about the best way to use it and in the process provide the best yield to their crops successfully (Kumar et al., 2022). Further, small business persons in the rural areas can take advantage of the electronic commerce system to market their products to the various users in the domestic market as well as the export market hence improving their revenues thus economically transforming their regions. Also, 5G will develop the opportunities of rural entrepreneur by providing the needed access to digital tools, to cloud, to training programs, etc. Decision makers in rural areas can attend sessions on new products through online mode, can make deals with partners online and can operate their firms by availing cloud computing systems. With the entrepreneurial development opportunities created by 5G, it will be instrumental in the boosting the rural economy and the creation of new employment opportunities in the rural areas of India (Mitra&Saha, 2021).

Enhancing Digital Services

Education, healthcare and eCommerce is the major sector in which this business is expected to thrive by serving the needs of the people through technological advances. Mr. Mugwisi also noted that through 5G technology, access to important digital services such as education, health, as well as commerce will also be availed in the rural areas. The capacity to access higher quality education and health care in rural regions is still limited by infrequent and, at times, unreliable internet connection. Students and patients at many rural schools and health care centers themselves are prefixed to avail digital services due to lack infrastructure. 5G will allow rural schools to deliver quality education in form of online classes, video conferencing, distant learning facilities etc. Rural students will receive equal chances for the same resources as urban students, and thus be equally capable of enrolling to universities and vocational colleges without having to move to cities. Other than that, 5G will be useful in telemedicine and remote consultation services, so patients with no capability to travel for long distances will be able to access needed healthcare services in rural areas (Ericsson, 2023). As for e-business, 5G with positive effects of make rural customers tends to use the online shopping platform and services to purchase products which are provided by the platform and fulfil cheap and diversified. Hence, with better connections physical, rural dwellers will be able to engage in marketplace business, make use of e-commerce and digital money, and also get an advantage of having goods delivered to door steps. This will help in raising the quality of life in a rural area besides also open up the rural market to be exploited by various business ventures (TRAI, 2023).

Measures Launched by the Government in Aid of the Development of 5G in Rural India

The Indian government has launched initiatives like the BharatNet project to expand fiber optic networks in rural areas, subsidized spectrum costs for telecom companies, and introduced 5G testbeds in agricultural zones. Additionally, it offers grants for tech innovation to enhance rural connectivity, aiming to bridge the digital divide. This part reveals the measures taken by the Government of India to promote 5G in Rural India.

Policies and Programs for Rural Digital Inclusion

Using various polices and programs the Indian government has been quite aggressive in a pursuit of expanding the use of internet in rural areas. Digital India which was kick started in 2015 is a campaign to make India a digitally empowered nation with an objective of providing Internet access to each of the Indian citizens, including those in the rural areas. To facilitate this, new programs like BharatNet and National Optical Fibre Network (NOFN) have been planned out to connect countryside with high speed broadband network (Department of Telecommunications, 2023). The BharatNet is one of the flagship programmes under the Digital India initiative to link about 250,000 Gram Panchayats through optical fiber connectivity. Continuity from BharatNet will progress with roll out of 5G networks which will enabled rural centers to access next generation connectivity and digital services. This way, the government seems to enhance the internet connectivity in rural area to close down the so-called digital divide and make more the economy digitalized (Mitra&Saha, 2021).

Collaborations Between Telecom Companies and the Government

To ensure that 5G drive is successful in the rural areas of India, the government will have to work hand-in-glove with telecom service providers. The government has already laid down policies for 5G implementation by securing the spectrum for 5G services and promoting private sector participation in telecommunication sector investment. Some of the major telecom players of India are Reliance Jio, BhartiAirtel, and Vodafone Idea have already signed_statement to roll out 5G networks in the country and the key focus will be on connecting rural region as well. Apart from spectrum sale, the government has recently come up with the PLI scheme to encourage the manufacturing of telecom equipments in India which will help for the implementation of 5G infrastructures. This means that, if the central government is not to do left want in the new revolution, it needs to coordinate closely with private telecom operators concerning the expansion of 5G networks in the rural regions.

These stakeholders have the opportunity to foster innovation altogether, and thus design solutions to the problems of rural connectivity and to implement 5G technology in all parts of the country (Sundararajan, 2023). 5G technologiesoffer the prospect to transform the scenario of rural India in terms of internet connection, employment generation for small traders and businesspersons and effective delivery of online services including education, health care and e-commerce facilities. Thanks to the higher network speed, lower latencies and increased bandwidth, 5G can connect rural and urban environment and give rural communities a new way for economic growth. To make 5G works in rural India, government programmes like Digital India and BharatNet are also conducive to the partnership between telecom companies and the government for successful implementation of the technology. With 5G networks being rolled out in the nation, the options of connectivity and digital services for rural India have never been better, raising the hope for a better life for the country's inhabitants.

STRATEGIES TO USE 5G FOR MARKING THE ENTRY IN THE DIGITAL MARKETING IN RURAL INDIA

Thus, the day has come when Indian can proudly speak about the country's digital coverage when speaking about crops The majority of the population is still rural and to them it constitutes almost 2/3 of the total, thus still remains an untouched ground. This makes the use of digital marketing an agent of change with the arrival of the 5G technology as a way of reducing the rural-urban digital divide this is because. Specifically, this roadmap identifies the following phases that are necessary for the penetration of digital marketing in rural India using 5G: Infrastructure promotion; Digital enlightenment; Strategic marketing; Market enlargement.

Phase 1: Infrastructure Development

The basis of digital marketing's growth in penetrating rural India is built upon sturdy 5G structures. Internet connectivity remains wanting in the rural areas where currently only 37% of rural households have internet (Communication Ministry, 2023). Increased deployment of the 5G service in these areas is crucial for the advancement of the digital marketing landscape 5G offers internet speed, low delays and the prospect of serving a large number of gadgets since the environments in the marginalized physical lands of the rural setting are physical hindrances (Mitra and Saha, 2021). Currently, the two giants of telecom are Reliance Jio and BhartiAirtel which have expressed intent on penetrating 5G in the rural areas of the country. More awareness and a greater push should be made to provide rural consumers an opportunity to interact with the digital platforms. More specific funding priorities in 5G development should involve deploying new towers and improving the availability of spectrum resources, especially to increase the fiber-optic connectivity in rural areas for the future (Sundararajan, 2023).

Public Relations Between Government, Telecom Companies and Other Private Stakeholders

One therefore concludes that partnerships with the government and the telecom industry and other private players is essential in achieving broad 5G coverage. The Indian government has paid a significant amount of attention through having various developmental programs like BharatNet that operates under Digital India programme with purpose of connecting villages to the high-speed internet (Department of Telecommunications, 2023). These investments, along with private funds, are needed to build the scalable and inexpensive 5G network. Joint efforts of telcos locally with businesses as well as the regional governments will help the decrementation of costs and faster implementation of 5G. It remains possible for the government to take a crucial role by subsidizing, offering spectrums at lower prices, and encouraging the private sector to fund the construction of rural facilities. Moreover, the telecom companies can tie up with the local merchants to develop cost-sharing structures to facilitate the construction of infrastructures (Mitra and Saha, 2021).

Cost Cutting and Fast-Track Deployment: Public-Private Partnerships

PPP is the way through which government aims to minimize the expenditure of the physical infrastructure required for 5G deployment in rural regions. The issue of high initial cost is addressed by the fact that PPPs are collaborative efforts where the two sectors can develop resources, knowledge as well as investment. Through government direction, telecom companies can engage the rural cooperatives, micro finance organizations and agricultural organizations to provide 5G structures at relatively cheaper prices (Sundararajan, 203). Additionally, they foster local and new entrant business initiatives and application developments by extending PPP options for 5G for rural SMEs. This not only fast forwards the deployment path but also allow the countryside populations to be an active stakeholder in the digital economy revolution. These partnerships can be furtherpositively influenced if the 5G deployment is in sync with the government's rural developments' plan to ensure the population in the rural areas receives the benefits of digital assimilation.

Phase 2: As a Way of Enhancing Digitization of Learning Support to Enhance Digital Literacy and Awareness, the Following Areas Need to be Addressed

The subsequent major goal of getting consumers from the rural areas becoming better connected is also accomplished once the necessary framework is developed. Most of the rural people have low awareness of digital platforms and services, and they cannot optimally gain from 5G digital marketing. To introduce potential benefits of digital tools, e-commerce platforms, and mobile internet to rural people, grassroots campaigns are important (TRAI, 2023).

Such campaigns should actually aim at spreading the word to the effects that rural businesspersons, farmers, and consumers could use digital marketing for better business and improved living standards. Community inhabitants can participate in local NGO's, local government agencies and other related non-profit organizations to help in the organization of workshops, training sessions and awareness campaigns. The effectiveness of such initiatives largely depends upon the answer to the question, Fine: there must be a positive impact to show – be it access to agricultural market or to education through the Internet or in IT healthcare and so on Sundararajan, 2023.It's also important that citizens had local influencers and community leaders to look up to.

There is tendency in involving prominent opinion leaders in actual local communities of rural India in promotion of the concept of digital literacy. Farmers are more likely to be influenced and follow those they know or have reasons to respect, making local influencer indispensable for any form of marketing, particularly that done on the digital platform. Community participation may be improved immensely when the local panchayat leaders, school teachers and rural entrepreneurs are involved in the campaigns (Mitra&Saha, 2021).Such influencers can be useful in increasing the use of digital channels because they can record and present actual cases the digital marketing endeavors of rural residents. Examples of a farmer earning more money from selling his produce online or a rural artisan acquiring customers through e commerce should be emotionally appealing to other similar individuals who simply have to embrace change and innovation and should also embrace the use of these new technologies.

Integrating Culturally and Linguistically Relevant Support for Education

A key in achieving the goal of increasing the internet use in the rural areas of India is the multilingual and multicultural society. The population mostly resides in rural areas and communicates in multiple regional languages, meaning that copying what is being done in other countries will not work. Literature has indicated that learning materials should be culturally sensitive and relevant depending on a particular region as the language tends to vary hence affecting the total of usages (Department of Telecommunications, 2023).Telecom operators working in conjunction with local government and regulatory bodies should create localized content in regional languages to educate people about the usage. Such applications, video tutorials, and websites should be of clear and simple for getting through to the villages and helping people to learn about the importance of being outcome of a digital literacy crusade. Also the use of customs and tradition when conducting marketing on the digital platforms can enhance the appeal of the marketing to the rural people (Sundararajan, 2023).

Phase 3: Creating Rural-Focused Digital Marketing Communications

Therefore, for digital marketing to appeal to the consumer from the rural area, it has to be customized depending on the needs, wants, and beliefs of the consumers from the rural area. Rural India is a heterogeneous market which comprises of people from different lifestyles, income and consumption capabilities. Marketers require creating new content targeting the rural audience on the basis of regional characteristics of Tri-State consumers.For instance, digital marketing should target products and services that will be suitable for this area, which include; farming equipment, cheap and reliable health services, and schools. Brands need to employ culturally appropriated words, signages and simplified stories in marketing communication as a way of reaching the country folk. Further, appeal to the self-organised values of rural consumers such as drawing attention to research findings noting how many products and services increase productivity and save time (TRAI, 2023).

Taking Advantage of 5G in the Delivery of Interactive and Multimedia Enhanced Campaigns

Another benefit of 5G technology is enhanced multimedia communications, such as videos, VR and AR. These technologies can be used in a way that a digital marketing campaign can be developed to appeal and grab the attention of the rural consumer. For instance, video commercials on automobile use using the 5G technology to explain product features or application photographs with success stories of people from the rural areas in endorsing the digital platforms may go a long way (Ericsson 2023). Likewise, imaginations in VR and AR can help craft engaging marketing experiences in which rural consumers can navigate products or services, or experiment with agricultural practices in real life. For example, an AR campaign could help farmers see how new equipment could function on their land hence improving their decision making. These campaigns involving multimedia, backed up by 5G, imply that rural area digital marketing is likely to derive far greater benefit (Sundararajan, 2023).

Real Life Examples of Successful Campaigns That Disseminated Information to Rural Consumers

Several digital marketing campaigns have already been done showing the possibility of reaching out the rural consumer. For instance, the Digital Mandi – an e-commerce venture allowing farmers to have direct interaction with potential buyers— has benefited Indian rural farmers by providing them better price for their produce by eliminating intermediaries (Mitra&Saha, 2021). Another easy example is the 'Jaago Re' campaign that was developed by Tata Tea to not only sell their products, but also raise awareness of points that would appeal to rural Indians (Sundararajan, 2023).These campaigns draw the need to establish marketing content fit the rural consumers' characteristics, and their lifestyles. Thus, by adding 5G to such campaigns to boost the influence across such regions, business individuals can penetrate the rural regions in their digital marketing efforts.

Phase 4: Market Access Creates the Necessary Conditions for the Functioning of Electronic Commerce

Currently, e-commerce holds the possibility for enabling the producers and consumers in the rural productive zones, allowing them to benefit from the expanded markets. But the key to rural e-commerce is the availability of efficient and stable internet connection and connection only. When 5G is rolled out, all the rural leaders and artisans can be connected to the E-commerce platforms whereby they can market their products to the urban and global clients (Kumar et al., 2022).Applications of 5G such as high speed can be directed by rural e-commerce to live stream their products, fashion a video shop and buy and/or live control of inventory. These features can increase the convenience of managing and improve the position of rural enterprises in the context of customer relations. Also, 5G will enhance e-commerce consumption in the rural areas through the ability of rural consumers to make online orders, track their consignments and making payment through electronic means (TRAI, 2023).

Integrating and Coordinating Rural Producers and Artisans to Urban and Global Markets

Thus, one the main avenues of digital marketing aimed at the rural populations is to integrate producers and artisans of rural India into the larger markets. Due to 5G, the rural players are able to access the national and international markets in that they can sell their commodities online. For instance, the rural food producers who carve wooden figurines, which are admired by rural cultural consumers or to artisans who weave baskets out of locally sourced organic material, e-commerce enables them offer their products to the urban market consumers who appreciate anything hand made or have cultural value (Ericsson, 2023).Digital marketing tools can prove useful in branding strategies; it may be effective for rural business to post some of their products on their business social accounts, market places, and independent ''Website Store'. In the context of enabling rural businesses to better promote their products to the market, 5G inherently assumes a very significant part in raising the enterprises' market prospects and, in extension, possible revenues (Mitra&Saha, 2021).

5-G Technology-Backed Rural E-Commerce Success Stories

A number of e-commerce success stories are already present in rural India and acts as examples that digital platforms can greatly help in the economic development of the country. For instance, DeHaat, an online marketplace that primarily targets small farmers in selling their produce and acquiring buying and supply partners, has revolutionalized the agricultural value chain through giving small farmers direct market access and minimizing transaction costs (Kumar et al., 2022). In the same way, MadhurMandi has empowered rural weavers to sell their handmade textiles to the consumers in India providing them better incomes and has advocated traditional arts (Ericsson, 2023).The above afore-scenarios show how the ecosystems of e-commerce through 5G networks empower the growth of rural areas. In extending the applicability of digital marketing and e-commerce in rural settings, 5G holds potential to bringadditional benefits to rural entrepreneurs and/or businesses.

The roadmap for penetrating digital marketing in rural India using 5G involves four key phases: infrastructure construction, cultivating people's digital skills, deploying appropriate marketing models for the vast rural areas, and establishing e-commerce and market connection. In September of last year,

I stated that through expanding the current coverage of the 5G network, mobile companies are actively popularising digital literacy and engaging the consumers residing in rural areas of the country; while, through developing culturally-suitable marketing content, businessmen can foster the rural borrowers as well as the rural market mid-wives. 5G will also make it easier for rural producers to sell their products to a wider market via e-commerce – spurring economic growth and incorporation of the entire country especially the rural area.

MAJOR ISSUES IN CREATING 5G-POWERED DIGITAL MARKETING IN RURAL AREA OF INDIA

There are massive opportunities for 5G-powered digital marketing in rural India, albeit the following challenges. All these present technical, economical, social and cultural problems that call for their solution for the penetration of digital marketing in rural areas.

Technical Challenges

One of the biggest technical constraints of deploying 5G in rural India is the high cost and installation inconvenience that comes with establishing that infrastructure. However, in rural areas, the population is generally sparse which means that it will not be cost effective for the telecom operators to lay the required and necessary infrastructure. Installation of 5G towers, fiber-optic cables, and base stations demand sizeable capital expenditure, and here the possibility of quick return on investments is not so high because of limited market and low level of digitalization in rural area (Verma, 2022).Also, factor of topography in the rural areas with features of mountains, forests, or River inhibits the establishment of 5G networks. Other challenges include the ability of these remote regions to provide power back up which is crucial for the running of 5G. They do not only add to the cost but also delay the timeline for achieving full fived development and the areas that are still in the most rural and marginalized part of the country.

Problems related to Maintenance and Technical Support for ICT and computers are difficult to handle with in the rural areas.The fifth and final technical challenge, from the applicant's analysis, is how to maintain and support 5G infrastructure in rural areas once it has been initially set up. Most of the time, there is inadequate human resource qualified in the complex 5G technology when it comes to rural areas. Therefore, if there is a problem such as a faculty's network or equipment problems then addressing them will take some time. Few service stations and access to spare parts, and repair services, especially in rural areas, compound the situation, meaning that for long, consumers have to wait for their 5G services and digital platforms to come back online, greatly eroding trust in these services.Moreover, maintaining technical support and ensuring that the rural consumers who may not fully understand advanced technologies would not be able to navigate will add another challenge to the process. Current subscribe telecommunications companies and digital marketers need to devise means by which the rural end users may be provided proper customer service/technical support after accessing and interacting with the new digital platforms.

Economic Challenges

Another limitation to delivering 5G for digital marketing in rural region of India is the issue of expensive 5G enabled device like smartphone, tablet, smart gadgets etc. While 5G smartphones are beginning to become available on the market at a more affordable price point in urban locations, they are still unattainable for the majority of consumers in rural areas, who are likely to have a lower capacity for disposable income. These devices are costly and this plays a big issue when it comes to the implementation of 5G technology to support results of digital marketing to the rural population (Sharma & Patel, 2023). Moreover, consumers in rural areas may not replaced their current devices with those supporting 5G, especially if they do not find immedite use of the technology. This hesitation is normally accompanied by the challenge that most of the rural areas have not even experienced 4G network let alone embracing 5G, which makes the transition even more restless.

The importance of this aspect is to ensure that mobile consumers in rural areas consume affordable data plans.Besides, the cost of the communication gadgets, the most important and significant economic issue among the four H's of mobile commerce is the cost of accessing available data plans. Although 5G networks are said to deliver higher and faster speedancy and lower latency they are also likely to contribute to the use of larger data and this implies that users will be charged more. Hence, there is the need to make 5G data plans cheap for them to make some Yi importance in the rural areas where people are believed to earn less as compared to urban dwellers.

The following is an area of contention: telecom firms need to become profitable thus offering expensive data tariffs while at the same time their tariffs have to be cheap enough to be consumed by rural users. If 5G services continue with high prices they will not be accepted by consumers and the desired change through digital advancement in the rural areas will not be realized (Rao, 2023). This is true as 5G technology may not be set at affordable rates and thus charged on a user's data plan or Government may need to subsidize 5G to make it popular in rural areas.

Social and Cultural Barriers

The biggest challenge that affects the implementation of digital marketing strategy through 5G in rural areas of India is lack of digital literacy. It is, therefore, important to note that while more and more people in the rural areas own smart phones most of them do not have adequate skills and knowledge to offer them full abilities in digital platforms. This digital divide is deep-rooted by the increased use of new technologies such as the 5G, which can be a little overbearing for new users (Bhatt & Kumar, 2022).

There is also a problem of technological resistance in some of the rural communities in that, besides digital illiteracy, some people still do not want to embrace new technological tools. This results in rural consumers being hesitant to resort to digital platforms as these are often met with skepticism due to inferiority complex from traditional beliefs and learning that dictates that anything considered complex isn't necessary. Such cultural resistance can be seen in people's unwillingness to transact with e-commerce platforms, social media applications or engage in digital payments which are regulators for the execution of digital marketing strategies. As for the movable cultural barrier, it is important to launch detailed educational campaigns, popularize the fact that such a platform exists both at the community and national levels along with engaging local opinion leaders who could help increase confidence in digital technologies.

Privacy Disregard and Lack of Trust in Online Markets

Another challenge that hinders the adoption of 5G powered digital marketing in rural India is the annoyance attributed to privacy and the overall skepticism that characterizes internet based platforms. Some of the country consumers especially those in the rural areas may not have any idea how such channels work and may also be worried about their identity being hacked. All these concerns are compounded by low awareness of data privacy, identity theft, and cyber risks (Singh, 2023, p. 8).For instance, rural users are likely to be reluctant to reveal their personal information or pay details on online fora, threatening the attempts of firms to reach out for them through properly defined market segments or electronic commerce. It will take time and a combined effort from the government and private business makes to ensure that users in rural areas have the knowledge of protection of their data and ensure that the platforms have adequate security measures put in place to discourage malicious actors.

5G technology is still under the-development process and has the possibility to bring a major transformation in digital marketing scenarios in the rural area of India, however, the following issues has to be effectively solved to provide the proper implementation of 5G. Hardware and networking, software, and infrastructure constraints of manufacturing and installation, costs of devices and data service, and cultural resistance to the technology also remain major hurdles. But if policies are formulated with such issues in mind and, if government and private entities work in tandem besides extensive multi-faceted digital inclusion programs, these barriers can be effectively overcome to enable the potential of 5G in the digital development of rural India.

SOLUTIONS AND RECOMMENDATIONS TO DEALING WITH CHALLENGES

Having said that, it is clear that the barriers related to implementing 5G-driven digital marketing in rural India are not insurmountable. Campaign for awareness created by both the government and private companies along with an increase in partnership with communities could significantly increase the number of people going online and therefore the potential in the rural markets. The following solutions and recommendations provided the framework necessary to combat such challenges.

Building Up Partnership Between the Government and Private Sector

Preparation and active involvement of telecom companies appear to be among the most direct routes towards realizing 5G in areas of interest such as rural regions. Telecom providers can and must play a part in driving down the cost of 5G devices and data. It becoming cheap: It is possible to offer subsidies to specific 5G Smartphones to make them affordable by the rural consumers individually or in conjunction with the government. The firm also following the low-cost and high-speed data plan options would persuade the rural consumers to go for the 5G digital technology platforms and make the digital platforms more appealing and useful for them.Possible models for subsidization are akin to those that helped to launch the use of mobile telephony in India. It makes sense that spreading the introduction of 5G handsets together with cheap data packages allows telecom operators to make progress in the penetration of digital marketing into the rural areas.

Government Policies to Support 5G Investment on the Rural Areas

They note that rural development of 5G infrastructure largely depends on Government policies to encourage its implementation. The government could apply sales tax exemptions, subsidies, and concessional credit for telecoms and other private actor willing to build 5G networks in rural areas. More recent liberalization measures to facilitate easier foundation of new 5G towers and the establishment of fiber-optics in rural areas would continue to chip away at the cost of infrastructure (Rao, 2023).Also, the government could announce certain Tax exemptions, discounts on the license fee for telecom companies and marketers of digital products in certain discrete geographical areas of the country's rural zones that would be zoned as SEZs for 5G networks deployment. This approach would not only enhance the physical connectivity but also bring forth more venture capital into the rural areas that enhance the digitization that is central of any digital marketing.

Measures of Supporting the Development of Digital Literacy Among Rural Citizens

With the help of 5G for digital marketing to thrive in rural India, there is a big need to invest in digital awareness. Appropriate digital literacy initiatives under which aspiring users will learn about the application, accessibility of services, and utilization of digital technologies are vital in enhancing the usage rate. Such programs should be designed based on their existing understanding of health issues, lately the rural users might have poor English speaking skills and thecarliness might hamper their ability handle technological gadgets (Bhatt & Kumar, 2022).Telecom organizations, digital marketing firms, and educational establishments could join and create basic-level training programs that will reflect useful tech proficiency. These modules could be delivered through smart phones, local learning centers or through community education centers. First of all, these initiatives should be conducted in regional languages and supported by hand signals and multimedia graphics.

A Form of Using NGOs and Local Influencers to Foster Trust as a Strategy

Another important consideration that affects the use of ICT in the rural areas is trust, for an organization to gain the trust of the rural people they have to work with NGOs and other people who the people trust. Some of these NGOs can be engaged to facilitate the roll out of 5G and the awareness creation among the people by reaching out to the telecommunications firms, as they have a long standing working relationship with those in the rural area. Since there are remarkable results regarding the efficiency of using digital services, local leaders, teachers, and heads of villages can carry on about the increased use of digital services with great authority (Singh, 2023).The use of local influencer in marketing also increases the reliability of digital services and products in the remote areas. Therefore, turning trusted members of the community that are suspicious about new technologies into endorsers of digital literacy and 5G, the digital marketers can abate the resistance and increase people's engagement with new technologies for they benefit the digital economy.

EMERGING MARKETING CONCEPTS FOR RURAL MARKETING IN INDIA

The strong aspect of 5G as a real-time data technology offers the digital marketers in reaching rural consumers through localized marketing strategies. Through use of location data, marketers can therefore create an advertising message that suits the needs, wants and buying frequency of the rural populace. It enables business entities to pass on their advertisements, sales promotions, and recommendations with in line with community culture and preference.For instance, digital marketers can make use of 5G-connected tools to endorse products the farmers in rural areas require during planting seasons such as seeds, or equipment required for the season in their particular region. Likewise, marketers could include geo-targeted promotions for stores or services at the nearest distance, which could help them write more compelling campaigns for rural users (Verma, 2022).

Using Artificial Intelligence and Machine Learning to Offer Customized Marketing Content to Rural Consumers

5G when integrated with AI and ML can redefine the digital marketing in rural India by allowing personalised marketing. Marketing strategies using AI can generate information processing of huge data on customers' preferences, their buying behavior, and trend in the particular regions making it easier to have markets appealing to the rural customer. This kind of targeting can enhance the performance of online advertising initiatives raising the engagement and conversion frequency (Jain, 2023).For instance, AI and ML were applied to suggesting that relevant products for rural customers concerning their webpage visits and local demand. This kind of hyper-personalization can also help to build more confidence in digital media, since consumers believe that the information which is provided to them is relevant for them.

It will also promote affordable internet to the rural areas since through 5G technology, these entrepreneurs will be able to market their products and services online. Due to the availability of the internet rural artisans, farmers and business persons can use digital marketing to market their products to other rural consumers, urban consumers and the international market. By improving the connection speed and enabling detailed internet advertising, 5G can improve the enrolment of rural ventures and promote the economic growth of rural areas (Rao, 2023).Therefore, developing 5G in support of these governmental programs will not only be an asset but also contribute more significantly by amplifying existing financial assistance or a mentoring program to the rural entrepreneurs. As the rural run business get an opportunity to make orders and sell their good online, attend online courses and even meet with potential customers in e-commerce platforms, they can contribute to the growth of the overall digital economy.

Some other rural businesses have already adopted this concept with a lot of profitability towards theirско digital marketing strategies. For example, organic farming cooperatives located in rural Maharashtra manage to sell their products to urban consumers through digital platforms increasing their sales and brand outreach. Likewise handloom and handicraft enterprises in the states such as Odisha and Rajasthan have benefitted effectively by the use of social media and e-trading sites in targeting buyers locally as well as internationally (Sharma & Patel, 2023).These success stories reveal how 5G digital marketing with the help of 5G can revolutionalise rural businesses as they are provided with the means necessary to succeed in the digital economy. The meaning is that as soon the 5G network will be the more entrepreneurs in rural areas will be able to implement the above mentioned strategies that in turn will help the rural Indian economy to grow.It will also be important to note here that if the rural markets of India have to embrace 5G-powered digital marketing the following factors will need to be taken into account. These barriers require appearance of the strengthened PPP, increased community involvement, and appropriate market positioning and product promotion strategies that fit rural clients. Besides, using advanced 5G service technologies, the rural business grievers will be facilitated to bring new prospects in the rural areas of India. India Can overcome the rural-urban divide by the adoption of 5G, and working for the digital literacy of the country.

FUTURE OUTLOOK FOR 5G AND DIGITAL MARKETING IN RURAL INDIA

The current publication correlates the advancement of 5G networks with changed opportunities for digital marketing and rural economic development in Indian states. As the country advances into the integration of 5G, there are several trends, economic and global effects that will define the imminent future of rural development. The second part focuses on the possible positive outcome of the crisis and how the Indian case can be instructive for the other emerging markets.

Emerging Trends

The enhancement of various features of digital marketing will be mandatory with the help of AI, AR, VR, IoT, etc in the rural area of India. AI will provide hyper localization in marketing opportunities due to the detailed results of consumer analytics, content customization, and predictive consumer trend analysis through effective real-time analysis. It will enable marketers to demystify the consumer behaviour and demands enabling them to create better campaigns that will appeal to country consumers (Patel & Sharma, 2023).Application of AR and VR together with the support of 5G low latency and high bandwidth will create novel shopping experiences for the rural consumers. For instance, using AR in the form of apps help customers in visualizing products before they get to make a purchase helping in improving the current digital shopping experience. One can use virtual reality suppliers to create a virtual environment through which a consumer based in a remote region would get a feel of the physical outlets and shops.

The use of IoT technology will also go hand in hand with the rural economy. The improved connectivity through 5G make IoT devices useful in agriculture, livestock management and supply chain in the rural areas. The collected information by IoT devices can also be employed by marketers to appeal to the consumers with various products which they may require like farming machinery, fertilizer, irrigation equipment, etc (Verma, 2023).Driving Rural e-commerce and FinTech value additions through B5g and 6g Enabling 5G for Rural e-commerce and FinTech Applications

The Main Stream has indicated that adoption of 5G technology is expected to enhance the rural e-commerce and financial technology (FinTech) services. Greater availability will also enable the rural consumers to access the online platforms that will help them make purchases efficiently. This can be a great opportunity to start targeting localized similar to the rural markets since they can benefit from products like the agricultural tools, educational as well as health related goods and services (Jain, 2023).

The FinTech services will also be on the rise for the rural region once the 5G network is put in place. The increased speed of the connection will lead to increased reliability of transactions and the spread of online payments services and banking. By increasing their access to financial services, rural consumers will have improved financial intermediary tools such as, savings accounts, loans, and insurance which can all be done online (Singh, 2022). This will also encourage rural players and small business entities to access the appropriate financial instruments for enhancement and expansion of their activities.

Long-Term Economic Impact

One of the most dramatic impacts of 5G will be the impact on rural economies by giving farmers better access to markets and jobs. Current producers like farmers, weavers, artisans and small scale industrial units will be enabled to link up with consumers in urban and even global markets using e-commerce platforms underpinned by 5G. These connectivity improvements will lower the boundaries that have hitherto kept rural populations in sectional environments, spur the flow of trade and commerce (Rao, 2023).

Moreover, 5G will thus make employment available to the individuals within rural areas by spurring development of other industries that are digital based. For instance, the rural-based businesspersons can use the 5G to create businesses, which sell products online, market the products online or provide work from home solutions. They also may have regional offices in rural areas to create new jobs among employees of telecom companies and tech startups. The increased economic growth that shall follow will not only increase the quality of life within the rural areas but also will also contribute towards the elimination of the rural-urban income inequality.

Eradicating Rural-Urban Income Divide Through Technology

Now, one of the critical goals of using 5G in rural India is to bring the digital gap between rural and urban residents to a minimum. Rationalizing internet accessibility will enable consumers in rural areas to engage in e-commerce activities as the urban dwellers. Once rural consumers have web access, the digital services and e-commerce coverage, the education, and healthcare they receive, and their employment opportunities will improve. In the long run, this will help in reduction the inequality across the of distribution of wealth and resources within the country (Sharma & Patel, 2023).The economic implication of implementing 5G will go a long way to boost the development of a highly integrated economy across the nation. This role emphatically indicates that 5G will significantly narrow the rural-urban divide in India – a divide which was created by the specific architecture of the Indian economy – through connecting producers in rural areas to urban centres of demand and providing consumers in such areas with access to new products and services.

Global Implications

Honest endeavour made by India to opt for 5G use in the rural areas will form useful benchmark for all other developing nations seeking same solutions. It is evident that many countries in Africa, South East Asia and Latin American have large populations in their rural-urban populations in regions that are underdeveloped in terms of digital connectivity. It would be important for these countries emulate India's model of cultivating and implementing government policies in coordination with private sector and engaging community support systems (Singh, 2023).

The conclusions that may be drawn from all these include; the idea of public-private partnership, the role of government in developing the infrastructure, and the role of community-based approaches to local training. In this manner, by using India as a model, other developing countries learn from such experiences, and apply these modifications to their settings in order to advance toward adopting 5G, and digital marketing in rural regions.

Being one of the largest and ethnically diverse nations it has a great potential to be a pioneer of rural digitalization around the globe. With the help of utilizing the latest innovation of 5G and converting the opportunity to offer digital marketing along with economic development for rural areas in particular, India should proceed as an example for other countries. The outcome of the 5G rubric undertaking in India could hence spread the centrality of the promotion of digital deprived region across the globe with India in the front line (Rao, 2023).

In addition, the current leadership position in this regard, may open up export opportunities for India in the area of digital marketing, telecommunications and e-commerce to other developing countries. Indian tec glamorous h firms, telecom service, and digital marketers could take a step forward to serve their triangle services to the countries that are still now starting to provide 5G services in their rural areas? This not only would fortify India's market in the realm of technology but also would help others asian countries or countries in the third world to achieve its dream of turning into a digital state.

The rural India holding out a bright promise for the future of 5G and digital marketing. The rural digital marketing will improve to better efficiency, effectiveness, and interactivity since more sophisticated technologies such as AI, AR, VR, and IoT will be implemented. The growth in the use of IT in the delivery of goods and services and the continued growth of e-commerce and FinTech services will add value to the rural consuming and producing sector, spurring economic productivity, and employment. With the implementation of the digital economy for bridging the rural –urban economic divide in India, the potential of accruable economy of the rural sector can be fully harnessed. Additionally, the study reveals that India has useful insights for other emerging market democracies, rather than highlighting the developing country's challenges, the country emerges as a world leader in improving digital access in rural areas.

CONCLUSIONS

If properly harnessed, the use of 5G technology can have a large effect on the rural arena of digital marketing in India, effecting changes in how business, consumer and entrepreneurs engage in these regions. Another is the incidental benefit of the 5G network which is going to bring the digital gap between the rural and urban peoples of India by offering them much higher speed and efficient services at a lower rate and bandwidth. This increased connection will benefit SMBs, promote the delivery of digital services in sectors such as education and health, and drive rural e-commerce. Moreover, it is expected that 5G will pave way for other facilitating technologies like AI, AR/VR and the IoT, to be incorporated into digital marketing and make the consumption experience of marketers' messages in the rural area more personalized and interactive.However, achieving this kind of digital marketing in rural areas with 5G has a clear path to be followed. Major stages are infrastructure construction, digital illiteracy and awareness, rural market orientation, and e-business and market enabler. All of them require the active involvement of the government, telecom operators, commercial consumers, and local inhabitants. The policies on the part of the government, its involvement in setting up public-private partnerships is, undoubtedly, very important for ensuring that the costs are brought down and more and more units are deployed for use. Additionally, working with key stakeholders and opinion leaders from those areas will be inevitable for achieving trust and adaptation onto digital among rural consumers.

REFERENCES

Bhatt, S., & Kumar, R. (2022). Digital literacy and its impact on the adoption of emerging technologies in rural India. *Journal of Rural Technology*, *19*(1), 45–59.

Census of India. (2011). *Population Census 2011*. https://www.censusindia.gov.in/

Department of Telecommunications. (2023). *Digital India and BharatNet: Transforming rural connectivity*. Government of India.

Ericsson. (2023). *5G: The next leap for mobile communication*. Ericsson Whitepaper.

Ghosh, A. (2022). Digital marketing strategies for rural India: A path to rural transformation. *Journal of Rural Development*, *41*(2), 75–90.

Jain, P. (2023). Challenges in maintaining 5G networks in rural India: A technical perspective. *Indian Journal of Telecommunications*, *34*(3), 101–115.

Jain, P. (2023). The role of AI and machine learning in hyper-local marketing. *Marketing Tech Review*, *24*(2), 88–99.

KPMG. (2022). *Digital India: A roadmap for bridging the digital divide*. KPMG Report.

Kumar, A. (2021). Marketing in rural India: Overcoming challenges with innovative solutions. *Journal of Marketing*, *57*(3), 89–101.

Kumar, R., Mitra, P., & Saha, A. (2022). The impact of 5G technology on rural development: Opportunities and challenges. *Journal of Telecommunications and Digital Innovation*, *12*(2), 89–103.

Kumar, S., Sharma, R., & Gupta, A. (2023). Telemedicine and e-learning in rural India: A case study of 5G's impact on healthcare and education in Uttar Pradesh. *Journal of Rural Innovation*, *15*(2), 45–59.

Mehta, R., & Agrawal, P. (2022). Bridging the digital divide in India: The role of 5G in rural connectivity. *Telecommunications Policy*, *46*(1), 102101. DOI: 10.1016/j.telpol.2021.102101

Mitra, P., & Saha, A. (2021). 5G in India: A roadmap for the future. *India Telecom Journal*, *14*(3), 45–62.

National Statistical Office. (2021). *Annual Report on the Demographics of Rural India*. Government of India.

Rajaraman, I. (2020). The impact of digital marketing on economic growth: A case study of India. *Economic and Political Weekly*, *55*(30), 34–40.

Ramaswamy, K. (2023). Empowering rural artisans through 5G-enabled e-commerce in Tamil Nadu: A case study. *Indian Journal of Digital Commerce*, *8*(1), 101–115.

Rao, M. (2023). *Leveraging 5G to empower rural entrepreneurs: Case studies from India*. Economic Times of India, 67(4), 34-41.

Rao, M. (2023). *The economics of 5G in rural India: Ensuring affordability for all*. Economic Times of India, 67(4), 65-72.

Sharma, P., & Patel, A. (2023). Digital transformation and the future of rural marketing: A case study of 5G deployment in India. *Journal of Indian Telecom Policy*, *22*(3), 55–68.

Sharma, P., & Patel, A. (2023). The affordability gap: 5G-enabled smartphones and rural India's digital transformation. *Journal of Indian Telecom Policy*, *22*(3), 55–68.

Sharma, P., Singh, M., & Desai, S. (2022). 5G and the future of digital marketing in rural India: Insights from a pilot project in Maharashtra. *Journal of Digital Transformation*, *12*(4), 98–113.

Singh, V. (2023). Community engagement strategies for fostering trust in rural digital platforms. *Indian Journal of Cybersecurity*, *12*(2), 23–36.

Singh, V. (2023). Data privacy and trust issues in rural India's digital landscape. *Indian Journal of Cybersecurity*, *12*(2), 23–36.

Statista. (2023). *India: Internet penetration rate 2023*. https://www.statista.com/

Sundararajan, R. (2023). The potential of 5G for rural India: Bridging the digital divide. *Indian Journal of Rural Development*, *26*(1), 15–30.

Telecom Regulatory Authority of India (TRAI). (2023). *Telecom statistics report 2023*.

Telecom Regulatory Authority of India (TRAI). (2023). *The telecom statistics report*. https://www.trai.gov.in

Verma, N. (2022). Infrastructure challenges in the deployment of 5G networks in rural India. *Journal of Infrastructure Development*, *8*(3), 87–100.

Verma, N. (2022). The future of hyper-local marketing in rural India: A 5G perspective. *Journal of Infrastructure Development*, *8*(3), 112–125.

Chapter 18
Roadmaps for Deploying 5G and 6G in Rural Areas for Unlocking the Future of Wireless Connectivity and Intelligent Networks

Sendhilkumar Manoharan
https://orcid.org/0000-0001-5116-8696
Presidency Business School, Presidency College, Bengaluru, India

Dhanabalan Thangam
https://orcid.org/0000-0003-1253-3587
Presidency Business School, Presidency College, Bengaluru, India

Ravindran Kandasamy
https://orcid.org/0000-0001-6763-0216
Presidency Business School, Presidency College, Bengaluru, India

M. Monica
https://orcid.org/0000-0001-7361-0684
Department of MBA, Acharya Institute of Technology, Bengaluru, India

Nidhi Shukla
https://orcid.org/0009-0004-0359-7233
Presidency Business School, Presidency College, Bengaluru, India

ABSTRACT

Abstract: The transition from Beyond 5G (B5G) to 6G is a sea change in wireless communication technology, which is expected to upend industries such as transportation, healthcare, and manufacturing. By expanding upon the 5G network's current capabilities, B5G boosts mobile broadband, reduces power consumption, and maximizes spectrum usage. Advanced Internet of Things (IoT) devices and real-time applications are made possible by essential features like massive machine-type communications

DOI: 10.4018/979-8-3693-7788-8.ch018

(mMTC) and ultra-low latency. The advent of 6G in the 2030s heralds revolutionary improvements such as intelligent surfaces, AI-driven network automation, and terahertz transmission frequencies. Thanks to this development, previously unimaginable speeds and bandwidth will be available, allowing for a proliferation of networked devices with latency reduced to microseconds. While resolving issues with present networks, these advancements will provide state-of-the-art solutions for smart cities and digital twins and enhance existing networks' capacities.

INTRODUCTION

The next generation of wireless communication networks, known as 6G, and Beyond 5G (B5G) are huge leaps forward. B5G, which is commonly considered as an improvement over 5G, prepares the way for 6G, which is expected to come out in the 2030s. While 5G aims to improve speed, latency, and connectivity in existing 5G areas, 6G is expected to offer even more revolutionary capabilities, including enhanced machine-to-machine (M2M) connectivity, terahertz (THz) frequencies, and ultra-reliable low-latency communication (URLLC) (Zhang et al., 2021). By facilitating more efficient network operations, faster data processing, and deeper integration of Artificial Intelligence (AI) into network management, these innovations have the potential to radically alter industries such as healthcare, manufacturing, and transportation. Better energy economy, higher spectrum utilization, and much improved mobile broadband (eMBB) are the main characteristics of B5G in comparison to 5G. In addition, B5G is concerned with making the most of the current 5G network to facilitate new features such as edge computing, mMTC, and more advanced Internet of Things (IoT) devices. Faster data transmission is a necessary for real-time applications, and the upgrade from 5G to B5G improves the network architecture to achieve this goal (Mahmood et al., 2020).

With 6G, the next generation of wireless technology, we may anticipate revolutionary improvements, including speeds that are 100 times faster than 5G, latency that is less than a microsecond, and the capacity to manage enormous amounts of data generated by billions of devices. According to Dang et al. (2020), 6G will bring new technologies including holographic communications, intelligent surfaces, and THz communication frequencies. These innovations will offer ultra-high-speed data transfer with unparalleled bandwidth. By addressing many of the problems that present networks encounter, these capabilities will propel developments in domains like smart cities, digital twins, and AI-driven automation. The digital divide, the difference in internet access between rural and urban regions, is a serious problem that has persisted for a long time. Due to variables such as low population density, topographical obstacles, and a lack of investment in infrastructure, rural areas have fallen behind their urban counterparts in terms of the benefits enjoyed by telecommunications technology (World Bank, 2020). This digital divide has exacerbated socioeconomic disparities by limiting economic prospects for rural populations and making it harder for them to obtain healthcare and education.

When it comes to solving the developmental problems that rural communities encounter, connectivity is crucial. Better healthcare through telemedicine, more educational options through online learning platforms, and economic development through expanded market access for rural firms are all outcomes of increased broadband connectivity (ITU, 2021). Furthermore, precision farming is one way that advancements in agriculture, which are frequently propelled by IoT technologies, can increase sustainability and production. Nevertheless, these technologies have not been widely used in rural areas due to the weak connectivity. To close this digital gap, B5G and 6G are essential. The upgraded mobile broadband services might be brought to rural areas that would not be economically feasible to establish regular 5G networks, thanks to B5G's greater capacity and efficiency. 6G, on the other hand, can make sure that even rural and underdeveloped areas have the same coverage as cities thanks to its energy-efficient network architecture and sophisticated spectrum usage. Using capabilities such as network slicing, which enables the development of virtual networks customized to individual requirements, 6G has the potential to provide long-term, affordable connectivity options for those living in rural areas (Saad et al., 2020). Therefore, ensuring that everyone has equitable access to the advantages of digital technology is a moral and economic responsibility, and it is also a technological challenge, to deliver B5G and 6G in rural areas.

With an emphasis on the specific difficulties and possibilities of rural areas, this assessment intends to offer a thorough evaluation of the plans for rolling out B5G and 6G technologies in these settings. There is consensus on the advantages of B5G and 6G technologies, but there are several obstacles to deployment in rural areas. These include expensive infrastructure, physical obstacles, and a lack of financial incentives for private investment. Considering these challenges, it is critical to investigate new approaches and solutions to guarantee that rural areas do not fall behind in the information technology revolution. This evaluation will start by analyzing the present connection in rural areas and then move on to explore several aspects of establishing B5G and 6G in these locations. Subsequently, it will go into the B5G and 6G technical aspects that are especially pertinent to rural connectivity, including LPWANs, communication via satellite, and spectrum-sharing methods. The study will also go over some possible deployment strategies, such as community-based network models, public-private partnerships (PPPs), and government subsidies, that have demonstrated promise in tackling the logistical and financial obstacles of rural connectivity (Boulogeorgos et al., 2020).

In addition, the evaluation will focus on important case studies that showcase the effective implementation of B5G and 6G technologies in rural areas. The goal is to learn from these experiences and apply them to future deployments. Additionally, it will analyze how policy interventions and regulatory frameworks might help rural areas establish next-generation networks. At the end of the day, the evaluation will give lawmakers, telecoms, and other interested parties advice on how to take advantage of B5G and 6G prospects while avoiding the pitfalls of rural deployment.

CURRENT STATE OF CONNECTIVITY IN RURAL AREAS

As things are, rural areas' telecommunications infrastructure is light years behind that of major cities. The mobile connectivity that enables basic internet access, voice calls, and SMS services is provided by 4G LTE, which is still the dominating technology in most rural areas. Localized areas, like schools or community centers, in certain regions, especially in more industrialized nations, use Wi-Fi networks to offer internet access. The lack of cell towers and other network infrastructure means that coverage and signal strength are typically worse in rural locations. Compared to metropolitan surroundings, this gap causes data speeds and service dependability to be lower (Greenstein, 2020). Even though 4G LTE has greatly increased mobile internet access, it isn't always up to the task of handling data-intensive apps, particularly in rural areas that are more isolated. The high expense of installing infrastructure over large, sparsely inhabited areas prevents the widespread deployment of fiber optic cables, which provide high-speed, high-capacity internet, in rural places. The digital divide is worsened in poorer nations due to even lower internet penetration and restricted access to fixed and mobile broadband services (OECD, 2019). While Starlink and similar satellite internet services have become more popular in some areas, they are still out of reach financially for many people living in remote areas, and they aren't always reliable or have low throughput.

Low population density is a major obstacle to rural connectivity. Compared to metropolitan areas, where larger population densities and consumer demand contribute to faster profit recovery, telecom companies typically find that the ROI for establishing infrastructure in rural areas is too low to justify the expenses. Because of the greater distances that need to be covered and the natural geographic obstacles like rivers, forests, and mountains, the cost of building the essential infrastructure, like cell towers and fiber-optic cables, is disproportionately expensive in rural regions (Clifton et al., 2020). When it comes to rural connection, environmental and geographical elements are major determinants. The installation of infrastructure is complicated and costly in many rural regions because of the topography. It is extremely difficult to keep network infrastructure in places that experience extreme weather, including deserts or areas that are prone to flooding or landslides. Service interruptions and increased operational expenses are common outcomes of these physical impediments (ITU, 2020).

One of the main problems with current solutions is the expensive cost of establishing communication infrastructure in remote places. Because of the greater distances between population centers and the more challenging terrain, rural areas typically need more infrastructure investment per inhabitant. This is especially problematic for underdeveloped countries with limited national finances since it makes the cost of fiber-optic or 4G LTE infrastructure prohibitive (Kim et al., 2021). Additionally, private corporations are not incentivized to engage in connectivity initiatives that serve rural areas. Areas with higher demand and more potential for infrastructure sharing tend to be urban and suburban areas, since this is where telecom providers may expect to see the quickest returns on their investments. Subsidies and incentives offered by governments have tried to bridge this gap, but they have frequently failed to achieve the goal of giving coverage to all rural residents (Samarajiva, 2019).

Important B5G and 6G Features for Rural Areas

The substantial improvement in data rates and decrease in latency is one of the key improvements of 6G and Beyond 5G (B5G) technologies. According to Zhang et al. (2019), rural areas will soon be able to access real-time apps that were previously inaccessible due to 4G networks' latency levels of up to 50 ms. Applications that necessitate real-time data transfer and high-speed video streaming, including telemedicine and remote education, will find this development particularly pertinent. Low latency and high-speed connectivity allow for the distant provision of vital services, enhancing the quality of life and economic prospects for those living in rural areas, where access to healthcare facilities and educational institutions is often limited (Pan et al., 2021). For instance, patients in remote places can consult with urban hospital doctors using telemedicine, cutting down on the need for patients to travel great distances. Virtual classrooms, course materials, and exams can all be accessible to students in rural areas through remote education platforms, giving them the same opportunities as their urban peers. The usage of augmented and virtual reality (AR/VR) in education and healthcare is supported by the high data rates of B5G and 6G, which further enhance the learning and diagnostic experience (Saad et al., 2020).

Massive Connectivity

With B5G and 6G networks, enormous connectivity is a key characteristic that lets billions of devices connect at the same time. When it comes to smart agriculture, healthcare, and industrial IoT applications in rural areas, this is particularly important. With the use of Internet of Things (IoT) sensors, farmers can automate irrigation systems, keep tabs on livestock, and keep an eye on the soil thanks to precision agriculture techniques made possible by widespread connection. More efficient use of resources, higher crop yields, and less environmental impact can result from this (Gupta et al., 2020).

Using Internet of Things (IoT) devices, healthcare providers in rural areas may keep tabs on their patients in real time, allowing for early intervention in the event of an emergency. Through process automation and the provision of real-time data on equipment performance and environmental conditions, the utilization of the Internet of Things (IoT) in rural industries like mining and forestry can also lead to cost reductions and increased efficiency (Hossain et al., 2020).

Energy Efficiency

Because of the high cost or lack of accessibility to consistent power sources, energy efficiency takes on added importance in rural locations. Compared to earlier generations, B5G and 6G networks are more energy efficient, meaning they use less power overall. The use of reliable electrical grids or diesel generators to power base stations and other infrastructure is common in rural locations, making this an especially significant consideration. B5G and 6G networks have the potential to lessen the financial and ecological burdens of rural connection projects by decreasing energy usage (Mumtaz et al., 2021). Particularly important in remote places without easy access to charging infrastructure, 6G networks' ability to reduce power consumption means that low-power devices like smartphones and Internet of Things sensors can last longer between charges. This increases the overall reliability of these systems and decreases the maintenance costs, allowing sensors and other equipment to be deployed for longer periods of time in agricultural, healthcare, and industrial applications (Zhao et al., 2020).

Network Segmentation

One other new aspect of B5G and 6G networks that is very useful in rural regions is network slicing. By dividing networks into smaller, more manageable pieces, operators can build virtual networks that are perfect for certain applications by adjusting the latency, reliability, and bandwidth of each slice. Network slicing can guarantee that each rural area sector—from agriculture to healthcare to education—receives the amount of service needed for maximum performance, even if their connectivity needs may vary (Foukas et al., 2017). Some network slices may be more concerned with low-power, long-range Internet of Things (IoT) connectivity for sensor networks in precision agriculture, whereas other slices may be more concerned with high-speed, low-latency connectivity for video conferencing and data transfer in telemedicine. Rural networks can be made more efficient and cost-effective by offering tailored connection solutions through network slicing. This makes them more adapted to the different needs of rural areas and optimizes resource allocation.

CHALLENGES IN DEPLOYING B5G AND 6G IN RURAL AREAS

Economic Barriers

The significant capital expenditure (CapEx) linked with infrastructure development is a big obstacle to implementing B5G and 6G networks in rural areas. Investing massive sums of money to cover huge areas with little returns is common in these places because of the low population density. It is often not feasible for telecom providers to build towers, install fiber-optic lines, and purchase advanced equipment for rural areas due to financial considerations. There are still a lot of rural areas that don't have access to the internet or are severely underserved (Rural Telecommunications Group, 2020). Keeping such infrastructure in outlying locations up and running can be expensive, especially considering the likelihood of frequent maintenance and repairs being required because of environmental variables. There are a lot of expenses involved, and telecom carriers don't get much money back when they deploy in rural areas. Rural regions sometimes have lesser income per user compared to metropolitan areas. This is because urban areas have a higher population density and a greater demand for advanced digital services. Because of this, telecom businesses find it difficult to rationalize substantial investments. Digital services such as smart agriculture, remote healthcare, and education are in demand in rural regions, but the money made from them isn't always enough to pay for the infrastructure that must be put in place and kept up (Khan, 2022). One of the key reasons rural areas are still left out in the worldwide race for B5G and 6G connection is this ROI difficulty.

Geographical and Environmental Constraints

The implementation of telecommunications infrastructure can be made more difficult in rural areas due to the presence of difficult terrains including forests, mountains, and large open spaces. For example, logistically challenging and costly, requiring specialized equipment and labor, is the construction of network towers in hilly terrain. Plus, these types of terrains can make it harder for signals to travel, which means you'll need more base stations to cover all the ground (Zhao et al., 2021). Dense woods can also reduce network efficiency by blocking signals. A new type of difficulty arises in extremely open areas, as those in agricultural or desert regions. Traditional infrastructure, such as towers and fiber-optic cables, might become too costly to build when users are too far apart, even while the landscape itself is not difficult to traverse (Azar & Hassan, 2020). With a tiny number of customers covered by each tower, it may not be worthwhile to spend heavily in infrastructure in remote locations with sparse inhabitants. In addition to these geographical challenges, harsh environmental conditions make them worse. Telecom infrastructure is vulnerable to natural disasters including heavy rain, snowfall, and strong winds that often strike rural locations. For instance, according to Chowdhury and Imran (2020), certain places' freezing temperatures might cause equipment problems, whereas tropical climates with significant humidity can cause equipment to corrode. It can be costly to install more durable and weather-resistant equipment to address these environmental concerns.

Regulatory and Policy Issues

Another obstacle to the rollout of B5G and 6G in remote regions is the regulatory landscape in numerous nations. Policies and incentives that are tailored to rural deployments are often vague or nonexistent. Telecom companies are hesitant to invest in areas with uncertain financial returns unless they receive specific subsidies or backing from the government. This regulatory sluggishness has the potential to postpone the deployment of advanced networks, which is especially problematic in developing nations where rural residents already face barriers to digital services (Maitland, 2021). Another major problem with regulations is spectrum allotment. While spectrum is crucial for rolling up B5G and 6G networks, many areas' rules do not lend itself to rollouts in rural areas. In many countries, telecom operators are forced to pay exorbitant amounts in spectrum auctions to acquire the bandwidth they need. These high spectrum costs are a big turnoff for rural locations, which have reduced revenue potential (Hong et al., 2022). Also, rural locations don't always have enough bandwidth for their networks because they must compete with urban areas for spectrum. This makes rollout even slower.

POTENTIAL SOLUTIONS AND ROADMAPS

Public-Private Partnerships (PPP)

For rural areas, establishing B5G and 6G might be financially challenging. One viable strategy to overcome this is to form public-private partnerships (PPP). To encourage telecom providers to invest in rural infrastructure, governments might offer subsidies, tax incentives, and other types of financial support. To help defray the hefty deployment expenses, several nations' governments have established rural connectivity funds to aid underprivileged regions (Eisenberg & O'Brien, 2021). This makes rural deployments more possible by reducing the financial load on telecom carriers through government-private partnerships that co-finance infrastructure projects. As a bonus, governments can make it easier for telecom carriers to implement B5G and 6G in rural areas by reducing regulatory barriers. One way to cut deployment costs and encourage investment is to streamline the process of procuring licenses. Another is to reduce spectrum costs. A third is to provide access to publicly owned infrastructure, including utility poles or rooftops (Zhao et al., 2021).

Affordable Network Models

For rural installations, cheap network options are just as important as PPPs. B5G and 6G connection in rural areas can be made more affordable using community-based network architectures, mesh networks, and collaborations with local Internet Service Providers (ISPs). In instance, rural regions with low populations and challenging topographies are ideal for mesh networks. Azar and Hassan (2020) state that these networks can provide wider coverage than traditional tower-based networks since they are based on numerous small nodes that communicate with one other. Some rural areas have also found success with community-based network concepts, in which local communities own and run their own networks. Communities can take charge of local network implementation and management through these models, with the help of larger telecom carriers or government agencies for technical support. Chowdhury and

Imran (2020) found that these methods might assist reduce operational costs and make sure that networks were suited to the needs of rural communities by integrating local stakeholders.

Innovative Infrastructure Solutions

To get over the geographical problems of rural deployments, another possible alternative is to use technologies based on unmanned aerial vehicles (UAVs) and satellites. When land-based infrastructure is impractical owing to factors like high prices or environmental restrictions, satellite networks step in to fill the void. Recent developments in satellite technology have provided a practical substitute for conventional infrastructure by bringing high-speed internet to formerly unserved rural areas (Khan, 2022). In a similar vein, solutions based on unmanned aerial vehicles (UAVs), such drones outfitted with communications gear, can offer short-term network coverage in times of crisis or in locations where constructing permanent infrastructure is not an option. During disaster recovery or other infrastructure damage situations, UAVs can assist in connecting remote areas (Hong et al., 2022). In remote places, deploying small cells and edge computing is another creative option. Deploying small cells, which are base stations with minimal power consumption, in less populated areas improves coverage and capacity. The combination of tiny cells and edge computing allows telecom operators to enhance service delivery in remote places while reducing latency (Maitland, 2021).

Spectrum Sharing and Dynamic Spectrum Allocation

To overcome the difficulties caused by the limited supply of spectrum in rural areas, dynamic spectrum allocation and spectrum sharing are crucial. An efficient use of spectrum and a decrease in the demand for exclusive licenses are achieved by spectrum sharing, which permits numerous operators to utilize the same frequency band in distant areas or at different times. Zhao et al. (2021) found that this method can work wonders in rural regions due to the decreased bandwidth demand and the ability to distribute spectrum more efficiently without interference. In contrast, spectrum can be assigned in real-time according to demand through dynamic allocation. Because of this, operators in rural regions can make better use of available bandwidth by dynamically allocating spectrum to various applications or consumers. When it comes to rural deployments, dynamic allocation and spectrum sharing are both great options because they make better use of existing resources and cut down on costly spectrum auctions (Eisenberg & O'Brien, 2021).

CASE STUDIES AND PILOT PROJECTS

Numerous pilot programs and case studies have investigated the rollout of B5G and early-stage 6G technology in remote regions, with encouraging results across a range of industries. One example is the agricultural sector, where the benefits of improved connectivity for remote farms have been proven by the "5G Rural First" initiative in the United Kingdom. Orkney and Somerset farmers were able to utilize IoT devices for real-time livestock, soil condition, and machinery monitoring thanks to this program (Adams et al., 2020). By improving data gathering and processing, farmers were able to maximize agricultural productivity and animal health with the implementation of B5G technology. This research showcases the potential of 5G to increase agricultural productivity while decreasing operational expenses in rural areas.

A prime example of a successful telemedicine initiative is Finland's "6G Smart Healthcare" program. Advanced healthcare services were made available to remote rural areas through this pilot initiative that employed 6G capabilities. Doctors at urban hospitals could consult with patients in remote areas in real-time and conduct diagnostics through ultra-low latency networks and high-speed connections (Haapola et al., 2021). Continuous monitoring of chronic patients was also made possible by the incorporation of wearable health sensors linked to the 6G network, which improved patient outcomes while decreasing the frequency of hospital visits. This trial shows how 6G might open new possibilities for remote consultations and tele-surgeries, transforming healthcare access in underprivileged areas. The Chinese government has also conducted important case studies, such as the testing of B5G and 6G networks for use in rural schools. Students in far-flung schools in Sichuan province were able to use 5G connectivity to access top-notch digital learning materials through the "Smart Village Education" initiative (Zhang & Wang, 2021). Thanks to lightning-fast internet, these kids were able to participate in VR and AR-based interactive lessons with teachers in urban areas. By closing the digital gap, this rollout has demonstrated how B5G and 6G technologies may play a pivotal role in alleviating educational disparity in rural regions.

Lessons Learned

Several important lessons regarding the deployment of B5G and 6G technologies in rural areas can be gleaned from these case studies. To begin, for deployments to be a success, public-private collaborations are crucial. Collaborating with local farmers, telecom companies, and the UK government was a key component of the 5G RuralFirst project (Adams et al., 2020). It would be financially impossible for private enterprises to construct advanced networks in sparsely inhabited regions without such cooperation. Secondly, infrastructure must be flexible enough to accommodate specific regional requirements. By deploying edge computing locally, the Finnish healthcare initiative was able to decrease the latency needed for real-time medical diagnoses (Haapola et al., 2021). When planning infrastructure for rural areas, it is crucial to take geographical and environmental considerations into account. Lastly, the societal effects of these initiatives highlight the revolutionary possibilities of B5G and 6G. New economic opportunities are created for rural communities because of the improved connectivity given by these technologies, which increases access to important services like healthcare and education. The Smart Village Education initiative in China was able to close the achievement gap between urban and rural pupils by providing them with access to online learning resources (Zhang & Wang, 2021). Beyond only fostering economic growth, this highlights the far-reaching social advantages of rural connectedness.

FUTURE OPPORTUNITIES AND EMERGING TRENDS

Advanced Use Cases for Rural Areas

In the future, B5G and 6G networks will likely enable a plethora of innovative use cases, which will have a profoundly positive impact on rural residents' quality of life. Smart agriculture is one of the most well-known uses, and it will use robotics, artificial intelligence (AI) data analytics, and Internet of Things (IoT) devices to oversee massive agricultural operations. According to El-Sayed et al. (2022), farms will have access to precision agriculture tools including AI-driven pest management, drones to help monitor crops, and automated irrigation systems once 6G networks are up and running. Thanks to better resource management, these innovations have the potential to greatly increase output while decreasing environmental impact. Telehealth is another exciting area of potential use. More sophisticated remote diagnosis and treatments may be possible with the advent of 6G networks, expanding upon current telemedicine services. According to Khan et al. (2023), clinicians may be able to detect possible health problems before they worsen if they use wearable sensors that track vital signs in real-time in conjunction with diagnostic tools powered by artificial intelligence. In addition, 6G's ultra-reliable low-latency communication (URLLC) capabilities will pave the way for real-time robotic surgery. This means that doctors in big cities may now remotely execute complicated procedures, expanding access to modern medical care to even the most remote rural areas.

Digital literacy programs stand to gain from these new tech developments. To help rural residents make the most of their new high-speed internet connections, governments and non-profits should launch digital education programs. By facilitating access to B5G/6G apps, such as online marketplaces, online banking, and online government services, these programs will help underserved rural areas improve their economic situation and become more involved in the digital economy (El-Sayed et al., 2022).

Research and Development

Investigations into 6G technology remain in their infancy, but several promising developments are soon to be announced. The advancement of terahertz (THz) communication is a major focus of study. Rural deployments of THz waves are hindered by their susceptibility to environmental interference, despite their 100-fold data transmission rates compared to 5G networks (Rappaport et al., 2019). According to Khan et al. (2023), researchers are looking into new methods to make THz signals stronger and more resilient. One approach is to utilize AI to forecast and reduce signal disruptions. Another is to apply enhanced beamforming techniques. For the rollout of 6G in remote places with potentially inadequate power infrastructure, energy-efficient networking is another important field of study. To make 6G networks run on renewable energy sources like solar power, scientists are working on low-power communication technology (Khan et al., 2023). 6G networks would be easier to set up and keep running in places without consistent power sources if this were the case. Future rural deployments are likewise anticipated to heavily rely on satellite-based communication. To ensure constant connectivity even in remote, thinly inhabited regions, scientists are looking at 6G's potential integration with satellite networks (Rappaport et al., 2019). In rural areas without access to terrestrial infrastructure, which can be expensive or otherwise problematic to build, this could be a lifesaver in terms of getting people online.

Long-Term Vision

Global connection and equality are anticipated to be greatly enhanced in the long run by the deployment of 6G networks. Not only will 6G bring about technological improvements, but it will also have the ability to make digital resources and services more accessible to a wider audience. 6G might close the digital gap and include every area in the digital revolution by making high-speed, dependable internet available everywhere (Rappaport et al., 2019). 6G has the potential to enable rural populations to become more involved in the global economy, which is a positive development from a socioeconomic standpoint. El-Sayed et al. (2022) found that better internet connections could help small companies in rural areas sell their wares online, reach a wider audience, and compete on a worldwide basis. In addition,

6G's capacity to enable smart cities and smart villages has the potential to bring about more equal and sustainable development, eliminating economic disparities between rural and urban areas.

DISCUSSIONS AND IMPLICATIONS

Discussions

With the advent of B5G and 6G in rural regions comes a plethora of pertinent concerns regarding the potential of technology to close the ongoing digital gap. Although B5G and 6G hold great potential for enhancing speed, latency, and connectivity, the specific obstacles encountered by rural areas cast doubt on the possibility of these advancements being accessible to everyone. Technological feasibility, economic viability, and socio-political implications are three areas that this analysis brings to light that need more in-depth discussion. The difficulties with the infrastructure are one of the main points of discussion. Problems with infrastructure deployment for B5G and 6G are common in rural locations due to factors such as low population density, tough terrain, and undeveloped utilities (Parvez et al., 2020). Essential to 5G and 6G networks are dense networks of tiny cells or fiber-optic links; however, these may be prohibitively expensive or unfeasible for deployments in rural areas. But new technology, such as communications by satellite and UAV, may hold the key. While these technologies have the potential to open new avenues for connectivity in inaccessible or far-flung locations, it is still unclear whether or not they can be scaled effectively. How feasible it is financially to roll out B5G and 6G in less-populated regions is another pressing concern. Due to smaller profit margins, telecom companies are typically unwilling to engage in rural connection. Due to the lower population density in rural areas, operators get a far lower return on investment (ROI) compared to their urban counterparts. Existing digital connectivity disparities may be further exacerbated if telecom providers are not incentivized or subsidized by the government to prioritize rural areas. One potential answer is public-private partnerships (PPPs), which aim to bring together public and private sectors to address a common problem (Zhao et al., 2021). To guarantee a successful deployment, the policy and regulatory environment is equally crucial. Regulatory support for the adoption of new technologies is currently lacking in rural regions of many nations. This is especially true when it comes to issues like infrastructure sharing and spectrum allotment. To facilitate dynamic spectrum sharing and inexpensive licensing models, governments should enact regulations that make the spectrum available for flexible use (Alwis et al., 2021). These rules will make it easier for smaller rural operators to implement B5G and 6G, which would lead to more widespread adoption in areas that are now underserved. Precision agriculture, telemedicine, and educational access are just a few examples of the rural needs that might be greatly improved by technological breakthroughs like edge computing, internet of things integration, and network slicing. Particularly in the agricultural sector, these use cases have the potential to radically alter rural economies. The efficient deployment and maintenance of these technologies in situations with limited resources will determine their success, though.

Implications

There will be major societal, economic, and technological ramifications for rural areas if B5G and 6G are successfully deployed there. Rural areas might benefit economically from better connection since it boosts productivity, especially in agriculture. With the use of the Internet of Things (IoT) and 6G, precision farming technology might maximize crop yields while decreasing food waste and increasing food safety. enterprises in rural areas will also have better access to international markets, which will encourage more people to start enterprises there and lessen the number of people leaving rural areas for cities (Parvez et al., 2020). More equitable access to basic services is one social benefit of better rural connectivity. Through telemedicine, which allows for remote diagnosis, consultations, and even surgery, improved communication networks have the potential to radically alter healthcare in rural regions. Gaining access to digital learning platforms and virtual classrooms will level the playing field for rural students relative to their urban counterparts, which will have a big impact on education, another important issue (Zhao et al., 2021).

Not to mention the significant technical ramifications. Network infrastructure and service delivery will undergo more innovation as B5G and 6G deployments are pushed forward in rural areas. In remote areas without easy access to the grid or conventional networks, energy-efficient networks, renewable energy sources like solar, and cutting-edge satellite communications will be necessities. Other neglected areas around the world, especially in developing nations, might learn from similar deployments' successes. That is why there may be a worldwide movement toward more equitable and long-term technological solutions brought about by the rollout of B5G and 6G in less-populated regions (Alwis et al., 2021). Prioritizing the deployment of B5G and 6G networks in rural areas requires careful consideration of their environmental impact. Although these networks are anticipated to have better energy efficiency, there may be substantial environmental impacts, particularly in rural areas, due to the widespread use of tiny cells and the infrastructure needed for extensive Internet of Things connectivity. To lessen the severity of these effects, it will be required to incorporate environmentally friendly technology such renewable energy sources and engage in meticulous planning (Zheng et al., 2020).

Lastly, there are major ramifications for politics and regulation. To ease the rollout of 5G and 6G in remote regions, governments must do more, especially by establishing favourable regulatory climates. Regulatory frameworks, financial incentives, and subsidies all fall under this category, as do policies that promote sharing of spectrum and infrastructure as well as community participation in network management. It is imperative that these steps are taken to guarantee that rural communities do not fall behind in the upcoming technology revolution. Although there are many obstacles to overcome, the advantages of rolling out B5G and 6G in rural regions are enormous. The revolutionary potential of next-generation connectivity can be realized in rural areas by creatively resolving the economic, technical, and regulatory hurdles. By working together, public and private organizations can close the digital divide and bring rural areas into the digital economy. Furthermore, future endeavours to provide connectivity to unserved people globally can be informed by the lessons learnt from previous installations.

CONCLUSIONS AND FUTURE DIRECTIONS FOR RESEARCH

There are enormous possibilities and substantial obstacles associated with rolling out B5G and 6G networks to more remote locations. The digital divide is a persistent barrier to social and economic growth in many rural locations across the globe, but these next-generation technologies show great promise for significantly improving connection in rural areas. Critical rural applications like telemedicine, precision agriculture, and remote education would benefit greatly from B5G and 6G's greater data speeds, ultra-low latency, and capacity to connect many devices (Zhao et al., 2021). To reap these benefits, though, several critical obstacles must be removed. Regulatory concerns pertaining to spectrum allocation and rural policy frameworks, geographical obstacles that make network infrastructure installation more difficult, and the high expense of infrastructure deployment are also factors to consider (Parvez et al., 2020). Potential ways to address these obstacles include dynamic spectrum-sharing technologies, creative infrastructure models (such as networks based on satellites or UAVs), and public-private collaborations. However, governments, businesses, and communities must work together to put these solutions into action. The consequences of effectively deploying B5G and 6G in rural areas are extensive. From an economic perspective, they have the potential to revitalise rural industries, particularly those involved in agriculture and small business, by boosting production and opening up new markets. On a societal level, they have the potential to raise living standards by expanding availability of essential services like healthcare and education. However, the possible effects of widespread deployments on the environment, particularly in outlying regions where infrastructure development can disturb ecosystems, require serious thought (Alwis et al., 2021).

If we want to successfully deploy B5G and 6G in rural areas, we need to dig further into a few areas of research. Research into cheaper and more energy-efficient infrastructure is crucial for rolling out B5G and 6G in rural areas, which are important areas for future studies. Though they show promise, solutions like solar-powered tiny cells, mesh networks, and communication via satellite still need improvement before they can be scaled and maintained over the long haul. Another important area of study that needs to be addressed is finding affordable solutions for backhaul connectivity in rural areas (Zhao et al., 2021). Obtaining reasonably priced spectrum licenses can be a challenge for rural areas because to the scarcity and high cost of spectrum. Dynamic spectrum-sharing technology may provide an answer to this issue by enabling flexible resource sharing among numerous users. Nevertheless, additional investigation is required to create effective algorithms for spectrum sharing and establish legislative frameworks that would facilitate their broad implementation (Parvez et al., 2020).

Extreme weather is a common threat to network infrastructure in rural places. We need more studies on how to construct resilient networks that can endure disturbances like natural catastrophes and extreme weather. This involves researching ways to guarantee consistent and dependable connectivity using technologies like self-healing networks, distributed computing, and redundant network pathways. More research is required on rural-specific applications that address the special demands of these locations, in contrast to the extensive research on generic B5G and 6G applications, such as the Internet of Things (IoT) and edge computing. Key use cases include precision agriculture, remote healthcare, and distance learning. However, to make these technologies inexpensive and accessible to rural populations, they require development strategies that are specifically designed for them. The unique problems faced by rural businesses and ecosystems should be the subject of future Internet of Things (IoT) research (Alwis et al., 2021).

There should be more research into the economic models that support rural connection initiatives. Is there a way to arrange public-private partnerships so that private enterprises and governments have an incentive to engage in rural connectivity? When it comes to rural network management and maintenance, what part can local communities play? For B5G and 6G deployments in rural areas to be financially feasible, there must be study into legislative reforms that can assist rural connectivity. These reforms should include things like infrastructure sharing, tax incentives, and subsidies. More research into the potential ecological effects of deploying B5G and 6G in remote regions is necessary. Research on environmentally friendly technologies that lessen the impact of network infrastructure should be prioritized for the future. Particularly in rural areas with lower rates of digital literacy, it is important to investigate ethical concerns related to data privacy and digital inclusion. It will be vital to explore how to avoid exploiting or leaving these communities behind when B5G and 6G go out. When contrasted with metropolitan populations, rural communities frequently exhibit distinct requirements, aspirations, and constraints. To guarantee that B5G and 6G technologies are available, useful, and beneficial for rural communities, research into user-cantered design for rural users is crucial. This involves taking into consideration cultural variations in technology use and perception, limited hardware accessibility, and lower levels of digital literacy when creating user interfaces and applications (Zhao et al., 2021).

REFERENCES

Adams, A., Browne, D., & Murray, A. (2020). 5G RuralFirst: Trialing 5G in rural and remote environments. *IEEE Access : Practical Innovations, Open Solutions*, *8*, 34512–34523. DOI: 10.1109/ACCESS.2020.2972099

Alwis, L., Kalla, B., & Jamalipour, A. (2021). 6G Vision: Future Mobile Networks for Intelligent IoT and Ubiquitous Connectivity. *IEEE Internet of Things Journal*, *8*(10), 8765–8777.

Alwis, L., Kalla, B., & Jamalipour, A. (2021). 6G Vision: Future Mobile Networks for Intelligent IoT and Ubiquitous Connectivity. *IEEE Internet of Things Journal*, *8*(10), 8765–8777.

Alwis, L., Kalla, B., & Jamalipour, A. (2021). 6G Vision: Future Mobile Networks for Intelligent IoT and Ubiquitous Connectivity. *IEEE Internet of Things Journal*, *8*(10), 8765–8777.

Azar, H., & Hassan, M. (2020). Deploying wireless technologies in rural areas: Opportunities and challenges. *Telecommunications Review*, *23*(3), 145–160.

Boulogeorgos, A. A., Alexiou, A., Papadopoulou, E., Devoti, F., Manios, M., Passas, N., & Berberidis, K. (2020). Terahertz technologies to deliver optical network quality of experience in wireless systems beyond 5G. *IEEE Communications Magazine*, *58*(6), 69–75. DOI: 10.1109/MCOM.001.2000018

Chowdhury, A., & Imran, M. (2020). Telecommunication in rural areas: Challenges and opportunities for developing countries. *Journal of Digital Communications*, *12*(1), 75–92.

Clifton, J., Díaz-Fuentes, D., & Warner, M. E. (2020). The loss of public values when public shareholders go private: The neglect of telecommunications infrastructure. *Public Management Review*, *22*(5), 724–746. DOI: 10.1080/14719037.2019.1604792

Dang, S., Amin, O., Shihada, B., & Alouini, M. S. (2020). What should 6G be? *Nature Electronics*, *3*(1), 20–29. DOI: 10.1038/s41928-019-0355-6

Eisenberg, M., & O'Brien, L. (2021). Public-private partnerships in telecommunications: A pathway for rural connectivity. *International Journal of Telecommunications Policy*, *34*(4), 207–219.

El-Sayed, H., El-Ghany, M. A., & Shahin, S. (2022). Applications of 5G and 6G technologies in rural areas: A review. *IEEE Communications Magazine*, *60*(4), 55–61. DOI: 10.1109/MCOM.2022.0000112

Foukas, X., Patounas, G., Elmokashfi, A., & Marina, M. K. (2017). Network slicing in 5G: Survey and challenges. *IEEE Communications Magazine*, *55*(5), 94–100. DOI: 10.1109/MCOM.2017.1600951

Greenstein, S. (2020). The economic impact of broadband in rural areas. *Journal of Rural and Community Development*, *15*(3), 1–15.

Gupta, A., Porambage, P., & Rajatheva, N. (2020). 6G-enabled Internet of Things (IoT): Technologies, challenges, and applications. *IEEE Access : Practical Innovations, Open Solutions*, *8*, 124718–124745. DOI: 10.1109/ACCESS.2020.3007278

Haapola, J., Leinonen, P., & Jämsä, T. (2021). 6G for smart healthcare in rural areas: Case studies from Finland. *Journal of Telemedicine and Telecare*, *27*(6), 333–341. DOI: 10.1177/1357633X20985634

Hong, S., Zhao, Y., & Wu, J. (2022). Spectrum allocation strategies for B5G and 6G networks in rural areas. *IEEE Communications Magazine*, *60*(5), 123–129.

Hossain, E., Bhargava, V., & Saad, W. (2020). *6G: The next frontier*. Cambridge University Press.

International Telecommunication Union (ITU). (2021). Connecting rural communities: ITU global connectivity report. *ITU Publications*. https://www.itu.int

ITU. (2020). Measuring digital development: Facts and figures 2020. International Telecommunication Union. https://www.itu.int/en/ITU-D/Statistics/Documents/facts/FactsFigures2020.pdf

Khan, M. A., Sohail, M., & Asghar, A. (2023). Energy-efficient and THz-based 6G technologies for rural deployments. *IEEE Wireless Communications*, *30*(1), 22–28. DOI: 10.1109/MWC.2023.3101011

Khan, N. (2022). Innovative solutions for rural B5G and 6G deployments. *Wireless Technology Journal*, *19*(2), 98–112.

Kim, K., Hong, S., & Lee, H. (2021). Policy strategies for enhancing rural connectivity: Lessons from international cases. *Telecommunications Policy*, *45*(1), 102–115. DOI: 10.1016/j.telpol.2020.102115

Mahmood, N. H., Berardinelli, G., Pedersen, K. I., Mogensen, P. E., & Lopez, D. G. (2020). On the potential of full-duplex for 5G and beyond. *IEEE Communications Surveys and Tutorials*, *22*(2), 1302–1321. DOI: 10.1109/COMST.2020.2968287

Maitland, C. (2021). Regulatory frameworks for rural telecommunications: Global perspectives. *Journal of Telecommunications Policy*, *45*(6), 105–116.

Mumtaz, S., Suzuki, K., & Saad, W. (2021). Green communication in 6G: Enabling technologies, challenges, and future research directions. *IEEE Communications Magazine*, *59*(1), 91–97. DOI: 10.1109/MCOM.001.2000735

OECD. (2019). Bridging the rural digital divide. https://www.oecd.org/digital/bridging-the-rural-digital-divide.pdf

Pan, Y., Zhang, Y., & Saad, W. (2021). Applications of 6G in rural connectivity: A survey. *IEEE Access : Practical Innovations, Open Solutions*, *9*, 127452–127468. DOI: 10.1109/ACCESS.2021.3104057

Parvez, I., Rahmati, A., Guvenc, I., Sarwat, A., & Dai, H. (2020). A Survey on Low-Latency towards 5G: RAN, Core Network and Caching Solutions. *IEEE Communications Surveys and Tutorials*, *20*(4), 3098–3130. DOI: 10.1109/COMST.2018.2841349

Rappaport, T. S., Xing, Y., & MacCartney, G. R. (2019). Terahertz wireless communications: Challenges and opportunities. *IEEE Communications Magazine*, *57*(5), 2–9. DOI: 10.1109/MCOM.2019.1800219

Rural Telecommunications Group. (2020). Bridging the digital divide: The economics of rural broadband deployment. *Policy Brief.*

Saad, W., Bennis, M., & Chen, M. (2020). A vision of 6G wireless systems: Applications, trends, technologies, and open research problems. *IEEE Network*, *34*(3), 134–142. DOI: 10.1109/MNET.001.1900287

Saad, W., Bennis, M., & Chen, M. (2020). A vision of 6G wireless systems: Applications, trends, technologies, and open research problems. *IEEE Network, 34*(3), 134–142. DOI: 10.1109/MNET.001.1900287

Samarajiva, R. (2019). Digital transformation of rural areas: Policies for rural broadband. *Journal of Digital Inclusion*, *2*(2), 10–21.

World Bank. (2020). Bridging the rural-urban digital divide: Policy interventions and recommendations. *World Bank Reports*. https://www.worldbank.org

Zhang, J., Zhang, Y., & Zhao, L. (2019). Towards 5G and beyond: Enabling technologies and applications. *IEEE Communications Magazine*, *57*(10), 4–6. DOI: 10.1109/MCOM.2019.8890359

Zhang, Y., & Wang, Z. (2021). B5G and 6G-enabled smart village education: A case study from China. *IEEE Access : Practical Innovations, Open Solutions*, *9*, 52798–52809. DOI: 10.1109/ACCESS.2021.3069005

Zhang, Z., Chen, X., Lin, Y., Liang, X., Huang, H., Liu, S., & Yang, J. (2021). 6G wireless networks: Vision, requirements, architecture, and key technologies. *IEEE Vehicular Technology Magazine*, *16*(3), 28–38. DOI: 10.1109/MVT.2019.2921208

Zhao, M., Li, X., & Liu, Y. (2021). Challenges and Opportunities in 6G Networks: A Comprehensive Survey. *IEEE Communications Surveys and Tutorials*, *23*(3), 1201–1231.

Zhao, Z., Liu, X., & Saad, W. (2020). Energy-efficient wireless systems for 6G: Key challenges and research directions. *IEEE Communications Magazine*, *58*(4), 28–33. DOI: 10.1109/MCOM.001.190087

Zheng, K., Zhao, Y., & Liu, J. (2020). Spectrum Allocation and Pricing for 5G and Beyond: Survey and Research Directions. *IEEE Transactions on Wireless Communications*, *19*(10), 6417–6434.

Chapter 19
A Systematic Literature Review Unveiling the Societal Impact of 5G/6G Technologies on Bridging the Digital Divide and Enabling Social Transformation

S. Baranidharan
https://orcid.org/0000-0002-7780-4045
Christ University, Bangalore, India

Amirdha Vasani Sankarkumar
St. Francis College, India

G. Chandrakala
Dayananda Sagar University, India

K. Dhanalakshmi
Christ University, India

Chippy Mohan
https://orcid.org/0000-0001-8589-4944
Christ University, India

Melvin Victor
https://orcid.org/0000-0002-1499-075X
Christ University, India

Rahul V. Rindavanam
https://orcid.org/0009-0003-0332-8810
Christ University, India

R. Sembiyan
St. Joseph's College of Arts and Science, Cuddalore, India

ABSTRACT

This study examines the multifaceted societal impact of 5G and 6G mobile communication technologies. Employing a systematic literature review (SLR) of recent scholarly literature (2022-2024), A comprehensive search strategy was implemented across reputable academic databases (Scopus, Web of Science, IEEE Xplore) to identify relevant publications. Selected articles underwent rigorous critical appraisal and thematic analysis to synthesize key findings and recurring themes. The review reveals the potential of 5G/6G to bridge the digital divide, particularly in underserved regions, through strategic infrastructure development. Social transformation can be achieved by fostering equitable access to communication services and deploying transformative applications in critical sectors like agriculture and healthcare.

DOI: 10.4018/979-8-3693-7788-8.ch019

However, challenges persist regarding infrastructure limitations, equitable access, and sustainable development. Effective policy frameworks and robust managerial strategies for responsible deployment are crucial to maximize the positive impact of 5G/6G technologies.

INTRODUCTION

The Digital Divide: A Persistent Challenge

The digital divide, a persistent social and economic disparity, refers to the unequal access to information and communication technologies (ICTs) between individuals and communities. This conceptual framework underscores the disparities in opportunities and resources associated with ICT access. Limited access to communication infrastructure,devices, and digital literacy skills can exacerbate existing socio-economic inequalities, further marginalizing disadvantaged populations. The digital age, characterized by its relentless dependence on information and communication technologies (ICTs), has undeniably transformed the landscape of human interaction, work, and education. However, this progress has been fragmented, creating a persistent and significant challenge: the digital divide. This term signifies the unequal access to and utilization of ICTs across populations, leading to disparities with profound social and economic ramifications. At its core, the digital divide encompasses disparities in three key areas: Infrastructure: Availability and quality of internet connectivity. This includes limitations in broadband access and mobile network coverage, particularly in remote and underserved communities. Devices: Ownership of computers, smartphones, and other ICT devices necessary to participate in the digital world. Digital Literacy: Skills required to navigate the complexities of technology, including the ability to use software,search for information online, and understand cybersecurity threats.

The consequences of the digital divide are multifaceted and far-reaching. Individuals with limited access to ICTs face significant disadvantages in various aspects of life. Education: Difficulty accessing online learning resources and platforms, hindering educational attainment and career prospects. Employment: Limited opportunities to search for jobs or participate in the growing online job market. Healthcare: Difficulty accessing telehealth services or online health information. Social Participation: Exclusion from online communities and social networks, potentially leading to isolation and a sense of disenfranchisement. Economic Development: Reduced competitiveness of businesses and communities that lack access to digital tools and resources. Furthermore, the digital divide can exacerbate existing social and economic inequalities. Low-income communities and marginalized groups are often disproportionately affected by this gap. Addressing this challenge necessitates a comprehensive approach that focuses on infrastructure development, affordability of technology, and robust digital literacy programs. Only by tackling these issues can we ensure an inclusive digital future for all.

Impactful Real-Time Cases Showing How 5G/6G Technologies are Transforming Society

Remote Surgery in Rural India demonstrates how 5G's ultra-low latency enabled doctors in Mumbai to perform complex robotic surgeries on patients in remote villages, bridging critical healthcare gaps. The near-instantaneous response time and high-definition video streaming allowed surgeons to make precise movements with robotic instruments while receiving real-time tactile feedback, effectively expanding access to specialized medical care.

Smart Traffic Management in Barcelona showcases how 5G-powered sensors and AI systems analyze traffic patterns in milliseconds, automatically adjusting signal timings and rerouting vehicles. This has reduced average commute times by 25% and lowered vehicle emissions by coordinating thousands of connected vehicles, traffic lights, and pedestrian crossings in real-time across the city grid.

Augmented Manufacturing at Volkswagen's plant in Wolfsburg illustrates how 5G private networks support thousands of connected robots and AR-equipped workers simultaneously. Maintenance technicians wearing AR glasses receive instant visual guidance for repairs, while AI systems monitor production lines through high-definition cameras to detect defects in real-time, increasing efficiency by 40%.

Emergency Response in Tokyo highlights how 6G's enhanced capabilities support disaster management through interconnected drones, vehicles, and wearables. During earthquakes, the system processes massive amounts of sensor data to coordinate first responders, guide evacuations through augmented reality overlays, and maintain critical communications even when traditional infrastructure fails, significantly improving response times and saving lives.

Table 1. Statistical breakdown of the real-time cases showing measurable impacts

Case Study	Key Metrics	Before 5G/6G	After 5G/6G	Impact % Change
Remote Surgery (India)	- Number of rural patients served/month	120 patients	450 patients	+275% ↑
	- Average surgery wait time	45 days	12 days	-73% ↓
	- Network latency	200ms	<10ms	-95% ↓
	- Surgery success rate	82%	94%	+12% ↑
Smart Traffic (Barcelona)	- Average commute time	42 min	31.5 min	-25% ↓
	- Vehicle emissions (CO2 tons/year)	58,000	43,500	-25% ↓
	- Traffic congestion index	72%	48%	-24% ↓
	- Real-time data processing speed	1-2 sec	<0.1 sec	-95% ↓
VW Manufacturing	- Production efficiency	Baseline	+40% output	+40% ↑
	- Quality control detection rate	85%	98%	+13% ↑
	- Maintenance downtime	18 hrs/month	7 hrs/month	-61% ↓
	- Connected devices/sq km	1,000	100,000	+9900% ↑
Tokyo Emergency Response	- Average response time	18 min	8 min	-56% ↓
	- Data processing volume	1 TB/hour	50 TB/hour	+4900% ↑
	- Network reliability during disasters	65% uptime	99.999% uptime	+35% ↑
	- Lives saved (estimated/year)	850	1,420	+67% ↑

5G/6G Technologies as Catalysts for Social Transformation

The advent of 5G and 6G, with their enhanced capabilities and broader network reach, presents a compelling opportunity to bridge this digital divide. These advancements can be conceptualized as catalysts for social transformation, fostering greater equity and inclusion in the digital age. By examining the role of infrastructure development (Sawid et al., 2023; Moncayo-Martínez & Salcedo, 2023) and the potential for transformative applications in critical sectors like agriculture (Zhang et al., 2022), we can explore how 5G/6G can empower individuals and communities. The digital divide remains a formidable obstacle to achieving a truly connected and equitable society. However, the emergence of 5G and the looming arrival of 6G technologies offer a glimmer of hope. These advancements hold immense potential to act as catalysts for social transformation, fostering greater inclusion and empowering individuals and communities. Enhanced Network Reach and Capacity: 5G networks boast significantly faster speeds and broader coverage compared to their predecessors. This opens doors for improved connectivity in remote and underserved areas,potentially bridging the geographical divide that often fuels the digital divide. 6G promises even further advancements, potentially providing ubiquitous and ultra-reliable connectivity, further shrinking the gap in access. Transformative Applications: The capabilities of 5G and 6G extend beyond simply browsing the internet. These technologies pave the way for the development and deployment of transformative applications in critical sectors like education, healthcare, and agriculture. Imagine remote students attending virtual classes with minimal lag or rural communities accessing telemedicine consultations in real-time. These advancements berate the potential to level the playing field and empower individuals regardless of their location. Digital Literacy Initiatives: The widespread adoption of 5G/6G can act as a catalyst for digital literacy initiatives.Increased internet accessibility can facilitate the development and dissemination of online learning platforms and resources, empowering individuals to develop the skills necessary to navigate the digital world. Economic Opportunities: Improved connectivity can unlock new economic opportunities for individuals and communities. Businesses in remote areas can gain access to global markets, and individuals can explore online entrepreneurship or remote work possibilities. This fosters economic growth and empowers individuals to contribute to the digital economy. Social Inclusion: Bridging the digital divide promotes social inclusion by fostering greater online participation.Individuals with limited access can connect with online communities, access information, and engage in civic discourse, fostering a more connected and inclusive society. However, it's crucial to acknowledge that technological advancements alone cannot dismantle the digital divide. Effective policy measures, infrastructure investments, and targeted digital literacy programs are essential to ensure equitable access and maximize the social transformative potential of 5G/6G technologies. By working collaboratively, governments,telecommunication companies, and civil society organizations can harness the power of these technologies to create a more inclusive and prosperous future for all.

Focus of this Review

This review leverages these conceptual frameworks to analyze the social, practical, and managerial implications of 5G/6G technologies. We will investigate how these advancements can bridge the digital divide, fostering equitable access to communication services, education, healthcare, and economic opportunities. Additionally, we will examine the practical considerations for infrastructure development and the managerial strategies necessary for sustainable implementation.Ultimately, this review aims to illuminate the transformative potential of 5G/6G technologies and their role in shaping a more connected and equitable future. The unrelenting evolution of mobile communication technologies necessitates a rigorous examination of their societal impact, particularly their potential to address the persistent challenge of the digital divide. As we transition from 4G to 5G and anticipate the imminent arrival of 6G, a paradigm shift is unfolding. This transformation extends beyond mere increases in network speeds and bandwidth; it redefines the very fabric of our interconnected world. This review delves into recent scholarly literature (2022-2024) to explore 5G and 6G technologies through a conceptual lens, examining their potential to bridge the digital divide and reshape societies on a global scale.

Conceptual Framework: 5G/6G Technologies as Catalysts for Social Transformation and The Enduring Challenge of the Digital Divide. The rapid advancement of mobile communication technologies has revolutionized the way we interact, communicate, and conduct business. This evolution, from 4G to 5G and now towards 6G, has been driven by increasing user demands for faster speeds, lower latency, and greater network capacity. This paper will delve into the key developments, challenges, and potential applications associated with this technological journey. The Transition from 4G to 5G - The transition from 4G to 5G marked a significant leap in mobile communication capabilities. 5G networks, characterized by higher data rates, lower latency, and greater network capacity, have enabled a wide range of new applications and services. As noted by Akhtar and Hassan (2021), the integration of terrestrial and non-terrestrial networks (TaNTIN) has been a key factor in the development of 5G and beyond. The Promise of 6G and Beyond 5G, the anticipated 6G network promises to deliver even more transformative capabilities. Chataut et al. (2024) highlight the potential of 6G networks to drive the AI revolution through advanced technologies and applications. Polymeni et al. (2023) emphasize the impact of 6G-IoT technologies on the development of agriculture 5.0, demonstrating the potential of these networks to revolutionize various industries. Challenges and Opportunities The journey from 4G to 6G is not without its challenges. Salahdine et al. (2023) discuss the technical and economic challenges associated with the development and deployment of 6G networks. However, the potential benefits of these networks, including improved connectivity, enhanced efficiency, and new business opportunities, far outweigh the challenges. Implications for Smart Cities The evolution of mobile communications has profound implications for the development of smart cities. As Chauhan (n.d.) points out, the transition from 4G to 6G can significantly impact the functionality and efficiency of smart city infrastructure. By enabling advanced applications and services, 6G networks can contribute to the creation of more sustainable, resilient, and connected urban environments. The evolution of mobile communications from 4G to 6G represents a remarkable technological achievement. As we continue to explore the potential of these networks, it is essential to address the associated challenges and capitalize on the opportunities they present. By understanding the key developments, applications, and implications of this journey, we can better prepare for the future of mobile communications and its impact on society. The digital divide, a persistent social and economic disparity, refers to the unequal access to and utilization of information and communication technologies (ICTs) between individuals and communities. This conceptual framework underscores the disparities in opportunities and resources associated with ICT access. Limited access to communication infrastructure,devices, and digital literacy skills can exacerbate existing socioeconomic inequalities, further marginalizing disadvantaged populations. The advent of 5G and 6G, with their enhanced capabilities and broader network reach, presents a compelling opportunity to bridge this digital divide. These advancements can be conceptualized as catalysts for social transformation, fostering greater equity and inclusion in the digital age. By examining the role of infrastructure development (Sawid et al., 2023; Moncayo-Martínez & Salcedo, 2023) and the potential for transformative applications in critical sectors like agriculture (Zhang et al., 2022), we can explore how 5G/6G can empower individuals and communities. This review leverages these conceptual frameworks to analyze the social, practical, and managerial implications of 5G/6G technologies. We will investigate how these advancements can bridge the digital divide, fostering equitable access to communication services, education, healthcare, and economic opportunities. Additionally, we will examine the practical considerations for infrastructure development and the managerial strategies necessary for sustainable implementation.Ultimately, this review aims to illuminate the transformative potential of 5G/6G technologies and their role in shaping a more connected and equitable future.

STATEMENT OF THE RESEARCH PROBLEM

The rapid advancement of mobile communication technologies, particularly the transition from 4G to 5G and the looming arrival of 6G, necessitates a comprehensive exploration of their societal impact, specifically their ability to address the enduring challenge of the digital divide. This research transcends a purely technical evaluation of network speed and bandwidth increases. Instead, it adopts a socio-economic perspective, focusing on the potential of 5G and 6G technologies to bridge the digital divide and promote equitable access to information and communication technologies (ICTs) on a global scale. By critically analyzing recent scholarly literature (2022-2024), this study aims to illuminate the multi-faceted implications of these advancements, encompassing social transformation, practical infrastructure considerations, and managerial strategies for sustainable implementation. Ultimately, the research seeks to contribute to the ongoing dialogue on harnessing technological progress to create a more inclusive and connected future for all.

METHODOLOGY

This study adopted a systematic literature review (SLR) methodology to investigate the multifaceted societal impact of 5G and 6G mobile communication technologies. The review focused on recent advancements, encompassing scholarly literature published between 2022 and 2024. To ensure comprehensive coverage, a search strategy was developed utilizing controlled vocabulary and keywords relevant to 5G/6G and their social or economic implications. Reputable academic databases, including Scopus, Web of Science, IEEE Xplore, and ACM Digital Library, were queried to identify relevant publications.

A two-step data analysis process was employed. First, a rigorous critical appraisal was conducted on each selected article. This appraisal aimed to extract key findings, methodological approaches, and theoretical frameworks relevant to the research question. The critical appraisal involved a close examination of the research design, data collection methods, and the authors' conclusions. Following the critical appraisal, a thematic analysis was conducted to identify recurring themes and categorize the reviewed literature based on their focus. This thematic analysis facilitated the identification of key areas of research and discussion concerning the multifaceted implications of 5G/6G technologies. Examples of potential themes included infrastructure challenges and opportunities for bridging the digital divide, the potential for social transformation through equitable access and novel applications, and managerial strategies for sustainable implementation and responsible development.

RESEARCH QUESTIONS

- To what extent can 5G/6G technologies bridge the digital divide, particularly in developing regions, and what infrastructure investments and policy frameworks are necessary to facilitate this process?

- How can 5G/6G applications in critical sectors like agriculture and healthcare contribute to social transformation, and what are the potential challenges and opportunities associated with ensuring equitable access to these advancements?
- What sustainable management strategies can telecommunication companies and technology firms implement to ensure the responsible development and deployment of 5G/6G technologies, considering factors like energy efficiency and electronic waste disposal?

LITERATURE REVIEW

This review synthesizes recent scholarly literature (2022-2024) on the potential of 5G and beyond (6G) mobile networks to promote environmental sustainability in sectors like agriculture, forestry, biodiversity monitoring, and water management. The rapid advancement of wireless communication technologies, particularly the transition from 5G to 6G, marks a pivotal transformation in global connectivity and digital infrastructure. As emphasized by Banafaa et al. (2023), 6G technology introduces unprecedented requirements, targets, and opportunities that extend far beyond traditional communication paradigms. The integration of these technologies has become increasingly crucial in addressing societal challenges across various sectors, including healthcare, agriculture, and urban development (Kumar et al., 2024; Majumdar et al., 2024). Recent studies have highlighted the multifaceted impact of these technologies on sustainable development. Beltozar-Clemente et al. (2023) and Cabanillas-Carbonell et al. (2023) demonstrate how 5G networks contribute significantly to achieving Sustainable Development Goals, particularly in economic growth and poverty reduction. The implementation of these advanced networks has shown remarkable potential in bridging the digital divide, as evidenced by Alizadeh et al. (2024) in their analysis of the National Broadband Network in Australia. In the context of infrastructure development, Dao et al. (2023) and Elamassie & Uysal (2023) emphasize the importance of exploring neglected aspects of 6G implementation, including underwater communications and free space optical communication for backhaul technology. This technological evolution has also catalyzed innovations in various sectors, as documented by Zhang et al. (2022) and Tomaszewski & Kołakowski (2023), particularly in smart agriculture, forestry, and environmental monitoring. However, the deployment of these technologies faces significant challenges, particularly in developing regions and rural areas. Sawad et al. (2023) and Moncayo-Martínez & Salcedo (2023) highlight the critical issues surrounding backhaul infrastructure and fiber optic deployment in underserved communities. Additionally, Kaur et al. (2023) address the crucial aspect of energy efficiency in 5G heterogeneous networks, emphasizing the need for sustainable implementation strategies. As progress toward the 6G era, the integration with emerging technologies such as Low Earth Orbit (LEO) networks presents both opportunities and challenges (Shayea et al., 2024). The evolution of these communication systems continues to reshape our understanding of connectivity, as outlined in comprehensive surveys by Vaezi et al. (2022) and Zhou et al. (2023), pointing toward a future where seamless, universal connectivity becomes increasingly achievable. This comprehensive analysis explores the current state, challenges, and future prospects of 5G/6G technologies, focusing on their implementation across various sectors and their potential to drive societal transformation while addressing existing technological and socioeconomic disparities.

5G Applications for Environmental Sustainability within the Technological Landscape

Several studies explore the application of 5G for environmental benefit. Tomaszewski and Kołakowski (2023) highlight the promise of 5G in smart agriculture, forestry, biodiversity monitoring, and water management. Their work identifies network capacity, latency, and security as key challenges for successful implementation. van Hilten and Wolfert (2022) conducted a similar review focusing on 5G's role in the agri-food sector, pinpointing opportunities for improved resource management, precision farming, and real-time monitoring. While the focus is on 5G, some studies explore the broader technological context. Beltozar-Clemente et al. (2023) conducted a systematic review examining the contribution of 5G networks to achieving Sustainable Development Goal 8 (Decent Work and Economic Growth). Their findings suggest that 5G can create new job opportunities and improve economic growth, potentially leading to indirect environmental benefits. Otlyvanska et al. (2024) and Blanutsa (2024) discuss the social and geographic implications of adopting 5G and future network technologies like 6G. These studies emphasize the importance of considering equity and accessibility when deploying these technologies for environmental applications.While the environmental impact of these technologies was not a central focus in all studies, some offered valuable insights. Majumdar et al. (2024) explored the potential for achieving sustainable agriculture through 5G-powered solutions, suggesting possibilities for reduced power consumption through specific network configurations. Although not their primary focus, studies by Adanvo et al. (2022) and Verma et al. (2023) discussed network efficiency strategies and applications with broader implications for environmental sustainability.

Studies Peripheral to Environmental Focus

Several studies included in the search results were not directly relevant to the environmental focus of this review. Kaur et al. (2023) reviewed energy-efficiency schemes for 5G base stations, a crucial aspect for overall network sustainability but not directly addressing environmental applications. Bastos et al. (2024) explored digitalization in the forest industry,which could have environmental implications but was not their primary focus. Alizadeh et al. (2024) investigated the socio-spatial patterns of the National Broadband Network in Australia, not directly related to environmental sustainability.Shen et al. (2023) and Malekzadeh (2023) focused on ecological modeling and IoT deployment in cellular networks,respectively, which are not specific to environmental applications of 5G. Ray (2021) discussed the future roadmap for 6G technology, which is outside the scope of this review on current and emerging environmental applications. The reviewed literature suggests that 5G technologies have significant potential to contribute to environmental sustainability in agriculture, forestry, biodiversity monitoring, and water management. However, challenges related to network capacity, latency, security, and equitable access need to be addressed. Future research should explore specific applications and the integration of 5G with other technologies to optimize environmental outcomes.

6G Network Integration, Innovation, Challenges and Considerations

Several studies explored the drivers of integration and innovation shaping 6G network development. Zhou et al. (2023) introduced the novel concept of aerospace integrated networks as a key enabler for 6G, highlighting its potential to significantly enhance performance and broaden overall network coverage. Shayea et al. (2024) investigated the integration of 5G, 6G, and the Internet of Things (IoT) with Low Earth Orbit (LEO) networks, identifying both the opportunities and challenges associated with this convergence. Plastras et al. (2024) examined the role of Non-Terrestrial Networks (NTNs) in connecting remote IoT devices within the 6G framework, with a particular focus on optimizing energy efficiency. Banafaa et al. (2023) provided a comprehensive overview of 6G technology, outlining its requirements, targets,applications, challenges, and potential benefits. The reviewed literature acknowledged the various challenges accompanying advancements in 5G and 6G technologies.Oughton et al. (2022) emphasized the critical role of policy decisions in ensuring the affordability of universal broadband access for existing 4G and 5G networks. Amini et al. (2024) raised concerns regarding data management and accessibility within cellular networks, highlighting a crucial consideration for future technological development. A few studies deviated from the core themes of 5G and 6G advancements. Allam and Takun (2022) presented a broader discussion on smart cities, which was not directly related to the technological aspects of this review. Pooja and Kumar (2023) examined modulation schemes within a 5G context, providing a technical analysis but not exploring broader network development. The reviewed literature provided compelling evidence of significant progress in 5G network capabilities and the ongoing development of 6G technologies. Integration with aerospace and non-terrestrial networks holds promise for enhanced performance and broader coverage. However, challenges related to affordability, data management, and environmental impact require continued attention from researchers and policymakers. Further research is necessary to explore the full potential of these technologies for sustainable applications across various sectors. Looking beyond immediate 5G deployment, Zhang et al. (2022) investigate the transformative potential of 6G technologies. Their review explores how 6G's enhanced capabilities, particularly in terms of ultra-low latency and high bandwidth, can revolutionize critical sectors like agriculture. This research suggests that 6G has the potential to unlock new opportunities for sustainable and efficient agricultural practices, potentially contributing to improved food security in developing nations.

Bridging the Rural-Urban Divide: Latency Reduction Strategies, Infrastructure Challenges and Opportunities in Developing Regions

A critical barrier to widespread 5G adoption in developing countries is the lack of robust backhaul infrastructure. Sawid et al. (2023) conducted a comprehensive review of this challenge, highlighting the limitations of traditional fiber optic deployment and exploring alternative backhaul solutions. Their work underscores the need for innovative approaches tailored to the specific economic and geographic realities of developing regions. Kumar et al. (2024) address the specific challenge of latency in rural areas. Their work examines strategies for reducing latency in 5G and beyond 5G (B5G) networks deployed in these regions. This research is crucial for enabling applications like smart healthcare, which rely on real-time data transmission and can significantly improve healthcare outcomes in remote locations. Moncayo-Martínez and Salcedo (2023) provide valuable insights into optimizing fiber optic deployment in small rural communities. Their research, focusing on the case of Mexico, offers practical solutions for overcoming infrastructure limitations and maximizing the reach of advanced communication networks. This case study serves as a valuable model for developing countries seeking to bridge the digital divide.

Emerging Infrastructure Needs: Underwater Communication

Dao et al. (2023) shift the focus to an often-overlooked aspect of future communication infrastructure: underwater communication networks. Their work highlights the critical, yet underdeveloped, role of underwater infrastructure in supporting the full potential of 6G applications across diverse domains. This research emphasizes the need for further investment in research and development to ensure that underwater communication capabilities keep pace with advancements in terrestrial 6G technologies. The reviewed literature underscores the critical role of infrastructure development in unlocking the transformative potential of 5G and 6G technologies. While challenges related to backhaul solutions and rural connectivity persist in developing countries, innovative approaches and targeted investments hold promise for bridging the digital divide.Furthermore, advancements in 6G offer exciting prospects for industries like agriculture, with the potential to contribute to global food security. Continuous research into infrastructure optimization, including the exploration of underwater communication networks, is essential to ensure that the benefits of 5G and 6G reach all corners of the globe. This review emphasizes the need for a multi-pronged approach that combines technological innovation, strategic infrastructure development, and equitable access policies to maximize the positive impact of these advancements on a global scale.

IMPLICATIONS

Societal Transformation through Technological Advancement

The widespread adoption of 5G and 6G presented significant opportunities to bridge the digital divide. As highlighted by Sawid et al. (2023), innovative backhaul solutions could overcome infrastructure limitations in developing regions. This fostered equitable access to essential communication services, potentially improving access to education, healthcare, and economic opportunities (Moncayo-Martínez & Salcedo, 2023). Furthermore, 6G's enhanced capabilities, as explored by Zhang et al. (2022), could lead to transformative applications in critical sectors like agriculture. This had the potential to revolutionize food production practices and contribute to global food security. Additionally, Kumar et al. (2024) discussed strategies for reducing latency in rural areas, paving the way for advancements in telemedicine and improved public health outcomes.

Practical Considerations for Infrastructure Development

Unlocking the potential of these technologies hinged on robust infrastructure development. The reviewed literature emphasized the need for significant investment in both traditional fiber optic deployment, optimized for rural areas (Moncayo-Martínez & Salcedo, 2023), and exploration of alternative backhaul solutions (Sawid et al., 2023). Dao et al.(2023) further highlighted the often-overlooked aspect of underwater communication networks, crucial for supporting the full potential of 6G applications. To facilitate these crucial developments, policymakers needed to create frameworks that incentivized infrastructure investment in developing countries and promoted equitable access (societal implication).

Managerial Strategies for Sustainable Implementation

Telecommunication companies and technology firms developed strategic plans that addressed the specific infrastructure challenges of developing regions. This necessitated exploring alternative backhaul solutions and collaborating with local governments to optimize network deployment. Additionally, a focus on sustainable practices throughout the development and deployment lifecycle was paramount (managerial implication). Research into energy-efficient network configurations and responsible electronic waste disposal was crucial for long-term environmental sustainability.

CONCLUSION

This review examined recent scholarly literature (2022-2024) on the evolving landscape of 5G and 6G mobile communication technologies. The analysis focused on the social, practical, and managerial implications associated with their development and deployment. The findings reveal significant potential for 5G and 6G technologies to bridge the digital divide and empower a more connected future. By overcoming infrastructure limitations in developing regions, as explored by Sawid et al. (2023) and Moncayo-Martínez & Salcedo (2023), these advancements can promote equitable access to communication services,education, healthcare, and economic opportunities. Additionally, 6G's enhanced capabilities, as discussed by Zhang et al.(2022), hold promise for transformative applications in critical sectors like agriculture, potentially contributing to global food security. However, unlocking this potential requires a multifaceted approach. Strategic infrastructure development is paramount,with a focus on optimizing both traditional fiber optic deployment (Moncayo-Martínez & Salcedo, 2023) and exploring alternative backhaul solutions (Sawid et al., 2023). Furthermore, as highlighted by Dao et al. (2023), investment in underwater communication networks is crucial to fully realize the potential of 6G applications. To facilitate these developments, policymakers must create frameworks that incentivize infrastructure investment and equitable access,particularly in developing regions. From a managerial perspective, telecommunication companies and technology firms need to develop strategic plans that address the specific infrastructure challenges of these regions. Collaboration with local governments and exploration of alternative backhaul solutions are key considerations. Sustainability must also be a central focus throughout the development and deployment lifecycle. Research into energy-efficient network configurations and responsible electronic waste disposal practices is essential for long-term environmental sustainability. 5G and 6G technologies offer a transformative opportunity to bridge the digital divide, empower societies,and promote global progress. By prioritizing infrastructure development, equitable access, and sustainable practices,researchers, policymakers, telecommunication companies, and technology firms can work together to unlock the full potential of these advancements and create a more connected and sustainable future.

REFERENCES

Adanvo, V. F., Mafra, S., Montejo-Sánchez, S., Fernández, E. M. G., & Souza, R. D. (2022). Buffer-aided relaying strategies for two-way wireless networks. *Sustainability (Basel), 14*(21), 13829. DOI: 10.3390/su142113829

Ahokangas, P., & Aagaard, A. (2024). *The changing world of mobile communications* (Ahokangas, P., & Aagaard, A., Eds.)., DOI: 10.1007/978-3-031-33191-6

Akhtar, M. W., & Hassan, S. A. (2021). TaNTIN: Terrestrial and non-terrestrial integrated networks-A collaborative technologies perspective for beyond 5G and 6G. *Internet Technology Letters*. Advance online publication. DOI: 10.1002/itl2.274

Alimi, I. A., Patel, R. K., Muga, N. J., Pinto, A. N., Teixeira, A. L., & Monteiro, P. P. (2021). Towards enhanced mobile broadband communications: A tutorial on enabling technologies, design considerations, and prospects of 5G and beyond fixed wireless access networks. *Applied Sciences (Basel, Switzerland), 11*(21), 10427. DOI: 10.3390/app112110427

Alizadeh, T., Helderop, E., Grubesic, T. H., & Ferrers, R. (2024). The multi-technology footprint of the National Broadband Network in Australia: Exploring the urban-regional divide and Socio-spatial patterns for inequality. *International Regional Science Review, 47*(2), 159–181. DOI: 10.1177/01600176231168025

Allam, Z., & Takun, Y. R. (2022). *Rethinking smart cities*. DOI: 10.4337/9781803926803

Amini, M., Stanica, R., & Rosenberg, C. (2024). Where are the (cellular) data? *ACM Computing Surveys, 56*(2), 1–25. DOI: 10.1145/3610402

Banafaa, M., Shayea, I., Din, J., Hadri Azmi, M., Alashbi, A., Ibrahim Daradkeh, Y., & Alhammadi, A. (2023). 6G mobile communication technology: Requirements, targets, applications, challenges, advantages, and opportunities. *Alexandria Engineering Journal, 64*, 245–274. DOI: 10.1016/j.aej.2022.08.017

Bastos, T., Teixeira, L. C., & Nunes, L. J. R. (2024). Forest 4.0: Technologies and digitalization to create the residual biomass supply chain of the future. *Journal of Cleaner Production, 467*(143041), 143041. DOI: 10.1016/j.jclepro.2024.143041

Beltozar-Clemente, S., Iparraguirre-Villanueva, O., Pucuhuayla-Revatta, F., Sierra-Liñan, F., Zapata-Paulini, J., & Cabanillas-Carbonell, M. (2023). Contributions of the 5G network with respect to decent work and economic growth (Sustainable Development Goal 8): A systematic review of the literature. *Sustainability (Basel), 15*(22), 15776. DOI: 10.3390/su152215776

Blanutsa, V. I. (2024). Contours of future directions for geographic study of 6G wireless communications. *Regional Research of Russia, 14*(1), 98–107. DOI: 10.1134/S2079970523600348

Cabanillas-Carbonell, M., Pérez-Martínez, J., & Zapata-Paulini, J. (2023). Contributions of the 5G network with respect to poverty (SDG1), systematic literature review. *Sustainability (Basel), 15*(14), 11301. DOI: 10.3390/su151411301

Chataut, R., Nankya, M., & Akl, R. (2024). 6G networks and the AI revolution-exploring technologies, applications, and emerging challenges. *Sensors (Basel)*, *24*(6), 1888. DOI: 10.3390/s24061888 PMID: 38544151

Chauhan, D. (n.d.). *Roadmap from 4G to 6G in smart cities impact and challenges*. doi:DOI: 10.5815/ijwmt.202.0.0

Dao, N.-N., Tu, N. H., Thanh, T. T., Bao, V. N. Q., Na, W., & Cho, S. (2023). Neglected infrastructures for 6G—Underwater communications: How mature are they? *Journal of Network and Computer Applications*, *213*(103595), 103595. DOI: 10.1016/j.jnca.2023.103595

Elamassie, M., & Uysal, M. (2023). Free space optical communication: An enabling backhaul technology for 6G non-terrestrial networks. *Photonics*, *10*(11), 1210. DOI: 10.3390/photonics10111210

Kaur, P., Garg, R., & Kukreja, V. (2023). Energy-efficiency schemes for base stations in 5G heterogeneous networks: A systematic literature review. *Telecommunication Systems*, *84*(1), 115–151. DOI: 10.1007/s11235-023-01037-x

Kumar, A., Gaur, N., & Nanthaamornphong, A. (2024). Improving the latency for 5G/B5G based smart healthcare connectivity in rural area. *Scientific Reports*, *14*(1), 6976. DOI: 10.1038/s41598-024-57641-7 PMID: 38521842

Kumar, N. (2023, November). Comparative analysis of Different Modulation Schemes for LMLC in 5G Scenario. In *2023 International Conference on Computing, Communication, and Intelligent Systems (ICCCIS)* (pp. 1057-1061). IEEE.

Majumdar, P., Mitra, S., Bhattacharya, D., & Bhushan, B. (2024). Enhancing sustainable 5G powered agriculture 4.0 : Summary of low power connectivity, internet of UAV things, AI solutions and research trends. *Multimedia Tools and Applications*. Advance online publication. DOI: 10.1007/s11042-024-19728-1

Malekzadeh, M. (2023). Enabling efficient and reliable IoT deployment in 5G and LTE cellular areas for optimized service provisioning. *The Journal of Supercomputing*, *79*(2), 1926–1955. DOI: 10.1007/s11227-022-04722-x

Moncayo-Martínez, L. A., & Salcedo, A. (2023). Optimising the optic fibre deployment in small rural communities: The case of Mexico. *Axioms*, *12*(3), 269. DOI: 10.3390/axioms12030269

Otlyvanska, G., Connolly, N., Walsh, G., & Dooley, J. (2024). Studying the adoption of 5G and future networks for social inclusion: An innovation systems transitions perspective for networks-as-a-service. In *IFIP Advances in Information and Communication Technology* (pp. 137–149). Transfer, Diffusion and Adoption of Next-Generation Digital Technologies., DOI: 10.1007/978-3-031-50204-0_12

Oughton, E. J., Comini, N., Foster, V., & Hall, J. W. (2022). Policy choices can help keep 4G and 5G universal broadband affordable. *Technological Forecasting and Social Change*, *176*(121409), 121409. DOI: 10.1016/j.techfore.2021.121409

Plastras, S., Tsoumatidis, D., Skoutas, D. N., Rouskas, A., Kormentzas, G., & Skianis, C. (2024). Non-Terrestrial Networks for energy-efficient connectivity of Remote IoT devices in the 6G era: A survey. *Sensors (Basel)*, *24*(4), 1227. DOI: 10.3390/s24041227 PMID: 38400391

Polymeni, S., Plastras, S., Skoutas, D. N., Kormentzas, G., & Skianis, C. (2023). The impact of 6G-IoT technologies on the development of agriculture 5.0: A review. *Electronics (Basel), 12*(12), 2651. DOI: 10.3390/electronics12122651

Ray, P. P. (2021). A perspective on 6G: Requirement, technology, enablers, challenges and future road map. *Journal of Systems Architecture, 118*(102180), 102180. DOI: 10.1016/j.sysarc.2021.102180

Salahdine, F., Han, T., & Zhang, N. (2023). 5G, 6G, and Beyond: Recent advances and future challenges. *Annals of Telecommunications -. Annales des Télécommunications, 78*(9–10), 525–549. DOI: 10.1007/s12243-022-00938-3

Sawad, I., Nilavalan, R., & Al-Raweshidy, H. (2023). Backhaul in 5G systems for developing countries: A literature review. *IET Communications, 17*(6), 659–669. DOI: 10.1049/cmu2.12578

Shayea, I., El-Saleh, A. A., Ergen, M., Saoud, B., Hartani, R., Turan, D., & Kabbani, A. (2024). Integration of 5G, 6G and IoT with Low Earth Orbit (LEO) networks: Opportunity, challenges and future trends. *Results in Engineering, 23*(102409), 102409. DOI: 10.1016/j.rineng.2024.102409

Shen, J., Zhu, W., Peng, Z., & Wang, Y. (2023). Improving landscape ecological network connectivity in urbanizing areas from dual dimensions of structure and function. *Ecological Modelling, 482*(110380), 110380. DOI: 10.1016/j.ecolmodel.2023.110380

Tomaszewski, L., & Kołakowski, R. (2023). Mobile services for smart agriculture and forestry, biodiversity monitoring, and water management: Challenges for 5G/6G networks. *Telecom, 4*(1), 67–99. DOI: 10.3390/telecom4010006

Vaezi, M., Azari, A., Khosravirad, S. R., Shirvanimoghaddam, M., Azari, M. M., Chasaki, D., & Popovski, P. (2022). Cellular, wide-area, and non-terrestrial IoT: A survey on 5G advances and the road toward 6G. *IEEE Communications Surveys and Tutorials, 24*(2), 1117–1174. DOI: 10.1109/COMST.2022.3151028

van Hilten, M., & Wolfert, S. (2022). 5G in agri-food - A review on current status, opportunities and challenges. *Computers and Electronics in Agriculture, 201*(107291), 107291. DOI: 10.1016/j.compag.2022.107291

Verma, H., Chauhan, N., & Awasthi, L. K. (2023). A Comprehensive review of 'Internet of Healthcare Things': Networking aspects, technologies, services, applications, challenges, and security concerns. *Computer Science Review, 50*(100591), 100591. DOI: 10.1016/j.cosrev.2023.100591

Zhang, F., Zhang, Y., Lu, W., Gao, Y., Gong, Y., & Cao, J. (2022). 6G-enabled smart agriculture: A review and prospect. *Electronics (Basel), 11*(18), 2845. DOI: 10.3390/electronics11182845

Zhou, D., Sheng, M., Li, J., & Han, Z. (2023). Aerospace integrated networks innovation for empowering 6G: A survey and future challenges. *IEEE Communications Surveys and Tutorials, 25*(2), 975–1019. DOI: 10.1109/COMST.2023.3245614

KEY TERMS AND DEFINITIONS

Digital Divide: The unequal access to and utilization of information and communication technologies (ICTs) between individuals, communities, and countries. This disparity can encompass limitations in infrastructure, devices, and digital literacy skills.

5G/6G Technologies: The latest generations of mobile communication technologies offering significantly faster speeds, broader network coverage, and enhanced capabilities compared to previous generations.

Social Transformation: A fundamental change in the social fabric of a society, often driven by technological advancements. In this context, it refers to the potential of 5G/6G to empower individuals and communities, improve access to essential services, and foster a more connected world.

Infrastructure Development: The process of creating and maintaining physical structures necessary to support 5G/6G networks. This includes fiber optic cables, cell towers, and other related equipment.

Equitable Access: The principle of ensuring that everyone has a fair and equal opportunity to access and utilize 5G/6G technologies and their associated benefits. This may require targeted policies and initiatives to address existing disparities.

Sustainable Development: A development approach that meets the needs of the present without compromising the ability of future generations to meet their own needs. In the context of 5G/6G, this includes considerations for energy efficiency, responsible e-waste disposal, and minimizing environmental impact.

Managerial Strategies: The plans and actions implemented by telecommunication companies and technology firms to ensure the responsible development and deployment of 5G/6G technologies. This can involve strategies for network optimization, resource allocation, and collaboration with stakeholders.

Compilation of References

3rd Generation Partnership Project. (2024). *3GPP specifications for 5G and beyond*. Retrieved from https://www.3gpp.org/

A. Mary, A. Y. Begum, and D. G. V., (2023) "A Novel IoT Framework and Device Architecture for Efficient Smart City Implementation," *2023 7th International Conference on Trends in Electronics and Informatics (ICOEI)*, Tirunelveli, India, pp. 420-426.

Aayog, N. I. T. I. (2021). *Transforming Indian Education through Effective Resource Management*. Government of India. https://www.niti.gov.in

Accenture. (2022). *Revolutionizing Education with AI: The Role of 5G Networks in Higher Education*. Accenture Insights. https://www.accenture.com/in-en/insights

Adams, A., Browne, D., & Murray, A. (2020). 5G RuralFirst: Trialing 5G in rural and remote environments. *IEEE Access : Practical Innovations, Open Solutions*, *8*, 34512–34523. DOI: 10.1109/ACCESS.2020.2972099

Adams, J. (2023). Grants for startups in vertical farming to boost rural economic growth. *Agricultural Economics Journal*, *12*(4), 214–222.

Adanvo, V. F., Mafra, S., Montejo-Sánchez, S., Fernández, E. M. G., & Souza, R. D. (2022). Buffer-aided relaying strategies for two-way wireless networks. *Sustainability (Basel)*, *14*(21), 13829. DOI: 10.3390/su142113829

Admin. (2024, March 20). *Digital transformation in livestock protection – Livetec*. Livetec. https://www.livetecsystems.co.uk/leading-the-change-digital-transformation-livestock-protection/

Agarwal, S. (2022). *Centralized Planning and Bureaucratic Challenges in Indian Higher Education*. Education Policy Journal, 25(2), 45-58. https://educationpolicyjournal.org

Agrawal, U., Dave, P., Tiwari, A., Upadhyay, V., Sankhe, M. S., & Shrivastav, S. (2022). Implementation of 5G technology in rural education of India. *IEEE International Conference on Broadband and Social Sustainability (IBSSC)*. doi:DOI: 10.1109/IBSSC56953.2022.10037466

Ahad, A., Jiangbina, Z., Tahir, M., Shayea, I., Sheikh, M. A., & Rasheed, F. (2024). 6G and Intelligent Healthcare: Taxonomy, technologies, open issues and future research directions. *Internet of Things : Engineering Cyber Physical Human Systems*, *25*, 101068. DOI: 10.1016/j.iot.2024.101068

Ahmed, M. A., Majeed, A. S., Dawood, Y. S. (2023). A Survey of 6G Mobile Systems, Enabling Technologies, and Challenges. *International journal of electrical and electronic engineering and telecommunications 12*(1). doi:DOI: 10.18178/ijeetc.12.1.1-21

Ahmed, R., & Khan, S. (2020). Social networks and the digital divide: A pathway to rural community resilience. *Rural Connectivity Journal*, *13*(2), 57–68.

Ahmed, R., & Sharma, P. (2022). The Role of Smart Classrooms in Modern Education: A 5G Perspective. *Journal of Educational Technology*, *34*(2), 123–134. https://jedutech.org

Ahokangas, P., & Aagaard, A. (2024). *The changing world of mobile communications* (Ahokangas, P., & Aagaard, A., Eds.)., DOI: 10.1007/978-3-031-33191-6

Akbar, M. S., Hussain, Z., Sheng, Q. Z., & Mukhopadhyay, S. (2022). *6G Survey on Challenges, Requirements, Applications, Key Enabling Technologies, Use Cases, AI integration issues and Security aspects*. 10.48550/arXiv.2206.00868.

Akhtar, M. W., & Hassan, S. A. (2021). TaNTIN: Terrestrial and non-terrestrial integrated networks-A collaborative technologies perspective for beyond 5G and 6G. *Internet Technology Letters*. Advance online publication. DOI: 10.1002/itl2.274

Akhtar, M. W., Hassan, S. A., Ghaffar, R., Jung, H., Garg, S., & Hossain, M. S. (2020). The shift to 6G communications: Vision and requirements. *Human-Centric Computing and Information Sciences*, *10*(53), 53. Advance online publication. DOI: 10.1186/s13673-020-00258-2

Akyildiz, I. F., Han, C., & Nie, S. (2018). Combating the Distance Problem in the Millimeter Wave and Terahertz Frequency Bands [June.]. *IEEE Communications Magazine*, *56*(6), 102–108. DOI: 10.1109/MCOM.2018.1700928

Akyildiz, I. F., Kak, A., & Nie, S. (2020). 6G and beyond: The future of wireless communications systems. *IEEE Access : Practical Innovations, Open Solutions*, *8*, 133995–134030. DOI: 10.1109/ACCESS.2020.3010896

Al-Ansi, A., Al-Ansi, A. M., Muthanna, A., & Koucheryavy, A. (2024). Blockchain technology integration in service migration to 6g communication networks: A comprehensive review. *Indonesian Journal of Electrical Engineering and Computer Science*, *34*(3), 1654–1664. DOI: 10.11591/ijeecs.v34.i3.pp1654-1664

Alexander, B. (2016). *Volha, Kalach., Alexander, Kicherer., Gustav, Murawski., Tim, Waizenegger., Matthias, Wieland. (2014). Policy-Framework - Eine Methode zur Umsetzung von Sicherheits-Policies im Cloud-Computing. Dilip, Krishnaswamy., Ramki, Krishnan., Diego, R., Lopez., Steven, Wright., Norival, Figueira.* Policy Architecture and Framework for NFV Infrastructures.

Al-Fuqaha, A., Guizani, M., Mohammadi, M., Aledhari, M., & Ayyash, M. (2015). Internet of Things: A survey on enabling technologies, protocols, and applications [Fourthquarter.]. *IEEE Communications Surveys and Tutorials*, *17*(4), 2347–2376.

Ali-Hassan, H. (2009). Social capital theory. In *Handbook of Research on Social Capital Theory*. DOI: 10.4018/978-1-60566-659-4.ch024

Alimi, I. A., Patel, R. K., Muga, N. J., Pinto, A. N., Teixeira, A. L., & Monteiro, P. P. (2021). Towards enhanced mobile broadband communications: A tutorial on enabling technologies, design considerations, and prospects of 5G and beyond fixed wireless access networks. *Applied Sciences (Basel, Switzerland), 11*(21), 10427. DOI: 10.3390/app112110427

Alizadeh, T., Helderop, E., Grubesic, T. H., & Ferrers, R. (2024). The multi-technology footprint of the National Broadband Network in Australia: Exploring the urban-regional divide and Socio-spatial patterns for inequality. *International Regional Science Review, 47*(2), 159–181. DOI: 10.1177/01600176231168025

Allam, Z., & Takun, Y. R. (2022). *Rethinking smart cities*. DOI: 10.4337/9781803926803

Alonistioti, N., Cardona, N., Waldemar, P., & Trivisonno, R. (2024). Establishing an open platform to foster 6G innovation and research. *IEEE Communications Magazine, 62*(1), 8–10. Advance online publication. DOI: 10.1109/MCOM.2024.10401872

Al-Shareeda, M., Al-Phryggy, M., Al-Kashani, A., Al-Jaber, H., & Al-Moussawi, D. (2024). Design and implementation of novel 6G cellular communication architecture using 3D modeling in real environments. doi:DOI: 10.54517/cte.v2i2.2668

Alston, P. (1979). The United Nations' Specialized Agencies and Implementation of the International Covenant on Economic, Social and Cultural Rights. *Columbia Journal of Transnational Law, 18*, 79.

Altaf Khattak, S. B., Nasralla, M. M., & Rehman, I. U. (2022). The Role of 6G Networks in Enabling Future Smart Health Services and Applications. *2022 IEEE International Smart Cities Conference (ISC2)*, Pafos, Cyprus, 2022. doi:DOI: 10.1109/ISC255366.2022.9922093

Álvaro, V. R., Petteri, K., Silvio, M., & Harish, V. (2023). The Role of AI on 6G MAC. doi:DOI: 10.36227/techrxiv.23708310.v1

Alwis, L., Kalla, B., & Jamalipour, A. (2021). 6G Vision: Future Mobile Networks for Intelligent IoT and Ubiquitous Connectivity. *IEEE Internet of Things Journal, 8*(10), 8765–8777.

Alzubaidi, L., Al-Sabaawi, A., Bai, J., Dukhan, A., Alkenani, A. H., Al-Asadi, A., Alwzwazy, H. A., Manoufali, M., Fadhel, M. A., Albahri, A. S., Moreira, C., Ouyang, C., Zhang, J., Santamaría, J., Salhi, A., Hollman, F., Gupta, A., Duan, Y., Rabczuk, T., & Gu, Y. (2023). Towards risk-free trustworthy artificial intelligence: Significance and requirements. *International Journal of Intelligent Systems, 2023*(1), 4459198. https://www.hindawi.com/journals/ijis/2023/4459198/. DOI: 10.1155/2023/4459198

Al-Zubi, M. M., & Alouini, M. (2023). End-to-End modelling and simulation of NLOS sub-6 GHz backhaul via diffraction for internet connectivity of rural areas. *IEEE Open Journal of the Communications Society, 4*, 3102–3114. DOI: 10.1109/OJCOMS.2023.3330078

Amazon India. (2021). *Amazon Saheli empowers women artisans and entrepreneurs*. https://www.amazon.in/saheli

Amini, M., Stanica, R., & Rosenberg, C. (2024). Where are the (cellular) data? *ACM Computing Surveys, 56*(2), 1–25. DOI: 10.1145/3610402

Amos, J. (2023, August 3). Embracing Change: Overcoming resistance during digital transformation - Industry leading ICT services, support. *soma technology group*. https://www.soma.com.au/blog/resistance-during-digital-transformation/

Anderson, J., & Rainie, L. (2012). The impact of the internet on education: A survey of teachers and technology use. Pew Research Centre. https://www.pewresearch.org

Andrew W, Gao J, Mullan S, Campbell N, Dowsey AW and Burghardt T (2020a). Visual identification of individual Holstein-Friesian cattle via deep metric learning. arXiv preprint arXiv, 2006.09205.

Andrew W, Greatwood C and Burghardt T (2019). Aerial animal biometrics: individual Friesian cattle recovery and visual identification via an autonomous with onboard deep inference. arXiv preprint arXiv, 1907.05310.

Andrew, W., Greatwood, C., & Burghardt, T. (2017). Visual localisation and individual identification of Holstein Friesian cattle via deep learning. Proceedings of the IEEE *International Conference on Computer Vision Workshops*, 2850–2859. DOI: 10.1109/ICCVW.2017.336

Andrews, J. G., Buzzi, S., Choi, W., Hanly, S. V., Lozano, A., Soong, A. C., & Zhang, J. C. (2014). What will 5G be? *IEEE Journal on Selected Areas in Communications*, *32*(6), 1065–1082. DOI: 10.1109/JSAC.2014.2328098

Andrew, W., Greatwood, C., & Burghardt, T. (2020b). Fusing animal biometrics with autonomous robotics: drone-based search and individual id of Friesian cattle. *Proceedings of the IEEE/CVF Winter Conference on Applications of Computer Vision Workshops* 1, 38–43. DOI: 10.1109/WACVW50321.2020.9096949

Appalachian Regional Commission. (2023). *Broadband initiatives*. Retrieved from https://www.arc.gov/programs/broadband-initiatives/

Appinventiv. (2023). 6G Technology: Revolutionizing Connectivity for the Future. Retrieved from https://appinventiv.com/blog/6g-technology-transforming-enterprises/

ArunKumar. S., Sivakami, K., VijayaKarthik, S. V., & Deepa, S. M. (2022). 6G With Smart Healthcare, Haptic Communication. In D. Satishkumar, G. Prabhakar, & R. Anand (Eds.), *Handbook of Research on Design, Deployment, Automation, and Testing Strategies for 6G Mobile Core Network* (pp. 303-329). IGI Global Scientific Publishing. doi:DOI: 10.4018/978-1-7998-9636-4.ch015

Asma India. (2023). Noteworthy Digital Initiatives that are Redefining Rural Education in India. Retrieved from https://www.asmaindia.in/blog/noteworthy-digital-initiatives-that-are-redefining-rural-education-in-india/

Awad, A. I. (2016). From classical methods to animal biometrics: A review on cattle identification and tracking. *Computers and Electronics in Agriculture*, *123*, 423–435. DOI: 10.1016/j.compag.2016.03.014

Azar, H., & Hassan, M. (2020). Deploying wireless technologies in rural areas: Opportunities and challenges. *Telecommunications Review*, *23*(3), 145–160.

Azari, , Solanki, S., Chatzinotas, S., Kodheli, O., Sallouha, H., Colpaert, A., Mendoza Montoya, J. F., Pollin, S., Haqiqatnejad, A., Mostaani, A., Lagunas, E., & Ottersten, B. (2022). Evolution of non-terrestrial networks from 5G to 6G: A survey. *IEEE Communications Surveys and Tutorials, 24*(4), 2633–2672. DOI: 10.1109/COMST.2022.3199901

Bahaj, A. S. (2011). Generating electricity from the oceans. *Renewable & Sustainable Energy Reviews, 15*(7), 3399–3416. DOI: 10.1016/j.rser.2011.04.032

Baker, L. (2023). The financial burden of initial investments in direct marketing and vertical farming. *Journal of Agribusiness, 15*(2), 57–65.

Banafaa, M., Ibraheem, S., Jafri, D., Marwan, H. A., Alashbi, A., Daradkeh, Y., & Alhammadi, A. (2023). 6G Mobile Communication Technology: Requirements, Targets, Applications, Challenges, Advantages, and Opportunities. *Journal of Network and Computer Applications, 4*(1), 270–280. DOI: 10.1016/j.jnca.2022.103

Banafaa, M., Shayea, I., Din, J., Hadri Azmi, M., Alashbi, A., Ibrahim Daradkeh, Y., & Alhammadi, A. (2023). 6G mobile communication technology: Requirements, targets, applications, challenges, advantages, and opportunities. *Alexandria Engineering Journal, 64*, 245–274. DOI: 10.1016/j.aej.2022.08.017

Bansal, K. (2023). *Bridging the Gap: How 5G Can Revolutionize Rural Education in India*. Educational Innovations Journal, 21(4), 78-89. https://eduinnova.com

Bansal, S. (2023). Data Privacy Challenges in the Age of 5G and 6G. *Journal of Information Security, 12*(3), 78–89. https://journalofinfosecurity.com

Barbedo, J. G. A., Koenigkan, L. V., Santos, T. T., & Santos, P. M. (2019). A study on the detection of cattle in UAV images using deep learning. *Sensors (Basel), 19*(24), 5436. DOI: 10.3390/s19245436 PMID: 31835487

Barbosa Pereira, C., Dohmeier, H., Kunczik, J., Hochhausen, N., Tolba, R., & Czaplik, M. (2019). Contactless monitoring of heart and respiratory rate in anesthetized pigs using infrared thermography. *PLoS One, 14*(11), e0224747. DOI: 10.1371/journal.pone.0224747 PMID: 31693688

Barron, U. G., Corkery, G., Barry, B., Butler, F., Mcdonnell, K., & Ward, S. (2008). Assessment of retinal recognition technology as a biometric method for sheep identification. *Computers and Electronics in Agriculture, 60*(2), 156–166. DOI: 10.1016/j.compag.2007.07.010

Barrutia, J. M., & Echebarria, C. (2021). Effect of the COVID-19 pandemic on public Managers' attitudes toward digital transformation. *Technology in Society, 67*, 101776. DOI: 10.1016/j.techsoc.2021.101776 PMID: 34642513

Barua, T., & Mohiuddin, K. (2024) "Dynamic Disaster Management with Real-Time IoT Data Analysis and Response," *2024 International Conference on Automation and Computation (AUTOCOM)*, Dehradun, India, pp. 142-147.

Basri, Y. M., Taufik, T., Yasni, H., Putri, R. I., Zahra, N. I., Rahmadani, N., Aslin, M., & Amri, T. (2023). *Improving company performance in rural areas: Institutional theory approach*. KnE Social Sciences., DOI: 10.18502/kss.v8i13.13750

Bastos, T., Teixeira, L. C., & Nunes, L. J. R. (2024). Forest 4.0: Technologies and digitalization to create the residual biomass supply chain of the future. *Journal of Cleaner Production, 467*(143041), 143041. DOI: 10.1016/j.jclepro.2024.143041

Basu, D., Ghosh, G., & Datta, R. (2022). 6G for Industry 5.0 and Smart CPS: A Journey from Challenging Hindrance to Opportunistic Future. *2022 IEEE Silchar Subsection Conference (SILCON). Silchar, India, 2022*, 1–6. DOI: 10.1109/SILCON55242.2022.10028927

Bazilian, M., Nussbaumer, P., Rogner, H., Brew-Hammond, A., Foster, V., Pachauri, S., & Howells, M. (2012). Energy access scenarios to 2030 for the power sector in sub-Saharan Africa. *Utilities Policy, 20*(1), 1–16. DOI: 10.1016/j.jup.2011.11.002

Beard, V. A., Harmsen, A. L., & Hinton, C. (2016). The Digital Divide and Its Implications for Rural Areas. *Journal of Rural Studies, 44*, 125–133. DOI: 10.1016/j.jrurstud.2016.01.009

Beck, F., & Martinot, E. (2004). Renewable Energy Policies and Barriers. In C. Cleveland (Ed.), Encyclopedia of Energy (Vol. 5, pp. 365-383). Elsevier.

Beltozar-Clemente, S., Iparraguirre-Villanueva, O., Pucuhuayla-Revatta, F., Sierra-Liñan, F., Zapata-Paulini, J., & Cabanillas-Carbonell, M. (2023). Contributions of the 5G network with respect to decent work and economic growth (Sustainable Development Goal 8): A systematic review of the literature. *Sustainability (Basel), 15*(22), 15776. DOI: 10.3390/su152215776

Bertani, R. (2012). Geothermal power generation in the world 2005–2010 update report. *Geothermics, 41*, 1–29. DOI: 10.1016/j.geothermics.2011.10.001

Bhattacharya, P. (2020). The impact of technology on rural healthcare in India. *Journal of Health Policy, 45*(3), 113–127. DOI: 10.1234/jhp.2020.45.3.113

Bhattacharya, P., & Mehta, S. (2022). Participatory governance and digital inclusion in rural India. *Journal of Governance and Policy Studies, 7*(1), 22–39.

Bhattacharyya, A., Agrawal, S., Rath, H. K., & Pal, A. (2019). Towards democratizing e-health under constrained ICT infrastructure through 'A-REaLiST' solution. *European Conference on Networks and Communications (EuCNC)*. doi:DOI: 10.1109/EuCNC.2019.8802034

Bhattacharyya, S. C. (2006). Energy access problem of the poor in India: Is rural electrification a remedy? *Energy Policy, 34*(18), 3387–3397. DOI: 10.1016/j.enpol.2005.08.026

Bhatt, S., & Kumar, R. (2022). Digital literacy and its impact on the adoption of emerging technologies in rural India. *Journal of Rural Technology, 19*(1), 45–59.

Bianchi, M., Bava, L., Sandrucci, A., Tangorra, F., Tamburini, A., Gislon, G., & Zucali, M. (2022). Diffusion of precision livestock farming technologies in dairy cattle farms. *Animal, 16*(11), 100650. DOI: 10.1016/j.animal.2022.100650 PMID: 36220000

Blanutsa, V. I. (2024). Contours of future directions for geographic study of 6G wireless communications. *Regional Research of Russia, 14*(1), 98–107. DOI: 10.1134/S2079970523600348

Boulogeorgos, A. A., Alexiou, A., Papadopoulou, E., Devoti, F., Manios, M., Passas, N., & Berberidis, K. (2020). Terahertz technologies to deliver optical network quality of experience in wireless systems beyond 5G. *IEEE Communications Magazine*, *58*(6), 69–75. DOI: 10.1109/MCOM.001.2000018

Brovarone, E. V., & Cotella, G. (2020). Improving Rural Accessibility: A multilayer approach. *Sustainability (Basel)*, *12*(7), 2876. DOI: 10.3390/su12072876

Brown, A. (2023). The role of social media marketing in direct-to-consumer agriculture. *Digital Marketing in Agriculture*, *8*(1), 33–40.

Brown, D. L. (1980). Rural electrification: The U.S. experience in international perspective. *Comparative Technology Transfer and Society*, *3*(1), 46–69. DOI: 10.1353/ctt.1980.0005

Byanyuma, M., Kalolo, S., Mrutu, S. I., Nyakyi, C., & Sam, A. (2013, July). Affordable broadband connectivity for rural areas. In *2013 Pan African International Conference on Information Science, Computing and Telecommunications (PACT)* (pp. 62-65). IEEE.

Cabanillas-Carbonell, M., Pérez-Martínez, J., & Zapata-Paulini, J. (2023). Contributions of the 5G network with respect to poverty (SDG1), systematic literature review. *Sustainability (Basel)*, *15*(14), 11301. DOI: 10.3390/su151411301

California Public Utilities Commission. (2023). *California Advanced Services Fund.* Retrieved from https://www.cpuc.ca.gov/industries-and-topics/internet-and-phone/california-advanced-services-fund

Carlos, S. (2023, October 25). José, Cunha., Guillermo, Candela., Santiago, Rodriguez., Xosé, Ramón, Sousa., Frederico, Branco. (2023). 4. The 6G Ecosystem as Support for IoE and Private Networks: Vision, Requirements, and Challenges. *Future Internet*, *15*(11), 348. Advance online publication. DOI: 10.3390/fi15110348

Census of India. (2011). *Population Census 2011.* https://www.censusindia.gov.in/

CEPT. (2023). Spectrum policy and regulation. https://www.cept.org/spectrum/

Chaccour, C., Al-Kadri, M., Saad, W., & Bennis, M. (2022). Can 6G be the ultimate wireless communication network? A contemporary review and discussion on future directions. *IEEE Internet of Things Journal*, *9*(1), 1–24. DOI: 10.1109/JIOT.2021.3074006

Chandra, S., & Dwivedi, S. (2020, February). Simulation analysis of high gain and low loss antenna with metallic ring and spoof SPP transmission line. In 2020 7th International Conference on Signal Processing and Integrated Networks (SPIN) (pp. 328-331). IEEE.

Chandra, G., Kumar, R., & Prasad, K. D. V. (2024). Securing digital authentication with cryptographic innovations in intrinsic verification systems. *Journal of Discrete Mathematical Sciences and Cryptography*, *27*(no. 2-A), 317–328.

Chandra, G., Kumar, R., & Prasad, K. D. V. (2024). Securing Digital Authentication with Cryptographic Innovations in Intrinsic Verification Systems. *Journal of Discrete Mathematical Sciences and Cryptography*, *27*(no. 2-A), 317–328.

Chandra, S., & Dwivedi, S. (2019, March). Comparative analysis of reconfigurable patch antenna array for different liquid crystal substrates. In *2019 URSI Asia-Pacific Radio Science Conference (AP-RASC)* (pp. 1-4). IEEE.

Chandra, S., & Dwivedi, S. (2021, April). Graphene based Radiation Pattern reconfigurable antenna. In *International Conference for convergence in Technology* (pp. 2-4).

Chaoub, A., Giordani, M., Lall, B., Bhatia, V., Kliks, A., Mendes, L., Rabie, K., Saarnisaari, H., Singhal, A., Zhang, N., Dixit, S., & Zorzi, M. (2022, February). 6G for Bridging the Digital Divide: Wireless Connectivity to Remote Areas. *IEEE Wireless Communications*, *29*(1), 160–168. Advance online publication. DOI: 10.1109/MWC.001.2100137

Chataut, R., Nankya, M., & Akl, R. (2024). 6G networks and the AI revolution—Exploring technologies, applications, and emerging challenges. *Sensors (Basel)*, *24*(6), 1888. [REMOVED HYPERLINK FIELD]DOI: 10.3390/s24061888 PMID: 38544151

Chaudhary, P., & Patel, R. (2021). *Bridging the Digital Divide in Higher Education: The Role of 5G Networks*. Digital Education Review, 10(4), 56-72. https://digitaledu.org

Chaudhary, P., Gupta, R., & Singh, M. (2022). Sustainable Resource Management in Higher Education: Leveraging IoT and 5G Networks. *Journal of Environmental Innovation*, *11*(2), 78–89. https://envinn.org

Chaudhary, R. (2022). Leveraging IoT and 6G in Indian agriculture. *Indian Agricultural Journal*, *29*(1), 72–90.

Chaudhary, S. (2020). Challenges of rural connectivity in India. *Journal of Rural Studies*, *34*(2), 45–60. DOI: 10.1080/10345672.2020.137628

Chauhan, D. (n.d.). *Roadmap from 4G to 6G in smart cities impact and challenges*. doi:DOI: 10.5815/ijwmt.202.0.0

Chauhan, R. (2023). Bridging India's digital divide: The potential of 5G and beyond in rural areas. *Telecom News Journal*. Retrieved from https://www.telecomnewsjournal.in/bridging-india-digital-divide

Chauhan, D., Mewada, H., Gondalia, V., Almalki, F. A., Patel, S., Modi, H., Kavaiya, S., Trivedi, Y., & Mujlid, H. M. (2024). Balancing Technological Innovation and Environmental Sustainability: A Lifecycle Analysis of 6G Wireless Communication Technology. *Sustainability (Basel)*, *16*(15), 6533. DOI: 10.3390/su16156533

Chaurey, A., Ranganathan, M., & Mohanty, P. (2004). Electricity access for geographically disadvantaged rural communities—Technology and policy insights. *Energy Policy*, *32*(15), 1693–1705. DOI: 10.1016/S0301-4215(03)00160-5

Chen, P. (2023). The impact of LED lighting on plant growth in vertical farming. *Journal of Sustainable Agriculture*, *14*(3), 201–209.

Chen, S., Liu, Y. J., Wang, W., Poor, H. V., & Cui, S. (2018). Wireless Communications in the Era of Big Data [Sept.]. *IEEE Communications Magazine*, *56*(9), 14–20.

Chen, S., Sun, S., & Kang, S. (2020). System integration of terrestrial mobile communication and satellite communication—The trends, challenges and key technologies in B5G and 6G. *China Communications*, *17*(12), 156–171. DOI: 10.23919/JCC.2020.12.011

Chen, S., Wang, H., Xu, X., & Zhou, X. (2020). Artificial intelligence for 6G: Challenges and opportunities. *IEEE Communications Surveys and Tutorials*, *23*(1), 40–57. DOI: 10.1109/COMST.2020.3031481

Chen, X., Chen, Z., Xu, W., & Chen, L. (2021). Vision, requirements, and network architecture of 6G mobile network beyond 2030. *IEEE Network*, *35*(4), 203–211. DOI: 10.1109/MNET.011.2000557

Chen, X., & Liu, H. (2024). Workforce Challenges in Rural 6G Deployment. *Journal of Wireless Communications*, *19*(2), 98–107. DOI: 10.1007/s11277-023-09123-1

Chen, X., Liu, Y., & Wang, J. (2024). Security and Privacy Challenges in 6G Networks. *IEEE Communications Magazine*, *62*(1), 36–42. DOI: 10.1109/MCOM.2024.1234567

Chiang, M., & Zhang, T. (2016). Fog and IoT: An overview of research opportunities [Dec.]. *IEEE Internet of Things Journal*, *3*(6), 854–864. DOI: 10.1109/JIOT.2016.2584538

Choudhary, A., & Desai, N. (2021). E-governance in rural India: Role of next-generation networks. *Indian Journal of Digital Policy*, *14*(3), 76–90.

Choudhury, T. (2023). *IoT-Enabled Campuses: Redefining Infrastructure in Higher Education*. Smart Education Review, 18(3), 233-245. https://smartedureview.com

Choudhury, P., & Banerjee, A. (2021). Economic challenges in rural B5G deployment. *Indian Journal of Telecommunication Economics*, *5*(3), 40–53.

Choudhury, R. (2022). Infrastructure Deficit in Indian Higher Education Institutions: An Analysis. *The Journal of Educational Research*, *31*(4), 122–137. https://joredu.com

Chowdhury, A., & Imran, M. (2020). Telecommunication in rural areas: Challenges and opportunities for developing countries. *Journal of Digital Communications*, *12*(1), 75–92.

Cisco. (2020). Cisco Annual Internet Report (2018–2023) *White Paper*. https://www.cisco.com/c/en/us/solutions/collateral/executive-perspectives/annual-internet-report/white-paper-c11-741490.html

Cisco. (2022). *Revolutionizing Education with VR and AR: How 5G Enhances Interactive Learning Experiences*. Cisco Insights. https://www.cisco.com

Clark, E. (2023). Technological literacy and rural farming communities: Bridging the gap. *Journal of Rural Studies*, *11*(5), 72–80.

Clifton, J., Díaz-Fuentes, D., & Warner, M. E. (2020). The loss of public values when public shareholders go private: The neglect of telecommunications infrastructure. *Public Management Review*, *22*(5), 724–746. DOI: 10.1080/14719037.2019.1604792

CRY. (2023). Education In Rural India: Challenges, Opportunities, Initiatives. Retrieved from https://www.cry.org/blog/education-in-rural-india/

Dang, S., Amin, O., Shihada, B., & Alouini, M. S. (2020). What should 6G be? *Nature Electronics*, *3*(1), 20–29. DOI: 10.1038/s41928-019-0355-6

Dao, N.-N., Tu, N. H., Thanh, T. T., Bao, V. N. Q., Na, W., & Cho, S. (2023). Neglected infrastructures for 6G—Underwater communications: How mature are they? *Journal of Network and Computer Applications*, *213*(103595), 103595. DOI: 10.1016/j.jnca.2023.103595

Das, A. K., Mohan, P. M., Panda, S., & Mishra, A. K. (2021). IoT-enabled smart city solutions for traffic management and urban development [Apr.]. *IEEE Internet of Things Journal*, *8*(7), 5744–5756.

Das, L., & Sharma, V. (2022). Women's health and telemedicine: Challenges and opportunities in rural India. *Healthcare Access Journal*, *18*(1), 121–136.

Das, M. (2022). *Panchayati Raj Institutions in India*. Galore International Journal of Applied Sciences and Humanities., DOI: 10.52403/gijash.20220402

Das, P., Ray, S., Sadhukhan, D., & Govil, C. G. (2024). 6G Communication Technology for Industry 5.0. In Ghosh, U., Mhlanga, F., & Rawat, D. B. (Eds.), *Secure and Smart Cyber-Physical Systems*., DOI: 10.1201/9781003376712-5

Das, S. (2021). Cost considerations in advanced network deployment for rural regions. *Telecom Studies Quarterly*, *14*(2), 85–92.

Das, S. L., Prakash, O., Naidu, V. J., Das, S., Chowdhury, S., & Somassoundaram, T. *6G applications and use cases. In 6G Communication Network*. CRC Press.

Dass, P., Ujjwal, S., Novotný, J., Zolotavkin, Y., Laaroussi, Z., & Köpsell, S. (2024). Addressing privacy concerns in joint communication and sensing for 6G networks: Challenges and prospects. arXiv:2405.01742

Davis, F. D. (1989). Perceived usefulness, perceived ease of use, and user acceptance of information technology. *Management Information Systems Quarterly*, *13*(3), 319–340. DOI: 10.2307/249008

Davis, M. (2023). Enhancing farmer-consumer interactions through technology in farmers' markets. *Agricultural Marketing Review*, *7*(2), 114–120.

De Alwis, C., Kumar, P., Pham, Q. V., Dev, K., Kalla, A., Liyanage, M., & Hwang, W. J. (2022). Towards 6G: Key technological directions. *ICT Express*. Advance online publication. DOI: 10.1016/j.icte.2022.10.005

Deepa, R., Sivasamy, A., Selvam, S. (2024). Intelligent networking unleashed: exploring AI and ML for enhanced 6G wireless communication systems. doi:DOI: 10.58532/V3BINC3P8CH3

Demirbas, A. (2009). Progress and recent trends in biofuels. *Progress in Energy and Combustion Science*, *33*(1), 1–18. DOI: 10.1016/j.pecs.2006.06.001

Department of Telecommunications. (2023). *Digital India and BharatNet: Transforming rural connectivity*. Government of India.

Devapitchai, J. J., Dheenadhayalan, K., Rajeshkumar, L., Meena, M. R., Soundarya, M., Sowmiya, M. S., & Manickam, T. (2024). Applications of Artificial Intelligence Techniques in Modern Banking Sectors. In *Blockchain Applications for Smart Contract Technologies* (pp. 279–298). IGI Global. DOI: 10.4018/979-8-3693-1511-8.ch014

Devapitchai, J. J., Krishnapriya, S. V., Karuppiah, S. P., & Saranya, S. (2024). Using AI-Driven Decision-Making Tools in Corporate Investment Planning. In *Generative AI for Transformational Management* (pp. 137–160). IGI Global. DOI: 10.4018/979-8-3693-5578-7.ch006

Digital Empowerment Foundation. (2023). *Bridging the Digital Divide in India's Education System.* https://defindia.org

Digital Empowerment Foundation. (2023). *The Digital Divide in India and How 5G Can Bridge the Gap in Education.* DEF Publications. https://www.defindia.org

Dilger, A. C., Schroeder, A. L., & Moseley, W. M. (2016). Barriers to global implementation of current and development of new performance-enhancing technologies in meat production. *Animal Frontiers : the Review Magazine of Animal Agriculture*, *6*(4), 50–55. DOI: 10.2527/af.2016-0044

Dively, M. (2003). Social capital in rural communities.

Dixit, S., Bhatia, V., Khanganba, S. P., & Agrawal, A. (2022). *Key considerations to achieve 5G and B5G connectivity in rural areas. In 6G: Sustainable Development for Rural and Remote Communities.* Springer Nature Singapore.

Dogra, A., Jha, R. K., & Jain, S. (2020). A survey on beyond 5G network with the advent of 6G: Architecture and emerging technologies. *IEEE Access : Practical Innovations, Open Solutions*, *9*, 67512–67547. DOI: 10.1109/ACCESS.2020.3031234

Drishtiias. (2023, August 9). *State of Elementary Education in Rural India2023.* Retrieved from Drishtiias.com: https://www.drishtiias.com/daily-updates/daily-news-analysis/state-of-elementary-education-in-rural-india-2023

Dwivedi, S. (2023, August). Partial Ground-Inserted H-shaped Antenna for Beyond 5G Applications. In International Conference on ICT for Sustainable Development (pp. 313-321). Singapore: Springer Nature Singapore.

Dwivedi, S. (2023, June). Design Demonstration of Frequency Reconfiguration Antenna using Varactor diode with SiO 2. In 2023 international conference on computer, electronics & electrical engineering & their applications (IC2E3) (pp. 1-5). IEEE.

Dwivedi, S. (2023, October). Elliptical Split Ring Structure Reconfigurable Antenna for Advance Communication. In International Conference on Microelectronics, Electromagnetics and Telecommunication (pp. 11-22). Singapore: Springer Nature Singapore.

Dwivedi, S. (2024, March). Design of triangular MIMO antenna for 5G and beyond. In 2024 11th International Conference on Signal Processing and Integrated Networks (SPIN) (pp. 40-44). IEEE.

Dwivedi, S. (2022). Design and Analysis of Metamaterial Waveguide Antenna for broadband applications. In *ICT Analysis and Applications: Proceedings of ICT4SD 2022* (pp. 417–423). Springer Nature Singapore.

Dwivedi, S. (2022). *ReconfigurableArray antenna*. Intech Open., DOI: 10.5772/intechopen.106343

Dwivedi, S. (2023). *Metamaterials-Based Antenna for 5G and Beyond*. IGI Global. DOI: 10.4018/978-1-6684-8287-2.ch001

Dwivedi, S. (2023). Metamaterials-based antenna for 5G and beyond. In *Metamaterial Technology and Intelligent Metasurfaces for Wireless Communication Systems* (pp. 1–35). IGI Global.

Dwivedi, S. (2024). *Antenna in Sixth Generation: Design*. Implementation, Specifications and Applications, Intech Open.

Dwivedi, S. (2024). *Metamaterial Antennas for Wireless Communication Systems*. IGI Global publications., DOI: 10.4018/979-8-3693-2599-5.ch002

Dwivedi, S. "Proposed Design for Beyond 5G Antenna for Upgraded Applications with review," , 2023 *14th International Conference on Computing Communication and Networking Technologies (ICCCNT)*. DOI: 10.1109/ICCCNT56998.2023.10307737

Dwivedi, Y. K., Ismagilova, E., Hughes, D. L., Carlson, J., Filieri, R., Jacobson, J., & Kumar, V. (2021). Setting the future of digital and social media marketing research: Perspectives and research propositions. *International Journal of Information Management*, *59*, 102168. Advance online publication. DOI: 10.1016/j.ijinfomgt.2020.102168

Edo, J. (2023a, September 26). The Interplay Between Culture and Technology: a Complex Relationship. *Medium*. https://medium.com/@JacobsEdo/the-interplay-between-culture-and-technology-a-complex-relationship-e46f9e31fd79

Edo, J. (2023b, September 24). The influence of culture on technology adoption and design. *Medium*. https://medium.com/@JacobsEdo/the-influence-of-culture-on-technology-adoption-and-design-e8ea61fbdac1

Eisenberg, M., & O'Brien, L. (2021). Public-private partnerships in telecommunications: A pathway for rural connectivity. *International Journal of Telecommunications Policy*, *34*(4), 207–219.

Ekeland, A. G., Bowes, A., & Flottorp, S. (2010). Effectiveness of Telemedicine: A Systematic Review of Reviews. *International Journal of Medical Informatics*, *79*(11), 737–747. DOI: 10.1016/j.ijmedinf.2010.08.006 PMID: 20884286

Elamassie, M., & Uysal, M. (2023). Free space optical communication: An enabling backhaul technology for 6G non-terrestrial networks. *Photonics*, *10*(11), 1210. DOI: 10.3390/photonics10111210

El-Badawy, H. M., Ahmed, H. A. S., Zainud-Deen, S. H., & Malhat, H. A. E. A. (2023, May). B5G/6G Network Planning For Study Case In Knowledge City Area As Model For Smart Cities. In 2023 40th National Radio Science Conference (NRSC) (Vol. 1, pp. 191-200). IEEE.

El-Sayed, H., El-Ghany, M. A., & Shahin, S. (2022). Applications of 5G and 6G technologies in rural areas: A review. *IEEE Communications Magazine*, *60*(4), 55–61. DOI: 10.1109/MCOM.2022.0000112

Ericsson. (2022). *5G in Higher Education: How IoT is Transforming Campus Resource Management*. Ericsson Reports. https://www.ericsson.com

Ericsson. (2023). *5G: The next leap for mobile communication*. Ericsson Whitepaper.

Ericsson. (2023). *6G Connectivity: Enabling a Hyper-Connected Education Ecosystem*. Ericsson White Papers. https://www.ericsson.com

Erika. (2024, March 13). *Blockchain and UI/UX design: Ensuring security and user trust*. The Data Scientist. https://thedatascientist.com/blockchain-and-ui-ux-design-ensuring-security-and-user-trust/

European Commission. (2021). *Horizon 2020: Smart education initiatives for rural areas*. https://ec.europa.eu/horizon2020

European Commission. (2022). Infrastructure sharing in telecoms. https://ec.europa.eu/digital-strategy/our-policies/infrastructure-sharing_en

Evans, R. (2023). Market competition in direct-to-consumer agricultural sales. *Journal of Market Economics*, *9*(4), 311–319.

Ezeigweneme, C. A., Umoh, A. A., Ilojianya, V. I., & Adegbite, A. O. (2024). Review of telecommunication regulation and policy: Comparative analysis USA and Africa. *Computer Science & IT Research Journal*, *5*(1), 81–99.

Facts and Figures 2023 - Internet use. (2023, October 10). https://www.itu.int/itu-d/reports/statistics/2023/10/10/ff23-internet-use/

FCC. (2023). Spectrum and Infrastructure for 6G Development. Retrieved from FCC.gov

Federal Communications Commission (FCC). (2023). Spectrum management. https://www.fcc.gov/spectrum-management

Federal Communications Commission. (2022). *Local partnerships and broadband expansion*. Retrieved from https://www.fcc.gov/local-partnerships-broadband

Feng, W., Li, Y., Jin, D., Su, L., & Chen, S. (2016). Millimetre-Wave backhaul for 5G networks: Challenges and solutions. *Sensors (Basel)*, *16*(6), 892. DOI: 10.3390/s16060892 PMID: 27322265

Fernandes, D., Lynch Jr, J. G., & Netemeyer, R. G. (2014). Financial literacy, financial education, and downstream financial behaviors. *Management Science, 60*(8), 1861-1883.3. Singh, Neena (2007) "Bridging the Digital Divide in India: Challenges and Opportunities", World Libraries, Vol. 17, No. 1.

Ferrag, M. A., Shu, L., Yang, X., Derhab, A., Maglaras, L., & Janicke, H. (2019). A Systematic Literature Review on Privacy-Preserving Blockchain-Based Distributed Machine Learning. *IEEE Access : Practical Innovations, Open Solutions*, *7*, 87665–87682.

Ferran, C. (2019). *Angelos, Mimidis., Nestor, Bonjorn., José, Soler*. Policy Framework Prototype for ONOS., DOI: 10.1109/NETSOFT.2019.8806691

Fischer, P. (2017). Bridging rural constraints: Do rural entrepreneurs exploit online social capital differently than urbanized entrepreneurs?

Fontana, R. J. (2002). A brief history of UWB communications. Multispectral. com, Multispectral Solutions, Inc., www. multispectral. com/history. html

Fontana, R., Ameti, A., Richley, E., Beard, L., & Guy, D. (2002, May). Recent advances in ultra wideband communications systems. In 2002 IEEE Conference on Ultra Wideband Systems and Technologies (IEEE Cat. No. 02EX580) (pp. 129-133). IEEE.

Foster, G. (2023). Incentives for urban agriculture: Promoting vertical farming in metropolitan areas. *Urban Development Journal, 17*(3), 196–204.

Foukas, X., Patounas, G., Elmokashfi, A., & Marina, M. K. (2017). Network slicing in 5G: Survey and challenges. *IEEE Communications Magazine, 55*(5), 94–100. DOI: 10.1109/MCOM.2017.1600951

Freeman, R. E., Harrison, J. S., Wicks, A. C., Parmar, B. L., & de Colle, S. (2010). *Stakeholder theory and strategic management.* Cambridge University Press., DOI: 10.1017/CBO9780511815768.005

Friedner, S., & House, R. (2018). 5G infrastructure requirements in the UK. LS telcom UK. Retrieved from https://www.computer.org/digital-library/magazines/it/rural-remote-connectivity-challenges

Fuentes, S., Viejo, C. G., Tongson, E., & Dunshea, F. R. (2022). The livestock farming digital transformation: Implementation of new and emerging technologies using artificial intelligence. *Animal Health Research Reviews, 23*(1), 59–71. DOI: 10.1017/S1466252321000177 PMID: 35676797

Galani, A., Koutsouris, N., Tsagkaris, K., Demestichas, P., Fuentes, B., Vazquez, C. G., & Nguengang, G. (2012, December). A policy based framework for governing Future networks. In 2012 [IEEE.]. *IEEE Globecom Workshops*, •••, 802–806.

Gao, S., & Li, X. (2024). Environmental Challenges in Deploying B5G and 6G Networks. *Journal of Environmental Management, 250*, 109–123. DOI: 10.1016/j.jenvman.2023.109123

Garcia, T. (2023). Infrastructure development for rural market access. *Journal of Development Economics, 10*(4), 219–227.

Gaska, T., Watkin, C., & Chen, Y. (2015). Integrated modular avionics-past, present, and future. *IEEE Aerospace and Electronic Systems Magazine, 30*(9), 12–23. https://ieeexplore.ieee.org/abstract/document/7321122/. DOI: 10.1109/MAES.2015.150014

Gaspard, H., & Baker, P. M. A. (2022). Innovation and Digital Connectivity : Comparative Policy Approaches for Connecting Rural Communities in the United States and Canada. *Choices, 37*(3), 1–9.

Gatherer, A., Sengupta, C., & Sen, S. P. (2023). *6G radio access implementation: Challenges and technologies.* Signals and Communication Technology., DOI: 10.1007/978-3-031-37920-8_19

Gautam, R. S., Rastogi, S., Rawal, A., Bhimavarapu, V. M., Kanoujiya, J., & Rastogi, S. (2022). Financial Technology and Its Impact on Digital Literacy in India: Using Poverty as a Moderating Variable. *Journal of Risk and Financial Management, 15*(7), 311. DOI: 10.3390/jrfm15070311

Geetharamani, G., & Aathmanesan, T. (2019). Metamaterial inspired THz antenna for breast cancer detection. *SN Applied Sciences, 1*(6), 595. DOI: 10.1007/s42452-019-0601-6

Ghassemian, M., Vukobratovic, D., Papadopoulos, C., An, X., Chatzimisios, P., Aijaz, A., Li, P., Meseguer Valenzuela, A., Shikh-Bahaei, M., Mihovska, A., Bartzoudis, N., Walker, R., Prasad, R. V., & Saghezchi, F. B. (2023). Standardisation landscape for 6G robotic services. *IEEE Conference on Standards for Communications and Networking (CSCN).* doi:DOI: 10.1109/CSCN60443.2023.10453144

Ghosh, A. (2022). Digital marketing strategies for rural India: A path to rural transformation. *Journal of Rural Development, 41*(2), 75–90.

Ghosh, A., Ratasuk, R., Mondal, B., Mangalvedhe, N., & Thomas, T. (2019). LTE-advanced: Next-generation wireless broadband technology. *IEEE Wireless Communications, 17*(3), 10–22. DOI: 10.1109/MWC.2010.5490974

Ghosh, M., & Gupta, P. (2021). Overcoming technical barriers in rural network expansion. *Digital India Journal of Connectivity, 9*(1), 12–23.

Gilani, S. M. M., Sultan, M. T., Shuai, Z., & Kabir, A. (2020). Long range backhaul microwave connectivity in wireless sensor networks via a new antenna designed for ISM 2.4 GHz band. *International Journal of Antennas and Propagation, 2020*, 1–14. DOI: 10.1155/2020/1789612

Giordani, M., Polese, M., Roy, A., Castor, D., & Zorzi, M. (2019). A Tutorial on Beam Management for 3GPP NR at mmWave Frequencies [Firstquarter.]. *IEEE Communications Surveys and Tutorials, 21*(1), 173–196. DOI: 10.1109/COMST.2018.2869411

Giovanetti, V., Decandia, M., Molle, G., Acciaro, M., Mameli, M., Cabiddu, A., Cossu, R., Serra, M., Manca, C., Rassu, S., & Dimauro, C. (2017). Automatic classification system for grazing, ruminating and resting behaviour of dairy sheep using a tri-axial accelerometer. *Livestock Science, 196*, 42–48. DOI: 10.1016/j.livsci.2016.12.011

Gismalla, , Azmi, A. I., Salim, M. R. B., Abdullah, M. F. L., Iqbal, F., Mabrouk, W. A., Othman, M. B., Ashyap, A. Y. I., & Supa'at, A. S. M. (2022). Survey on device to device (D2D) communication for 5GB/6G networks: Concept, applications, challenges, and future directions. *IEEE Access : Practical Innovations, Open Solutions, 10*, 30792–30821. DOI: 10.1109/ACCESS.2022.3160215

Global Wind Energy Council (GWEC). (2022). Global Wind Report 2022. https://gwec.net/global-wind-report-2022/

Global Wind Energy Council. (2021). Global wind report 2021. https://gwec.net/global-wind-report-2021/

Global Wind Energy Council. (2022). *Global Wind Report 2022.* Retrieved from https://gwec.net/global-wind-report-2022/

Gonzalez Viejo, C., Fuentes, S., Torrico, D., & Dunshea, F. (2018). Non-contact heart rate and blood pressure estimations from video analysis and machine learning modelling applied to food sensory responses: A case study for chocolate. *Sensors (Basel), 18*(6), 1802. DOI: 10.3390/s18061802 PMID: 29865289

Government of Canada. Statistics Canada. (2022, October 18). *Average cost to prevent or detect cyber security incidents by industry and size of enterprise.* https://www150.statcan.gc.ca/t1/tbl1/en/tv.action?pid=2210013101

Government of India. (2021). *DIKSHA: Digital infrastructure for teachers and students.* https://diksha.gov.in

Government of India. (2021). *Startup India initiative: Supporting entrepreneurship in rural India.* https://startupindia.gov.in

Greener, S. (2022). Digging for acceptance theory. *Interactive Learning Environments, 30*(4), 587–588. Advance online publication. DOI: 10.1080/10494820.2022.2062170

Greenstein, S. (2020). The economic impact of broadband in rural areas. *Journal of Rural and Community Development, 15*(3), 1–15.

Gryech, I., Vinogradov, E., Saboor, A., Bithas, P. S., Mathiopoulos, P. T., & Pollin, S. (2024). A systematic literature review on the role of UAV-enabled communications in advancing the UN's sustainable development goals. *Frontiers in Communications and Networks, 5*, 1286073. DOI: 10.3389/frcmn.2024.1286073

GSMA. (2022). *Network Slicing in 5G: Customizing Network Environments for Education.* GSMA White Paper. https://www.gsma.com

Gunkel, D. J. (2003). Second Thoughts: Toward a Critique of the Digital Divide. *New Media & Society, 5*(3), 499–522. DOI: 10.1177/146144480354003

Guo, H., Liu, J., Tang, J., & Zhang, J. (2020). Edge AI and 5G: The First Wave of Driving Applications [Feb.]. *IEEE Journal on Selected Areas in Communications, 38*(2), 246–252.

Guo, H., Liu, X., & Li, Y. (2022). Artificial intelligence and machine learning in 6G wireless networks: Opportunities and challenges. *IEEE Access : Practical Innovations, Open Solutions, 12*, 25034–25048. DOI: 10.1109/ACCESS.2022.1230567

Gupta, A., Kumar, S., & Patel, R. (2022). *Adaptive Learning Technologies in Higher Education: A 5G Perspective.* Educational Technology Quarterly, 15(4), 56-72. https://edutechq.org

Gupta, A., Singh, V., & Kumar, S. (2022). *AR/VR in Education: Transforming Learning in the Digital Era.* Virtual Learning Quarterly, 15(1), 45-59. https://vlq.org

Gupta, U., & Dwivedi, S. (2021, June). Frequency reconfigurable antenna using metamaterial split ring resonators for smart applications. In 2021 international conference on electrical, communication, and computer engineering (ICECCE) (pp. 1-5). IEEE.

Gupta, A., & Jha, R. K. (2021). A survey of 5G network: Architecture and emerging technologies. *IEEE Access : Practical Innovations, Open Solutions, 3*, 1206–1232. DOI: 10.1109/ACCESS.2021.3059431

Gupta, A., Kumar, S., & Sharma, V. (2020). The future of telecommunications: A look at Beyond 5G (B5G) and 6G technologies. *Telecom Research Journal, 12*(3), 245–259.

Gupta, A., Kumar, S., & Sharma, V. (2022). Digital divide and its impact on rural India: The role of mobile networks. *Telecom Policy Journal, 17*(3), 89–110.

Gupta, A., & Mehta, V. (2023). Ethical Implications of AI in Education: A 5G Perspective. *Journal of Educational Ethics, 15*(2), 34–47. https://educationalethics.com

Gupta, A., Porambage, P., & Rajatheva, N. (2020). 6G-enabled Internet of Things (IoT): Technologies, challenges, and applications. *IEEE Access : Practical Innovations, Open Solutions*, *8*, 124718–124745. DOI: 10.1109/ACCESS.2020.3007278

Gupta, A., & Singh, R. (2021). Rural digital inclusion and the promise of B5G and 6G. *Digital Divide Journal*, *18*(4), 233–245.

Gupta, R. R., Kumar, A., & Gadhavi, D. J. (2020). Implementation of B5G and 6G networks for enhanced rural connectivity [Sept.]. *IEEE Communications Magazine*, *58*(9), 42–47.

Gustavsson, U., Frenger, P., Fager, C., Eriksson, T., Zirath, H., Dielacher, F., Studer, C., Parssinen, A., Correia, R., Matos, J. N., Belo, D., & Carvalho, N. B. (2021). Implementation challenges and opportunities in beyond-5G and 6G communication. *IEEE Journal of Microwaves*, *1*(1), 86–100. Advance online publication. DOI: 10.1109/JMW.2020.3034648

Haapola, J., Leinonen, P., & Jämsä, T. (2021). 6G for smart healthcare in rural areas: Case studies from Finland. *Journal of Telemedicine and Telecare*, *27*(6), 333–341. DOI: 10.1177/1357633X20985634

Halachmi, I., Guarino, M., Bewley, J., & Pastell, M. (2019). Smart animal agriculture: Application of real-time sensors to improve animal well-being and production. *Annual Review of Animal Biosciences*, *7*(1), 403–425. DOI: 10.1146/annurev-animal-020518-114851 PMID: 30485756

Han, J., Li, H., & Lin, X. (2019). Spectrum sharing in 5G and beyond: Recent advances and future challenges [Dec.]. *IEEE Communications Magazine*, *57*(12), 168–174.

Hansen, M. F., Smith, M. L., Smith, L. N., Salter, M. G., Baxter, E. M., Farish, M., & Grieve, B. (2018). Towards on-farm pig face recognition using convolutional neural networks. *Computers in Industry*, *98*, 145–152.

Harris, N. (2023). Implementing rural vertical farming initiatives to promote fresh food access. *Journal of Agricultural Innovation*, *16*(4), 103–111.

Hatem, I.-K., Laroui, M., Alfaqawi, M., Magnouche, A., Moungla, H., & Afifi, H. (2023). 6G-edge support of Internet of Autonomous Vehicles: A survey. *European Transactions on Telecommunications*. Advance online publication. DOI: 10.1002/ett.4918

Heather, H. (2019). 5G Mobile Broadband: Spectrum Challenges for Rural Regions. Social Science Research Network, doi: DOI: 10.2139/SSRN.3427548

Hegge, K., & Agrawal, R. (2023) "Wireless and Solar-Powered Multipurpose Robot for Industrial Automation: A Sustainable Solution," *2023 8th International Conference on Communication and Electronics Systems (ICCES)*, Coimbatore, India, pp. 146-151.

Heo, G., & Doh, I. (2024). Blockchain and differential privacy-based data processing system for data security and privacy in urban computing. *Computer Communications*, *222*, 161–176. Advance online publication. DOI: 10.1016/j.comcom.2024.04.027

Hilbert, M. (2016). The bad news is that the digital access divide is here to stay: Domestically installed bandwidths among 172 countries for 1986–2014. *Telecommunications Policy*, *40*(6), 567–581. DOI: 10.1016/j.telpol.2016.01.006

Hoffmann, T., & Solarte-Vasquez, M. C. (2022). The Estonian e-residency programme and its role beyond the country's digital public sector ecosystem. *CES Derecho*, *13*(2), 184–204.

Hong, S., Zhao, Y., & Wu, J. (2022). Spectrum allocation strategies for B5G and 6G networks in rural areas. *IEEE Communications Magazine*, *60*(5), 123–129.

Horrigan, J. B. (2020). Baltimore's Digital Divide: Gaps in Internet Connectivity and the Impact on Low-Income City Residents. *The Abell Report*, *33*, 4.

Hossain, E., Bhargava, V., & Saad, W. (2020). *6G: The next frontier*. Cambridge University Press.

Hosseini, S. A., Nguyen, D. T., Garcia, N., & Tran, D. C. (2021). A comprehensive survey of 6G networks: Applications, challenges, trends, and opportunities. *IEEE Access : Practical Innovations, Open Solutions*, *9*, 43253–43273. DOI: 10.1109/ACCESS.2021.3072101

Huang, L., Li, S., Zhu, A., Fan, X., Zhang, C., & Wang, H. (2018). Non-contact body measurement for Qinchuan cattle with LiDAR sensor. *Sensors (Basel)*, *18*(9), 3014. DOI: 10.3390/s18093014 PMID: 30205607

Huang, W., Zhu, W., Ma, C., & Guo, Y. (2020). Weber texture local descriptor for identification of group-housed pigs. *Sensors (Basel)*, *20*(16), 4649. DOI: 10.3390/s20164649 PMID: 32824819

Huawei. (2023). *Advancements in 5G and the Path to 6G: Enabling Edge Computing and IoT in Education*. Huawei White Papers. https://www.huawei.com

Hussain, N., Dixit, S. K., & Gupta, A. K. (2023). A Novel Approach of Unsupervised Feature Selection Using Iterative Shrinking and Expansion Algorithm. *Journal of Interdisciplinary Mathematics*, *26*(3), 519–530. DOI: 10.47974/JIM-1678

Hyongsoon, K. (2010). *Eunyoung, Lee*. Policy Analysis on Resolving Digital Divide in Rural Areas.

IEA. (2022). *World Energy Outlook 2022*. International Energy Agency.

IEEE. (2024). *Bridging the Future: From 5G Advanced to 6G Development*. Retrieved from IEEE Testbed.

India Education Report. (2022). *India's Growing Student Population and the Demand for Scalable Solutions in Higher Education*. Indian Ministry of Education. https://www.education.gov.in

India, C. (2023). *Empowering Indian Education through 5G and AR/VR Technology*. Cisco White Papers. https://www.cisco.com

Innovation, Science and Economic Development Canada & Digital Transformation Service Sector. (2022, April 11). *High-Speed Access for all: Canada's connectivity strategy*. https://ised-isde.canada.ca/site/high-speed-internet-canada/en/canadas-connectivity-strategy/high-speed-access-all-canadas-connectivity-strategy

Institute of Electrical and Electronics Engineers. (2023). *IEEE 802.11 and IEEE 802.15 working groups*. Retrieved from https://standards.ieee.org/standard/802_11-2020.html

Intel. (2023). *Predictive Analytics and 5G: Creating Scalable Solutions for Higher Education*. Intel Education Insights. https://www.intel.com

International Energy Agency. (2020). World Energy Outlook 2020. https://www.iea.org/reports/world-energy-outlook-2020

International Energy Agency. (2021). Renewables 2021: Analysis and forecast to 2026. https://www.iea.org/reports/renewables-2021

International Organization for Standardization. (2023). *ISO/IEC JTC 1/SC 6: Telecommunications and information exchange between systems*. Retrieved from https://www.iso.org/committee/45362.html

International Renewable Energy Agency (IRENA). (2020). Renewable capacity statistics 2020. https://www.irena.org/publications/2020/Mar/Renewable-Capacity-Statistics-2020

International Renewable Energy Agency (IRENA). (2021). Renewable Capacity Statistics 2021. https://www.irena.org/publications/2021/Mar/Renewable-Capacity-Statistics-2021

International Renewable Energy Agency. (2021). *Renewable Capacity Statistics 2021*. Retrieved from https://www.irena.org/publications/2021/Mar/Renewable-Capacity-Statistics-2021

International Telecommunication Union (ITU). (2021). Connecting rural communities: ITU global connectivity report. *ITU Publications*. https://www.itu.int

International Telecommunication Union (ITU). (2023). Universal service and access. https://www.itu.int/en/ITU-T/universal-service/Pages/default.aspx

International Telecommunication Union. (2020). *Technologies for Network 2030: New technologies for an evolved network ecosystem*. Retrieved from https://www.itu.int/en/ITU-T/tech/Pages/default.aspx

International Telecommunication Union. (2023, September 12). *Press release*. ITU. https://www.itu.int/en/mediacentre/Pages/PR-2023-09-12-universal-and-meaningful-connectivity-by-2030.aspx

Inzillo, V., Garompolo, D., & Giglio, C. (2024). Enhancing smart city connectivity: A multi-metric CNN-LSTM beamforming-based approach to optimize dynamic source routing in 6G networks for MANETs and VANETs. *Preprints*. DOI: 10.20944/preprints202407.1161.v1

IRENA. (2016). Renewable Energy Benefits: Measuring the Economics. International Renewable Energy Agency. https://www.irena.org/publications/2016/Jan/Renewable-Energy-Benefits-Measuring-the-Economics

Islam, S. R., Kwak, D., Kabir, M. H., Hossain, M., & Kwak, K. S. (2015). The Internet of Things for health care: A comprehensive survey. *IEEE Access : Practical Innovations, Open Solutions*, *3*, 678–708. DOI: 10.1109/ACCESS.2015.2437951

ITU. (2020). Measuring digital development: Facts and figures 2020. International Telecommunication Union. https://www.itu.int/en/ITU-D/Statistics/Documents/facts/FactsFigures2020.pdf

ITU. (2021). *Understanding Latency in 5G Networks*. ITU Reports. https://www.itu.int

ITU. (2022). The next generation of intelligent networks. ITU Insights. https://www.itu.int

ITU. (2024). Innovations in Next-Generation Networks. Retrieved from ITU.int

ITU/UN tech agency. (2024, April 16). *Measuring digital development - Facts and Figures: Focus on Landlocked Developing Countries, April 2024 - ITU*. ITU. https://www.itu.int/hub/publication/d-ind-ict_mdd-2024-2/

Ivan, Armuelles, Voinov., Joaquin, Chung., Rajkumar, Kettimuthu., Borja, Bordel., Ramón, Alcarria., Tomás, Robles. (2022). Towards 6G Networks for Rural Environments: Vision for Improving Digital Inclusion through Open Source Hardware and Software. DOI: 10.1109/CONCAPAN48024.2022.9997739

Jain, A. D. (2023, August 24). *Providing last-mile internet connectivity to rural India*. Retrieved from tndindia.com: https://www.tndindia.com/providing-last-mile-internet-connectivity-to-rural-india/#:~:text=Internet%2Dpenetration%20in%20rural%20India,reliable%20access%20to%20the%20internet

Jain, P. (2023). Challenges in maintaining 5G networks in rural India: A technical perspective. *Indian Journal of Telecommunications*, *34*(3), 101–115.

Jain, P. (2023). The role of AI and machine learning in hyper-local marketing. *Marketing Tech Review*, *24*(2), 88–99.

Jebadurai, D. J. (2013). An overview of problems of rural entrepreneurs in India. *International Journal of Advanced Research in Management and Social Sciences*, *2*(7), 202–208.

Jebadurai, D. J., Sheela, M. V., Rajeshkumar, L., Soundarya, M., Meena, R., Manickam, T., & Manikandan, M. (2024). AI-Driven Decision-Making and Optimization in Modern Agriculture Sectors. In *Using Traditional Design Methods to Enhance AI-Driven Decision Making* (pp. 269–288). IGI Global. DOI: 10.4018/979-8-3693-0639-0.ch012

Jeng, J. J., Chang, H., & Chung, J. Y. (2003, June). A policy framework for business activity management. In EEE International Conference on E-Commerce, 2003. CEC 2003. (pp. 238-245). IEEE.

Jha, A. V., Appasani, B., Khan, M. S., Zeadally, S., & Katib, I. (2024). 6G for intelligent transportation systems: Standards, technologies, and challenges. *Telecommunication Systems*, *86*(2), 1–28. DOI: 10.1007/s11235-024-01126-5

Johnson, R. (2023). Blockchain technology and food supply chain transparency. *Agricultural Technology Review*, *6*(2), 188–195.

Jonek-Kowalska, I., Wolniak, R., Marinina, O., & Ponomarenko, T. (2023). Stakeholder theory. *Journal of Business Research*, *166*, 114104. Advance online publication. DOI: 10.1016/j.jbusres.2023.114104

Joseph, J. A., & Varghese, D. (2023). Role of community in rural tourism: A pilot study conducted at Kumbalgodu Panchayat, Karnataka. *Journal of Management Studies and Business Management*. DOI: 10.46610/JMSBM.2023.v04i02.002

Joshi, M., & Ramesh, K. (2021). Remote health monitoring through next-gen networks. *Healthcare Informatics Research*, *17*(1), 99–113.

Kapoor, S. (2020). Digital Literacy and its Implications for Rural India. *Journal of Digital Inclusion*, *15*(2), 101–115.

Kapoor, S., & Aggarwal, N. (2021). Addressing the digital literacy gap in India: Implications for rural development. *International Journal of Rural Studies*, *27*(2), 112–129.

Kaundinya, D. P., Balachandra, P., & Ravindranath, N. H. (2009). Grid-connected versus stand-alone energy systems for decentralized power—A review of literature. *Renewable & Sustainable Energy Reviews*, *13*(8), 2041–2050.

Kaur, H., & Kamal, T. (2023). Infrastructure Challenges in Rural 5G Deployment. *Telecommunications Policy*, *47*(3), 101–112. DOI: 10.1016/j.telpol.2023.103001

Kaur, P., Garg, R., & Kukreja, V. (2023). Energy-efficiency schemes for base stations in 5G heterogeneous networks: A systematic literature review. *Telecommunication Systems*, *84*(1), 115–151. DOI: 10.1007/s11235-023-01037-x

Khan, L. U., Yaqoob, I., Imran, M., Han, Z., & Hong, C. S. (2020). 6G wireless systems: A comprehensive survey. *IEEE Internet of Things Journal*, *8*(1), 400–418. DOI: 10.1109/JIOT.2020.3010833

Khan, M. A., Sohail, M., & Asghar, A. (2023). Energy-efficient and THz-based 6G technologies for rural deployments. *IEEE Wireless Communications*, *30*(1), 22–28. DOI: 10.1109/MWC.2023.3101011

Khan, N. (2022). Innovative solutions for rural B5G and 6G deployments. *Wireless Technology Journal*, *19*(2), 98–112.

Khan, S. (2023). Pesticide reduction in vertical farming: A controlled environment approach. *Environmental Agriculture Studies*, *14*(3), 298–305.

Khan, T., & Roy, V. (2021). Cybersecurity in B5G and 6G networks: Implications for rural India. *Indian Journal of Cybersecurity Policy*, *8*(4), 99–109.

Kharche, S., & Kharche, J. S. (2023). 6G intelligent healthcare framework: A review on the role of technologies, challenges, and future directions. *Journal of Mobile Multimedia*. doi:DOI: 10.13052/jmm1550-4646.1931

Kharche, S., & Dere, P. (2022). Interoperability Issues and Challenges in 6G Networks. *J.Mobile Multimedia*, *18*(5), 1445–1470. DOI: 10.13052/jmm1550-4646.1856

Kimberling, E. (2023, April 25). Most important change management deliverables in a digital transformation - Third stage consulting. *Third Stage Consulting*. https://www.thirdstage-consulting.com/most-important-change-management-deliverables-in-a-digital-transformation/

Kim, H., Kim, D., & Park, Y. (2021). Dynamic spectrum management for IoT and smart agriculture [Jan.]. *IEEE Internet of Things Journal*, *8*(2), 1020–1033.

Kim, H., Oh, J., & Park, S. (2020). 5G mobile networks: Performance and impact on future generations. *Communications Journal*, *18*(4), 323–345. DOI: 10.1007/s10207-021-00508-5

Kim, J., & Lee, S. (2022). The impact of digital connectivity on rural development: Evidence from developing countries. *Journal of Rural Studies*, *92*, 234–242.

Kim, K., Hong, S., & Lee, H. (2021). Policy strategies for enhancing rural connectivity: Lessons from international cases. *Telecommunications Policy*, *45*(1), 102–115. DOI: 10.1016/j.telpol.2020.102115

Kim, N., Kim, G., Shim, S., Jang, S., Song, J., & Lee, B. (2024). Key Technologies for 6G-Enabled Smart Sustainable City. *Electronics (Basel)*, *13*(2), 268. DOI: 10.3390/electronics13020268

Kohno, R., & Takizawa, K. (2005, September). Overview of research and development activities in NICT UWB consortium. In *2005 IEEE International Conference on Ultra-Wideband* (pp. 735-740). IEEE.

Koratagere Anantha Kumar, S., & Oughton, E. J. (2023). Techno-economic assessment of 5G infrastructure sharing business models in rural areas. *Frontiers of Computer Science*, *5*, 1191853.

KPMG. (2022). *Digital India: A roadmap for bridging the digital divide*. KPMG Report.

KPMG. (2023). *Higher Education in India: Challenges and Opportunities*. KPMG India Reports. https://home.kpmg.com

KPMG. (2024). *6G and the Future of Higher Education: Insights and Trends*. https://home.kpmg.com

KPMG. (2024). *The Future of 6G in Higher Education: Opportunities and Challenges*. https://home.kpmg.com

Kukulska-Hulme, A. (2009). Will mobile learning change language learning? *ReCALL*, *21*(2), 157–165. DOI: 10.1017/S0958344009000223

Kulkarni, A., Goudar, R. H., Rathod, V., Dinesh, P. M., & Hukkeri, G. S. (2024). *New directions for adapting intelligent communication and standardization towards 6G*. ICST Transactions on Scalable Information Systems., DOI: 10.4108/eetsis.5126

Kumar, R., Sharma, P., & Verma, A. (2021). *The Role of IoT in Optimizing Resource Utilization in Universities*. Smart Campus Review, 18(3), 123-137. https://smartcampus.org

Kumar, S. K. A., Ihita, G. V., Chaudhari, S., & Arumugam, P. (2022, January). A survey on rural internet connectivity in India. In 2022 14th International Conference on COMmunication Systems & NETworkS (COMSNETS) (pp. 911-916). IEEE.

Kumar, A. (2021). Marketing in rural India: Overcoming challenges with innovative solutions. *Journal of Marketing*, *57*(3), 89–101.

Kumar, A., Gaur, N., & Nanthaamornphong, A. (2024). Improving the latency for 5G/B5G based smart healthcare connectivity in rural area. *Scientific Reports*, *14*(1), 6976. DOI: 10.1038/s41598-024-57641-7 PMID: 38521842

Kumar, M., Patel, J., & Nguyen, S. (2021). E-commerce platforms in agriculture: Expanding the reach of farmers. *Digital Agriculture Journal*, *12*(1), 45–53.

Kumar, N. (2023, November). Comparative analysis of Different Modulation Schemes for LMLC in 5G Scenario. In *2023 International Conference on Computing, Communication, and Intelligent Systems (ICCCIS)* (pp. 1057-1061). IEEE.

Kumar, P., Chauhan, D. S., Srivastava, P. K., & Ahmad, F. (2021). Sustainable development in 6G: Challenges and future directions. *Wireless Personal Communications*, *120*(4), 2573–2591. DOI: 10.1007/s11277-021-08582-3

Kumar, R. (2022). Faculty Shortages and Their Impact on Indian Higher Education Quality. *Indian Journal of Academic Research*, *19*(3), 233–249. https://ijar.in

Kumar, R., & Mishra, A. (2021). The Impact of Remote Education on Higher Education in India: Insights from the Pandemic Era. *Indian Journal of Educational Research*, *12*(2), 98–112. https://ijer.in

Kumar, R., Mitra, P., & Saha, A. (2022). The impact of 5G technology on rural development: Opportunities and challenges. *Journal of Telecommunications and Digital Innovation*, *12*(2), 89–103.

Kumar, R., & Singh, M. (2022). Economic Barriers to 5G Deployment in Higher Education. *Indian Journal of Technology Policy*, *19*(1), 90–102. https://indiantechpolicy.org

Kumar, S. (2023). Beyond 5G and 6G: Emerging technologies and their implications. *Journal of Wireless Communications*, *18*(2), 112–126. DOI: 10.1234/jwc.2023.01802

Kumar, S., & Sharma, A. (2021). Data-Driven Resource Management in Indian Universities: Challenges and Prospects. *The Journal of Higher Education*, *14*(1), 78–92. https://higheredu.org

Kumar, S., Sharma, R., & Gupta, A. (2023). Telemedicine and e-learning in rural India: A case study of 5G's impact on healthcare and education in Uttar Pradesh. *Journal of Rural Innovation*, *15*(2), 45–59.

Lee, C., & Wang, H. (2023). Case studies in rural broadband expansion: Lessons learned from B5G and 6G pilot projects. *Rural Technology Journal*, *21*(4), 210–225. DOI: 10.6789/rtj.2023.02104

Lee, K. (2023). Training farmers in digital marketing and e-commerce tools. *Rural Development Studies*, *13*(4), 79–88.

Lemelin-Bellerose, S. (2023, September 7). *Deployment of broadband internet in rural and remote areas of Canada*. https://lop.parl.ca/sites/PublicWebsite/default/en_CA/ResearchPublications/202311E

Levin, J., & Kim, D. (2015). U.S. Patent No. 8,935,365. Washington, DC: U.S. Patent and Trademark Office.

Liao, X., You, W., Lv, J., Wang, M., Zhao, Y., & He, G. (2023). 6G Polymorphic Smart Network Architecture and Key Technologies. *2023 5th International Conference on Frontiers Technology of Information and Computer (ICFTIC)*, Qiangdao, China. doi:DOI: 10.1109/ICFTIC59930.2023.10455881

Lien, S., Chen, K., & Ding, Y. (2019). 5G new radio: Unveiling the essentials of the next generation mobile communication technology [Feb.]. *IEEE Communications Magazine*, *57*(2), 24–30.

Li, H. (2023). The role of automation in vertical farming: A case study on labor cost reduction. *Journal of Automated Systems in Agriculture*, *7*(3), 232–239.

Liu, J., Wang, T., & Zhang, Z. (2023). Leveraging IoT and 6G for sustainable agriculture. *IEEE Sustainable Agriculture and Systems Journal*, *25*(1), 23–33. DOI: 10.1109/SASJ.2023.456274

Liu, R., Huang, J., & Wang, Y. (2020). Advanced spectrum management for 5G and beyond: A comprehensive survey [Fourthquarter.]. *IEEE Communications Surveys and Tutorials*, *22*(4), 2151–2176.

Liu, R., Lin, H., Lee, H., Chaves, F., Lim, H., & Sköld, J. (2023). Beginning of the journey toward 6G: Vision and framework. *IEEE Communications Magazine*, *61*(10), 8–9. Advance online publication. DOI: 10.1109/MCOM.2023.10298069

Liu, Y., Peng, S., Zhang, M., Shi, S., & Fu, J. (2024). Towards secure and efficient integration of blockchain and 6G networks. *PLoS One*, *19*(4), e0302052. DOI: 10.1371/journal.pone.0302052

Liu, Y., Razman, M. R., Zakaria, S. Z. S., Ern, L. K., Hussain, A., & Chamola, V. (2024). Utilizing ubiquitous learning to foster sustainable development in rural areas: Insights from 6G technology. *Computers in Human Behavior*, *161*, 108418. DOI: 10.1016/j.chb.2024.108418

Liu, Z., Liu, J., & Wang, X. (2023). Terahertz Communication in 6G: The Future of High-Speed Data Transfer. *Journal of Telecommunications Research*, *18*(4), 112–126. https://www.jtrjournal.org

Li, X., Zhang, Q., & Wang, T. (2021). Ethical implications of 6G technology: A systematic review. *Journal of Information.Communication and Ethics in Society*, *19*(4), 487–502. DOI: 10.1108/JICES-04-2021-0034

Lu, Y., & Zheng, X. (2020). 6G: A survey on technologies, scenarios, challenges, and the related issues. *Journal of Industrial Information Integration*, *19*, 100158.

Lu, Y., & Zheng, X. (2020). 6G: A survey on technologies, scenarios, challenges, and the related issues. *Journal of Industrial Information Integration*, *19*, 100158. Salameh, A. I., & El Tarhuni, M. (2022). From 5G to 6G—challenges, technologies, and applications. *Future Internet*, *14*(4), 117.

Lwakatare, L. E., Raj, A., Crnkovic, I., Bosch, J., & Olsson, H. H. (2020). Large-scale machine learning systems in real-world industrial settings: A review of challenges and solutions. *Information and Software Technology*, *127*, 106368. https://www.sciencedirect.com/science/article/pii/S0950584920301373. DOI: 10.1016/j.infsof.2020.106368

Lyons, A. C., & Kass-Hanna, J. (2021). A methodological overview to defining and measuring "digital" financial literacy. *Financial Planning Review*, *4*(2), e1113. DOI: 10.1002/cfp2.1113

Maddineni, V. K., & Gummadi, V. A. K. (2024) "Deep Learning Approaches for Real-Time IoT Data Processing and Analysis," *2024 5th International Conference on Smart Electronics and Communication (ICOSEC)*, Trichy, India, pp. 757-764.

Maduranga, P. W., Tilwari, V., Rathnayake, R. M. M. R., & Sandamini, C. (2024). *AI-Enabled 6G Internet of Things: Opportunities, Key Technologies, Challenges, and Future Directions*. Retrieved from https://www.mdpi.com

Mahapatra, P., Gupta, R., & Mishra, S. (2021). Impact of telemedicine on healthcare accessibility in rural Odisha. *Journal of Health Informatics in Developing Countries*, *15*(1), 44–58.

Mahindra, T. (2023). *5G and Resource Management in Education: Creating Digital Campuses in India*. Tech Mahindra White Papers. https://www.techmahindra.com

Mahmood, N. H., Berardinelli, G., Pedersen, K. I., Mogensen, P. E., & Lopez, D. G. (2020). On the potential of full-duplex for 5G and beyond. *IEEE Communications Surveys and Tutorials*, *22*(2), 1302–1321. DOI: 10.1109/COMST.2020.2968287

Mahmoud, H. H. H., Amer, A. A., & Ismail, T. (2021). 6G: A comprehensive survey on technologies, applications, challenges, and research problems. *Transactions on Emerging Telecommunications Technologies*, *32*(4), e4233. DOI: 10.1002/ett.4233

Maitland, C. (2021). Regulatory frameworks for rural telecommunications: Global perspectives. *Journal of Telecommunications Policy*, *45*(6), 105–116.

Majumdar, P., Mitra, S., Bhattacharya, D., & Bhushan, B. (2024). Enhancing sustainable 5G powered agriculture 4.0 : Summary of low power connectivity, internet of UAV things, AI solutions and research trends. *Multimedia Tools and Applications*. Advance online publication. DOI: 10.1007/s11042-024-19728-1

Mak, C., & Hong, J. (2020). Creating learning organization 2.0: A contextualized and multi-stakeholder approach. *The Learning Organization*, *27*(3), 235–248. https://www.emerald.com/insight/content/doi/10.1108/TLO-01-2020-0020/full/html. DOI: 10.1108/TLO-01-2020-0020

Malekzadeh, M. (2023). Enabling efficient and reliable IoT deployment in 5G and LTE cellular areas for optimized service provisioning. *The Journal of Supercomputing*, *79*(2), 1926–1955. DOI: 10.1007/s11227-022-04722-x

Maligi, A. S., Bhavana, H. T., Sanjana, T., & Archana, H. R. (2022). Technological and Non-Technical Challenges Associated With 6G Networks. In Bharathy, A., & Alhadidi, B. (Eds.), *Challenges and Risks Involved in Deploying 6G and NextGen Networks* (pp. 167–176). IGI Global Scientific Publishing., DOI: 10.4018/978-1-6684-3804-6.ch011

Mansell, R. (2002). From digital divide to digital entitlements in knowledge societies. *Current Sociology*, *50*(3), 407–426.

Mansell, R. (2021). Inequality and digitally mediated communication: Divides, contradictions, consequences. *Oxford Review of Economic Policy*, *37*(3), 428–449. DOI: 10.13140/RG.2.2.23191.16805

Mao, Y., You, C., Zhang, J., Huang, K., & Letaief, K. B. (2017). A Survey on Mobile Edge Computing: The Communication Perspective [Fourthquarter.]. *IEEE Communications Surveys and Tutorials*, *19*(4), 2322–2358. DOI: 10.1109/COMST.2017.2745201

Marouan, M., Brambilla, M., Tagliaferri, D., Mazzucco, C., Debbah, M., Mach, T., Simeone, R., Mandelli, S., Frascolla, V., Lombardi, R., Magarini, M., Nicoli, M., & Spagnolini, U. (2021). *6G V2X Technologies and Orchestrated Sensing for Autonomous Driving*. 10.48550/arXiv.2106.16146.

Matinmikko-Blue, M., Yrjölä, S., & Ahokangas, P. (2024). Multi-perspective approach for developing sustainable 6G mobile communications. *Telecommunications Policy*, *48*(2), 102640. DOI: 10.1016/j.telpol.2023.102640

Mazur, A. (2019). Institutional support for the development of family-type households in rural areas. *Journal of Rural Economics and Social Studies*. DOI: 10.37128/2411-4413-2019-3-8

McKinsey. (2023). *The Future of Work and Digital Skills in India: Implications for Higher Education*. https://mckinsey.com

Mehta, R., & Agrawal, P. (2022). Bridging the digital divide in India: The role of 5G in rural connectivity. *Telecommunications Policy*, *46*(1), 102101. DOI: 10.1016/j.telpol.2021.102101

Meit, Y. (2020). *Digital India: Empowering rural citizens through e-governance*. Ministry of Electronics and Information Technology. https://digitalindia.gov.in

Mentsiev, A., Aygumov, T., & Amirova, E. (2023). DATA-DRIVEN DIGITAL TRANSFORMATION: CHALLENGES AND STRATEGIES FOR EFFECTIVE BIG DATA MANAGEMENT. *ResearchGate*. https://doi.org/DOI: 10.24412/1932-2321-2023-575-526-531

Mhlanga, D., & Moloi, T. (2020). COVID-19 and the digital transformation of education: What we are learning in South Africa. *Preprints*.https://doi.org/DOI: 10.20944/preprints202004.0195.v1

Mihovska, A., Vlahov, A., & Poulkov, V. (2024). 6G-based intelligent, context-aware, and trustworthy user-centric healthcare applications. *Wireless Telecommunications Symposium (WTS)*. doi:DOI: 10.1109/WTS60164.2024.10536689

Miller, J. (2023). Government subsidies for farmers in direct marketing initiatives. *Public Policy and Agriculture*, *5*(4), 143–149.

Ministry of Education. (2022). *Improving Accessibility and Quality in Indian Higher Education*. Government of India. https://www.education.gov.in

Ministry of Science and ICT. (2022). *K-education platform leveraging 6G: Empowering rural learning through advanced technologies*. South Korean Government. https://msit.go.kr

Mitra, P., & Saha, A. (2021). 5G in India: A roadmap for the future. *India Telecom Journal*, *14*(3), 45–62.

Modiba, F., & Kaye, S. (2023). EVALUATION OF INFORMATION AND COMMUNICATION TECHNOLOGIES (ICTs) TOOLS CONTRIBUTING TO RURAL DEVELOPMENT. *Russian Journal of Agricultural and Socio-Economic Sciences*, *142*(10), 19–29. DOI: 10.18551/rjoas.2023-10.03

MoE. (2021). *Survey of Digital Infrastructure in Indian Higher Education*. Ministry of Education, India. https://mhrd.gov.in

Mogavi, R. H. (2024, July). *ChatGPT in education: A blessing or a curse? A qualitative study exploring early adopters' utilization and perceptions*. Retrieved from Science Direct: https://www.sciencedirect.com/science/article/pii/S2949882123000270

Mohan, P. M., Das, A., & Panda, S. (2020). Analyzing the impact of edge computing on healthcare services. *IEEE Access : Practical Innovations, Open Solutions*, *8*, 203720–203729.

Moncayo-Martínez, L. A., & Salcedo, A. (2023). Optimising the optic fibre deployment in small rural communities: The case of Mexico. *Axioms*, *12*(3), 269. DOI: 10.3390/axioms12030269

Morris, L. (2023). Evaluating energy consumption in vertical farming: Environmental impacts and solutions. *Sustainable Energy Journal*, *19*(2), 113–121.

Morrison-Smith, S., & Ruiz, J. (2020). Challenges and barriers in virtual teams: A literature review. *SN Applied Sciences. SN Applied Sciences*, *2*(6), 1096. Advance online publication. DOI: 10.1007/s42452-020-2801-5

Müller, B. (2007). *Designing the Clean Development Mechanism: Operational and institutional issues*. Edward Elgar Publishing.

Mumtaz, S., Rodriguez, J., & Dai, L. (2018). When Mobile Crowd Sensing Meets Ubiquitous Computing: Vision and Challenges [Dec.]. *IEEE Communications Magazine*, *56*(12), 52–59.

Mumtaz, S., Suzuki, K., & Saad, W. (2021). Green communication in 6G: Enabling technologies, challenges, and future research directions. *IEEE Communications Magazine*, *59*(1), 91–97. DOI: 10.1109/MCOM.001.2000735

Murata. (2023). Technological Trends Leading to Beyond 5G/6G. Retrieved from https://article.murata.com/en-sg/article/communication-network-management-using-ai-and-ml

Mutambara, D., & Bayaga, A. (2021). Determinants of mobile learning acceptance for STEM education in rural areas. *Computers & Education*, *160*, 104010. Advance online publication. DOI: 10.1016/j.compedu.2020.104010

Nadeem, Q., Seddik, K. G., & Ghauri, S. A. (2022). Regulatory challenges in the adoption of 6G networks: A critical review. *IEEE Communications Standards Magazine*, *6*(1), 25–33. DOI: 10.1109/MCOMSTD.001.2000015

Nair, M. (2021). Financial inclusion and digital connectivity: Empowering women in rural India. *Journal of Economic Development Studies*, *19*(4), 207–225.

Najjuuko, C., Ayebare, G. K., Lukanga, R., Mugume, E., & Okello, D. (2021). *A survey of 5G for rural broadband connectivity. 2021 IST-Africa Conference*. IST-Africa.

Nanda, S., & Verma, T. (2022). *Upskilling the Workforce for 5G and Beyond*. Workforce Development Quarterly, 14(3), 112-130. https://workforcedev.org

Nanda, P. (2023). Access to Academic Resources in Rural Indian Colleges: Challenges and Solutions. *The Journal of Educational Research*, *29*(2), 201–216. https://joer.in

Nanda, R. (2022). Empowering rural communities through digital literacy: Challenges and strategies. *Journal of Rural Development Studies*, *7*(1), 21–38.

Nanda, S. (2023). AI-Driven Resource Allocation in Indian Universities: A Case Study Approach. *Journal of Higher Education Policy*, *29*(3), 189–201. https://johep.org

Nanda, S. (2023). Inclusive Education through 5G: Innovations and Challenges. *Journal of Digital Inclusion*, *19*(1), 45–61. https://digitalinclusion.org

Narula, K., Nagai, Y., & Pachauri, S. (2012). The role of decentralized distributed generation in achieving universal rural electrification in South Asia by 2030. *Energy Policy*, *47*, 345–357. DOI: 10.1016/j.enpol.2012.04.075

Narula, S. (2020). Empowering rural farmers through eNAM: A case study of digital agriculture in India. *Journal of Rural Development*, *39*(2), 23–37.

NASSCOM. (2021). *Meeting the Needs of a Growing Student Base: Challenges in Technical and Professional Programs*. NASSCOM Reports. https://www.nasscom.in

Nath, P., & Kumar, D. (2021). Empowering rural entrepreneurs through digital networks. *Journal of Economic Empowerment*, *13*(4), 44–61.

National Broadband Plan. (2021). *Challenges and solutions for local broadband deployment*. Retrieved from https://www.broadband.gov/plan

National League of Cities. (2022). *Local regulations and broadband deployment*. Retrieved from https://www.nlc.org/resource/local-regulations-and-broadband-deployment/

National Rural Electric Cooperative Association (NRECA). (2020). History of NRECA. https://www.electric.coop/our-organization/history

National Statistical Office. (2021). *Annual Report on the Demographics of Rural India*. Government of India.

Nayak, R. (2018). A conceptual study on digitalization of banking-issues and challenges in rural India. International Journal 5. of Management. *IT and Engineering*, *8*(6), 186–191.

Nayak, S., & Patgiri, R. (2021). *6G communication technology: A vision on intelligent healthcare*. Springer Communications in Computer and Information Science., DOI: 10.1007/978-981-15-9735-0_1

Neethirajan, S. (2023). Artificial intelligence and sensor technologies in dairy livestock Export: Charting a Digital transformation. *Sensors (Basel)*, *23*(16), 7045. DOI: 10.3390/s23167045 PMID: 37631580

NEP. (2020). *National Education Policy 2020*. Government of India. https://nep2020.gov.in

Neuronimbus. (2021). Beyond 5G: Exploring the future of connectivity with 6G technology. https://www.neuronimbus.com/blog/beyond-5g-exploring-the-future-of-connectivity-with-6g-technology/

Neuronimbus. (2022). Beyond 5G: Exploring the Future of Connectivity with 6G Technology. Retrieved from https://www.neuronimbus.com/blog/beyond-5g-exploring-the-future-of-connectivity-with-6g-technology

Nguyen, A. (2024). Policy and regulatory frameworks for next-generation connectivity: A global perspective. *Telecommunications Policy Review*, *32*(1), 75–90. DOI: 10.2345/tpr.2024.03201

Nguyen-Son, V., Trung, D., & Zhichao, S. (2022). The Key Trends in B5G Technologies, Services and Applications. Computer Networks and Communications. Retrieved from https://link.springer.com/article/10.1007/s11036-022-01999-3

Nguyen, T., & Patel, R. (2023). Space-efficient farming solutions: Vertical farming in urban settings. *Journal of Urban Agriculture*, *9*(1), 67–75.

Nguyen, V. L., Lin, P. C., Cheng, B. C., Hwang, R. H., & Lin, Y. D. (2021). Security and privacy for 6G: A survey on prospective technologies and challenges. *IEEE Communications Surveys and Tutorials*, *23*(4), 2384–2428. DOI: 10.1109/COMST.2021.3108618

Niehaves, B., & Plattfaut, R. (2010). *The Age-Divide in Private Internet Usage: A Quantitative Study of Technology Acceptance*. AIS Electronic Library (AISeL). https://aisel.aisnet.org/amcis2010/407

Niranjana, C., & Jijesh, J. J. Lakshmi, Manasa, B., & Vineeth, Kumar, P, K. (2024). Future outlook and challenges in deploying 6G. In *Futuristic Trends in Network & Communication Technologies Volume 3 Book 3*.doi:DOI: 10.58532/V3BINC3P7CH1

Nishikawa, D. K., Higuchi, T., & Kojima, Y. (2020). Rural broadband networks and their economic impact on remote areas [Mar.]. *IEEE Communications Magazine*, *58*(3), 58–63.

Nokia. (2021). Public-Private Partnerships in 5G and Beyond. Retrieved from Nokia

Nokia. (2022). *Enhancing Education through Reliable and Energy-Efficient 5G Networks*. Nokia Reports. https://www.nokia.com

Norp, T. (2018). 5G requirements and key performance indicators. *Journal of ICT Standardisation*, *6*(1), 15–30. DOI: 10.13052/jicts2245-800X.618

Nugent, J., & Jose, L. (2017). Stakeholder theory and technical communication academic programs. *Technical Communication (Washington)*, 11–30. Advance online publication. DOI: 10.4324/9781315232713-2

O'Connor, B. (2023). Funding for vertical farming research: Supporting sustainable agriculture. *Governmental Research Support Quarterly*, *11*(3), 91–98.

OECD. (2019). Bridging the rural digital divide. https://www.oecd.org/digital/bridging-the-rural-digital-divide.pdf

Office of the Auditor General of Canada. (2023, March). *Report 2—Connectivity in rural and remote areas*. https://www.oag-bvg.gc.ca/internet/English/parl_oag_202303_02_e_44205.html

Olejniczak, A., Blaszkiewicz, O., Cwalina, K. K., Rajchowski, P., & Sadowski, J. (2023). LOS and NLOS identification in real indoor environment using deep learning approach. *Digital Communications and Networks*. Advance online publication. DOI: 10.1016/j.dcan.2023.05.009

Olugboja, A., & Moghalu Agbakwuru, E. (2024). *Bridging healthcare disparities in rural areas of developing countries: Leveraging artificial intelligence for equitable access*. IEEE Advanced Computing and Data Science Applications. ACDSA., DOI: 10.1109/ACDSA59508.2024.10467443

Organisation for Economic Co-operation and Development (OECD). (2023). Universal service obligations in telecommunications. https://www.oecd.org/sti/telecom-universal-service-obligations.htm

Otlyvanska, G., Connolly, N., Walsh, G., & Dooley, J. (2024). Studying the adoption of 5G and future networks for social inclusion: An innovation systems transitions perspective for networks-as-a service. In *IFIP Advances in Information and Communication Technology* (pp. 137–149). Transfer, Diffusion and Adoption of Next-Generation Digital Technologies., DOI: 10.1007/978-3-031-50204-0_12

Oughton, E. J., Boch, E., & Kusuma, J. (2022). Engineering-Economic Evaluation of Diffractive NLOS Backhaul (e3nb): A Techno-economic Model for 3D Wireless Backhaul Assessment. *IEEE Access : Practical Innovations, Open Solutions*, *10*, 3430–3446. DOI: 10.1109/ACCESS.2022.3140421

Oughton, E. J., Comini, N., Foster, V., & Hall, J. W. (2022). Policy choices can help keep 4G and 5G universal broadband affordable. *Technological Forecasting and Social Change*, *176*(121409), 121409. DOI: 10.1016/j.techfore.2021.121409

Ouyang, Y., Zhang, Y., Wang, P. S., Liu, Y. Y., Qiao, W., Zhu, J., Liu, Y., Zhang, F., & Wang, S. (2023). 6G network business support system.

Pachauri, S., & Rao, N. D. (2013). Gender impacts and determinants of energy poverty: Are we asking the right questions? *Current Opinion in Environmental Sustainability*, *5*(2), 205–215. DOI: 10.1016/j.cosust.2013.04.006

Painuly, J. P. (2001). Barriers to renewable energy penetration; a framework for analysis. *Renewable Energy*, *24*(1), 73–89. DOI: 10.1016/S0960-1481(00)00186-5

Paliwal, R. C. (2018). Role of Panchayati Raj Institutions (PRIs) in rural development of Uttarakhand. *Journal of Rural Development*, *31*(1), 89–95. Advance online publication. DOI: 10.29070/31/57440

Pan, C., Ren, W., Wang, J., Elkashlan, M., Wang, J., & Hanzo, L. (2018). NOMA-aided Mobile Edge Computing: A Joint Optimization of Computation Offloading and Time Allocation [July.]. *IEEE Transactions on Wireless Communications*, *17*(7), 4568–4583.

Pandey, B. K., Paramashivan, M. A., Pandey, D., George, A. S., Agarwal, A., Dadheech, P., & Mahajan, D. A. (2024). A Recent Survey on 6G Vehicular Technology: Applications and Challenges. In Pandey, B. K., Pandey, D., & Ahmad, T. (Eds.), *6G Security Education and Multidisciplinary Implementation*, 337-350. IGI Global. 10.4018/979-8-3693-7421-4.ch019

Pandey, D., Goyal, S., Bhaumik, K., Suneja, S., Sharma, M., & Dadheech, P. (2024). A systematic review of security issues in 6G networks and communication. *Advances in Wireless Technologies and Telecommunication*, •••, 1–11. Advance online publication. DOI: 10.4018/979-8-3693-2931-3.ch001

Pandey, S., Samdanis, K., Costa-Pérez, X., & Guo, W. (2020). 6G Network Formation and Management for Multi-RAT Heterogeneous Environments [Nov.]. *IEEE Communications Magazine*, *58*(11), 38–44.

Pan, Y., Zhang, Y., & Saad, W. (2021). Applications of 6G in rural connectivity: A survey. *IEEE Access : Practical Innovations, Open Solutions*, *9*, 127452–127468. DOI: 10.1109/ACCESS.2021.3104057

Papaioannou, P., Pastellas, I., Tranoris, C., Karagiorgou, S., & Denazis, S. (2024). AI-fuelled Dimensioning and Optimal Resource Allocation of 5G/6G Wireless Communication Networks. *2024 IEEE International Mediterranean Conference on Communications and Networking (MeditCom)*, Madrid, Spain. doi:DOI: 10.1109/MeditCom61057.2024.10621426

Park, S., Jung, M. C., & Lee, J. Y. Enabling Sustainable Broadband Adoption in Rural Areas. In *Communication Research and Policy Conference*, 2013 (p. 12).

Parvez, I., Rahmati, A., Guvenc, I., Sarwat, A., & Dai, H. (2020). A Survey on Low-Latency towards 5G: RAN, Core Network and Caching Solutions. *IEEE Communications Surveys and Tutorials*, *20*(4), 3098–3130. DOI: 10.1109/COMST.2018.2841349

Pašti, I. A. (2022). Diffusion of Innovation theory. In *Handbook of Innovation Theory*. DOI: 10.4324/9781003109945-16

Patel, A. (2023). *Scaling Remote Learning Platforms with 5G in India*. Higher Education Innovations Journal, 21(2), 33-47. https://hei.org

Patel, R. (2024, January 12). How much does custom software development cost in 2024? *Space-O Technologies (Canada) Logo*. https://www.spaceo.ca/blog/software-development-cost/

Patel, J., & Joshi, M. (2021). Digital empowerment of rural women through B5G and 6G. *Journal of Gender and Technology*, *16*(2), 45–60.

Patel, J., Sharma, M., & Sen, A. (2021). Education and technology in rural India: Challenges and opportunities. *Indian Journal of Educational Technology*, *12*(3), 321–338.

Patel, S. (2024). Perspective of Digital Transformation in Rural India. [IJSR]. *International Journal of Science and Research (Raipur, India)*, *13*(4), 387–390. https://www.researchgate.net/publication/379756122_Perspective_of_Digital_Transformation_in_Rural_India/citation/download. DOI: 10.21275/SR24402140819

Patil, R., & Desai, N. (2021). Bridging the digital gap in rural India: Challenges and prospects. *Indian Journal of Communication and Development*, *22*(1), 76–91.

Patterson, C. & Canadian Sheep Federation. (2020). AgroLedger. In *Canadian Sheep Federation*. https://animalhealthcanada.ca/pdfs/5%20-%20CSF_AgroLedger_Presentation%20Council%20Forum%202020_Corlena%20Patterson.pdf

Paul, L. C., Ankan, S. S. A., Rani, T., Karaaslan, M., Hossain, M. N., Al-Gburi, A. J. A., Saha, H. K., & Alkurt, F. Ö. (2023). A Wideband Highly Efficient Omnidirectional Compact Antenna for WiMAX/Lower 5G Communications. *International Journal of RF and Microwave Computer-Aided Engineering*, *2023*, 1–10. DOI: 10.1155/2023/7237444

Paul, S. (2024). Investigating federated learning implementation challenges in 6G networks. *IEEE International Conference on Advances in Electrical, Computing, and Telecommunication (ICAECT)*. doi:DOI: 10.1109/ICAECT60202.2024.10469650

Pelton, J. N. (2020). The rise of small satellites and the emergence of a new satellite industry. *Space Policy*, *47*, 21–25. DOI: 10.1016/j.spacepol.2018.10.002

Plastras, S., Tsoumatidis, D., Skoutas, D. N., Rouskas, A., Kormentzas, G., & Skianis, C. (2024). Non-Terrestrial Networks for energy-efficient connectivity of Remote IoT devices in the 6G era: A survey. *Sensors (Basel)*, *24*(4), 1227. DOI: 10.3390/s24041227 PMID: 38400391

PLOS ONE. (2024). *Towards Secure and Efficient Integration of Blockchain and 6G Networks*. Retrieved from PLOS ONE

Podgorskaya, S., & Schitov, S. (2021). The role and importance of social capital in rural development. *E3S Web of Conferences*. DOI: 10.1051/e3sconf/202127308072

Polese, M., Giordani, M., Roy, A., Castor, D., & Zorzi, M. (2020). Toward End-to-End, Full-Stack 6G Terahertz Networks [Nov.]. *IEEE Communications Magazine*, *58*(11), 48–54. DOI: 10.1109/MCOM.001.2000224

Polymeni, S., Plastras, S., Skoutas, D. N., Kormentzas, G., & Skianis, C. (2023). The impact of 6G-IoT technologies on the development of agriculture 5.0: A review. *Electronics (Basel)*, *12*(12), 2651. DOI: 10.3390/electronics12122651

Porcino, D. (2003, June). UWB Regulations\& Coexistence: From the FCC First Report and Order to the Path to Approval in Europe. In *International Workshop on Ultra Wideband Systems*, Oulu, Finland.

Prasad, A., Sharma, A., & Kumar, R. (2022). Smart villages: Bridging the digital divide through B5G and 6G technologies. *Journal of Rural Studies*, *94*, 187–196. DOI: 10.1016/j.jrurstud.2022.10.003

Prasad, R., & Reddy, S. (2020). Spectrum allocation and regulatory frameworks for 6G in India. *Telecom Policy Review*, *16*(3), 64–77.

Premkumar, S., Hussain, N., & Chinamuttevi, V. S. (2023). A Novel RF-SMOTE Model to Enhance the Definite Apprehensions for IoT Security Attacks. *Journal of Discrete Mathematical Sciences and Cryptography*, *26*(3), 861–873. DOI: 10.47974/JDMSC-1766

Primmia, D. R., Mahabooba, M., Karpagam, J., Sharma, K., Singh, A., & Manoj, S. (2024). The Development of 6-G Technology in Integration with AI type of Synergy. *2024 4th International Conference on Advance Computing and Innovative Technologies in Engineering (ICACITE)*, Greater Noida, India. doi:DOI: 10.1109/ICACITE60783.2024.10616538

Qadir, J., Ali, A., & Saleem, A. M. (2021). Artificial intelligence-enabled edge computing for IoT: A comprehensive survey [Feb.]. *IEEE Internet of Things Journal*, *8*(4), 3037–3066.

Qadir, J., Ali, A., & Saleem, A. M. (2021). Artificial Intelligence-Enabled Edge Computing for IoT: A Comprehensive Survey [Feb.]. *IEEE Internet of Things Journal*, *8*(4), 3037–3066.

Qamar, F., Siddiqui, M. U. A., Hindia, M. N., Hassan, R., & Nguyen, Q. N. (2020). Issues, challenges, and research trends in spectrum management: A comprehensive overview and new vision for designing 6G networks. *Electronics (Basel)*, *9*(9), 1416. DOI: 10.3390/electronics9091416

Quy, V. K., Chehri, A., Quy, N. M., Han, N. D., & Ban, N. T. (2023). Innovative Trends in the 6G Era: A Comprehensive Survey of Architecture, Applications, Technologies, and Challenges. *IEEE Access : Practical Innovations, Open Solutions*, *11*. Advance online publication. DOI: 10.1109/ACCESS.2023.3269297

Rainer, L., Shafi, M., Shivanandan, G., Chandramouli, D., & Thiebaut, L. (2022). Perspectives on a 6G Architecture. 10.48550/arXiv.2210.03286

Rajaraman, I. (2020). The impact of digital marketing on economic growth: A case study of India. *Economic and Political Weekly*, *55*(30), 34–40.

Raja, S., Jebadurai, D. J., Ivan, L., Mykola, R. V., Ruslan, K., & Nadiia, P. R. (2024). Impact of Artificial Intelligence in Students' Learning Life. In *AI in Business: Opportunities and Limitations* (Vol. 2, pp. 3–17). Springer Nature Switzerland. DOI: 10.1007/978-3-031-49544-1_1

Raj, M., & Prakash, A. (2021). Bridging the digital literacy gap in rural India. *Education and Connectivity Research*, *10*(4), 77–89.

Ramanathan, P., Kapoor, N., & Desai, R. (2024). *Exploring the Potential of 6G Networks in Education: Emerging Trends and Applications*. Advanced Communication Journal, 22(1), 123-137. https://advcomm.org

Ramanathan, P., Kapoor, N., & Desai, R. (2024). *Holographic Communication in Education: Opportunities with 6G Networks*. Advanced Communication Journal, 22(1), 112-128. https://advcomm.org

Ramaswamy, K. (2023). Empowering rural artisans through 5G-enabled e-commerce in Tamil Nadu: A case study. *Indian Journal of Digital Commerce*, *8*(1), 101–115.

Ramesh, K., Patel, S., & Das, A. (2022). Cybersecurity Risks in 5G-Enabled Campuses. *Journal of Digital Security*, *9*(4), 67–81. https://digitalsecurityjournal.org

Ramraj, D., Choudhary, G., Dragoni, N., Lalwani, P., Khare, U., & Kundu, S. (2023). 6G Mobile Networks: Key Technologies, Directions, and Advances. *Telecom*, *4*(4). Advance online publication. DOI: 10.3390/telecom4040037

Rangan, S., & Erkip, E. (2023). Economic Barriers to B5G and 6G Adoption in Rural Areas. *Proceedings of the IEEE*, *101*(3), 366–385. DOI: 10.1109/JPROC.2014.2299397

Rani, S., Kumar, P., Kumar, S., Dar, S. A., & Prabhakar, P. (2024). ICT based applications to support rural development in India-A review. *International Journal of Bio-Resource and Stress Management*, *15*(2), 1–9.

Rao, M. (2023). *Leveraging 5G to empower rural entrepreneurs: Case studies from India*. Economic Times of India, 67(4), 34-41.

Rao, M. (2023). *The economics of 5G in rural India: Ensuring affordability for all*. Economic Times of India, 67(4), 65-72.

Rao, V., & Srinivasan, N. (2023). *Policy Frameworks for 5G Implementation in Education*. Policy Insights Journal, 11(3), 23-45. https://policyinsights.org

Rao, R., & Singh, K. (2020). Enhancing e-governance with 5G in rural India. *Government Innovation Journal*, *25*(3), 88–101.

Rappaport, T. S., Xing, Y., Kanhere, O., Ju, S., Madanayake, A., Mandal, S., & Singh, A. (2019). Wireless communications and applications above 100 GHz: Opportunities and challenges for 6G and beyond. *IEEE Access : Practical Innovations, Open Solutions*, *7*, 78729–78757. DOI: 10.1109/ACCESS.2019.2921522

Rappaport, T. S., Xing, Y., & MacCartney, G. R. (2019). Terahertz wireless communications: Challenges and opportunities. *IEEE Communications Magazine*, *57*(5), 2–9. DOI: 10.1109/MCOM.2019.1800219

Rappaport, T. S., Xu, Q., & Kanhere, S. (2021). 6G Wireless Systems: A Vision and Research Agenda. *IEEE Access : Practical Innovations, Open Solutions*, *9*, 22575–22592. DOI: 10.1109/ACCESS.2021.3060105

Rawal, N. (2008). Social inclusion and exclusion: A review. *Dhaulagiri*, *2*, 161–180. DOI: 10.3126/dsaj.v2i0.1362

Ray, P. P. (2021). A perspective on 6G: Requirement, technology, enablers, challenges and future road map. *Journal of Systems Architecture*, *118*(102180), 102180. DOI: 10.1016/j.sysarc.2021.102180

Reddy, M., & Tanguturi, R. (2022) "Improved Secure Communication Mechanism for IoT Platform Devices," *2022 International Conference on Smart Generation Computing, Communication and Networking (SMART GENCON)*, Bangalore, India, pp. 1-6.

Reddy, N. S., & Ashreetha, B. (2023) "Technologies for Comprehensive Information Security in the IoT," *2023 International Conference for Advancement in Technology (ICONAT)*, Goa, India, pp. 1-5.

Reevely, D. (2023, December 25). *Canadian farmers dream of blockchain sheep*. The Logic. https://thelogic.co/news/the-big-read/canadian-farmers-dream-of-blockchain-sheep/

REN21. (2021). Renewables 2021: Global status report. https://www.ren21.net/reports/global-status-report/

REN21. (2023). Renewables 2023 Global Status Report. REN21 Global Platform for Renewable Energy.

Report, D. I. (2023). *Digital Infrastructure in Rural Educational Institutions: Progress and Challenges*. Government of India. https://www.digitalindia.gov.in

ResearchGate. (2020). A Key 6G Challenge and Opportunity - Connecting the Remaining 4 Billions: A Survey on Rural Connectivity. Retrieved from https://www.researchgate.net/publication/337062133_A_Key_6G_Challenge_and_Opportunity_-_Connecting_the_Remaining_4_Billions_A_Survey_on_Rural_Connectivity

Roberts, C. (2023). Urban vertical farms and local grocery supply chains. *Journal of Food Supply Chain Management, 10*(4), 321–329.

Robinson, G. M. (2024). Building the rural future and alleviating rural poverty. In *Transforming Rural China* (pp. 225–252). Edward Elgar Publishing., DOI: 10.4337/9781803928586.00015

Roldan, C. S., Giraldo, G. A. M., & Santana, E. L. (2023). Sustainable development in rural territories within the last decade: A review of the state of the art. *Heliyon, 9*(7), e17555. Advance online publication. DOI: 10.1016/j.heliyon.2023.e17555 PMID: 37496929

Rural Telecommunications Group. (2020). Bridging the digital divide: The economics of rural broadband deployment. *Policy Brief.*

Rwanda Ministry of Health. (2022). *Telemedicine services in Rwanda: Progress and impact*. https://www.moh.gov.rw

Saad, W., Bennis, M., & Chen, M. (2020). A vision of 6G wireless systems: Applications, trends, technologies, and open research problems. *IEEE Network, 34*(3), 134–142. DOI: 10.1109/MNET.001.1900287

Saarnisaari., ... (2020). A 6G white paper on connectivity for remote areas. *arXiv preprint arXiv:2004.14699.*

Saarnisaari, H., Chaoub, A., Heikkilä, M., Singhal, A., & Bhatia, V. (2021). Wireless terrestrial backhaul for 6G Remote access: Challenges and low power solutions. *Frontiers in Communications and Networks, 2*, 710781. Advance online publication. DOI: 10.3389/frcmn.2021.710781

Sahita, R., Hahn, S., Chan, K., & McCloghrie, K. (2003). Framework policy information base (No. rfc3318).

Saidur, R., Rahim, N. A., Islam, M. R., & Solangi, K. H. (2011). Environmental impact of wind energy. *Renewable & Sustainable Energy Reviews, 15*(5), 2423–2430. DOI: 10.1016/j.rser.2011.02.024

Saimler, M., İckin, S., Bernini, G., Toumi, N., Diamanti, M., Papavassiliou, S., Zivkovic, M., Akgül, Ö. U., & Khorsandi, B. M. (2024). The Role of AI Enablers in Overcoming Impairments in 6G Networks. *2024 Joint European Conference on Networks and Communications & 6G Summit (EuCNC/6G Summit)*, Antwerp, Belgium. doi:DOI: 10.1109/EuCNC/6GSummit60053.2024.10597102

Salahdine, F., Han, T., & Zhang, N. (2023). 5G, 6G, and Beyond: Recent advances and future challenges. *Annales des Télécommunications, 78*(9), 525–549. DOI: 10.1007/s12243-022-00938-3

Salameh, A. I., & El Tarhuni, M. (2022). From 5G to 6G—Challenges, technologies, and applications. *Future Internet, 14*(4), 117. DOI: 10.3390/fi14040117

Salem, H., Kouzayha, N., El Falou, A., Alouini, M. S., & Al-Naffouri, T. Y. (2023). Exploiting hybrid terrestrial/LEO satellite systems for rural connectivity. *arXiv*. https://doi.org//arxiv.2311.02591DOI: 10.48550

Samarajiva, R. (2019). Digital transformation of rural areas: Policies for rural broadband. *Journal of Digital Inclusion, 2*(2), 10–21.

Samsukha, A. (2024, February 2). How much does it cost to build an AI app in 2024: a detailed guide. *Tech Blog | Mobile App, eCommerce, Salesforce Insights*. https://www.emizentech.com/blog/cost-to-build-an-ai-app.html

Sarabu, V. K. (2018). Rural development in India-a Way forward. Hanamkonda, FormerMember of-BoardofStudiesKakatiyaUniversity, Warangal.

Sarkar, A. (2020). Affordable technology for rural populations in B5G networks. *Telecom Access Journal, 11*(3), 56–65.

Sarma, T., & Mishra, K. (2021). COVID-19 and the Digital Divide in Indian Education: Lessons Learned. *Journal of Social Policy, 12*(3), 99–114. https://jsp.in

Sawad, I., Nilavalan, R., & Al-Raweshidy, H. (2023). Backhaul in 5G systems for developing countries: A literature review. *IET Communications, 17*(6), 659–669. DOI: 10.1049/cmu2.12578

Saxena, A. C. S., Kumar, R., Al-Farouni, M., Kumar, S., Karpagam, C. J., & Kavitha, M. (2024). Navigating 6G network into the latest AI technology for the fastest communication without losing data. *IEEE International Conference on Advances in Computing, Communication and Information Technology for Engineering (ICACITE)*. doi:DOI: 10.1109/ICACITE60783.2024.10616812

Scalise, P., Boeding, M., Hempel, M., Sharif, H., Delloiacovo, J., & Reed, J. (2024). A Systematic Survey on 5G and 6G Security Considerations, Challenges, Trends, and Research Areas. *Future Internet, 16*(3), 67. DOI: 10.3390/fi16030067

Shahzadi, S., Chaudhry, N. R., & Iqbal, M. (2023). A Novel 6G Conversational Orchestration Framework for Enhancing Performance and Resource Utilization in Autonomous Vehicle Networks. *Sensors (Basel), 23*(17), 7366. Advance online publication. DOI: 10.3390/s23177366

Sharma, T. (2023). *AR and VR Applications in Higher Education: A 5G Enabled Revolution*. Virtual Learning Review, 17(2), 89-102. https://vlr.org

Sharma, T., Kapoor, N., & Desai, P. (2023). *Economic and Ethical Challenges of 5G in Education*. Educational Technology Horizons, 8(2), 89-105.

Sharma, A., & Singh, R. (2020). Evolution of mobile networks from 2G to 5G. *Wireless Technology Journal, 33*(4), 345–369.

Sharma, A., Tyagi, V. V., Chen, C. R., & Buddhi, D. (2009). Review on thermal energy storage with phase change materials and applications. *Renewable & Sustainable Energy Reviews, 13*(2), 318–345. DOI: 10.1016/j.rser.2007.10.005

Sharma, J. (2022). Spectrum auction and rural accessibility in next-generation networks. *Journal of Regulatory Telecom Studies, 19*(2), 102–115.

Sharma, M., Sharma, K., Dwivedi, S., Singh, D. K., Gill, R., & Kumar, N. (2023, December). A multiple millimeter-wave 5G MIMO antenna including n257/n258/n259/n260/n261 FR2-bands with high diversity performance. In *2023 International Conference on Power Energy, Environment & Intelligent Control (PEEIC)* (pp. 279-284). IEEE.

Sharma, P., & Patel, A. (2023). Digital transformation and the future of rural marketing: A case study of 5G deployment in India. *Journal of Indian Telecom Policy, 22*(3), 55–68.

Sharma, P., & Patel, A. (2023). The affordability gap: 5G-enabled smartphones and rural India's digital transformation. *Journal of Indian Telecom Policy, 22*(3), 55–68.

Sharma, P., Singh, M., & Desai, S. (2022). 5G and the future of digital marketing in rural India: Insights from a pilot project in Maharashtra. *Journal of Digital Transformation, 12*(4), 98–113.

Sharma, S., Kushwaha, D. S., & Misra, M. (2020). Towards improved resource provisioning for IoT-enabled cloud systems [Dec.]. *IEEE Internet of Things Journal, 7*(12), 11900–11914.

Sharma, S., Popli, R., Singh, S., Chhabra, G., Saini, G. S., Singh, M., Sandhu, A., Sharma, A., & Kumar, R. (2024). The Role of 6G Technologies in Advancing Smart City Applications: Opportunities and Challenges. *Sustainability (Basel), 16*(16), 7039. DOI: 10.3390/su16167039

Sharma, T., Chehri, A., & Fortier, P. (2021). Review of optical and wireless backhaul networks and emerging trends of next generation 5G and 6G technologies. *Transactions on Emerging Telecommunications Technologies, 32*(3), e4155. DOI: 10.1002/ett.4155

Shayea, I., El-Saleh, A. A., Ergen, M., Saoud, B., Hartani, R., Turan, D., & Kabbani, A. (2024). Integration of 5G, 6G and IoT with Low Earth Orbit (LEO) networks: Opportunity, challenges and future trends. *Results in Engineering, 23*(102409), 102409. DOI: 10.1016/j.rineng.2024.102409

Sheikh, Y. A. (2023). *Role of Panchayati Raj Institutions towards rural development: A study with special reference to Pulwama District of Jammu and Kashmir*. VIDYA - A Journal of Gujarat University., DOI: 10.47413/vidya.v2i2.181

Shekhar, S., & Ashreetha, B. (2023). Performance Analysis of Energy Efficiency and Security Solutions of Internet of Things Protocols. *IJEER, 11*(2), 442–450. DOI: 10.37391/ijeer.110226

Shen, J., Zhu, W., Peng, Z., & Wang, Y. (2023). Improving landscape ecological network connectivity in urbanizing areas from dual dimensions of structure and function. *Ecological Modelling, 482*(110380), 110380. DOI: 10.1016/j.ecolmodel.2023.110380

Shimaa, A., Hussein, H., & Kim, H. (2022). Vision and research directions of 6G technologies and applications. *Computers & Electrical Engineering, 99*, 107805. DOI: 10.1016/j.compeleceng.2022.107805

Shin, H., Park, S., Kim, L., Kim, J., Kim, T., Song, Y., & Lee, S. (2024). The future service scenarios of 6G telecommunications technology. *Telecommunications Policy*, *48*(2), 102678. DOI: 10.1016/j.telpol.2023.102678

Shi, W., Cao, J., Zhang, Q., Li, Y., & Xu, L. (2016). Edge computing: Vision and challenges. *IEEE Internet of Things Journal*, *3*(5), 637–646. DOI: 10.1109/JIOT.2016.2579198

Signé, L., & Fox, L. (2022, June 28). Overcoming the barriers to technology adoption on African farms. *Brookings*. https://www.brookings.edu/articles/overcoming-the-barriers-to-technology-adoption-on-african-farms/

Sihombing, H. C. (2021). A 5g Rural, Remote, and Difficult Areas Hope: A Review. *Teknokom*, *4*(1), 21–29.

Silayo, D., & Bano, W. (2019), https://www.researchgate.net/publication/351371349_Digital_Initiatives_in_Rural_India/citation/download

Simpson, L., Daws, L., & Wood, L. (2003). More than just an internet connection: Building rural social capital through public access. *Rural Society*, *13*(2), 113–125. Advance online publication. DOI: 10.5172/rsj.351.13.2.113

Singh, M., & Kapoor, A. (2023). *The Role of Predictive Analytics in Higher Education: Leveraging 5G for Real-Time Decision Making*. Analytics in Education, 10(3), 67-81. https://analedu.org

Singh, A. (2023). Climate control in vertical farming: Achieving year-round production. *Agricultural Science and Technology Review*, *18*(3), 194–200.

Singhal, V. K. (2015). An overview of Panchayati Raj Institutions in India. *Social Science Research Network*. https://doi.org/DOI: 10.2139/SSRN.2692135

Singh, L., & Raj, V. (2021). Infrastructure challenges in 5G and beyond: A focus on rural India. *Infrastructure Development Journal*, *22*(1), 33–47.

Singh, M. K., Singh, R., Singh, N., & Yadav, C. S. (2022). Technologies Assisting the Paradigm Shift from 5G to 6G. In Dutta Borah, M., Singh, P., & Deka, G. C. (Eds.), *AI and Blockchain Technology in 6G Wireless Network. Blockchain Technologies*. Springer., DOI: 10.1007/978-981-19-2868-0_1

Singh, M., & Mehta, A. (2023). AI and 5G: Transforming Administrative Processes in Universities. *Journal of Educational Administration*, *29*(4), 190–202. https://eduadmin.org

Singh, P., & Bhatt, R. (2021). Digital literacy and community awareness through advanced connectivity in rural India. *Journal of Rural Education and Empowerment*, *22*(3), 143–159.

Singh, V. (2023). Community engagement strategies for fostering trust in rural digital platforms. *Indian Journal of Cybersecurity*, *12*(2), 23–36.

Singh, V. (2023). Data privacy and trust issues in rural India's digital landscape. *Indian Journal of Cybersecurity*, *12*(2), 23–36.

Sinha, P., Kandasamy, M., Shanmugam, R., & Nageswari, P. (2024). *6G communication challenges. 6G Communication Network: Architecture*. Security and Applications.

Siwiak, K., & Mc, D. (2004). *Keown, "Ultra-Wideband Radio Technology*. John Wiley & Sons, Ltd. DOI: 10.1002/0470859334

Skouby, K. E. (2021). 5G business models & trends in rural communication. *Journal of Multimedia*. Advance online publication. DOI: 10.13052/jmm1550-4646.17139

Smith, J., & Jones, A. (2022). Mobile applications for direct farm-to-consumer sales. *Journal of Digital Agriculture*, *6*(1), 123–130.

Smith, R., & Johnson, T. (2022). Challenges and opportunities in rural broadband deployment. *International Journal of Rural Development*, *29*(3), 45–59. DOI: 10.5678/ijrd.2022.02903

Sohel, K. (2013). *Jon, Boone., Robert, Lambardia*. Communication Policy Frame.

Sovacool, B. K. (2012). The political economy of energy poverty: A review of key challenges. *Energy for Sustainable Development*, *16*(3), 272–282. DOI: 10.1016/j.esd.2012.05.006

Sovacool, B. K., & D'Agostino, A. L. (2012). A comparative analysis of renewable electricity support mechanisms for Southeast Asia. *Energy*, *44*(1), 189–204.

Sovacool, B. K., & Drupady, I. M. (2012). *Energy Access, Poverty, and Development: The Governance of Small-Scale Renewable Energy in Developing Asia*. Ashgate Publishing.

Spiceworks. (2021). What is a 6G network? Definition, features, and benefits. https://www.spiceworks.com/tech/networking/articles/what-is-6g/

Srinivas, V., & De, T. (2023) "Wi-Fi Router Signal Coverage Position Prediction System using Machine Learning Algorithms," *2023 International Conference on Sustainable Computing and Smart Systems (ICSCSS)*, Coimbatore, India, pp. 253-258.

Srinivas, V., & De, T. (2023) "Wi-Fi Router Signal Coverage Position Prediction System Using Machine Learning Algorithms," *2023 International Conference on Sustainable Computing and Smart Systems (ICSCSS)*, Coimbatore, India, pp. 253-258.

Statista Research Department. (2022). *Higher Education Institutions in India: Statistics and Analysis*. Statista. https://www.statista.com

Statista. (2023). *India: Internet penetration rate 2023*. https://www.statista.com/

Statista.com. (2024, February 27). *Rural and urban population in India from 2018 to 2022*. Retrieved from Statista.com: https://www.statista.com/statistics/621507/rural-and-urban-population-india/

Stewart, P. (2023). University partnerships in agricultural innovation and vertical farming education. *Journal of Educational Partnerships*, *8*(4), 311–318.

Suleiman, A., & Al-Turjman, F. (2020). *6G applications and standards: An overview*. Springer Communications in Computer and Information Science., DOI: 10.1007/978-3-030-69431-9_4

Sundararajan, R. (2023). The potential of 5G for rural India: Bridging the digital divide. *Indian Journal of Rural Development*, *26*(1), 15–30.

Suri, T., & Jack, W. (2016). The long-run poverty and gender impacts of mobile money. *Science*, *354*(6317), 1288–1292. DOI: 10.1126/science.aah5309 PMID: 27940873

Talagala, N. (2022, March 4). Data as the new oil is Not enough: Four Principles for Avoiding Data Fires. *Forbes*. https://www.forbes.com/sites/nishatalagala/2022/03/02/data-as-the-new-oil-is-not-enough-four-principles-for-avoiding-data-fires/?sh=4c7e5061c208

Talwar, S., Himayat, N., Nikopour, H., Xue, F., Wu, G., & Ilderem, V. (2021). 6G: Connectivity in the era of distributed intelligence. *arXiv: Information Theory*. 10.48550/arXiv.2110.07052

Tarhini, A., Hone, K., Liu, X., & Tarhini, T. (2017). Examining the moderating effect of individual-level cultural values on users' acceptance of E-learning in developing countries: A structural equation modeling of an extended technology acceptance model. *Interactive Learning Environments*, *25*(3), 306–328. DOI: 10.1080/10494820.2015.1122635

Telecom Regulatory Authority of India (TRAI). (2022). *Annual Report on the State of Rural Connectivity in India*. Retrieved from https://www.trai.gov.in/

Telecom Regulatory Authority of India (TRAI). (2023). *Telecom statistics report 2023*.

Telecom Regulatory Authority of India (TRAI). (2023). *The telecom statistics report*. https://www.trai.gov.in

Telecom Regulatory Authority of India. (2023, September 29). *Consultation paper on digital transformation through 5G ecosystem* (Consultation Paper No. 24/2023). Mahanagar Door Sanchar Bhawan. https://www.trai.gov.in

The Economic Times. (2023, July 27). *UN cautions schools against 'excessive' use of tech in classrooms, says it can impede learning*. Retrieved from The Economic Times: https://economictimes.indiatimes.com/magazines/panache/un-cautions-schools-against-excessive-use-of-tech-in-classrooms-says-it-can-impede-learning/articleshow/102175016.cms?from=mdr

The Economic Times. (2024, April 24). *AI is touching lives of 9 in 10 internet users: report*. Retrieved from The Economic Times: https://economictimes.indiatimes.com/tech/technology/ai-is-touching-lives-of-9-in-10-internet-users-report/articleshow/109601626.cms?from=mdr

The Economic Times. (2024, August 21). *Improved internet quality ahead: New subsea cable routes to widen India's data highway*. Retrieved from The Economic Times: https://economictimes.indiatimes.com/industry/telecom/telecom-news/improved-internet-quality-ahead-new-subsea-cable-routes-to-widen-indias-data-highway/articleshow/112663013.cms?from=mdr

The Economic Times. (2024, June 28). *80% maths teachers in India, Middle East falter on basic mathematical questions: Study*. Retrieved from The Economic Times: https://economictimes.indiatimes.com/industry/services/education/80-maths-teachers-in-india-middle-east-falter-on-basic-mathematical-questions-study/articleshow/111341070.cms?from=mdr

The Print. (2022, November 4). *Only 34% Indian schools have internet access, less than 50% have functional computers, shows data*. Retrieved from The Print: https://theprint.in/india/education/only-34-indian-schools-have-internet-access-less-than-50-have-functional-computers-shows-data/1196870/

The Times of India. (2023, Dec 6). *Internet connectivity in government schools increased from 5*. Retrieved from The Times of India: https://timesofindia.indiatimes.com/education/internet-connectivity-in-government-schools-increased-from-5-5-to-24-2-in-five-years/articleshow/105788392.cms

Thompson, L. (2023). Zoning policies and the promotion of local food systems. *Journal of Agricultural Policy*, *5*(3), 83–89.

Thota, S., & Govind, T. R. (2024). 6G: Opportunities and challenges. Towards Wireless Heterogeneity in 6G Networks, 1-17.

Tilwari, V., Sharma, D., Solanki, S., & Chakraborty, N., & P., W. M. (2024). A multicriteria aware multipath routing method to increase the QoS for future 6G networks. *IEEE Conference Proceedings*. https://doi.org/DOI: 10.1109/CSNT60213.2024.10546037

Times of India. (2024, February 28). *Over half of 821mn internet users from rural India: Study*. Retrieved from Times of India: https://timesofindia.indiatimes.com/business/india-business/over-half-of-821mn-internet-users-from-rural-india-study/articleshow/108057081.cms

Times of India. (2024, January 19). *Taught, but thought? Latest survey of rural education again shows India's demographic advantage isn't a sure thing*. Retrieved from Times of India: https://timesofindia.indiatimes.com/blogs/toi-editorials/taught-but-thought-latest-survey-of-rural-education-again-shows-indias-demographic-advantage-isnt-a-sure-thing/

Tiwari, S., Lane, M., & Alam, K. (2016). Does broadband connectivity and social networking sites build and maintain social capital in rural communities? *arXiv: Computers and Society*.

Tognisse, I. S., & Degila, J. (2021). Rural technology adoption and use model in rural Africa: A predictive approach to telephony acceptance. [IJEA]. *International Journal of E-Adoption*, *13*(1), 36–55.

Tomaszewski, L., & Kołakowski, R. (2023). Mobile services for smart agriculture and forestry, biodiversity monitoring, and water management: Challenges for 5G/6G networks. *Telecom*, *4*(1), 67–99. DOI: 10.3390/telecom4010006

Townsend, S. (2022, October 24). *What is resource management and why is it important?* Planview. https://www.planview.com/resources/guide/resource-management-software/resource-management-leverage-people-budgets/

TRAI. (2022). *Digital Access in India: A Statistical Overview*. Telecom Regulatory Authority of India. https://trai.gov.in

Türk, U., Dolciotti, M., Östh, J., Reggiani, A., & Nijkamp, P. (2020). Social capital, rurality, and accessibility: A comparative study between Turkey and Italy. DOI: 10.1007/978-981-15-1435-7_23

UGC. (2022). *Faculty Availability in Indian Higher Education Institutions*. University Grants Commission Reports. https://ugc.ac.in

Ukhurebor, K., & Nwankwo, W. (2019). Investigating the Performance of Point to Multipoint Microwave Connectivity across Undulating Landscape during Rainfall. *Journal of Nigerian Society of Physical Sciences*, *1*(3), 103–115. Advance online publication. DOI: 10.46481/jnsps.2019.16

Ulman, M., Šimek, P., Masner, J., Kogut, P., Löytty, T., Crehan, P., & Sabbatini, T. (2020). Towards Future Oriented Collaborative Policy Development for Rural Areas and People. *AGRIS On-Line Papers in Economics and Informatics, 12*(1), 111–124.

UNESCO. (2021). *Promoting Transparency and Accountability in Higher Education Resource Management*. UNESCO Reports. https://www.unesco.org

UNESCO. (2021). The role of information and communication technologies in rural development. https://en.unesco.org/news/role-information-and-communication-technologies-rural-development

UNESCO. (2022). *The GIGA project: Connecting schools and promoting digital literacy in rural areas*. https://unesco.org/giga

United Nations. (2016). *Leaving no one behind: The imperative of inclusive development*. United Nations Department of Economic and Social Affairs.

Vaezi, , Azari, A., Khosravirad, S. R., Shirvanimoghaddam, M., Azari, M. M., Chasaki, D., & Popovski, P. (2022). Cellular, wide-area, and non-terrestrial IoT: A survey on 5G advances and the road toward 6G. *IEEE Communications Surveys and Tutorials, 24*(2), 1117–1174. DOI: 10.1109/COMST.2022.3151028

Vaidya, S., & Patil, R. (2022). Community empowerment through social networks: The role of B5G and 6G technologies. *Journal of Social Inclusion and Digital Connectivity, 10*(4), 79–99.

van Bakel, M., & Horak, S. (2024). Social capital theory. In *Handbook of Social Capital Theory*. DOI: 10.4337/9781035308767.ch33

Van der Horst, D. (2007). NIMBY or not? Exploring the relevance of location and the politics of voiced opinions in renewable energy siting controversies. *Energy Policy, 35*(5), 2705–2714. DOI: 10.1016/j.enpol.2006.12.012

Van Dijk, J. A. (2012). *The Evolution of the Digital Divide - The Digital Divide turns to Inequality of Skills and Usage* (pp. 57–78). https://doi.org/DOI: 10.3233/978-1-61499-057-4-57

van Dijk, J. A. (2006). Digital divide research, achievements and shortcomings. *Poetics, 34*(4-5), 221–235. DOI: 10.1016/j.poetic.2006.05.004

Van Dijk, J. A. G. M. (2006). *The network society: Social aspects of new media* (2nd ed.). SAGE Publications.

Van Dijk, J. A. G. M. (2020). *The Digital Divide: The Internet and Social Inequality*. Polity Press.

van Hilten, M., & Wolfert, S. (2022). 5G in agri-food - A review on current status, opportunities and challenges. *Computers and Electronics in Agriculture, 201*(107291), 107291. DOI: 10.1016/j.compag.2022.107291

Vanelli-Coralli, A., Guidotti, A., Foggi, T., Colavolpe, G., & Montorsi, G. (2020). 5G and beyond 5G non-terrestrial networks: Trends and research challenges. *IEEE 5G World Forum (5GWF)*. doi:DOI: 10.1109/5GWF49715.2020.9221119

Vankadara, Y., Pathak, P., Ashok, P., & Pillai, S. (2024). 6G revolution: Redefining communication systems. *IEEE International Conference on Smart Sustainable Energy Solutions (ICSSES)*. doi:DOI: 10.1109/ICSSES62373.2024.10561292

Vassilakopoulou, P., & Hustad, E. (2021). Bridging Digital Divides: A Literature Review and Research Agenda for Information Systems Research. *Information Systems Frontiers*, *25*(3), 955–969. DOI: 10.1007/s10796-020-10096-3 PMID: 33424421

Vasylkivskyi, M., Horodetska, O., Klymchuk, B. & Hovorun, V. (2023). Strategies of technological development of hardware of infocommunication radio networks. *Informatsiini Tekhnologii ta Kompiuterna Inzheneriia*. doi:DOI: 10.31649/1999-9941-2023-56-1-83-91

Venkatesh, V., & Davis, F. D. (2000). A theoretical extension of the technology acceptance model: Four longitudinal field studies. *Management Science*, *46*(2), 186–204. DOI: 10.1287/mnsc.46.2.186.11926

Verma, H., Chauhan, N., & Awasthi, L. K. (2023). A Comprehensive review of 'Internet of Healthcare Things': Networking aspects, technologies, services, applications, challenges, and security concerns. *Computer Science Review*, *50*(100591), 100591. DOI: 10.1016/j.cosrev.2023.100591

Verma, N. (2022). Infrastructure challenges in the deployment of 5G networks in rural India. *Journal of Infrastructure Development*, *8*(3), 87–100.

Verma, N. (2022). The future of hyper-local marketing in rural India: A 5G perspective. *Journal of Infrastructure Development*, *8*(3), 112–125.

Verma, P. (2021). Language inclusivity in digital content for rural India. *Journal of Social Connectivity*, *8*(2), 67–79.

Verma, R., Verma, K., Verma, J., Singh, T., & Bisen, P. (2024). Indian rural development: A review of technology and society. *SN Social Sciences*, *4*(2), 42. Advance online publication. DOI: 10.1007/s43545-024-00831-x

Vetrithangam, D., Komala, C. R., Sugumar, D., Rao, Y. M., Karpagavalli, C., & Kiruthiga, B. (2024). AI's Role in 6G Security Machine Learning Solutions Unveiled. In Pandey, D., Pandey, B., & Ahmad, T. (Eds.), *Security Issues and Solutions in 6G Communications and Beyond* (pp. 26–44). IGI Global Scientific Publishing., DOI: 10.4018/979-8-3693-2931-3.ch003

Violino, B. (2023, May 19). *Mitigating the hidden risks of digital transformation*. CIO. https://www.cio.com/article/191398/mitigating-the-hidden-risks-of-digital-transformation.html

Viswanathan, H., & Mogensen, P. (2020). Communications in the 6G era. *IEEE Access : Practical Innovations, Open Solutions*, *8*, 57063–57074. DOI: 10.1109/ACCESS.2020.2974747

Voinov, I. A., Chung, J., Kettimuthu, R., Bordel, B., Alcarria, R., & Robles, T. (2022, November). Towards 6G networks for rural environments: vision for improving digital inclusion through open source hardware and software. In 2022 IEEE 40th Central America and Panama Convention (CONCAPAN) (pp. 1-6). IEEE.

Vo, N. N., Le, T. A., & Tran, N. H. (2018). Energy-Efficient 5G Green Network with Wireless Power Transfer for IoT Devices. *IEEE Access : Practical Innovations, Open Solutions*, *6*, 17160–17173.

Wang, F., Wijaya, T. T., Habibi, A., & Liu, Y. (2022). Predictors Influencing Urban and Rural Area students to Use Tablet Computers as Learning Tools: Combination of UTAUT and TTF Models. *Sustainability (Basel)*, *14*(21), 13965. DOI: 10.3390/su142113965

Wang, X., Ma, X., & Niu, Y. (2020). Security and privacy issues in 6G: A comprehensive survey. *IEEE Network*, *34*(4), 16–23. DOI: 10.1109/MNET.001.1900467

Wang, X., Mei, J., Cui, S., Wang, C. X., & Chen, X. (2023). Realizing 6G: The operational goals, enabling technologies of future networks, and value-oriented intelligent multi-dimensional multiple access. *IEEE Network*, *37*(1), 10–17. Advance online publication. DOI: 10.1109/MNET.001.2200429

Wang, Y. (2023). The use of big data analytics in optimizing vertical farming. *Journal of Agricultural Data Science*, *9*(2), 211–218.

Wang, Y., Zhao, T., & Liu, S. (2023). Enhancing digital inclusion through 6G networks in rural areas. *Journal of Digital Society*, *29*(2), 52–67.

Wargo, E., Chellman, D. C., Budge, K., & Canfield, K. D. (2021). On the digital frontier: Stakeholders in rural areas take on educational technology and schooling. *Journal of Research on Technology in Education*, *53*(2), 140–158. Advance online publication. DOI: 10.1080/15391523.2020.1760753

Warschauer, M. (2003). *Technology and social inclusion: Rethinking the digital divide*. MIT Press.

Wellalage, M. P., (2024). *Towards Secure and Efficient Integration of Blockchain and 6G Networks.* Retrieved from https://journals.plos.org

White, G., Barbu, V., & Williams, A. L. (2020). Innovative applications of 5G technology in the healthcare industry. *IEEE Access : Practical Innovations, Open Solutions*, *8*, 155234–155246.

Williams, D. (2023). Community Supported Agriculture (CSA): A model for sustainable local food systems. *Journal of Sustainable Food Systems*, *14*(2), 147–154.

World Bank. (2019). Hydropower development and resettlement. https://www.worldbank.org/en/topic/hydropower/overview

World Bank. (2020). Bridging the rural-urban digital divide: Policy interventions and recommendations. *World Bank Reports*. https://www.worldbank.org

World Bank. (2020). *Digital dividends: How to close the digital divide in rural areas*. https://www.worldbank.org/en/news/feature/2020/03/24/digital-dividends-how-to-close-the-digital-divide-in-rural-areas

World Bank. (2021). The Role of Renewable Energy in Rural Development. https://www.worldbank.org/en/topic/renewableenergy/publication/renewable-energy-and-rural-development

World Bank. (2021). World Development Report 2021: Data for a Better World. The World Bank. Bhattacharyya, S. C. (2013). Rural Electrification Through Decentralised Off-grid Systems in Developing Countries. Springer.

World Bank. (2022). Digital divide in Indian education. World Bank Education Series. Retrieved from https://www.worldbank.org

World Bank. (2023). Public-private partnerships for broadband infrastructure. https://www.worldbank.org/en/topic/ict/brief/public-private-partnerships-for-broadband

World Health Organization. (2016). Burning Opportunity: Clean Household Energy for Health, Sustainable Development, and Wellbeing of Women and Children. https://www.who.int/airpollution/publications/burning-opportunity/en/

Xiao, D. D., Wang, X. Y., Lu, L., Shi, N. X., Liu, C., Zhang, T., & Sun, T. (2023). 6G Architecture Design: From Overall, Logical and Networking Perspective. *IEEE Communications Magazine*, *61*(7), 158–164. Advance online publication. DOI: 10.1109/MCOM.001.2200326

Xia, X., Zhang, S., Zhang, Y., & Wu, L. (2021). Beyond 5G (B5G) Wireless Networks: Key Technologies, Applications, and Challenges. *IEEE Network*, *35*(4), 40–46. DOI: 10.1109/MNET.011.2100130

Yaacoub, E., & Alouini, M. S. (2019). A key 6G challenge and opportunity—connecting the remaining 4 billions: a survey on rural connectivity. *arXiv preprint arXiv:1906.11541*.

Yaacoub, E., & Alouini, M. S. (2020). A key 6G challenge and opportunity—Connecting the base of the pyramid: A survey on rural connectivity. *Proceedings of the IEEE*, *108*(4), 533–582. DOI: 10.1109/JPROC.2020.2976703

Yadoo, A., & Cruickshank, H. (2012). The role for low carbon electrification technologies in poverty reduction and climate change strategies: A focus on renewable energy mini-grids with case studies in Nepal, Peru and Kenya. *Energy Policy*, *42*, 591–602.

Yang, S., Li, S., Chen, W., & Zhao, Y. (2024). A redactable Blockchain-Based data management scheme for agricultural product traceability. *Sensors (Basel)*, *24*(5), 1667. DOI: 10.3390/s24051667 PMID: 38475202

Yoshihisa, H. (2022). Defining 6G: Beyond the conventional boundaries of wireless communication. *International Journal of Wireless Networks*, *28*(3), 321–338. DOI: 10.1007/s11276-021-02656-0

You, X., Huang, Y., Liu, S., Wang, D., Ma, J., Zhang, C., Zhan, H., Zhang, C., Zhang, J., Liu, Z., Li, J., Zhu, M., You, J., Liu, D., Cao, Y., He, S., He, G., Yang, F., Yang, L., & Gao, W. (2023). Toward 6G TKμ Extreme Connectivity: Architecture, Key Technologies and Experiments. *IEEE Wireless Communications*, *30*(3), 86–95. Advance online publication. DOI: 10.1109/MWC.004.2200482

You, X., Wang, C., Huang, J., Gao, X., Zhang, Z., Wang, M., Huang, Y., Zhang, C., Jiang, Y., Wang, J., Zhu, M., Sheng, B., Wang, D., Pan, Z., Zhu, P., Yang, Y., Liu, Z., Zhang, P., Tao, X., & Liang, Y.-C. (2021). Towards 6G Wireless Communication Networks: Vision, Enabling Technologies, and New Paradigm Shifts [Feb.]. *Science China. Information Sciences*, *64*(1), 1–74. DOI: 10.1007/s11432-020-2955-6

Yrjölä, S., Matinmikko-Blue, M., & Ahokangas, P. (2023). Developing 6G visions with stakeholder analysis of the 6G ecosystem. *EuCNC & 6G Summit*. doi:DOI: 10.1109/EuCNC/6GSummit58263.2023.10188379

Yunus, N. A. M., Hanapi, Z. M., & Kamarudin, S. (2024, July). 6G on the Horizon: Technologies, Requirements, Trends, and Potential Techniques. In *2024 International Conference on Electrical, Computer and Energy Technologies (ICECET* (pp. 1-6). IEEE.

Zaidi, A., Athanasiou, G., Mahmood, K., Teyeb, O., Medbo, J., & Wikström, G. (2020). Integration of Terrestrial and Non-Terrestrial Networks in 6G: Requirements, Challenges, and Directions [Sept.]. *IEEE Communications Standards Magazine*, *4*(3), 49–55.

Zhang, Y., Li, L., & Yang, J. (2023). Dynamic spectrum sharing for next-generation wireless networks. IEEE Communications Magazine, 61(1), 45-52. https://doi.org/ International Telecommunication Union (ITU). (2022). Global spectrum management. https://www.itu.int/en/ITU-R/terrestrial/fmd/Pages/default .aspxDOI: 10.1109/MCOM.2023.XXXXX

Zhang, C., Patras, P., & Haddadi, H. (2019). Deep learning in mobile and wireless networking: A survey. *IEEE Communications Surveys and Tutorials*, *21*(3), 2224–2287. DOI: 10.1109/COMST.2019.2904897

Zhang, F., Zhang, Y., Lu, W., Gao, Y., Gong, Y., & Cao, J. (2022). 6G-enabled smart agriculture: A review and prospect. *Electronics (Basel)*, *11*(18), 2845. DOI: 10.3390/electronics11182845

Zhang, J., Zhang, Y., & Zhao, L. (2019). Towards 5G and beyond: Enabling technologies and applications. *IEEE Communications Magazine*, *57*(10), 4–6. DOI: 10.1109/MCOM.2019.8890359

Zhang, X. (2023). Real-time environmental monitoring in vertical farms through IoT sensors. *Journal of AgriTech Innovation*, *17*(2), 183–192.

Zhang, Y., & Wang, X. (2023). Standardization and Interoperability in B5G and 6G Networks. *International Journal of Wireless Information Networks*, *30*(1), 12–23. DOI: 10.1007/s10776-023-00567-8

Zhang, Y., & Wang, Z. (2021). B5G and 6G-enabled smart village education: A case study from China. *IEEE Access : Practical Innovations, Open Solutions*, *9*, 52798–52809. DOI: 10.1109/ACCESS.2021.3069005

Zhang, Y., Zhao, M., & Liu, S. (2020). Exploring the impact of 5G technology on education: Opportunities and challenges. *Journal of Educational Technology & Society*, *23*(4), 42–54. https://www.jstor .org/stable/10.2307/26924001

Zhang, Z., Xiao, Y., Ma, Z., Xiao, M., Ding, Z., Lei, X., & Li, H. (2020). 6G wireless networks: Vision, requirements, architecture, and key technologies. *IEEE Vehicular Technology Magazine*, *14*(3), 28–41. DOI: 10.1109/MVT.2020.3003744

Zhao, M., Li, X., & Liu, Y. (2021). Challenges and Opportunities in 6G Networks: A Comprehensive Survey. *IEEE Communications Surveys and Tutorials*, *23*(3), 1201–1231.

Zhao, Z., Liu, X., & Saad, W. (2020). Energy-efficient wireless systems for 6G: Key challenges and research directions. *IEEE Communications Magazine*, *58*(4), 28–33. DOI: 10.1109/MCOM.001.190087

Zhao, Z., Li, Z., & Zhang, H. (2021). The road to 6G: Ten critical issues. *IEEE Communications Magazine*, *59*(3), 68–73. DOI: 10.1109/MCOM.001.2000397

Zheng, K., Zhao, S., & Xu, X. (2020). Quality-of-experience assurance in B5G/6G networks [Sept./Oct.]. *IEEE Network*, *34*(5), 190–197.

Zheng, K., Zhao, Y., & Liu, J. (2020). Spectrum Allocation and Pricing for 5G and Beyond: Survey and Research Directions. *IEEE Transactions on Wireless Communications*, *19*(10), 6417–6434.

Zhou, D., Sheng, M., Li, J., & Han, Z. (2023). Aerospace integrated networks innovation for empowering 6G: A survey and future challenges. *IEEE Communications Surveys and Tutorials*, *25*(2), 975–1019. DOI: 10.1109/COMST.2023.3245614

Zong, B., Fan, P., Zhang, A., Li, C., & Shen, X. (2019). Evolution of mobile communications: From 1G to 5G. *IEEE Communications Magazine*, *57*(4), 58–64. DOI: 10.1109/MCOM.2019.1800411

About the Contributors

Thirupathi Manickam, currently working as an assistant professor in the Department of Professional Studies at Christ University, Bangalore, one of the leading institutions accredited in NIRF, NBA, and NAAC. He has more than 7 years of teaching and Research experience. He has been teaching and guiding the B.Com, B.Com (CA) BBA, M.Com students in various colleges in Karnataka & Tamil Nadu like Acharya Bangalore B-School, Sambhram Academy of Management Studies, Acharya Institute of Graduate Studies from Bengaluru & Don Bosco College (Arts & Science) Dharmapuri Tamil Nadu. He has secured an M.Com (General) & M.Phil from Thiruvalluvar University- Vellore and Ph.D in Alagappa University- Karaikudi in E-Commerce & Digital Payments. He also cleared TN-SET & K-SET eligibility examinations which are conducted Tamil Nadu & Karnataka. He was the Resource person in National level seminar on "Influence of Digital Marketing in Development of Rural Youth" Organised by Don Bosco College, Dharmapuri- Tamil Nadu. He has Organized seminar and Workshops on "Entrepreneurial Development, SPSS & AMOS, Plagiarism Free Report Reporting, Report Writing techniques, Environmental Ecosystem, Intellectual Property rights and Business Innovation and Replead Consumer Protection Act-2019". His areas of expertise are Financial Accounting, Corporate Accounting, Financial Management, Management Accounting, Taxation, Business Statistics, E-Business, E-Business and Computerized Accounting. Research Perspectives He has 68 citations and 4 h-Index and has published 28 research papers in Scopus, Web of Science, UGC-CARE, and UGC-approved and leading international journals. He has also presented 15 papers at national and international conferences. He has also participated in over 50 seminars, conferences, FDP & workshops at National and International Levels. He is a Reviewer in several journals, including Scopus and Web of Science Indexed Journal, and an Editor in "International Journal of Commerce and Management Research". He is holding professional membership in "Education Research and development Association".

Sriram Ananthan, a former banker turned entrepreneur. I'm also a coach, a professor, and a writer, driven by a singular goal: to use my experiences to guide and motivate young professionals who may just need a little nudge in the right direction. With a background rooted in my PhD studies in Business Management, I've not only applied those learnings in my professional journey, but I'm also determined to translate my life experiences into words for the benefit of all. As an avid student and analyzer of philosophy, the themes in my current books and upcoming projects revolve around it, delving into topics that will help expand your mind to the boundless possibilities of the world.

S. Gopalakrishnan presently working as head-dept. Of commerce, and professor, department of commerce & management, east point group of institutions, bengaluru. He has 26 years of experience including 12 years of industry and 14 years of commerce & management teaching experience. He was holding responsibilities of r&d head, vice principal in his previous assignments and he has got an international experience of 1 year worked in nigeria as lecturer-coordinator at university of uyo. His industry experience includes holding positions like senior manager - education & training at pentagon global solutions-chennai, academics head at aptech limited-chennai, hod for department of management. He had conducted several communication and personality development workshops in various institutes. He is a holder of a patent and authored more than 25 publications in indexed, national level and international journals. He has also written 3 booksin "strategic management and healthcare" recommended for mba syllabus bangalore university. He was awarded with a research project for 8 lakhs on "solid waste management" by central government department icssr (indian council for social sciences research, new delhi). He acted as a resource person for several conferences and seminars addressing the students and faculty members in various topics of social sciences. He was also involved in several university assignments like online teaching, webinars, curriculum design, mass communication through tv/internet He organized several international, national conferences, seminars, workshops and faculty developments programmes sponsored by aicte, vtu and several other bodies. His professional association includes, iste & bma.

Chandan A. Chavadi, a distinguished academician with a Ph.D. from Karnatak University, MBA in Marketing, and B.E. in Electronics and Communication, is the Dean and Professor at Presidency Business School. With 37 research papers, 1 book, 2 book reviews, and 6 book chapters published, he has made significant contributions to management. His work, "Impact of Customer Characteristics on Retail Format Selection," demonstrates leadership in cutting-edge research. Chavadi has also played a pivotal role in institutional development, serving as a principal presenter during NAAC accreditation cycles and contributing significantly to autonomous system implementation for the MBA department.

***

Bharath Booshan is a dedicated academician with a strong foundation in corporate experience, holding a Doctorate in Management Science and an MBA in Marketing. His academic expertise spans general management, marketing, public relations, branding, consumer behavior, and neuromarketing, supported by a graduate background in psychology, economics, and sociology. Dr. Booshan's research interests have led him to publish papers in numerous reputable journals, contributing valuable insights to his fields of study. His professional background includes successful roles in financial institutions and media with organizations like ICICI Securities, HDFC Bank, ING Vysya Bank, IndusInd Bank, Infomedia-Network 18 Group, and Gettit Group. With extensive experience in wealth management and customer relationship management, Dr. Booshan combines industry knowledge with academic insight, making significant contributions to both fields.

Shabista Booshan is an accomplished academician with a blend of corporate experience, specializing in general management, event management, human resources, marketing, public relations, branding, consumer behavior, and neuromarketing. She holds a Doctorate in Management, a Postgraduate Diploma in Public Relations, and an MBA in Human Resources, complemented by an undergraduate degree in English literature, psychology, and journalism. Her passion for research has led to numerous publications in reputable journals. Dr. Shabista has also worked on a variety of impactful projects and assignments with major media channels and newspapers, including ETC, Zee TV, Rajasthan Patrika, and Dakshin

Bharat. Her professional journey also encompasses substantial experience in journalism and event management, adding depth to her academic and practical expertise.

Lakshmidevi C.G. is an educator, researcher, and social advocate with expertise in Earth Sciences and Sociology. Holding a PhD in Earth Sciences from Pondicherry University and an MA in Sociology, she has also served as a CSIR Senior Research Fellow. Her international research experience includes a 6-month internship at Université de Poitiers, France. Currently, Dr. Lakshmidevi serves as Principal of Mahe Co-operative College of Higher Education and Technology, fostering social responsibility and community engagement among students.

Smrity Dwivedi is working as an Assistant Professor in the department of Electronics Engg., IIT (BHU), Varanasi. She is Senior Member of IEEE, Fellow of IETE, Member of IET, LMISC, Guest Editor in Elsevier, Life member of VEDA society. She completed her doctoral work from IIT (BHU), VNS, INDIA in RF and Microwaves. She worked as a research fellow (postdoctoral type work) from IIT-BHU, INDIA. She has published many research papers in international journals, international conferences and national conferences. She is also published many chapters in reputed books.

Raghunandan G received his PhD in Management from the Bangalore University, Bangalore in 2019. He currently works with the Department of Commerce, CHRIST (Deemed to be University), Bangalore, India. His area of interest is into risk management in financial service firms. He is also the coordinator for the Chartered Institute for Securities and Investment professional certificate programme at CHRIST (Deemed to be University).

Vimala Govindaraju is an Experienced Lecturer with a demonstrated history of working in the education industry. Skilled in Research, Leadership, Strategic Planning, Organizational Development, Management and Training. Education professional with a Doctor of Philosophy (Ph.D.) in Human Communication from University Putra Malaysia.

Dankan Gowda V is currently working as an Assistant Professor in the Department of Electronics and Communication Engineering at BMS Institute of Technology and Management in Bangalore. Previously, he worked as a Research Fellow at ADA DRDO and as a Software Engineer at Robert Bosch. With a total experience of 14 years, including teaching and industry, Dr. Gowda has made significant contributions to both academia and research. He has published over 120 research papers in renowned international journals and conferences. In recognition of his innovative work, Dr. Gowda has been granted six patents, including four from Indian authorities and two international patents. His research interests primarily lie in the fields of IoT and Signal Processing, where he has conducted workshops and handled industry projects. Dr. Gowda is passionate about teaching and strives to create an engaging and effective learning environment for his students. He consistently seeks opportunities to enhance his knowledge and stay updated with the latest advancements in his field.

Rajiv Gurugopinath is an Assistant Professor of Law at Ramaiah College of Law, Bengaluru. He is registered as an advocate on the rolls of the Karnataka State Bar Council. He graduated with a degree of B.A. LL.B. from B.M.S. College of Law, Bengaluru. He completed his Master of Laws (LL. M.) in Constitutional and Administrative Law from the University Law College, Bangalore

University, Bengaluru and wrote a Thesis on Constitutional Implications of the appointment of Judges of the Higher Judiciary in India – A Critical Evaluation. He also a holder of Master of Business Laws (M. B. L.) from the National Law School of India University, Bengaluru. Sri. Rajiv Gurugopinath, is currently pursuing his Ph.D., from the School of Law, Vel Tech Rangarajan Dr.Sagunthala R&D Institute of Science and Technology, Avadi, Chennai on the topic "Right to Health - A Study on the Legal Framework and Its Implementation with Special Reference to Bangalore Urban District." Sri. Rajiv Gurugopinath, has been a part of many National and International seminars, conferences, symposiums where he has presented and published various papers. He has also been a resource person at various national and international seminars. Sri. Rajiv Gurugopinath, has time and again been invited to deliver various guest lectures at reputed colleges, also adjudge many National Level Moot Court competitions and various Law Events like, Mock Trial, Legal Debate, etc. Sri. Rajiv Gurugopinath, has a diverse experience both as an advocate and as an academician. Sri. Rajiv Gurugopinath, tries to imbibe a healthy mixture of both practice as well as academics and therefore is a strong advocate of holistic legal education. His areas of Interest include Constitutional Law, Administrative Law, Jurisprudence, Law of Crimes, Business Laws, Intellectual Property Law, Health Laws and Medical Laws among others.

Hendra is a distinguished academic and professional leader, currently serving as the Chair of Research and Community Service Development at Politeknik Wilmar Bisnis Indonesia. In addition to his administrative role, he lectures in International Marketing Management, bringing a wealth of knowledge and a Doctorate in Management Science to the classroom. Dr. Hendra's dedication to education and community empowerment is demonstrated through impactful projects, such as the ecoprinting batik initiative in Timbang Jaya Village, which fosters sustainable local enterprise. His commitment to lifelong learning is reflected in his diverse certifications, including TESOL and Digital Marketing, which he shares with students to inspire and equip future entrepreneurs.

Kiran Hiremath is finance faculty with interest in Behavioural Finance. Her academic qualifications include – PhD (Finance) from Tumkur University, MMS from Symbiosis Institute of Management, Pune, affiliated to Pune University She worked in the industry for 6 years before getting into teaching profession. She has 21 years' teaching experience. The courses she has specialized in are - Managerial Accounting, Financial Management, Business Statistics, Quantitative Techniques and Operations Research, Risk Management and Derivatives and Corporate Taxation. She has also presented many research papers at national and international conferences including conferences held at IIM – K, Accounting Research Institute, Kuala Lumpur, Malaysia, ICBAI, IISc, Bangalore, 7th Indian Finance Conference at IIM-B, ICBM 2020 (Sri Lanka). A paper titled 'Is Retirement Planning a Priority for Individuals? A Study in Bengaluru City', won 'Best Paper Award' at International Conference conducted by ISME, Business School in February 2019. Apart from conducting regular research, teaching and guiding students, she also takes care of inviting professionals from industry to interact with students.

Devapitchai Joel Jebadurai is currently working as Associate Professor in the Department of MBA,St.Joesph's College of Engineering, OMR, Chennai, Tamil Nadu, India. He has published more than 40 research articles in a peer reviewed journals.

Dhanalakshmi K (Beulah, pen name), Assistant Professor, School of Business and Management, Christ University, specialized in Marketing, Human Resources, and Entrepreneurship. She has completed her Ph.D. at Bharathiar University. Her research interests are in customer relationship management, consumer behavior, entrepreneurship, education management, and SDG as well as on education management. She has authored more than six books. She is the recipient of the Global Jury Award 2021 by the Wadhwani Foundation, National Entrepreneurship Network; and two Best Paper awards in an International Conference organized in Bangalore in 2014 and 2016, respectively. She has published more than twenty papers publications and seven patent publications. She is a Fellow Member of Eduoxia Research Center, a Fellow Member of Eduoxia Research University, She is a lifetime member of the Teaching and Education Research Association (TERA); Member of International Association of Academic Plus Corporate Society (IAAC); Member of EAE Research Centre, a part of Global Research Network. She has served as an Editor / Editorial Board Member for Journals, at Bangalore University affiliated colleges. She has also served as an External Examiner / Panel Member / Moderator / Question Paper Setter for the colleges in Bangalore. She has also served as Chairperson and Resource Person in various National and International Conferences organized in India.

Kandasamy Ravindran is an Associate Professor at Presidency Business School in Bangalore, with 15 years of academic and research experience. He has published over 30 articles in national and international journals and holds a patent in the marketing domain. Under his mentorship, one candidate has completed a Ph.D., and five research scholars are currently pursuing their Ph.D. degrees. He has received a major research project grant from ICSSR valued at 15 lakhs. Additionally, Dr. Ravindran has received several awards and serves as a reviewer and editor for reputable journals.

Amit Kohli, Post Doctorate Research at the University of Arizona, USA, Ph.D., India, and two Master's degrees. He has more than 20 + years of academic and industrial experience. He has been bestowed the Best teacher award and young scientist award. He has about 50 research papers to his credit. His papers are published in various Global Journals of repute such as Taylor & Francis, Springer, Inderscience, Emerald, etc. He has participated in various international conferences held in countries such as the UK, Italy, etc. He is a publisher of two books. He has two patent designs and one utility patent to his recognition.

Ravi Kumar, an academician with a Ph.D. in Management Science from Anna University, serves as an Associate Professor at Presidency College, Bangalore, since October 2023. With over 15 years of teaching experience, he has held positions at CMR University and KIT-Kalaignar Karunanidhi Institute of Technology. He specializes in Marketing and Human Resources, covering subjects like Security Analysis, HR Management, and Marketing Management. Dr. Kumar has presented at numerous conferences, published research in esteemed journals, and authored several books. He is recognized for his teaching excellence and commitment to student development.

L. Rajeshkumar, Assistant Professor in the Department of MBA. He has graduated BBA from Aditanar College of Arts and Sciences, MBA, and M.Phil from Manonmaniam Sundarnar University in Tirunelveli. In December 2012, he qualified for the UGC-NET examination. Then 2018, he received his Ph.D with a marketing specialization especially in FMCG. He started his career at Manonmaniam Sundaranar University College, Govindaperi, in September, 2012. After that, he moved to DMI. St. Eugune University, Zambia, for one year in the 2014–15 academic year. In 2015, he began his Ph.D. studies at Manonmaniam Sundarnar University in Tirunelveli. In July 2018, he submitted his thesis and started his journey at the Aarupadai Veedu Institute of Technology, Chennai, in February 2018. Then, in February 2020, he shifted his work to TNJFU - Fisheries Business School, Tamilnadu Dr.J. Jayalalithaa Fisheries University, OMR Campus, Chennai. After that, in October 2022, he joined as an assistant professor in the department of MBA at St.Joseph's College of Engineering, Chennai. Overall, he has 11 years of experience, including 3 years of research and 8 years of teaching. He has specialized in Marketing and

Sendhil Kumar works as an Assistant Professor at Presidency Business School, Bengaluru, India. He is serving as IIC Convener for Presidency College. He is enrolled in the "Certificate Programme in Leadership and Change Management at IIM Indore". He completed his Ph.D. degree in management at Bharathiar University, Tamil Nadu, India. He completed an MBA from Anna University Chennai and an MSc (Psy) from Annamalai University. His research interests are organizational behavior, startups, small business management, and artificial intelligence in management. He has authored more than 50 research articles in Scopus and UGC care, and 12 proceedings presented at various professional conferences and venues. He has received the Best Paper award at International and National conferences. He has received two funding supports from the Indian Council of Social Science Research (ICSSR, New Delhi). He has been serving as a Reviewer for two journals, one from IGI Publisher and the other from Sage Publications. He has more than 23+ years of experience in Industry and Academics. His area of expertise is in the field of AI tools in Research and Digital Marketing and served as a resource person in various conferences, FDPs, and MDPs. He is the certified trainer of Personality Grooming by Gillette.

Chippy Mohan is an accomplished Assistant Professor at Christ University, India, known for her academic expertise and contributions to the field of [specific field/discipline, e.g., Finance, Marketing, HR, etc.]. With a strong background in HR, Dr. Chippy has established herself as a prolific researcher and educator. She has authored numerous publications in reputed national and international journals, showcasing her commitment to advancing knowledge in her domain. Her research interests include [list specific interests or topics, e.g., behavioral finance, digital marketing, employee engagement, etc.], and she actively participates in conferences, seminars, and workshops to share insights and learn from global peers.

Deepika N is an Assistant Professor in the Department of Social Work at Acharya Institute of Graduate Studies, Bengaluru. With a keen interest in rural development and community engagement, she has over two years of teaching experience and currently, she is pursuing a Ph.D. with a focus on researching tribal livelihood and development in the Western Ghats region of Karnataka, India. Deepika has extensive field experience, having served as Field Operations Head for an NGO focused on organic farming, empowering farmers, and providing skill enhancement training for rural and tribal women. She has worked on Self Help Group (SHG) projects in Sri Lanka, organized activities to improve farmers' livelihoods, and participated in workshops on self-sustainability. Her contributions also include reviewing child protection policies and assisting in business proposal submissions for international funding. A dedicated scholar and academician, Deepika has presented papers on human rights violations against transgender individuals at the Karnataka State Human Rights Commission and published a book on "Gender-Based Violence and Rights." She has attended numerous national and international conferences, consistently contributing to the academic and practical aspects of social work. Her previous roles include overseeing operations in Uttara Kannada District, conducting training programs for the Siddi tribe farmers, and ensuring efficient field operations and reporting functions.

Muddasir Ahamed Khan N is an accomplished academician and professional with over 14 years of experience in education and industry. Currently serving as an Associate Professor in the Department of Management at Acharya Institute of Graduate Studies, Bengaluru, I am renowned dedication to teaching and research. I Hold Ph.D. in Management, and holds an MBA in Marketing. I also published extensively in reputable journals, contributing to the fields of FMCG, advertising trends, and green marketing, alongside authoring textbooks on strategic and marketing management. My career reflects a commitment to academic excellence, mentoring, and curriculum innovation. As I am actively participated in numerous conferences and faculty development programs, consistently striving to stay at the forefront of education and industry trends.

N. Raghu is an Associate Professor at Presidency Business School in Bangalore, with 15 years of academic and research experience. He has published over 10 articles in national and international journals. He guided several student dissertations and research projects under his mentorship. Additionally he has been an IQAC coordinator for presidency College.

Gowher Amrin Naaz is having the Academic qualifications of Masters in Literature, Psychology and Education. Presently working as an Academic Coordinator in Presidency College, Bengaluru, India. She is having 18 years of Academic and Education Experiences. Presently she is pursuing Ph.D in the Domain of Education. She has presented papers in various National and International Conferences.

Swetha Appaji Parivara is an Associate Professor at Presidency College (Autonomous), Bengaluru, with nearly two decades of experience in academia. Her extensive career spans prestigious state universities and institutions, where she has made significant contributions to the field of finance and analytics. Dr. Swetha holds a Ph.D. in Finance and is a certified Data Science professional, with expertise in Python, R, and Tableau. Her scholarly work includes presenting and publishing research in national and international forums, including Scopus-indexed journals. In addition to her academic role, Dr. Swetha is a freelance Data Science professional and a corporate trainer, with a focus on finance and data science domains. Her diverse interests also extend to finance, accounting, and tourism, reflecting her multifaceted expertise.

Perumal Revathi is an Assistant Professor at Knowledge Institute of Technology, Salem. She completed MBA (Finance and Marketing) in SNS College of Technology, she is a Gold Medalist in MBA..also completed M.Com(Banking) in Madurai Kamaraj University through Distance Education. Currently pursuing Ph.D. With 13 years of Teaching Experience for MBA Students in engineering college, She has published 2 national journals, 1 international journal and done 10 paper presentation in National and International conferences.

KDV Prasad, works as a Faculty of Research in Symbiosis Institute of Business Management, Hyderabad; Symbiosis International (Deemed University), Pune, India. Professor Prasad holds Masters in Computer Applications and Masters in Software Systems from BITS, Pilani; MBA (Human Resources), IGNOU, New Delhi.. Professor Prasad Possess PhD in Business Management (Kanpur

University); and PhD in Business Administration (RTM Nagpur University. He is AIMA certified Management Teacher, Fellow, World HR Board, Carlton Advanced Management Institute, USA. Professor Prasad Published over 100 articles in scopu/WOS indexed journals, and 3 books.

R. Ramesh is currently an Associate Professor at the Knowledge Institute of Technology in Salem, Tamil Nadu, India. He is also a recognized research supervisor at Anna University. With 15 years of experience in teaching and research, Dr. Ramesh has published several papers in both national and international journals.

R. Saranya is an Assistant Professor at Knowledge Institute of Technology, Salem. She holds a B.E. in Computer Science and Engineering from SRM Valliammai Engineering College and an M.E. in Computer Science and Engineering from Madha Engineering College. With 8 years of experience in the engineering teaching profession, she has published 10 papers in international journals and conferences. She is certified in Full Stack using Java through Wipro TalentNext and also certified faculty Universal Human Value - II.

Mirshad Rahman is an accomplished social work educator with extensive experience in teaching, research, and administration. He is currently an Assistant Professor and the Coordinator of the Internal Quality Assurance Cell (IQAC) at Acharya Institute of Graduate Studies in Bengaluru. In addition to these roles, Dr. Rahman manages key institutional responsibilities, including coordinating NIRF and AISHE, as well as overseeing the Entrepreneurial Development Cell and the Institutional Innovation Council. His academic involvement extends to his role as an Academic Counselor at IGNOU's Regional Centre in Cochin, Kerala. Dr. Rahman holds a Ph.D. in Social Work from Pondicherry Central University, along with a Master of Social Work from Amrita Vishwa Vidyapeetham University and a Master of Arts in Psychology from IGNOU. Dr. Rahman has over seven years of teaching and six years of research experience. His scholarly contributions include five articles and two book chapters in Scopus-indexed international publications, and four articles in UGC Care-listed journals. His research interests focus on experiential learning, student welfare, public-private partnerships in child welfare, child protection policies, and the impact of technology on social work practice. He is particularly interested in networking governance and how digital tools can enhance both child protection efforts and school social work. In addition to his academic and professional roles, he is an active member of professional associations, including the Kerala Association of Professional Social Workers, the Indian Society of Professional Social Work, and the Indian Counselors Association, where he serves as an executive member of the Kerala Chapter.

Aruna Rani, Multitalented and compassionate professional in Accounting, Finance, and Quantitative Techniques, with more than two decades of academic experience, dedicated to fostering student success. Skilled in designing objective-based lesson plans and developing formative and summative assessments to track progress. Adept at differentiating instruction for diverse learners and maintaining an inclusive, engaging classroom environment. Experienced in multicultural settings, with strong communication and collaboration skills for case study analysis. Proficient in educational technology and committed to continuous professional development. Demonstrates adaptability and creative problem-solving, alongside a solid understanding of educational standards and ethical teaching practices.

Keshvi Rastogi, a third-year graduation student, is currently pursuing BCom Honours in Christ University, Bangalore. It is one of the leading institutions in Bangalore, Karnataka, and is a NIRF and NAAC Accredited University. She has demonstrated a keen interest and aptitude in the field of commerce and finance. Throughout her academic journey, she has consistently excelled in coursework, showcasing a strong grasp of various topics. Her commitment to academic excellence is further reflected in her active participation in various research projects, internships, and competitive fests, where she has applied theoretical knowledge to practical scenarios. Keshvi is known for her analytical skills, attention to detail, and ability to work collaboratively in team settings. Her dedication to continuous learning and professional growth makes her a promising candidate for contributing valuable insights and innovative solutions in the realm of commerce and finance.

Gowtham S is an Assistant Professor Department of Social Work, Acharya Institute of Graduate Studies, Bengaluru. He Post Graduated in Social Work from Department of Studies and Research in Social Work, Tumukur University, Tumukur also qualified Karnataka State Eligibility Test(K-SET) conducted by Mysore University, Mysore, Karnataka. He had work experience in national level projects as counsellor in NTQL – NIMHANS and as Project Technical Officer in NCDIR – ICMR, Bengaluru. His attended many workshops related social work education and he successfully completed online certificate course on 'Community Wellbeing for Youth' conducted by LGBRIMH, Tezpur Assam.

Shabana S is an accomplished Assistant Professor in the Department of Commerce at Bharathamatha College of Arts and Science, Palakkad where she coordinates the PG Department of Commerce, serves as Vice President of the Internship Activity Coordination, and leads the Anti-Ragging Cell and Social Action Club. Holding a Ph.D. in Commerce with a specialization in Human Resources from Bharathiar University, she also completed an MBA in HR, M.Phil., and M. Com in Finance, reflecting her dedication to academic excellence. Her technical skills include MS Office, Tally, SPSS, and a PGDCA, enhancing her role as an educator and researcher. With over five years as a Skill Development Executive under Kerala's Additional Skill Acquisition Programme, she has trained students in communicative English and IT skills. Dr. Shabana has published 15 articles in Scopus-indexed, Web of Science, peer-reviewed, and UGC Care-listed journals.

S. B. Inayath Ahamed is presently associated as Assistant Professor in Kalasalingam Business School, Kalasalingam Academy of Research and Education (Deemed to be University), Anand Nagar, Krishnankoil, Tamil Nadu. He holds the position of Programme Coordinator of MBA . He had cleared UGC NET –JRF/SRF and SET respectively and did his Ph.D from Periyar University, Salem as JRF/SRF Scholar. He has 4 years of Research Experience and Teaching Experience each. His areas of expertise include Supply Chain Management and Marketing Management. He had published many research articles in International and national journals of Repute. He is presently the Innovation Brand Ambassador of MHRD & AICTE at KARE. He had completed FDP at IIM Ahmedabad.

Anil Sharma is working as Assistant Professor with Department of Computing Science, Amity School of Engineering & Technology, Amity University, Noida, India. He holds a M.Tech and PhD degree in Computer Science & Engineering from Netaji Subhas University of Technology and Guru Gobind Singh Indraprastha University, Delhi, India respectively. He has an experience of 3 years in IT industry as ERP consultant and more than 14 years as academician in India and abroad. He has worked with Debre Tabor University Ethiopia, East Africa on United Nation Project. He has published and presented 34 research papers in reputed international journals and conferences, including 15 SCI/ESCI publications. He has authored two textbooks on Data Management and holds three Indian patents. He has delivered several one-week workshops and seminars in universities in India and abroad. He has chaired and co-chaired technical sessions in international conferences of repute. He is a reviewer for reputed Scopus, Web of Science and SCI journals including International Journal of System Assurance Engineering and Management (Springer), International Journal of Intelligent Computing and Cybernetics (Emerald Publishing), IET Collaborative Intelligent Manufacturing (Wiley Publication), Journal of Sensors (Hindawi), Journal of Intelligent & Fuzzy Systems (IOS Press), International Journal of Services, Economics and Management (Inderscience), Wireless Personal Communication (Springer), Frontiers of Oncology and many more. He is a member and reviewer for professional body Internet Society Delhi Chapter. His area of research is Machine Learning, Information Retrieval, NLP, Semantic Web technologies, Bioinformatics & Swarm based optimization.

Geetanjali Sharma is a distinguished professional with an impressive academic journey and a wealth of experience. Holding a doctorate in Development Economics, a post-graduate degree in Economics, and an MBA, she brings a multifaceted perspective to her role. With a rich teaching career spanning over two decades across various esteemed management institutes, she has honed her expertise in education. Currently positioned as the Head of Learning for all MA Programs at JainOnline, Dr. Geetanjali plays a pivotal role in shaping the academic landscape. Her passions extend beyond the classroom. With a keen interest in social, economic, and development issues, she actively engages in research and analysis, often sharing her insights through reviews published in The Economic Times. Dr. Geetanjali firmly believes that true productivity flourishes at the intersection of emotional well-being and intellectual acumen. This philosophy underscores her approach, advocating for a harmonious equilibrium between emotions and intelligence to unlock heightened levels of individual and collective productivity. Her global voyages have endowed her with a unique cross-cultural perspective, offering profound insights into the dynamics of diverse societies. These experiences have not only enriched her understanding of various cultures but have also granted her a nuanced comprehension of global socio-economic intricacies.

Julie Sunil has a doctorate in Marketing Management with 17 years of experience in corporate and academic fields. She is involved in professional courses aimed at building confidence among management aspirants to compete and excel in leadership roles. She has published and presented articles in reputed journals and is best known for writing multi-disciplinary case studies designed for training MBA professionals on handling the complexity in decision making. She is currently working as Associate Professor at Presidency Business School, Bangalore.

Uma C Swadimath is a keen Learner, an enthusiast, a team player and a staunch disciplinarian. Ph. D in Economics, Professor with over 23 years of experience in teaching and facilitating Management subjects is currently working with CMS Business School, JAIN (Deemed to-be University). She has been appointed as the Chief Superintendent for conducting MBA / M.Com examination of Bharathiar University, Examination in charge of Bengaluru City University, a Board of Examiner for multiple reputed institutions in Bangalore urban including the likes of BMS Institute of Management Studies, Ramaiah Institute of Management Studies, New Horizon College of Engineering, Bengaluru City University et al. She had also been appointed as Chief Superintendent for conducting the MBA examination of

Bengaluru City University. Has been awarded as the' Best Faculty" twice. She has published 25 articles in National and International Journals and presented several papers in Conferences. As a core member of Autonomous Committee, formulated Curriculum for Autonomous MBA Course.

Dakshinamurthy with a Master's in Engineering Design and an MBA from PSG Institute of Management, coupled with over a decade in academia and three years of industry experience, he possesses a unique skill set that spans across crucial areas in today's industries. His academic career underscores a deep commitment to fields such as Technological Licensing, Entrepreneurship, Innovation, Strategic Management, International Business, and Economics. His dual expertise allows him to effectively bridge the gap between theoretical knowledge and practical application, offering valuable guidance to startups and advising companies on optimizing operations and navigating complex market landscapes. Beyond academia and consulting roles, Dakshinamurthy has actively pursued innovation through tangible projects, including the development of a prototype app designed to facilitate alumni mentorship for current college students.

T. Suganya is an Assistant Professor at Knowledge Institute of Technology Salem. She completed her MBA with the specialization of Finance and Marketing. She has 5 years of experience in Academia along with 2 years of experience in Secondary Market, Commodities and futures market operations. To Enrich her knowledge in various disciplines she has attended many FDPs and Seminars in her tenure of Experience. She also published a patent on "Device to identify financial loan Eligibility of a person".

T. N. Seenivasan is an Assistant Professor at Gnanamani college of Technology, Namakkal. he holds a BBM (Business Management) from Gobi Arts and Science College, MBA from VLB Janakiammal college of Arts and Science and an Ph.D from Bharathiar University.5 years of experience in Industry (Banking and Insurance) 12 years of experience in teaching profession, he has published 5 papers in international journals and conferences.

Dhanabalan Thangam presently working as an Associate Professor at Presidency Business School, Bengaluru, India. Earlier he worked as Post a doctoral researcher in Konkuk School of Business, Konkuk University, Seoul, Korea South. He received his Ph.D. degree in Management from Alagappa University, Tamilnadu, India. His current research interests are marketing, small business management, and artificial intelligence in management fields.

P. Veeramanikandan was born in Ramanathapuram, Tamilnadu, India, in 1982. He received B.E degree in Electrical and Electronics Engineering and M.E degree in Power Electronics and Drives both from Anna University, Chennai in 2010 and 2012, respectively and the Ph.D. degree in Electrical Engineering from Anna University, Chennai in 2021. He presented more than 25 research papers in various national and international Journals and Conferences. He has 12 years of teaching experience. He is currently working as an Associate Professor in the Department of Electrical & Electronics Engineering at Dhanalakshmi Srinivasan College of Engineering and technology, Mahabalipuram, Chennai, Tamilnadu, India. His research interests are in the area of Power Electronics, Solar MPPT, Multilevel Inverter and Soft computing techniques.

Bangaru Venugopal is a senior Professor of Law, Currently the Professor and Dean at Vel Tech Rangarajan Dr. Sagunthala R&D Institute of Science and Technology, earlier worked in the Madaras University. Sir has guided many scholars and Attended many national and international seminars, conferences and symposiums. Sir has been the resource person for many instutions of national and international repute.

Index

Symbols

A

B

C

D

E

F

H

I

L

T

U

W

www.ingramcontent.com/pod-product-compliance
Lightning Source LLC
LaVergne TN
LVHW060008031026
849161LV00025B/2363